The Physics of SiO$_2$

and Its Interfaces

THE PHYSICS OF SiO$_2$

AND ITS INTERFACES

Proceedings of the International Topical Conference on
THE PHYSICS OF SiO$_2$ AND ITS INTERFACES
held at the **IBM** Thomas J. Watson Research Center
Yorktown Heights, New York
March 22-24, 1978

Editor
SOKRATES T. PANTELIDES
IBM, Yorktown Heights

PERGAMON PRESS
New York/Oxford/Toronto/Sydney/Frankfurt/Paris

Pergamon Press Offices:

U.S.A.   Pergamon Press Inc., Maxwell House, Fairview Park,
                    Elmsford, New York 10523, U.S.A.

U.K.   Pergamon Press Ltd., Headington Hill Hall, Oxford OX3,
                    OBW, England

CANADA   Pergamon of Canada, Ltd., 75 The East Mall,
                    Toronto, Ontario M8Z 5W3, Canada

AUSTRALIA   Pergamon Press (Aust) Pty. Ltd., 19a Boundary Street,
                    Rushcutters Bay, N.S.W. 2011, Australia

FRANCE   Pergamon Press SARL, 24 rue des Ecoles,
                    75240 Paris, Cedex 05, France

FEDERAL REPUBLIC   Pergamon Press GmbH, 6242 Kronberg/Taunus,
OF GERMANY   Frankfurt-am-Main, West Germany

Library of Congress Cataloging in Publication Data

International Topical Conference on the Physics of
    SiO2 and Its Interfaces, Yorktown Heights,
    N. Y., 1978.
    The physics of SiO2 and its interfaces.

    Includes index.
    1. Silica--Congresses. 2. Surfaces (Physics)--
Congresses. 3. Surface chemistry--Congresses.
I. Pantelides, Sokrates T. II. Title.
QC585.75.S55T57 1978   546'.683'2 78-18205
ISBN 0-08-023049-0

Printed in the United States of America

FOREWORD

Silicon dioxide (SiO_2) and silicon are the heart and soul of the semiconductor industry which has given us the myriad of electronic devices that have become so much part of our lives. In fact, the quality and properties of thermal SiO_2, as compared with those of the oxides that form on other semiconductors, are largely responsible for the dominance of Si in semiconductor technology. SiO_2 is also all around us in a variety of other forms: The earth's rocks are composed predominantly of SiO_2; sand is quartz, one of the crystalline forms of SiO_2; most glasses are primarily amorphous SiO_2, with other substances in small quantities imparting characteristic properties and color. For these and other reasons SiO_2 has been the object of extensive studies and research by engineers, physicists, chemists, crystal growers, metallurgists, geologists, mineralogists and others. Interaction among these groups of scientists has been minimal, however, even though the problems are often quite similar, if not identical. Typically, research on amorphous films of SiO_2 grown on Si for electronic devices is reported in different journals and at different conferences from research on crystalline quartz, research on glasses, research on high-pressure polymorphs of SiO_2, etc. Finally, despite its technological importance, SiO_2 has not been a very popular material among solid-state physicists. The reason for the latter is perhaps the fact that SiO_2 is useful mainly in the amorphous form, and amorphous materials have only recently been studied extensively. Amorphous Si, Ge, As, Se, and the chalcogenide glasses have been more popular for basic experimental and theoretical physics research, at the expense of SiO_2. Even in its crystalline form, SiO_2 has not attracted many physicists, largely because of very complicated crystal structures. Characteristically, detailed energy band calculations for SiO_2 have only appeared in the last few years, and many questions are still unanswered.

The purpose of organizing the "International Topical Conference on the Physics of SiO_2 and Its Interfaces" was primarily to bring together the diverse groups of scientists who have been studying SiO_2 from different points of view and focus on the basic properties of the material which transcend specific applications. Thus, this conference was to address, for example, the fundamental properties of chemical impurities in both amorphous and crystalline SiO_2, but not the specific effects that particular impurities may have on either MOS devices or glasses. At the same time, the purpose of the conference was to stimulate solid-state physicists to pay more attention to this important and interesting material. For this reason, the conference was scheduled for March 22-24, 1978, the week preceding the popular solid-state March meeting of the American Physical Society. Since the March meeting was in Washington, D.C., an east coast location was chosen. The IBM Thomas J. Watson Research Center in Yorktown Heights, New York, where extensive research on SiO_2 has been carried out, was an apppropriate place where a laboratory atmosphere would be provided. An Organizing Committee and an International Advisory Committee were formed (the names of committee members appear on page ii). The American Physical Society and IBM agreed to sponsor it. Financial support was granted by IBM Corp., by IBM World Trade Corp. (A/FE and E/ME/A) to assist with travel expenses of foreign scientists, and also by the Office of Naval Research, the Advanced Research Projects Agency (ARPA), and the National Science Foundation for travel assistance to participants from both the United States and abroad.

The response to the initial announcement was very satisfactory. Over 600 people from all over the world indicated their interest and their desire to receive more information. Final registration exceeded expectations. Though the initial goal was to have only about 150 participants, registration was actually not cut off until the capacity of the IBM auditorium (265 seats) was exceeded. I had the painful task of turning down a large number of scientists who simply could not be physically accommodated. 263 scientists from 13 countries attended.

The highlight of the program was Sir Nevill Mott's keynote invited paper on optical and transport properties. The session was chaired by P. W. Anderson who shared with Mott and also with J. H. Van Vleck the 1977 Nobel Prize in Physics. Because of the large number of contributed papers that were received, the number of invited papers was kept to a minimum in order to accommodate as many diverse points of view and subjects as possible. Thus, in addition to Mott's paper, there was a review paper on impurities in amorphous thermal thin films of SiO_2 by D. J. DiMaria and a review paper on impurities in α-quartz and fused silica by D. L. Griscom. Finally, there was an invited paper by R. B. Laughlin on recent calculations on the electronic structure of the Si-SiO_2 interface. The contributed papers covered a wide range of properties, including transport properties, tunneling, electronic structure, structural properties, impurities in both amorphous films and quartz, vibrational and thermal properties and other topics. An entire session containing 10 papers was devoted to the stoichiometry of the Si-SiO_2 interface and the question of the existence of an SiO_x interface layer. In order to allow ample time for discussion after each oral presentation, 29 papers were presented by poster.

This volume contains the papers that were presented at the conference. They were submitted in camera-ready form and were reviewed at the time of the conference. Reviewers were asked to concentrate on detecting errors of substance and other major deficiencies, as opposed to style, because of the inflexibility of the camera-ready format and the desire for speedy publication. In those few cases where disagreements between authors and reviewers could not be settled at the time of the conference, protracted refereeing was avoided in favor of speedier publication. All papers have been included in the volume.

Thanks for the success of this conference are due to the Office of Naval Research, the Advanced Research Projects Agency, the National Science Foundation, IBM Corp., and IBM WTC A/FE and E/ME/A for their financial contributions, to the IBM T. J. Watson Research Center for providing the necessary facilities, to the members of the conference committees for their suggestions, and to the members of the organizing committee who helped plan the technical program.

Sokrates T. Pantelides
April 1978.

CONTENTS

CHAPTER I: TRANSPORT PROPERTIES AND TUNNELING

Electronic Properties of Vitreous Silicon Dioxide - N. F. Mott 1

Small Polaron Formation and Motion of Holes in a-SiO$_2$ -
R. C. Hughes and D. Emin 14

Field Dependent Hole Transport in Amorphous SiO$_2$ -
F. B. McLean, H. E. Boesch, Jr., and J. M. McGarrity 19

Exciton Transport in SiO$_2$ - Z. A. Weinberg and G. W. Rubloff..................................24

High-Electric Field Transport of Electrons in SiO$_2$ - D. K. Ferry..................................29

Electron Emission from Silicon Dioxide Into Vacuum - P. M. Solomon 35

Electron Transport in Silicon Oxynitride -
A. V. Rzhanov, K. P. Mogilnikov, and V. A. Gritsenko 40

The Nature of Electron Tunneling in SiO$_2$ - M. Av-Ron,
M. Shatzkes, T. H. DiStefano, and I. B. Cadoff..................................46

**Evidence for a Band Tail on the Conduction Band Edge of
Thermal SiO$_2$ from Photon-Assisted Tunneling Measurements** -
A. Hartstein, Z. A. Weinberg, and D. J. DiMaria..................................51

CHAPTER II: ELECTRONIC STRUCTURE AND SPECTRA

Electronic Structure of Crystalline and Amorphous SiO$_2$
D. J. Chadi, R. B. Laughlin, and J. D. Joannopoulos 55

**Electronic Structure of α-Quartz and the Influence of Some
Local Disorder: A Tight-binding Study** - R. N. Nucho and
A. Madhukar..................................60

**Electronic Structure Investigations of Two Allotropic Forms of
SiO$_2$: α-Quartz and β-Cristobalite** - I. P. Batra 65

Band Structures and Electronic Properties of SiO$_2$ - W. B. Fowler,
P. M. Schneider, and E. Calabrese 70

K X-ray Spectra of Amorphous and Crystalline SiO$_2$ - C. Senemaud
and M. T. Costa Lima 75

The Optical-Absorption Spectrum of SiO$_2$ - S. T. Pantelides 80

Inelastic Electron Scattering in SiO$_2$ - A. E. Meixner,
P. M. Platzman, and M. Schlüter..................................85

**Electronic Structure of SiO$_2$ from Electron Energy Loss
Spectroscopy** - J. Olivier, P. Faulconnier and R. Poirier 89

The Absorption and Photoconductivity Spectra of Vitreous SiO$_2$ -
A. Appleton, T. Chiranjivi, and M. Jafaripour-Ghazvini..................................94

Calculated and Measured Auger Lineshapes in SiO$_2$ - D. E. Ramaker, J. S. Murday, N. H. Turner, G. Moore, M. G. Lagally, and J. Houston...99

Is Silicon Dioxide Covalent or Ionic? - W. A. Harrison 105

Chemical Bond and Related Properties of SiO$_2$ - K. Hübner...............................111

Topological Effects on the Band Structure of Silica - M. F. Thorpe.......................116

CHAPTER III: THERMAL AND STRUCTURAL PROPERTIES

Heat Pulse Experiments on Vitreous SiO$_2$ in the Temperature Range 2.5-300K - W. Block, M. Meissner, and K. Spitzmann.......................122

Thermal Conductivity of SiO$_2$ - B. H. Armstrong 128

Neutron Diffraction by Vitreous Silica - A. C. Wright and R. N. Sinclair 133

Raman Spectra and Atomic Configurations of Vitreous Silica - S. W. Barber 139

Electrostriction and Piezoelectricity of Thermally Grown SiO$_2$ Films - K. Misawa, A. Moritani, and J. Nakai 144

Critical Need for S(k,ω) Determinations in Amorphous SiO$_2$: Calculation of Physical Properties Via Frozen Liquid Phonons - E. Siegel 149

Properties of Localized Silicon-Dioxide Clusters in Layers of Disordered Silicon on Silver - Cheol Jung Kim, K. Shu, H. Oona and S. O. Sari 155

CHAPTER IV: DEFECTS AND IMPURITIES IN THERMAL SiO$_2$

The Properties of Electron and Hole Traps in Thermal SiO$_2$ Layers Grown on Silicon - D. J. DiMaria 160

Dynamic Behavior of Mobile Ions in SiO$_2$ Layers - M. W. Hillen 179

Photo-injection Studies of Traps in HCl/H$_2$O Oxides - J. Dorosti and C. R. Viswanathan 184

Photodepopulation of Electrons Trapped in SiO$_2$ on Sites Related to As and P Implantation - R. F. DeKeersmaecker, D. J. DiMaria, and S. T. Pantelides..189

Chemical State of Phosphorus in Deposited SiO$_2$(P) Films - A. N. Saxena and R. A. Powell...195

Spectroscopic and Structural Properties of Nitrogen-Doped Low-Temperature SiO$_2$ Films - G. W. Anderson, W. A. Schmidt, and J. Comas 200

Some Observations of Defects in Amorphous SiO$_2$ Films - E. A. Irene 205

Measurement of Hydrogen Profiles in SiO$_2$ by a Nuclear Reaction Technique - D. D. Allred, C. W. White, G. J. Clark, B. R. Appleton, and I. S. T. Tsong..210

Interaction of Dissolved Molecular Hydrogen with a Vitreous Silica Host - J. Vitko, C. M. Hartwig, and P. L. Mattern......................215

Hydrogen in SiO$_2$ Films on Silicon - A. G. Revesz..............................222

ESR Centers and Charge Defects Near the Si/SiO$_2$ Interface - E. H. Poindexter, E. R. Ahlstrom, and P. J. Caplan......................227

CHAPTER V: DEFECTS AND IMPURITIES IN α-QUARTZ AND FUSED SILICA

Defects and Impurities in α-Quartz and Fused Silica - D. L. Griscom......................232

A Germanium Tri-Hydrogen Center in α-Quartz - F. C. Laman and J. A. Weil..253

Electron Paramagnetic Resonance Studies on Al Centers in Vitreous Silica - K. L. Brower..258

Oxygen-Associated Trapped Hole Centers in High-Purity Fused Silica - M. Stapelbroek and D. L. Griscom......................263

A Model for Point Defects in Silica - G. N. Greaves......................268

Auger Spectra of SiO$_2$ Surface Defect Centers - K. Schwidtal......................273

Vibrational and Electronic Spectroscopy of Ion-Implantation-Induced Defects in Fused Silica and Crystalline Quartz - G. W. Arnold......................278

Raman Studies of Structural Defects in Vitreous SiO$_2$ - F. L. Galeener, J. C. Mikkelsen, Jr., and N. M. Johnson......................284

Cathodoluminescence Studies of SiO$_2$ - Na, Cl, Ge, Cu, Au, and Oxygen Vacancy Results - C. E. Jones and D. Embree......................289

Anomalous Dielectric Absorption in SiO$_2$-Based Glasses - M. A. Bösch......................294

XPS Study of Sodium Oxide in Amorphous SiO$_2$ - B. W. Veal and D. J. Lam......................299

Modification of SiO$_x$ - S. R. Ovshinsky, K. Sapru, and K. Dec......................304

Intrinsic Surface Phonons in Porous Glass - C. A. Murray and T. J. Greytak......................309

Positronium-Surface Interaction in the Pores of Vycor Glass - S. M. Kim and W. J. L. Buyers......................314

Ion Irradiation and Stored Energy in Vitreous SiO$_2$ - M. Antonini, A. Manara, and P. Lensi......................316

CHAPTER VI: ELECTRONIC STRUCTURE OF THE Si-SiO$_2$ INTERFACE

Electronic States of Si-SiO$_2$ Interfaces - R. B. Laughlin,
J. D. Joannopoulos, and D. J. Chadi 321

The Defect Structure of the Si-SiO$_2$ Interface, A Model Based on Trivalent Silicon and Its Hydrogen "Compounds" - C. M. Svensson 328

Electronic Structure of a Model Si-SiO$_2$ Interface - F. Herman,
I. P. Batra and R. V. Kasowski 333

CHAPTER VII: THE STOICHIOMETRY OF THE Si-SiO$_2$ INTERFACE

Continuous-Random-Network Models for the Si-SiO$_2$ Interface -
S. T. Pantelides and M. Long 339

Studies of the Si-SiO$_2$ Interface by MeV Ion Scattering -
L. C. Feldman, I. Stensgaard, P. J. Silverman
and T. E. Jackman 344

**Transmission Electron Microscopy of Microstructural Defects in
Si-SiO$_2$ Systems - Si Clusters in SiO$_2$ Film** - J. J. Chen
and T. Sugano 351

**A High-Resolution Electron Microscopy Study of the Si-SiO$_2$
Interface** - O. L. Krivanek, D. C. Tsui, T. T. Sheng,
and A. Kamgar 356

Structure of the Si-SiO$_2$ Interface by Internal Photoemission -
T. H. DiStefano 362

Auger Sputter Profiling Studies of the Si-SiO$_2$ Interface -
C. R. Helms, N. M. Johnson, S. A. Schwarz, and W. E. Spicer 366

Auger Analysis of the SiO$_2$/Si Interface of Ultrathin Oxides -
J. F. Wager and C. W. Wilmsen 373

Studies of Si/SiO$_2$ Interfaces and SiO$_2$ by XPS - T. Hattori and
T. Nishina 379

X-ray Photoelectron Spectroscopy of SiO$_2$-Si Interfacial Regions -
S. I. Raider and R. Flitsch 384

**Chemical Structure of the Transitional Region of the SiO$_2$/Si
Interface** - F. J. Grunthaner and J. Maserjian 389

**MOS Solar Cell as a Tool to Study the Transition Region
Associated with Ultra Thin Films of SiO$_x$** - R. Singh, K. Rajkanan
and J. Shewchun 396

CHAPTER VIII: INTERFACE PROPERTIES

Initial Stages of SiO$_2$ Formation on Si(111) - R. S. Bauer,
J. C. McMenamin, H. Petersen, and A. Bianconi 401

The Si-SiO$_2$ Interface and Localization in the Inversion Layer -
M. Pepper 407

Metastabilities at the Si-SiO$_x$ Interface - C. T. White and K. L. Ngai 412

Photocapacitance Probing of Si-SiO$_2$ Interface States - E. Kamieniecki
and R. Nitecki 417

**Transient Capacitance Measurements of Electronic States at the
SiO$_2$-Si Interface** - N. M. Johnson, D. J. Bartelink, and M. Schulz 421

**Interface States Resulting from a Hole Flux Incident on the
SiO$_2$/Si Interface** - J. M. McGarrity, P. S. Winokur, H. E. Boesch, Jr.,
and F. B. McLean 428

**The Influence of the pH on the Surface State Density at the SiO$_2$-Si
Interface** - N. F. de Rooij and P. Bergveld 433

**The Si-SiO$_2$ Interface: Oxide Charge, Electron Affinity and Fast
Surface States** - L. A. Kasprzak and A. K. Gaind 438

Lateral Nonuniformities (LNU) of Oxide and Interface State Charge -
N. Zamani and J. Maserjian 443

**Temperature Dependence of Relaxation of Injected Charge at the
Polycrystalline-Silicon/SiO$_2$ Interface** - T. W. Hickmott 449

Effects of Ultra-thin SiO$_x$ in Conducting M-I-S Structures -
T. E. Sullivan, R. Childs, J. Ruths, and S. J. Fonash 454

Confirmation of Hydrogen Surface States at the Si-SiO$_2$ Interface -
B. Keramati and J. N. Zemel 459

Electrical Properties of SiO$_2$-Si Interface for Deformed Si Surfaces -
K. Murty, B. Lalevic, B. W. Lee, H. Suga, and S. Weissman 464

Shear Strength of Metal-SiO$_2$ Contacts - S. V. Pepper 470

List of Participants 475

Author Index.. 487

xi

CHAPTER I

TRANSPORT PROPERTIES AND TUNNELING

ELECTRONIC PROPERTIES OF VITREOUS SILICON DIOXIDE

N. F. Mott
Cavendish Laboratory, Madingley Road, Cambridge, U.K.

ABSTRACT

The types of self-trapped holes and excitons that may exist in vitreous SiO_2 are discussed. Similarities and differences between this material and the chalcogenide semiconductors are pointed out.

1. MOBILITY EDGES AND POLARONS

The study of amorphous semiconductors during the last decade has provided concepts which give a fair understanding of charge transport in these materials, though there are still many points that are controversial. The aim of this paper is to apply these concepts to the electrical and optical properties of vitreous silicon dioxide. I have published two papers on this subject;[1,2] here I shall summarise the ideas presented there, bring them up to date and revise a few of them.

The conduction and valence bands of non-crystalline systems are thought to have tails of localized states (traps), separated from extended states by a sharp energy[3] denoted by E_c and called the mobility edge.[4]. The firmest evidence for a mobility edge for electrons in a random field comes from the study of impurity bands and inversion layers.[5] In semiconductors, the evidence for silicon and germanium deposited in a silane glow discharge is in my view good,[6] but there is doubt about the extent to which the disorder producing the edge in the conduction band is due to hydrogen incorporated in the network. Nagels et al.[7] also give evidence for a range of order 0.1 eV of localized states in the valence band of As_2Te_3, though Emin and coworkers[8] have suggested that these results are due to polaron formation, the arguments to the contrary being set out by Mott, Davis and Street.[9] In vitreous SiO_2, measurements[10] of the drift mobility μ_D of electrons show no evidence for a mobility edge, the mobility decreasing with increasing temperature. Another group of materials for which this is true is the liquid rare gases.[11] I believe that in the conduction bands of such materials the range ΔE of localized states is very small ($<<$ kT), and that this will be so when the wave functions are s-like at each atom. Unfortunately, calculations[12] of ΔE are confined to simple models appropriate to an impurity band, and have not been carried out for a continuous random network appropriate to a glass.

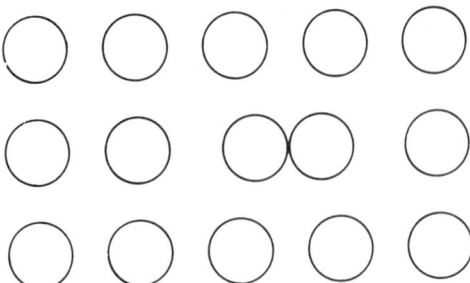

Fig. 1.   Molecular polaron formed by positive hole in a crystalline solid.

Turning now to the valence band, SiO_2 in common with the chalcogenides has[13] a comparatively narrow (2-3 eV) non-bonding valence band formed from oxygen lone-pair orbitals, separated from and above a wider band formed from Si-O bonding orbitals. Holes in this narrow band may or may not form "small" polarons, though this in my view is unlikely to occur in conduction bands. Polarons are of two types, "molecular" and "dielectric", and here I shall discuss only the former. The best-known examples are the V_K centres in alkali halides and self-trapped holes in solid and liquid rare gases.[14] What is envisaged is shown in Fig. 1; the "hole" on one atom forms a bond with a neighbouring atom (in, for instance, solid argon), and both are displaced to form Ar_2^+. It is well known[15,16,17] that for this kind of polaron (unlike the dielectric kind) the electron-phonon interaction has to exceed a certain value before there is any polaron formation at all. The argument is as follows. To displace an atom or pair of atoms as in Fig. 1 by a distance q requires an elastic energy Aq^2 (cf. Fig. 2). The displacement will also produce a potential well for the carrier (the hole in our case). In three dimensions the depth of such a hole must attain a certain critical value q_o before a bound state occurs (unless $V(r)$ falls off as $-C/r$, which is not the case here). It will then decrease, first as const. $(q-q_o)^2$ and then linearly. Thus, the electronic energy will appear as in Fig. 2. A minimum will normally occur, but if the energy at the minimum is positive, no stable polaron can be formed. If the energy is negative, it can; but as pointed out by Mott and Stoneham,[17] there will be a time delay τ of order given by $1/\tau \sim \omega \exp(-w/kT)$ (or $\omega \exp(-w/\frac{1}{4}\hbar\omega)$ at low T) in the formation of the polaron. Here w is the energy shown in Fig. 2, ω the phonon frequency.

One has therefore to ask whether, in a given material, polarons are formed or not. I

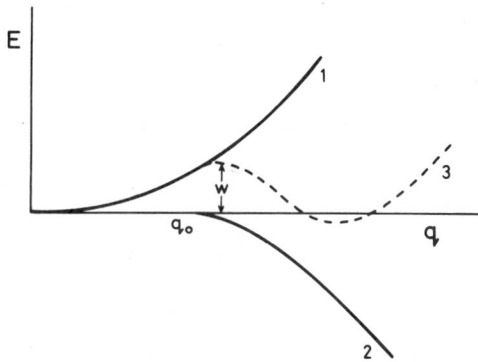

Fig. 2.　　Energy of a molecular polaron as shown in Fig. 1 as a function of displacement q. (1) is the kinetic energy Aq^2, (2) the potential energy of the carrier in the potential well, and (3) the total energy.

shall not here repeat the arguments given elsewhere[9,18] that they are not formed in arsenic-based glasses or amorphous selenium, but, following the work of Hughes,[19,20] I believe they are in the valence band of SiO_2. Since the polaron is formed from a hole in a lone-pair oxygen band, which does not take part in bonding with silicon, I believe the process is likely to be due to a moving together of two oxygen atoms, as in Fig. 1. Lattice dynamics investigations due to Striefler and Barsch[21] show that the effective charge on the oxygen ion is about 0.47 e, so the pair will have bonding similar to that of O_2, not O_2^{3-}; I am grateful to Dr. A. M. Stoneham for pointing this out.

I must now ask whether the non-crystalline nature of the material affects polaron formation. Of course, if a carrier is trapped in any localized state, there is bound to be some distortion of the network around it, which becomes smaller as the radius of the state increases. In any optical transition involving a localized state, this *must* produce a Stokes shift. This will, however, be small, and is usually neglected when the radius of the state is large, as in a donor centre in Si:P for example. I believe that this is normally true also for band-edge localized states due to disorder. I do not think that, within the radius of the orbital, the phase of the wave function changes in a random way from atom to atom; if it did, disorder would produce a major change in the wave functions, and, since much kinetic energy would be present in the extended state, polaron formation would be considerably easier. On the contrary, within a band-edge localized state the wave function is much the same as for the crystal, the phase on each atom or bond changing slowly and becoming random in a distance comparable with the radial extension of the state. If this is so, we expect for the valence band

(1) A "mobility edge", say ~ 0.1 eV from the band edge, as found by Nagels et al.[7] for As_2Te_3.

(2) Formation of polarons, if at all, *after* a certain delay.

I have attempted[1] to describe Hughes' "prompt" mobility and his smaller long-term mobility in terms of this model. I suppose that the prompt mobility with activation energy 0.17 eV is due to excitation to a mobility edge at high temperatures, going over to variable range hopping at low T so that the activation energy 0.37 eV is due, in my view, to the formation of molecular polarons. Hughes, on the other hand, assigns it to trapping by defects. But this does not seem compatible with the finding of McLean et al.[22] that this activation energy too falls off at low T (to 0.06 eV), which is just what we expect for polarons.

2. DEEP STATES IN THE GAP

In deep levels, in crystalline as for amorphous materials, we should normally expect a large Stokes shift. Figure 3 shows the configuration diagrams for ground and excited states of a centre giving a deep level. PQ is the absorption frequency and $h\Delta\nu$ the (large) breadth of the line. RS is the emission frequency, if the transition is radiative. The following recombination mechanisms are possible:

(a) Radiation from R to S, with a transition probability ~ 10^8 s^{-1}.

(b) A multiphonon transition from R, exciting the vibrations from A to B and, in the limit of low temperatures, having a rate per unit time[9] ~ $\omega \exp(-\gamma E/\hbar\omega)$, with $\gamma \sim 1$ and E as shown in Fig. 3.

(c) Excitation over the crossing point X with probability per unit time ~ $\omega \exp(-W/kT)$.

(d) If P lies above X, then according to Dexter, Klick and Russell[22] energy will be dissipated by interaction with phonons until the energy X is reached, and then there is a high probability of cross-over to the lower curve. Just how large this probability is has not as yet been calculated, though estimates (unpublished) by Stoneham suggest ~ 99% in some cases. If so, very weak radiation with $h\nu = RS$ is still expected.

In amorphous materials deep states can either be due to defects (e.g., dangling bonds) or to impurities (e.g., Na^+ bound to a negative non-bridging oxygen). The latter is a neutral centre; an isolated dangling bond can give up an electron, acting as a deep donor, or accept one. In either case it becomes charged and without spin.

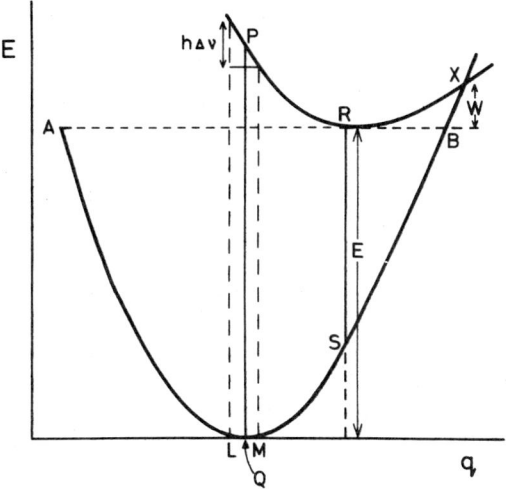

Fig. 3. Configuration diagram of a carrier in a deep level.

Such centres play a considerable role in this article, and I shall make a few remarks about them.

    (a)   A dangling bond carries an unpaired spin when neutral, but no spin when negatively or positively charged.

    (b)   In crystals dangling bonds can only exist in pairs, giving no spin in the ground state, *unless* one of the atoms has threefold co-ordination, as in As_2Se_3. In amorphous materials they are always possible. In SiO_2 an oxygen dangling bond is a "non-bridging oxygen". A silicon dangling bond is another possibility.

There is a marked difference between the absorption spectrum to be expected from charged and from neutral centres. A neutral centre will always, in principle, give a series of broadened lines leading to a series limit (as, for instance, in an F-centre). For a charged centre, this is not so, and in the excited state, if there is one, a carrier should be weakly bound in a state with a large radius (which should cut down the probability of radiative and multiphonon recombination). We have to ask whether the former statement is true for neutral centres in non-crystalline systems. I think it is, though I shall not attempt to prove it here. In SiO_2 the low dielectric constant and "smooth" conduction band makes it highly probable that such bound excited states exist with binding energy $\sim$ 1.5 eV. Thus, the absorption edge due to Na^+ with a non-bridging oxygen must be of this type, and is really a broad line as in Fig.

3 (cf. Sigel[24]). The absorption of light by such centres should not lead to photoconduction, except perhaps at high temperatures.

For chalcogenides Street and Mott[25] and Mott, Davis and Street (MDS)[9] proposed
that at such dangling bonds, if neutral or positively charged, covalent bonds would be
formed with the lone-pair orbital of a neighbouring chalcogen, the chalcogen being
displaced thereby. What happens is illustrated in Fig. 4, for amorphous Se. Denot-
ing the centres in their charged states by D^-, D° and D^+, these authors supposed
that the energy gained by forming a D^+ is so great that the reaction

$$2D^\circ \rightarrow D^+ + D^-$$

is exothermic. All states are thus positively or negatively charged. Following
Anderson[26] it is shown that such a model pins the Fermi energy. Kastner, Adler and
Fritsche (KAF)[27] have also discussed this model, calling the centre C_3^+, C_3° and C_1^- in
its three charge states. They point out that the formation of a pair does not decrease
the number of bonds, so that in a quenched chalcogenide glass the concentration can
be quite high.

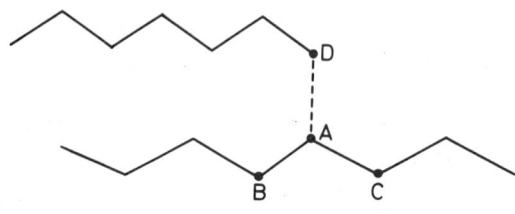

Fig. 4.   Showing a chain end (dangling bond) in a-Se. If it
is negative (D^-) the centre has no unpaired spin
and atom D forms no bond with the next chain. If
it is positive (D^+) a covalent bond between D and
A is formed by two electrons, DA contracts to the
length AB or AC and A is three-fold co-ordinated.
For the neutral centre AD carries one σ^*
(antibonding) and two σ, and the contraction will
be less.

Street, Searle and Austin (for refs., see Street[28]) first gave evidence that charged
centres are responsible for photoluminescence, and Mott and Street[18] show that

under different circumstances both D^+ and D^- can fulfill this role. In either case there is a large Stokes shift as in Fig. 3.

This brings me to the difference between the MDS and KAF models. Starting from D^+ (C_3^+), and adding a σ^* electron, KAF believe that this moves freely between the three bonds AB, AC, AD (Fig. 4), producing only the small distortion of an extended centre. In their model the neutral centre is threefold co-ordinated. Two electrons, however, in their model as in ours, can be localized on one bond, pushing A and D apart. MDS, however, believe that *one* σ^* electron can do this too, so D° is not threefold co-ordinated. Both models are possible in principle. If A and D are pushed apart, the elastic energy is Aq^2 and the energy gained is bound to begin as $-Bq^2$, so it depends on which is larger. The evidence in favour of the MDS hypothesis is that, according to Street[27] and Mott and Street[18], D^- and D^+ can both give photoluminescence with comparable Stokes shifts.

The question whether dangling bonds with this property exist in SiO_2 is of interest. It will *not* be true that the energy to form them is small, since the Si-O bond energy is much larger than anything we can expect from O-O bonds. Perhaps therefore we should expect them to occur in thermally or anodically grown oxide, but to disappear on annealing. In collaboration with Pepper[5,29] I have discussed their role in the oxide near a Si/SiO_2 interface, both as regards providing the "slow" surface states and the random charges to produce Anderson localization in the inversion layer.

3. EXCITONS

It is sometimes stated that excitons do not exist in non-crystalline materials. However, in my view it is certain that in their lowest state an electron and hole are coupled together by a Coulomb attraction. For chalcogenides and for amorphous selenium the work of Pai and Enck[30] on the photoelectric quantum efficiency gives the clearest evidence; in the absence of a sufficient field or temperature the electron and hole do not separate from each other, but recombine without contributing to a photocurrent. This does not, however, mean that an exciton *line* necessarily appears in the absorption spectrum; the spectrum of states in which the electron and hole are bound together is not properly understood, particularly in the case when ΔE is greater than the exciton binding energy. The broadening by random fields, as described by Dow and Redfield[31] as leading to an Urbach tail, may be widely applicable.

In chalcogenides an interesting point, if the model of Street, Searle and Austin (for refs., see 28) for photoluminescence is accepted, is that while the dangling bonds when excited return to the ground state by emitting radiation, this is not so for the excitons; as far as we know, radiation only occurs from defects. The difference between the two cases on this model is that the excited D^+ or D° centre consists of a D centre together with a *weakly* bound electron (or hole) in a large orbit, and this will not contribute to the distortion. An exciton, on the other hand, will relax into a state with two σ and *two* σ^* electrons, *not* spread out in space; one is the excited electron, the other promoted from the new bond. These σ^* electrons cannot both go on to the bond AB in Fig. 5 but may go onto a neighbouring bond AC and expand it or even break it, and if so produce a D⁻. There is thus more distortion in this case and so the possibility arises of recombination by the mechanism of Dexter, Klick and Russell.

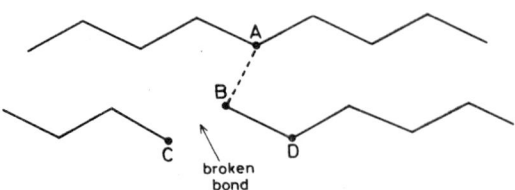

Fig. 5.   Suggested model for a self-trapped exciton.   AB forms a bond with two σ electrons but two σ^* anti-bonding electrons sit on the bonds BC, BC, expanding them.   If both sit on one bond, that bond will be broken.

I turn now to the question of excitons in SiO_2. The calculated binding energy of an electron-hole pair in this material is from 1.5 to 2 eV, and it seems to me very unlikely that any broadening of an absorption line due to disorder is as great as this, especially as there is no observable mobility edge in the conduction band.   Thus I believe that the low frequency side of that part of the absorption edge which is not due to impurities or to defects *must* be due to the formation of electron-hole pairs. If the transition is optically forbidden, it can be allowed with the help of phonons, or disorder in the glass.   I shall use the calculations of Chelikowsky and Schlüter,[13] giving a band structure as in Fig. 6b.   Excitons for such a case are discussed by Dexter and Ross.[32]   I think one can then reasonably assume that the peak at 10.4 eV is due to an *allowed* exciton formed from the states A,B, that this is an exciton of Frenkel type to be thought of as the excitation of an oxygen ion, and that its bonding energy is sufficiently great for direct absorption to occur for the frequencies giving the indirect transition C to A, though the latter might possibly be responsible for part of the tail.   Chelikowsky and Schlüter[18] state that the first direct vertical transition is dipole-forbidden, but this does not mean that a Frenkel exciton will necessarily be forbidden, and will certainly be allowed for some frequency in the exciton band, probably for the peak at 10.4 eV.

I would like now to outline the properties of an exciton of this kind.

(a)   An exciton is mobile; it may well have a band-width of 1-2 eV, though in crystals optical transitions are only allowed to the k = 0 state.   In glasses, however, transitions to all states should be allowed to some extent.

9

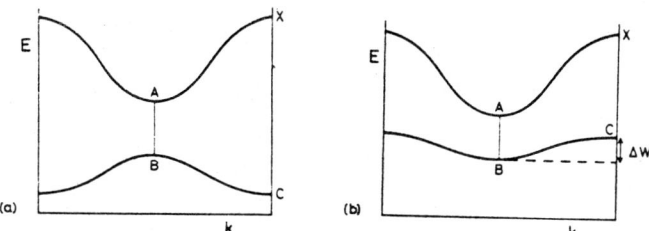

Fig. 6.   Suggested band structures (a) a material with a
direct gap, (b) with an indirect gap, as in SiO_2.

(b)   If holes can be self-trapped, so probably can excitons. If so, the configuration diagram should be as in Fig. 7, with a delay in self-trapping due to the
barrier w. The phonon broadening $h\Delta\nu$ of Fig. 3 is absent.

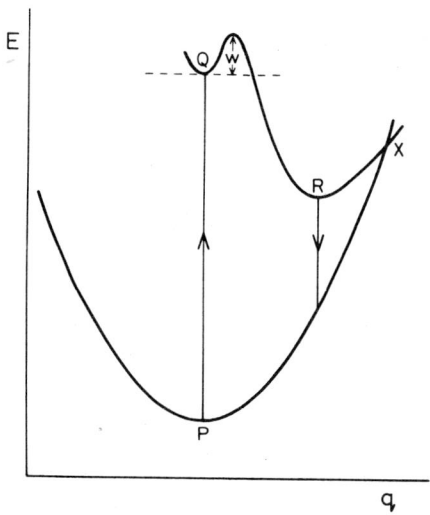

Fig. 7. Configurational diagram for an exciton.

(c)   Excitons can only produce photoelectrons by dissociating at impurities or
electrodes, or perhaps in very strong fields.

(d)   Recombination from Q to P is coherent with the incident radiation (in spite of the A-coefficient, see Heitler[34]), unless the exciton has exchanged energy with a phonon. If so, fluorescence is expected very near the absorption frequency, and difficult to distinguish from Rayleigh scattering. This kind of radiation is sometimes called that from a "phonon-dressed free exciton".[35] It is not known if it exists in SiO_2. In alkali halides it has recently been observed by several authors.[36] The temperature dependence enables an estimate of w in Fig. 7 to be given and is of order 20 meV.

There is, however, another possibility by which excitons can produce photoelectrons. This is, if the indirect gap AC of Fig. 6b overlaps the exciton band. If this is so, ΔW in Fig. 6 must be greater than the bonding energy of the direct exciton. The high photoelectric quantum efficiency observed by DiStefano and Eastman[37] in SiO_2 and also the results presented by Weinberg and Rubloff in this volume make it, in my view, very probable that this occurs. Thus, the true gap must be at about 9 eV, so that the direct-gap exciton can dissociate.

It is an interesting question to ask how the energy of the exciton is converted into phonons. If the exciton did not dissociate, we could certainly appeal to the mechanism of Dexter, Klick and Russell. If it does dissociate, and the hole is self-trapped before they recombine, it is not so clear that one can, and fluorescence with a large Stokes shift might well occur. The same should occur under excitation by high energy radiation. The problem merits experimental investigation.

If this model is correct, an indirect exciton formed from an electron at A and hole at C must exist below the band edge. I believe it might diffuse a long way before recombining and is probably the kind of exciton which diffuses through 1000 Å of oxide in the paper by Weinberg and Rubloff. It is even possible that a triplet exciton is formed, which would be more stable still.

It will be noted that these effects can only exist if the band form is as in Fig. 6b, and the gap is not direct. They are impossible if the gap is direct, as in the work of Fowler, Schneider and Calabrese, reported in this volume.

Selection rules are not expected to be rigidly obeyed in a glass, and absorption by the indirect exciton should be observable. To show that this is probably so, I reproduce some direct absorption spectra of vitreous silica (type IV spectrosil) due to Appleton and co-workers.[37] Figure 8 shows some absorption coefficients α for both glass (a) and crystal (b). The part below 8 eV is structure sensitive and due to defects or impurities. The steep part between 8 and 8.4 eV may be an Urbach edge to the direct exciton, but as it does not show photoconduction the hypothesis that it is the indirect exciton is attractive. It is very little dependent on T for the glass, strongly so for the crystal as we expect. Above 9 eV (where the results are due to Philipp,[38] we think that the band width of the moving direct exciton is being observed.

Figure 9 (due to Appleton) shows some typical photocurrents for blown films with aluminum electrodes in sandwich configuration. The important point here is that the structure-sensitive absorption at about 8 eV does show photoconductivity; if it is due to defects, they must be charged. The drop in the photocurrent at the true edge

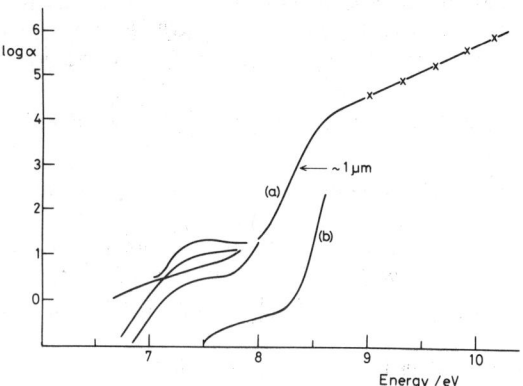

Fig. 8.   Absorption coefficient α of thin films of SiO_2, (a) vitreous, (b) crystalline. The structure-sensitive tail is for various specimens of thickness $\sim 100\ \mu$. The crosses refer to Philipp's data[33] (A. Appleton).

occurs, I believe, because here most of the radiation forms indirect excitons for which the efficiency for dissociation is small.

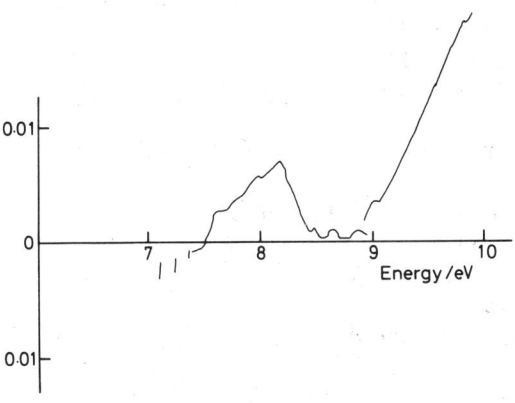

Fig. 9.   Photoconductive yield (arbitrary units) of vitreous silica, sandwich electrodes (A. Appleton).

If charged centres are present, the question arises of their nature. They could be

D^+, D^- centres (charged, non-bridging oxygens), but if they are to be identified with the "slow states" of the Si/SiO interface their energies must be higher in the gap.[1] Greaves[39] has suggested that a silicon dangling bond may be an E' centre, but will normally lose an electron to form a D^- (non-bridging oxygen), and that the latter, contrary to my earlier assumption, may give states low in the gap.

Finally, I sum up my conclusions from the experimental evidence.

1.   The gap AC in Fig. 6b is at about 9 eV and indirect.

2.   The peak at 10.4 eV is a direct gap exciton, perhaps of Frenkel type, and it overlaps the gap. Its width from 9 eV upwards is that of the exciton band, the dipole-allowed state lying at 10.4 eV and the lowest state near 9 eV.

3.   The indirect exciton lies between 8 and 9 eV, and has a long lifetime and large diffusion distance.

4.   If an electron and hole produced in any way form an indirect (stable) exciton *before* the hole is self-trapped, recombination by the mechanism of Dexter *et al.* is expected.   If the hole is self-trapped first, this is much less likely and radiation is to be expected, unless recombination through surface states or defects occurs.   The radiation should show a large Stokes shift.

REFERENCES

1.   N. F. Mott, *Adv. Phys.* **26**, 363 (1977).
2.   N. F. Mott (1977) *Electronic Processes in Glasses*, ed.   P. H. Gaskell, Taylor & Francis, London, p. 101.
3.   N. F. Mott, *Adv. Phys.* **16**, 49 (1967).
4.   M. H. Cohen, H. Fritzsche and S. R. Ovshinsky, *Phys. Rev. Lett.* **22**, 1065 (1969).
5.   N. F. Mott, M. Pepper, S. Pollitt, R. H. Wallis and C. J. Adkins, *Proc. R. Soc. A* **345**, 169 (1975).
6.   W. E. Spear, *Adv. Phys.* **26**, 811 (1977).
7.   P. Nagels, R. Callaerts and M. Denayer (1974) *Proc. 5th Int.   Conf. on Amorphous and Liquid Semiconductors*, eds. J. Stuke and W. Brenig, Taylor & Francis, London, p. 867.
8.   D. Emin, C. H. Seager and R. K. Quinn, *Phys. Rev. Lett.* **28**, 813 (1972).
9.   N. F. Mott, E. A. Davis and R. A. Street, *Phil. Mag.* **32**, 961 (1975).
10.   R. C. Hughes, *Phys. Rev. Lett.* **30**, 1333 (1973).
11.   S. H. Howe, P. G. Le Comber and W. E. Spear, *Solid State Commun.* **9**, 65 (1971).
12.   J. T. Edwards and D. J. Thouless, *J. Phys. C* **5**, 807 (1972).
13.   J. R. Chelikowsky and M. Schlüter, *Phys. Rev. B* **15**, 4020 (1977).
14.   A. M. Stoneham (1975) *Theory of Defects in Solids*, Clarendon Press, Oxford.
15.   Y. Toyozawa, *Prog. Theor. Phys. Osaka* **26**, 29 (1961).

16. D. Emin, *Adv. Phys.* **22**, 57 (1973).
17. N. F. Mott and A. M. Stoneham, *J. Phys. C* **10**, 3391 (1977).
18. N. F. Mott and R. A. Street, *Phil. Mag.* **36**, 33 (1977).
19. R. C. Hughes, *Appl. Phys. Lett.* **26**, 436 (1975).
20. R. C. Hughes, *Phys. Rev. B* **15**, 2012 (1977).
21. M. E. Streifler and G. R. Barsch, *Phys. Rev. B* **12**, 4553 (1975).
22. F. B. McLean, H. E. Boesch and J. M. McGarrity, *IEEE Trans. Nucl. Sci.* **23**, 1506 (1976).
23. D. L. Dexter, C. C. Klick and G. A. Russell, *Phys. Rev.* **100**, 603 (1955).
24. G. H. Sigel, *J. Non-Cryst. Solids* **13**, 372 (1973/74).
25. R. A. Street and N. F. Mott, *Phys. Rev. Lett.* **35**, 1293 (1975).
26. P. W. Anderson, *Phys. Rev. Lett.* **34**, 953 (1975).
27. M. Kastner, D. Adler and H. Fritzsche, *Phys. Rev. Lett.* **37**, 1504 (1976).
28. R. A. Street, *Adv. Phys.* **25**, 397 (1976).
29. M. Pepper, this volume.
30. D. M. Pai and R. C. Enck, *Phys. Rev. B* **11**, 5163 (1976).
31. J. D. Dow and D. Redfield, *Phys. Rev. B* **5**, 594 (1972).
32. D. L. Dexter and R. S. Knox (1965) *Excitons*, No. 25, Interscience Tracts on Physics and Astronomy, Wiley, New York.
33. W. Heitler (1954) *The Quantum Theory of Radiation*, 3rd edn.
34. D. Pudewill, F. J. Himpsel, V. Saile, N. Schwentner and M. Skibowski, *J. Chem. Phys.* **65**, 5226 (1976).
35. I. L. Kuusmann, P. Kh. Liblik, G. G. Liid'ya, N. E. Lushchik, Ch. B. Lush-chik and T. A. Soovich, *Sov. Phys. Solid State* **17**, 2312 (1976); T. Hayashi, T. Ohata and S. Koshino, *J. Phys. Soc. Japan* **43**, 347 (1977); G. Guillot, E. Mercier and A. Nouailhat, *J. de Physique* **38**, L-495 (1977).
36. T. H. DiStefano and D. E. Eastman, *Solid State Commun.* **9**, 2259 (1971).
37. A. Appleton, this volume.
38. H. R. Philipp, *J. Non-Cryst. Solids* **8-10** 627 (1972).
39. N. Greaves, this volume.

SMALL POLARON FORMATION AND MOTION OF HOLES IN a-SiO$_2$[*]

R. C. Hughes and D. Emin
Sandia Laboratories, Albuquerque, New Mexico 87185

ABSTRACT

X-ray generated holes in SiO$_2$ are observed to be reduced to low mobility in times of the order of vibrational periods, 10^{-12} s. The temperature dependence, electric field dependence and magnitude of this mobility for times up to about 100 ns are consistent with those of hole-like small polarons. The circumstances which favor the occurrence of rapid small polaron formation are a large effective mass (narrow valence band), the presence of the long-range hole-lattice interaction characteristic of an ionic material and the presence of disorder, all of which are found in amorphous SiO$_2$. An alternative explanation involving trapping requires an extremely large localized state density and fortuitous temperature and field dependences of the hopping rates.

In a-SiO$_2$ as in numerous crystalline materials, there is a great disparity between electron and hole transport (1,2). Namely, in relatively pure a-SiO$_2$ the electrons move in a manner which is similar to that of high-mobility electrons in crystals to which a scattering picture is often applied (1). Holes, on the other hand, are relatively immobile (2,3), displaying the characteristics expected of small-polaron hopping motion. Exploiting the many order-of-magnitude difference between the electron and hole mobilities in a-SiO$_2$, one may utilize drift mobility measurements of uniformly (x-ray) generated electron-hole pairs in thin films to study the details of hole motion in a-SiO$_2$ (2,3). In particular, one can learn about the predominant process via which x-ray generated holes relax to form small polarons.

To begin, recall that the nature of the transformation of an unbound (dry) hole to a small polaron depends critically upon the details of the coupled carrier-lattice system (4-7). For example, consider a stationary carrier which couples with the vibrations of an ionic lattice solely via the long range interaction characteristic of a charge interacting with electric dipoles. In this case, as shown in Fig. 1a, concomitant with the polarizing of the lattice surrounding it, the carrier will always find it energetically advantageous to attain a characteristic radius, the magnitude of which decreases monotonically with an increase in the strength of the electron-lattice interaction (6). The time required for such polaron formation is determined by the time needed for the atoms of the lattice to displace to new equilibrium positions -- a time of the order of a picosecond. However, if one considers the case of a stationary carrier in a solid in which the carrier-lattice interaction is short-ranged, as is the deformation-potential interaction, one finds that there are but two distinct states possible of this system (4-6). As depicted in Fig. 1b, an initially unbound (nonpolaronic) carrier may remain unbound, not forming a polaron at all, or its wavefunction may shrink in size to a radius less than, or comparable to, an interatomic spacing, thereby forming a small polaron. Since in a solid

[*]Work supported by U.S. Department of Energy.

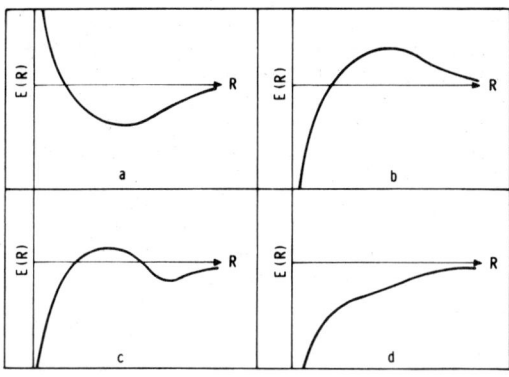

Fig. 1. The groundstate energy of
the system comprising an electron in
a deformable continuum, E(R), is
plotted against the spatial extent of
the electronic wavefunction, R, for
situations described in the text.

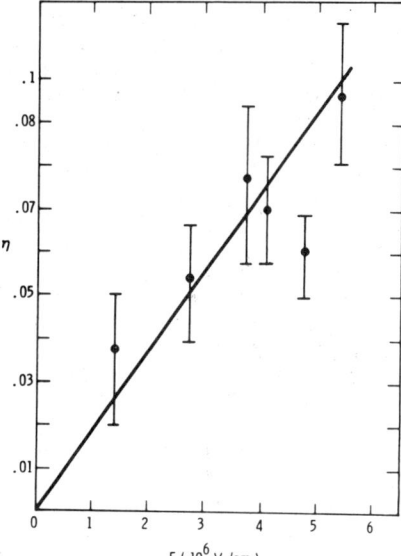

Fig. 2. The fraction of the thick-
ness of the SiO_2 layer traversed by holes in less than 10^{-8} s as a function
of the applied field. The SiO_2 layer is 3700 Å thick and the temperature
was 300°K.

these curves only have meaning when R exceeds an interatomic separation,
with a sufficiently weak carrier-lattice interaction only the nonpolaronic
situation will exist. However, with a stronger interaction both nonpolaronic and small-polaronic states will exist with the values of the physical
parameters determining which of the two states is the energetically preferred (stable) state. Furthermore, associated with the fact that only two
distinct situations can exist for a short range interaction is the fact that
an energy barrier must be traversed to convert a metastable nonpolaronic
carrier to a stable small polaron (4-7). The barrier exists because of
competition between the kinetic energy of the carrier and its potential
energy (associated with carrier-induced-atomic displacements) as a function
of the radius of the carrier's wavefunction. As the radius of the wavefunction shrinks the kinetic energy increases faster than the potential
energy decreases; but at a sufficiently small radius the situation is
reversed. This leads to the peak in the plot of the energy of the system
versus the radius of the carrier's wavefunction manifested in Fig. 1b (6).
Due to the presence of the barrier the conversion of a nonpolaronic carrier
to a small polaron is impeded. Thus the time required for this transformation can, depending on the physical parameters, greatly exceed a vibrational period. However, with an increase of the effective mass of the
carrier the kinetic energy, the barrier height, and concomitantly the small-
polaron formation time, are reduced.

In a material such as a-SiO2 the ionic character of the material supports a
long-range component of the interaction in addition to the short-range component. As depicted in Fig. 1 with the addition of a long range component
of the interaction the barrier is reduced (Fig. 1c) and for a sufficiently

strong long-range component even eliminated (Fig. 1d). Furthermore, the presence of energetic disorder also aids small-polaron formation by providing an additional premium for localization in a spatial region of low energy. For example, even with only a short-range carrier-lattice interaction the placing of the carrier in an increasingly deep coulombic well acts to reduce and ultimately eliminate the barrier: again, see the progression of Fig. 1b to Fig. 1d (6). Thus, three factors which are presumably present in a-SiO$_2$ all tend to reduce or eliminate a barrier to small-polaron formation. These three are: (a) the large effective mass of a hole in the rather narrow uppermost valence band, (b) the presence of a long-range polar type carrier-lattice interaction, and (c) the presence of disorder. In the absence of a detailed computation of these effects in a-SiO$_2$ one cannot comment on the magnitude or even existence of the barrier in this material. However, one can examine the drift mobility data for holes in a-SiO$_2$ to determine whether the small-polaron formation time is substantially longer than the minimum time for small-polaron formation (associated with the absence of a barrier) -- a time, of the order of a picosecond.

In our experiments (2) electron-hole pairs are generated uniformly throughout a thin film (typically $\leqslant 4 \times 10^3$ Å) of a-SiO$_2$ by the absorption of high energy x rays. With the application of a high electric field ($\geqslant 10^8$ V/m) the high-mobility electrons are swept out of the material in less than a nanosecond. The charge transport which occurs at later times is then due to holes. We envision the hole transport to occur in roughly three time domains. The first is an interval immediately after the hole is produced. During this period it has not formed a small polaron and hence moves with a relatively high mobility, say, 1 cm^2/V-s. After this period the carrier relaxes to form a small polaron and moves with a very much lower mobility; for example, it is observed to be $\sim 10^{-5}$ cm^2/V-s at 300°K (2). Finally, at still longer times the carrier encounters hard hops (presumably associated with "defects") and its motion is further impeded. It is only the initial stage of the carrier's motion which occurs too rapidly to be observed directly.

Nonetheless, at high fields it is observed that the total charge which has traversed the sample at a given time is somewhat more than is consistent with the hole possessing the low second stage mobility from the time of its inception (8). Thus the carrier must possess a higher mobility for some initial period. One may determine that fraction, η, of the total thickness of the sample which the carrier traverses before being slowed down. Figure 2 depicts η as a function of E. Despite the sizeable experimental error it is clear that the holes move only a small distance in their initial (high) mobility state. Employing the relation $\eta = \mu\tau E/d$, where μ is the initial hole mobility, τ is the lifetime of this high-mobility state, E is the strength of the applied electric field and d is the film thickness, then yields a value for $\mu\tau$ of $7 \pm 2 \times 10^{-13}$ cm^2/V. If one assumes a value of the mobility in the range 1 - 10 cm^2/V-s one finds that τ is 10^{-12} - 10^{-13} s. This is just the time which is expected for the formation of a small-polaron in the absence of a substantial delay time, i.e., in the absence of a significant barrier.

Alternatively it has been suggested (9) that after some initial high-mobility motion, rather than forming small-polarons, the holes are trapped in sufficiently large radius states so as to not force small-polaron formation. In particular, a localized state radius of ten interatomic spacings has been specifically suggested (7). Assuming a simple diffusion model one may relate the lifetime against trapping to the capture radius, R_t, and the localized state density, N_t, via the relation $\tau^{-1} = 4\pi D\, N_t\, R_t = 4\pi (kT/e)\, \mu\, R_t\, N_t$, where D is the diffusion constant; the Einstein relation was employed in obtaining the second equality. In our case one has that $N_t\, R_t = 5 \times 10^{12}$ cm^{-2}. At this point we note both (a) that the unexcited material is electronically neutral and (b) that the electron transport is not trap controlled. This implies that such localized states are neutral. However, neutral traps are well known to have capture radii which are typically 10^{-8} cm or less (10). Thus, for our experiment this implies a density of large radius states well in excess of 10^{20} cm^3. At such a density one might envision the overlap between such large radius states to be such as to yield extended states rather than the presumed localized states (11). Thus, unlike the small-polaron picture, this simple picture does not seem consistent with the experimental facts as we know them.

Furthermore the mobility which is directly measured in this (second) regime manifests magnitudes, temperature dependences and field dependencies which are all in accord with the predictions of small-polaron theory (2,12). Namely, the mobility is low and thermally activated at temperatures above one third the Debye temperature, breaking to a non-activated behavior at lower temperatures. Also consistent with the small-polaron model is the mild field dependence of the mobility at very high fields.

Still additional support for the small-polaron model emerges by comparing the jump rate associated with small-polaron diffusion with that associated with the nearest neighbor hopping of a hole about a substitutional aluminum atom, termed an A-center (13). Here, since the hole is well localized it may be viewed as simply being a bound small polaron. Again, as in the case of the measured hole mobility, one observes a break in the jump rate from a high-temperature activated behavior to a nonactivated behavior below one third the Debye temperature (2). Indeed, the jump rate associated with the motion of a small-polaron hole about an aluminum center is very similar to that which we ascribe to the diffusion of a small polaron in a-SiO$_2$. It is interesting to note that EPR studies of the A-center show that the bound small polaron is not of the V_K-center variety (14).

Finally, it should be noted that at longer times, in the third temporal regime, the transits become dispersive (2,3) with the transit times being associated with an activation energy (.37 eV) which is about 0.2 eV larger than that of the drift mobility (0.16 eV). Again, suggestive of small-polaronic behavior is the fact that at below about one third the Debye temperature the transit time departs from an Arrhenius behavior. The dispersive behavior suggests that after having moved a sufficient distance a significant fraction of the small-polarons begin to encounter hard hops. These especially difficult jumps may be associated with either defects in the material or with the wide fluctuations of the Si-O-Si bond angle which could significantly affect small-polaron inter-oxygen hopping motion.

REFERENCES

(1) R. C. Hughes, Hot electrons in SiO$_2$, Phys. Rev. Lett. 35, 449 (1975).
(2) R. C. Hughes, Time-resolved hole transport in a-SiO$_2$, Phys. Rev. 15, 2012 (1977).
(3) F. B. McLean, H. E. Boesch, Jr., and J. M. McGarrity, Hole transport and recovery characteristics of SiO$_2$ gate insulators, IEEE Trans. Nuc. Sci. NS-23, 1506 (1976).

18

(4) Y. Toyozawa, Self-trapping of an electron by the acoustical mode of lattice vibration. I, Prog. Theo. Phys. 26, 29 (1961).

(5) D. Emin, On the existence of free and self-trapped carriers in insulators: An abrupt temperature-dependent conductivity transition, Ad. Phys. 22, 57 (1973).

(6) D. Emin and T. Holstein, Adiabatic theory of an electron in a deformable continuum, Phys. Rev. Lett. 36, 323 (1976).

(7) N. F. Mott and A. M. Stoneham, The lifetime of electrons, holes and excitons before self-trapping, J. Phys. C: Solid State Phys. 10, 3391 (1977).

(8) R. C. Hughes, High field electronic properties of SiO_2, Solid-State Electronics 21, 251 (1978).

(9) N. F. Mott, Silicon dioxide and the chalcogenide semiconductors; similarities and differences, Adv. Phys. 26, 363 (1977).

(10) A. G. Milnes, (1973) Deep Impurities in Semiconductors, Wiley-Interscience, New York.

(11) N. F. Mott and E. A. Davis (1971) Electronic Processes in Non-Crystalline Materials, Clarendon Press-Oxford.

(12) D. Emin, Phonon-assisted transition rates I. Optical-phonon-assisted hopping in solids, Adv. Phys. 24, 305 (1975).

(13) W. J. De Vos and J. Volger, Dielectric relaxation phenomena in smoky quartz, Physica 47, 13 (1970).

(14) J. A. Weil, The aluminum centers in α-quartz, Rad. Eff. 26, 261 (1975).

FIELD-DEPENDENT HOLE TRANSPORT IN AMORPHOUS SiO$_2$

F. B. McLean, H. E. Boesch Jr., and J. M. McGarrity
Harry Diamond Laboratories, Adelphi, Maryland   20783

ABSTRACT

Field-dependent studies of the delayed, dispersive radiation-induced hole transport in thin amorphous films of SiO$_2$ have been carried out for a series of temperatures between 79 and 293 K. The data are analyzed assuming that polaron hopping between randomly distributed localized sites is the microscopic charge transfer mechanism. The average hopping distance in the field direction is found to be 9 $\pm$ 2 A for fields in the range of 1 to 6 MV/cm.

INTRODUCTION

Photo- or radiation-induced transient hole conductivity in amorphous SiO$_2$ films used as the gate insulators in metal-oxide-semiconductor (MOS) structures has been the subject of active interest over the past few years (Ref. 1-4). The general physical picture of the charge response that has emerged is the following. When a MOS structure is irradiated, hole-electron pairs are created in the SiO$_2$ film. Depending upon the applied electric field across the SiO$_2$, a fraction of the radiation-generated carriers escapes initial recombination (Ref. 5-7). Those electrons escaping initial recombination are rapidly swept out of the oxide. Most of the holes remain behind near their point of generation and cause negative flat-band voltage or threshold voltage shifts in the MOS structure. Over a period of time (milliseconds at room temperature, thousands of seconds at cryogenic temperature), the holes execute a relatively slow, temperature- and field-activated transport through the oxide and are removed at an interface (at the Si interface under positive gate bias). This hole transport is highly dispersive in nature, taking place over many decades in time. Although the origin of the dispersion is still a topic of active investigation, we have found that the stochastic nature of the transport, including the features of dispersion and universality with respect to temperature and field, is well described by the biased continuous time random walk (CTRW) model developed by Scher and Montroll (Ref. 8). In their model, the carriers are presumed to move by hopping between localized sites, and the dispersion is attributed to a broad distribution of hopping times or intersite transfer integrals (as would be the case for a random spatial distribution of sites). The microscopic intersite transfer mechanism consistent with both the thermal and the optical data along with the stochastic features of the transport is that of small polaron hopping between the localized sites (Ref. 2).

This paper presents the results of a field-dependent investigation of the delayed dispersive hole transport. The charge relaxation following pulsed ionizing irradiation was studied as a function of field from 1 to 6 MV/cm for a series of temperatures from 79 K to room temperature. These studies shed further light on the hopping nature of the transport. In particular, analysis of the results in terms of the semiclassical polaron hopping mobility yields values for the average hopping distance in the field direction and for the zero field activation energy.

SAMPLES AND EXPERIMENTAL TECHNIQUES

Thin film SiO_2 capacitors, grown as a part of a radiation hardening effort, were supplied by Hughes Aircraft Corporation. The samples consisted of wet-grown SiO_2 layers deposited on <100> n- and p-type Si substrates with vapor-deposited Al gate electrodes. The wet-process SiO_2 was grown with pyrogenic H_2O at 950 C to 965 A thickness and annealed in N_2 at 925 C for 20 min. These samples are especially suitable for transport studies because, as shown by late time measurements, less than 2 percent of the holes are permanently trapped in the oxide.

The experiments were carried out by using the electron linear accelerator (LINAC) at the Armed Forces Radiobiology Research Institute. The LINAC produced a nominal 13 MeV electron beam with a pulse width of 4 µs, with approximately 30 krad (SiO_2) per pulse delivered to the sample. The samples were irradiated in an evacuated sample holder which had provisions for electron beam dosimetry and for control of sample temperature from 80 to 500 K. Charge relaxation, or transport, was measured by monitoring the shift in the flat-band voltage in the capacitance-voltage (C-V) characteristic between the pre-irradiation value and the value at logarithmically increasing time intervals after the radiation pulse from 0.2 ms to 800 s, using a fast high frequency C-V measurement apparatus (Ref. 2). The flat-band voltage shift $\Delta V_{FB}(t)$ is proportional to the time-dependent first spatial moment (relative to the metal-oxide interface) of the radiation induced charge distribution in the oxide. As the charge transports and is collected at the electrodes, the flat-band voltage relaxes towards its preirradiation value.

RESULTS

Figure 1 shows a typical set of field-dependent relaxation data, in this case for positive gate bias and for T = 79 K. The data are normalized to the initial shift $\Delta V_{FB}(0^+)$ in the flat-band voltage immediately after the radiation pulse. The data are plotted in the negative direction because the shifts are negative, indicative of net positive charge induced in the oxide. For the data in Fig. 1, $\Delta V_{FB}(0^+)$ = -4.5 V.

There are several points to note about the data. First, for fields less than 3 MV/cm, there is little recovery of the flat-band voltage from the initial shift, indicating little or no transport. The initial shift under negative bias is also negative and within ∿ 5 percent of the value under positive bias. Since electrons are very mobile (Ref. 1-4) in the SiO_2 and are swept out under either bias before 10^{-4} s, this result is consistent with uniform generation of the holes throughout the SiO_2 and, at low temperature and field, the holes remain frozen very near their generation point for the time scale of the experiment (800 s).

Second, transport at 79 K is evidently strongly field activated. At 3 MV/cm, some transport and recovery begin to take place on the time scale of the observations, and as the field is increased above 3 MV/cm, the charge transport increases strongly until, at 6 MV/cm, there is almost complete recovery of ΔV_{FB} at 800 s, indicating that most of the positive charge has been collected at the electrode.

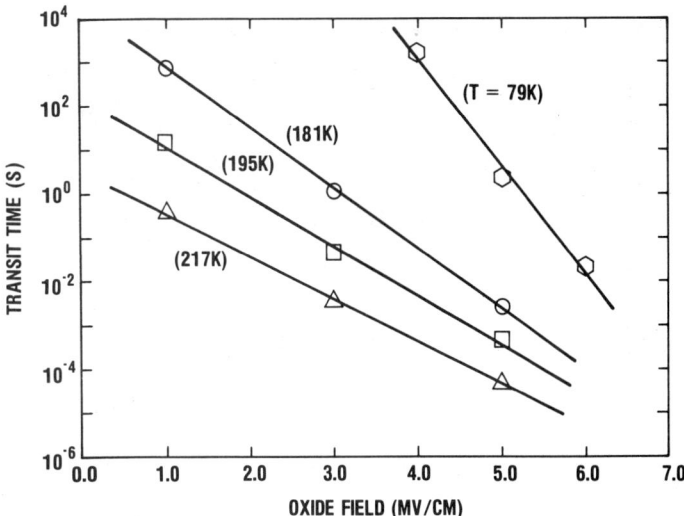

Fig. 2. Hole transit time versus electric field in the oxide using the time at which half recovery occurs as a measure of the transit time.

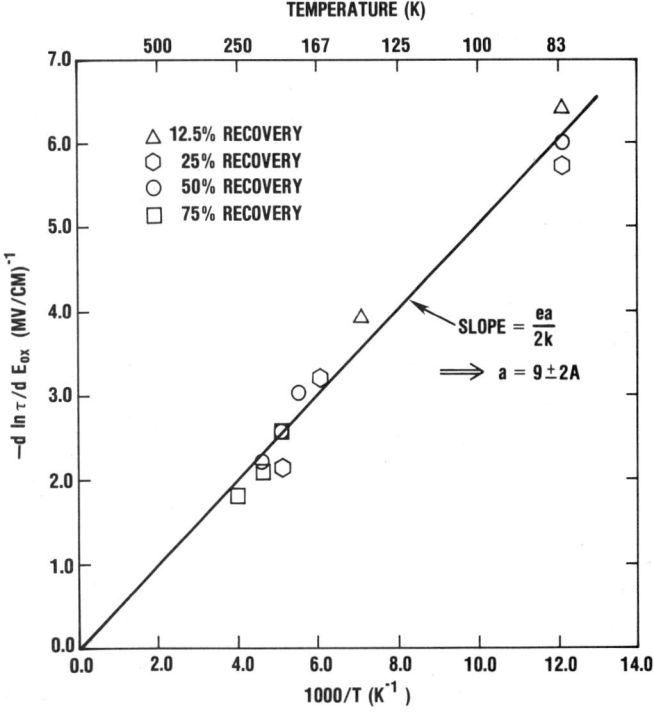

Fig. 3. The field slopes versus 1000/T for times τ corresponding to various fractions of recovery. The solid line is fitted through the points for half recovery.

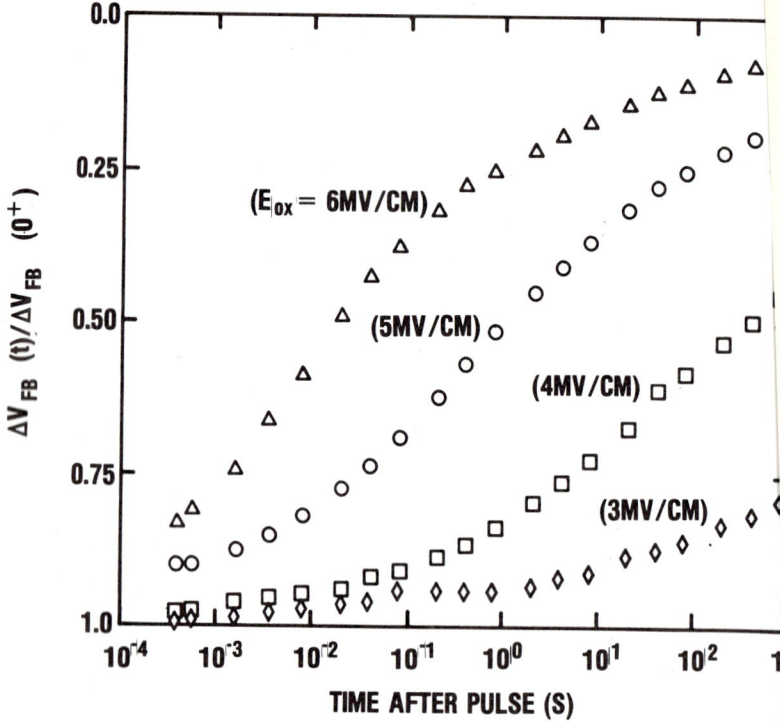

Fig. 1. Flat-band voltage recovery data at T = 79 K. The data a
normalized to the initial flat-band voltage shift (-4.5 V)
immediately after the radiation pulse.

From the data in the field range of 4 to 6 MV/cm, it is evident that t
recovery, or hole transport, is very dispersive in time, taking place
least 8 decades. Note also the universal appearance of the data, name
changing the field affects the time scale of the transport without sig
cantly changing the shape of the recovery curve. If the data are repl
in terms of a single scaled time, for example, the time $t_{\frac{1}{2}}$ at which hal
recovery occurs ($\Delta V_{FB}(t)/\Delta V_{FB}(0^+) = 0.5$), then the recovery curves very
trace out a single "universal" curve.

Finally, in the recovery curves for 3 and 4 MV/cm, there appears to be
tively small linear (on a semilog plot) component of recovery that cons
about 10 percent of the recovery at 800 s. From low temperature relaxa
studies on films of varying thickness, we have ascertained that this co
which is nearly field and temperature independent, is due to removal of
holes within approximately 50 A of the Si interface by electrons tunnel
from the silicon. In carrying out the analysis of the field dependence
transport discussed below, we subtracted out the effect of this tunnelin
component.

DISCUSSION

Field-dependent data such as shown in Fig. 1 were taken for a series of temperatures between 79 K and room temperature. The data are analyzed in terms of the semiclassical small polaron hopping mobility, $\mu(E) = \mu(0) \sinh(\beta eEa/2)/(\beta eEa/2)$, where E is the oxide electric field, $\beta = (kT)^{-1}$, a is an average hopping distance and $\mu(0)$ is the low field mobility. Taking the transit time $t_{\frac{1}{2}}$ as a measure of the average inverse mobility and assuming the case of large argument, $\beta eEa/2 \gg 1$, which is true for the fields and temperatures here, we have $t_{\frac{1}{2}}(E) = t_{\frac{1}{2}}(0) \exp(-\beta eEa/2)$. In Fig. 2, we plot log $t_{\frac{1}{2}}$ versus E with temperature as a parameter. The exponential dependence of the recovery with field is evident, with the magnitude of the slope increasing as the temperature is lowered. If the semiclassical hopping formula is applicable, the field slope should be ea/2kT.

Figure 3 shows the field slopes plotted versus 1/T. The circles are the values of the slopes taken from Fig. 2 for the times required for half recovery. The straight line is a fit through these points and the origin. According to the hopping mobility formula the line yields a value of 9 A for the average hopping distance. Also shown in Fig. 3 are points taken from other plots similar to Fig. 2 where other fixed fractions of charge recovery (12.5, 25, and 75 percent) were used as a measure of the hole transport speed. Only transit time versus field plots with at least three field points are included. Though there is some scatter in these points, all the data are consistent within experimental resolution with the straight line through the half recovery points. From the uncertainties in the data and the analysis, we estimate error limits on the average hopping distance of ± 2 A.

Although not discussed in this paper, an average hopping distance of 9 A is consistent with the reduction in activation energy we have observed with increasing field in the high-temperature activated hopping regime. The activation energy in this regime can be described within experimental accuracy as $\Delta(E) = \Delta_o - eEa/2$ with the zero field activation energy $\Delta_o = 0.65 \pm 0.05$ eV and an average hopping distance of $a = 10 \pm 3$ A.

Combining the field-dependent data reported here with the temperature characteristics of the transport, we conclude that the hole transport process in amorphous SiO_2 in that of small polaron hopping between randomly distributed localized sites. The hole transit times are well described by the semiclassical, large argument hopping mobility with an average hopping distance in the field direction of 9 ± 2 A.

REFERENCES

1. F. B. McLean, G. A. Ausman, H. E. Boesch, Jr., J. McGarrity, Application of Stochastic Hopping Transport to Hole Conduction in Amorphous SiO_2, J. Appl. Phys. 47, 1529 (1976).
2. F. B. McLean, H. E. Boesch, Jr., and J. M. McGarrity, Hole Transport and Recovery Characteristics of SiO_2 Gate Insulators, IEEE Trans. Nucl. Sci. NS-23, 1506 (1976).
3. R. C. Hughes, Time-Resolved Hole Transport in a-SiO_2, Phys. Rev. B 15, 2012 (1977).
4. O. L, Curtis, Jr., and J. R. Srour, The Multiple-Trapping Model and Hole Transport in SiO_2, J. Appl. Phys. 48, 3819 (1977).
5. O. L. Curtis, Jr., J. R. Srour, and K. Y. Chiu, Hole and Electron Transport in SiO_2 Films, J. Appl. Phys. 45, 4506 (1974).
6. G. A. Ausman, Jr., and F. B. McLean, Electron-Hole Pair Creation Energy in SiO_2, Appl. Phys. Lett. 26, 173 (1975).
7. H. E. Boesch, Jr., and J. M. McGarrity, Charge Yield and Dose Effects in MOS Capacitors at 80 K, IEEE Trans. Nucl. Sci. NS-23, 1520 (1976).
8. H. Scher and E. W. Montroll, Anomalous Transit-Time Dispersion in Amorphous Solids, Phys. Rev. B 12, 2455 (1975).

EXCITON TRANSPORT IN SiO$_2$*

Z. A. Weinberg and G. W. Rubloff
IBM Thomas J. Watson Research Center, Yorktown Heights, NY 10598

ABSTRACT

Positive charge accumulates at the Si-SiO$_2$ interface in MOS structures when illuminated by VUV photons under conditions where neither holes nor photons can reach the interface. The transport of energy from the metal side of the oxide (where the photons are absorbed) to the Si-SiO$_2$ interface is attributed to diffusion of neutral excitons. The reduction of the effect by ion implantation is consistent with this picture. Models are discussed for excitons which might diffuse across a 1000-2000 Å oxide. In addition, the field dependence of photoconductivity suggests that the SiO$_2$ band gap is $\sim$ 9.3 eV.

INTRODUCTION

We have been studying the accumulation of positive charge at the Si-SiO$_2$ interface in MOS structures caused by vacuum ultraviolet (VUV) photon illumination at energies where the SiO$_2$ is strongly absorbing. These results are particularly interesting because they appear to give new information related to the transport and decay of excitons in SiO$_2$.[1] In this paper, we present new experimental results and discuss models to explain the exciton transport and decay and their implications for understanding a number of other observations in MOS structures.

The experimental phenomenon and its possible causes are illustrated schematically in Fig. 1. Monochromatic VUV light incident on the MOS structure passes through the thin metal gate and is strongly absorbed in the SiO$_2$ within a few hundred Å of the metal; this results in the accumulation of positive charge at the Si-SiO$_2$ interface, which is measured by shifts of capacitance-voltage curves. Although a variety of techniques (e.g. high field stressing, ionizing radiation, high energy electron bombardment) are known to produce positive charge at the Si-SiO$_2$ interface, the experimental approach discussed here is particularly valuable in that it allows us to rule out the usual explanations for this effect involving holes, photons, and impact ionization (see Fig. 1). This has been accomplished by an extensive set of experiments which include the dependence of the effect on photon energy, oxide thickness, and applied negative bias. An initial discussion of these conclusions has been given previously.[1]

We have concluded that the effect arises from transport of energy through the oxide in an electrically neutral form, most likely an exciton. If the diffusion length of the excitons is sufficiently long, excitons can reach the Si-SiO$_2$ interface to produce the positive charge by various mechanisms.

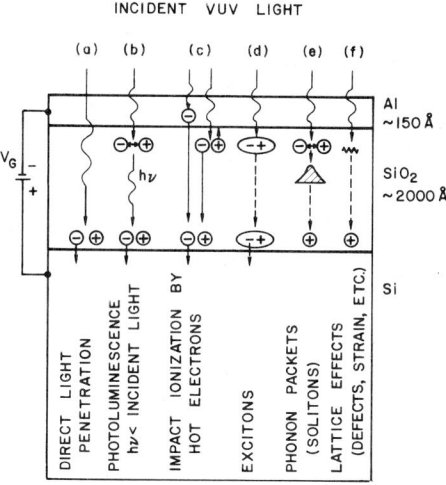

Fig. 1. Schematic illustration of possible mechanisms to produce positive charges at the Si-SiO$_2$ interface. The incident photons (wiggly lines) are strongly absorbed by the SiO$_2$. Under negative gate bias (V$_G$) the positive charge can be a result of photon processes including light penetration to the interface (a) and photoluminescence (b), hot electron effects (c), or propagation of neutral excitations (indicated by dashed lines) such as excitons (d), localized, phonon packets (e), or other lattice excitations (f).

EXPERIMENTAL RESULTS

In this section we present a few of our recent results.

1. We have measured the optical transmission of thin thermally-grown oxide membranes ($\sim$ 500 Å thick produced by etching a window in the silicon substrate). This yields an absorption length of 140 $\pm$ 20 Å at 10.2 eV, which rules out light penetration to the interface (explanation (a)). We have also established an upper limit of 10^{-3} for the photoluminescence yield at 10.2 eV, which rules out explanation (b).

2. Since exciton transport through the oxide should be suppressed by traps or defects, we have performed measurements which show that an ion-implantation layer in the oxide quenches the effect. Arsenic was chosen as the implanted species because it traps both electrons and holes[2] and therefore is expected to be a recombination site for excitons. The relative reduction of positive charge accumulation at the Si-SiO$_2$ interface by the implantation is shown in Fig. 2. The charge trapped in the implanted region of the oxide during the VUV illumination was almost entirely removed by a second illumination at photon energies below 5.5 eV with the MOS capacitor shorted[3] (using a deuterium source and a cutoff filter); this second illumination removed only a fraction of the interface charge, which was measured by calibration runs on unimplanted samples. Figure 2 shows that a layer of 1 × 10^{13} As/cm^2 sufficiently deep in the oxide (centroid $\sim$ 470 Å from the Aℓ) removes the effect completely. Because the exciton motion is

random in space, this implies a capture cross-section smaller than 10^{-13} cm^2.

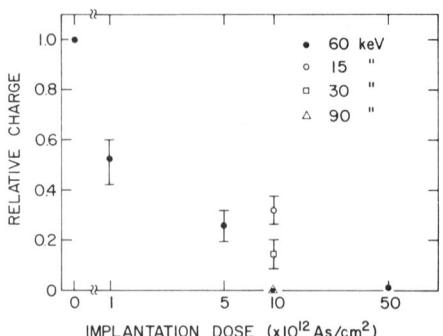

Fig. 2. The effect of ion implantation (arsenic) on the amount of positive charge accumulation at the Si-SiO$_2$ interface after illumination at 10.2 eV for 15 min. with $V_G = -7$ volts. The relative charge is normalized by calibration samples (cut from an unimplanted portion of the wafer) which are designated by zero implantation dose in the figure. The optical absorption length of the 10.2 eV light is 140 ± 20 Å (see text), and the centroids of the implantation, measured by the photo I-V technique, are located at 165, 260, 470, 600 Å away from the Al gate for 15, 30, 60, 90 KeV implantation energy, respectively (Ref. 17). The SiO$_2$ thickness was 1400 Å.

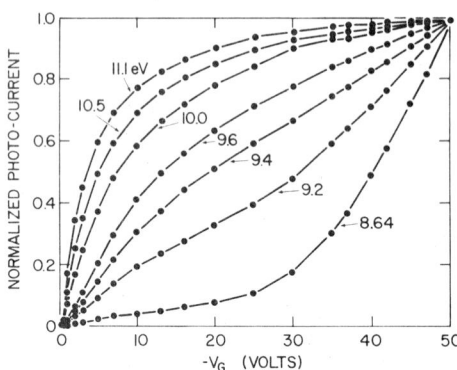

Fig. 3. A comparison of the voltage dependence of photoconductivity at several photon energies. The photocurrents are all normalized (to their value at 50 volts) to fit within the same scale. The oxide (1440 Å) was thermally grown on 2Ω-cm p-type silicon.

3.   The field dependence of photoconductivity is interesting because it measures the separation of free electrons and holes, including those produced by exciton dissociation. The voltage dependence of photoconductivity at several photon energies, shown in Fig. 3, displays a striking resemblance to that of internal photoemission. This similarity is not surprising because in both cases an electron is separated from a positive charge. In internal photoemission experiments, a change in curvature occurs when the photon energy equals the barrier height.[4] The change in curvature between 9.2 and 9.4 eV in Fig. 3 suggests that the band gap is ~ 9.3 eV, a value consistent with previous work.[5] The stronger field dependence at lower photon energy would then be attributed to field ionization of the excitons.

DISCUSSION

To explain the effect, namely the accumulation of positive charge at the Si-SiO$_2$ interface under VUV illumination and negative bias, we envision the following processes. A fraction of the photons absorbed in the oxide produce electron-hole pairs which are not separated by the applied electric field. These electron-hole pairs, or excitons, must have a substantial binding energy (~ 1 eV, typical of many insulators), so that they are not immediately field-ionized by the applied electric field. The diffusion length of the excitons must be several hundred Å.[1] Such values are consistent with the diffusion distances before self-trapping for excitons in solid rare gases and alkali halides.[6]

The effect is observed over a considerable range of photon energies (~9-13 eV)[1] near and above the one-electron band gap (E$_g$ ~ 9.3 eV). It seems likely that excitons of excitation energy E$_{exc}$ > E$_g$ will decay rapidly into band-like excitations; of these, the free electrons and holes which are sufficiently separated in space (to prevent recombination back into an exciton) will then contribute to photoconductivity but not to the observed effect. However, some of the excitons with E$_{exc}$ > E$_g$ may decay quickly by phonon emission into excitons with E$_{exc}$ < E$_g$; these lower energy excitons would then have considerably longer lifetimes. In this picture, it seems most likely that energy transport across the oxide occurs by rather strongly-bound excitons below band gap.

There are three candidates for such an exciton below E$_g$, as follows: (a) If the band gap is indirect[7] and ~ 9.3 eV, an indirect exciton with E$_{exc}$<E$_g$ may be generated from the initial exciton by phonon emission; these excitons would appear only weakly in the optical absorption spectrum of SiO$_2$ (optically allowed only by phonon coupling or disorder). (b) Triplet excitons (spin aligned) might be generated by nonradiative decay from the initial singlet excitons. (c) The initial exciton, during its nonradiative recombination process, may relax into a relatively stable state which is coupled to a strong local lattice deformation that can travel with the exciton (a form of mobile self-trapped excitons). The required lattice deformation may be provided by the phonon emission during the initial stages of the relaxation.

In all three cases, the excitons would be below band gap and have significantly longer lifetimes than the initial singlet excitons generated by the optical absorption at 9 eV and above. In this way, metastable exciton states might have energies which extend considerably (one or even several eV) below the band gap. Although these states would not be excited by optical absorption, they may be excited whenever hot carriers of sufficient energy and momentum are generated. Such a metastable exciton model is attractive because it might explain long exciton diffusion lengths and also several other observations in SiO$_2$. First, the creation of such metastable excitons by hot electrons could represent an efficient mechanism for energy loss at energies below ~ 9 eV and therefore explain the high dielectric strength of SiO$_2$ as well as the appearance of positive charge at the Si-SiO$_2$ interface observed in negative corona charging[8,9], high field stressing[10-12], and avalanche injection[13]. Second, since metastable exciton states are unlikely to exist at energies more than a few eV below band gap, this energy loss mechanism will not contribute in those experiments where a relatively low energy loss per unit distance is deduced, such as electron emission through thin SiO_2 layers into vacuum[14,15] and internal photoemission at low fields[4]. Finally, this model would provide a simple explanation for the oxide thickness dependence of the positive charge at the $Si-SiO_2$ interface observed in certain cases of bombardment of MOS structures by ionizing radiation or high energy electrons[16].

In conclusion, these experiments seem to provide strong evidence for exciton transport in SiO_2. They give novel information about excitonic states beyond what is usually inferred from optical data and band structure calculations.

ACKNOWLEDGMENT

The authors are grateful to E. Bassous for his cooperation in this work and appreciate illuminating discussions with D. Emin, R. C. Hughes, and N. F. Mott. In particular we are grateful for D. Emin's suggestion about triplet excitons.

REFERENCES

* This research was supported by the Defense Advanced Research Projects Agency and monitored by the Deputy for Electronic Technology, RADC, under Contract F19628-76-C-0249.

1. Z. A. Weinberg and G. W. Rubloff, *Appl. Phys. Lett.* **32**, 184 (1978).
2. D. J. DiMaria, *Paper-BA1*, this conference.
3. R. F. DeKeersmaecker, D. J. DiMaria, and S. T. Pantelides, *Paper-BA5*, this conference.
4. R. J. Powell, *J. Appl. Phys.* **41**, 2424 (1970).
5. T. H. DiStefano and D. E. Eastman, *Solid-State Commun.* **9**, 2259 (1971).
6. N. F. Mott, *Advances in Physics*, **26**, 363 (1977).
7. J. R. Chelikowsky and M. Schlüter, *Phys. Rev.* **B15**, 4020 (1977).
8. Z. A. Weinberg, W. C. Johnson, and M. A. Lampert, *J. Appl. Phys.* **47**, 248 (1976).
9. M. H. Woods and R. Williams, *J. Appl. Phys.* **47**, 1802 (1976).
10. M. Shatzkes and M. Av-ron, *J. Appl. Phys.* **47**, 3192 (1976).
11. P. M. Solomon and J. M. Aitken, *Appl. Phys. Lett.* **31**, 215 (1977).
12. D. J. DiMaria, Z. A. Weinberg, and J. M. Aitken, *J. Appl. Phys.* **48**, 898 (1977).
13. D. R. Young, private communication.
14. R. Poirier and J. Olivier, *Appl. Phys. Lett.* **21**, 334 (1972).
15. J. Peisner, G. Aszódi, M. Németh-sallay, and G. Forgács, *Thin Solid Films* **36**, 251 (1976).
16. T. P. Ma, *Appl. Phys. Lett.* **27**, 615 (1975).
17. D. J. DiMaria, D. R. Young, R. F. Dekeersmaecker, W. R. Hunter, and C. M. Serrano, to be published.

HIGH-ELECTRIC FIELD TRANSPORT OF ELECTRONS IN SiO$_2$

D. K. Ferry
Colorado State University, Fort Collins, Colorado 80523

ABSTRACT

The transport of electrons at high electric fields in SiO$_2$ has been calculated using an iterative solution of the path-variable form of the Boltzmann equation. Scattering due to the polar-optical modes and acoustic modes of the lattice were included. It is found that both the 0.153eV and 0.063eV polar optical modes are important for scattering of electrons. The "saturated" velocity is near 1.9×10^7cm/sec. The calculated velocity-field curves are in excellent agreement with experiment, yielding a best estimate of the polaron corrected conductivity mass of 1.3m. Time resolved polar runaway is observed with an absolute threshold field near 1×10^7V/cm, with the runaway field increasing for short times, such as may be appropriate for transport through thin oxides. Impact ionization is found to occur at fields near 1×10^7V/cm for realistic mass ratios.

INTRODUCTION

The transport of electrons in amorphous SiO$_2$ is of interest primarily because of its usage as a dielectric medium for use in metal-oxide-silicon semiconductor devices. Primary attention with regard to the transport properties has usually centered on the breakdown phenomena of this material. Hughes (1) has measured the transport properties of hot electrons as a function of the applied electric field and observed that the drift velocity begins to saturate at electric fields of the order of 8×10^5V/cm. He earlier pointed out that transport of electrons in a-SiO$_2$ was probably dominated by intrinsic phonon scattering above 200K, since the mean free path is so short in this material (Ref. 2), being essentially shorter than the ordering length.

Early attempts to treat scattering by the LO phonons in polar crystals and to determine the resultant electron mobility centered around techniques involving variational methods by Howarth and Sondheimer (3) or the use of a displaced Maxwellian approximation by Stratton (4) to facilitate solution of the Boltzmann equation. Recently, an iterative-integral technique was used (Ref. 5) with a single LO scatterer to determine the velocity of hot electrons in a-SiO$_2$. Thornber and Feynman (6) earlier utilized a quantum-mechanical path-variable technique and obtained qualitatively similar results. In the present paper, the calculations of Ref. 5 are extended to consider scattering by the major LO scatterers and to investigate the role played by polar runaway. Calculations for carrier ionization coefficients are presented to estimate the avalanche breakdown field of the oxide.

VELOCITY-FIELD RELATIONSHIP

In recent years, solutions of the semi-classical Boltzmann equation, which retain the full form of the collision integral, have been obtained using the Monte Carlo and path-variable iterative techniques (see, e.g., Price, Ref. 7). These methods are exact in the sense that no approximations to the physics, other than in the material model itself, are introduced. In particular, the iterative techniques utilized here build upon a path-variable formalism and offer a relatively rapid computational result.

The major scattering process in SiO_2 has previously been shown to be the high-energy LO phonon at 0.153eV (Ref. 8). Little is known of the details of the SiO_2 conduction-band structure, but low-field transport studies (Refs. 1,2) indicate that a polaron-corrected effective mass of $1.3m_0$ is appropriate to this material. Using this effective mass and the a value of $(\varepsilon_\infty^{-1}-\varepsilon_0^{-1})=0.143$, given by the dielectric dispersion analysis of Lynch (8), the velocity-field curve can be calculated. This is shown in Fig. 1 as the dashed curve labeled "a." The data of Hughes (1) is also shown for comparison. This curve appears to have the proper curvature, but gives too large a value of the velocity. A relatively saturated velocity of about 2.2×10^7 cm/sec is found at high fields. The curve is insensitive to the value utilized for the acoustic deformation potential in the range 0-15eV used for the calculations. A value of 7eV has been taken in the results reported here. The velocity-field curve is sensitive to the value of the effective mass used. For a reduced mass, $1.0m_0$, the dashed curve marked "b" in Fig. 1 is found. Not only is the velocity too large, but the curvature near the knee of the data at $5-8 \times 10^5$V/cm appears wrong. Using an effective mass smaller than $1.3m_0$ does not yield curves that agree well with the data.

A second strong LO phonon exists at an energy of 0.063eV (Ref. 8), although this phonon is considerably weaker than the high-energy mode. It is found that this mode contributes sufficient scattering to affect the velocity-field curve. From the known optical and static dielectric constants of SiO_2 and the above mentioned strength of the high-energy LO phonon, the coupling constant of the low-energy LO mode can be determined. This gives $(\varepsilon_{int}^{-1}-\varepsilon_0^{-1})=0.063$, where ε_{int} is the dielectric constant in the region between the two LO modes, and ε_0 in the equation above for the high energy LO mode is replaced by ε_{int} also. The velocity-field curve calculated with both LO modes is shown as the solid curve in Fig. 1. The fit to the data is remarkable in the region where theory and experiment overlap. Moreover, the curve appears to extrapolate to a low-field mobility of about $30cm^2$/V-sec in close agreement to the estimates made by Lynch (8). At high fields, the saturated velocity is about 1.9×10^7 cm/sec.

VELOCITY RUNAWAY

Velocity-runaway has been predicted in most calculations of the transport in polar material (Refs. 4,6). More explicitly, above a specific velocity, a stable velocity-field solution does not exist in calculations based upon energy loss considerations. In iterative calculations, the actual distribution function is calculated rather than the velocity, and the flow of particles to high energy is sensed by a build-up of the high-energy tail of the distribution function. One of the important aspects of the iterative technique is that each iteration corresponds to a specific, and well-defined, time step of the solution (Ref. 9). Thus, the transient behavior of the distribution function can be monitored effectively. For electric fields above about 10^7V/cm, the distribution function is stable for short times only. For longer times, the high

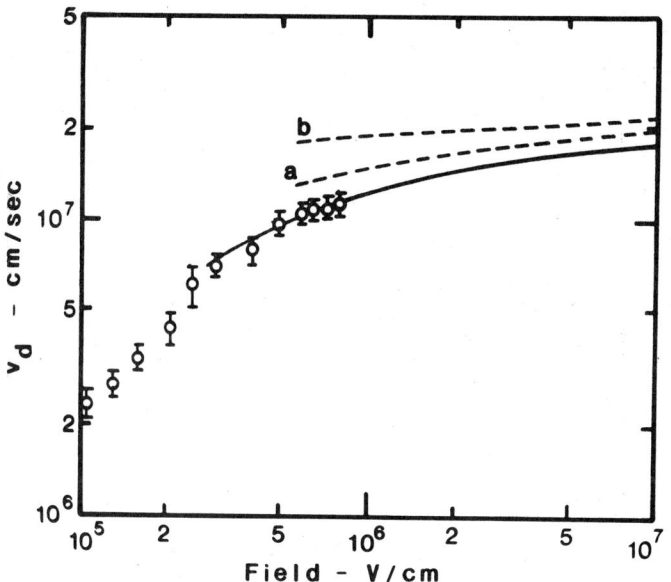

Fig. 1. Velocity as a function of electric field. The data is from Hughes (1) and the various curves are discussed in the text.

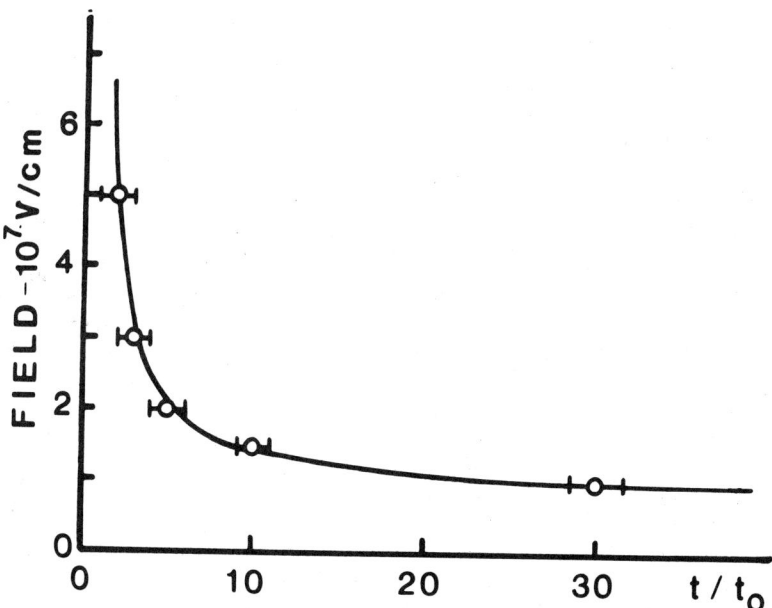

Fig. 2. Stability time before polar runaway as a function of electric field.

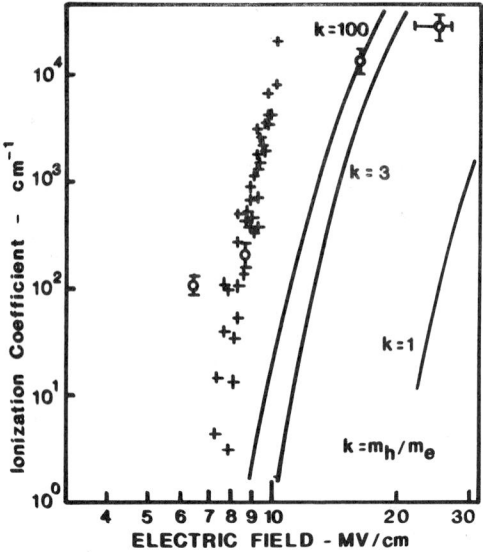

Fig. 3.   Impact ionization coefficient.   Data is from
Solomon and Klein (20) and Smith, et al., (21).

energy tail builds up and velocity runaway is encountered. This behavior is
shown in Fig. 2.   Shown in the figure is the approximate boundary between stable
and unstable regions. For values of the field and time that lie below the curve,
the distribution is stable. For values that lie above the curve, velocity runaway occurs. The normalized time t_o on the horizontal axis is related to the
minimum relaxation time for LO scattering, and in SiO_2 is about 10^{-15}sec. For
fields as low as $8-9\times10^6$V/cm, the distribution function is stable for very long
times. This agrees well with the calculations of Thornber and Feynman (6) in
which a steady-state calculation yields the maximum field for which a stable
velocity is found.

AVALANCHE BREAKDOWN

The dielectric strength of amorphous silicon dioxide film has an important
bearing on the performance and reliability of metal-oxide-silicon structures.
Dielectric breakdown has generally been attributed to thermal breakdown (Ref.
10) or velocity-runaway (Ref. 6), with only recent experimental data suggesting
intrinsic impact ionization as a mechanism (Ref. 11). Carrier multiplication
through impact ionization has been treated theoretically by several authors
(Refs. 12-16) and applied to SiO_2 by the present author (Ref. 17). In Fig. 3,
the theoretical impact ionization coefficient is shown as a function of the
electric field (solid curves) for several values of the electron-hole mass
ratio. These curves are based upon the Okuto and Crowell (18) analytical

expression for the detailed numerical calculations of Baraff (14). The effective mass ratio enters the calculation in its effect on the threshold energy for pair production (Ref. 19). The measured data from MOS devices measured by Solomon and Klein (20) is shown as the crosses and the laser data on bulk quartz measured by Smith, et al., (21) is shown as the circles.

DISCUSSION

With the short mean free paths experienced in this material and the short time between collisions, there arises concern over the use of the Boltzmann equation techniques. Thornber and Feynman (6) explicitly considered a full quantum mechanical treatment and obtained similar results for the case of a single LO scatterer. Moreover, the results obtained here agree well with the experimental data, so that is some confidence in the validity of the results. Both Barker (22) and Thornber (23) have pointed out that the scattering rates are modified in the presence of high electric fields. These effects can be expected to be important for fields above $2-5 \times 10^6$V/cm. Only preliminary calculations have been carried out with these effects included, but these tend to indicate the results of Fig. 1 are changed very little.

The calculations of impact ionization for realistic mass ratios indicate that avalanche breakdown could occur at fields of the order of $7-10 \times 10^6$V/cm. These are still above the experimental data, but are of the same magnitude as those for which velocity runaway occurs.

REFERENCES

(1) R. C. Hughes, Hot electrons in SiO_2, Phys. Rev. Letters 35, 449 (1975); High Field Electronic Properties of SiO_2, Sol. State Electr. 21, 251 (1978).
(2) R. C. Hughes, Charge-carrier transport phenomena in amorphous SiO_2: direct measurement of the drift mobility and lifetime, Phys. Rev. Letters 30, 1333 (1973).
(3) D. J. Howarth and E. H. Sondheimer, The theory of electron conduction in polar semiconductors, Proc. Roy. Soc. (London) A219, 53 (1953).
(4) R. Stratton, The influence of interelectronic collisions on conduction and breakdown in polar crystals, Proc. Roy. Soc. (London) A246, 406 (1958).
(5) D. K. Ferry, Electron transport at high fields in a-SiO_2, Appl. Phys. Letters 27, 689 (1975)
(6) K. K. Thornber and R. P. Feynman, Velocity acquired by an electron in a finite electric field in a polar crystal, Phys. Rev. B 1, 4099 (1970).
(7) P. J. Price, Calculation of hot electron phenomena, Sol.-State Electr. 21, 9 (1978).
(8) W. T. Lynch, Calculation of electric breakdown in quartz as determined by dielectric dispersion analysis, J. Appl. Phys. 53, 3274 (1972).
(9) H. D. Rees, Numerical solution of electron motion in solids, J. Phys. C 5, 641 (1972).
(10) J. J. O'Dwyer, Theory of dielectric breakdown in solids, J. Electrochem. Soc. 116, 239 (1969).
(11) M. Shatzkes, M. Av-Ron, and M. Anderson, On the nature of conduction and switching in SiO_2, J. Appl. Phys. 45, 2006 (1975).
(12) P. A. Wolff, Theory of electron multiplication in silicon and germanium, Phys. Rev. 95, 1415 (1954).
(13) W. Shockley, Problems related to p-n junctions in silicon, Sol.-State Electr. 2, 35 (1961).

(14) G. A. Baraff, Distribution function and ionization rates of hot electrons, Phys. Rev. 128, 2507 (1962).
(15) L. V. Keldysh, Concerning the theory of impact ionization in ionic semi-conductors, Zh. Eksperim i Teor. Fiz. 48, 1962 (1965) [Sov. Phys.-JETP 21, 1135 (1965)].
(16) W. P. Dumke, Theory of avalanche breakdown in InSb and InAs, Phys. Rev. 167, 783 (1968).
(17) D. K. Ferry, Impact ionization in silicon dioxide, Sol. State Commun. 18, 1051 (1976).
(18) Y. Okuto and C. R. Crowell, Ionization coefficients in semiconductors: a nonlocalized property, Phys. Rev. B 10, 4284 (1974).
(19) C. L. Anderson and C. R. Crowell, Threshold energies for electron-hole pair production by impact ionization in semiconductors, Phys. Rev. B 5, 2267 (1972).
(20) P. Solomon and N. Klein, Impact ionization in silicon dioxide at fields in the breakdown range, Sol. State Commun. 17, 1397 (1975).
(21) W. L. Smith, J. H. Bechtel, and N. Bloembergen, Picosecond laser-induced breakdown at 5321 and 3547Å: Observation of frequency-dependent behavior, Phys. Rev. B 15, 4039 (1977).
(22) J. R. Barker, Quantum transport theory of high-field conduction in semiconductors, J. Phys. C 6, 2663 (1973).
(23) K. K. Thornber, High-field electronic conduction in insulators, Sol.-State Electr. 21, 259 (1978).

ELECTRON EMISSION FROM SILICON DIOXIDE
INTO VACUUM

P.M. Solomon
IBM T.J. Watson Research Center Yorktown Heights, NY 10598.

ABSTRACT:

Electron emission from SiO_2 films into vacuum is investigated. Au-SiO_2-Si structures were stressed at fields of 8-10 MV/cm (Au +) and the electrons emitted through the thin Au electrode into vacuum were collected and their energy distribution measured. For a sample with SiO_2 and Au thicknesses of 300 and 120 Å respectively, electrons with kinetic energies of over 10eV were detected. The electron energy distributions had a Maxwellian tail with effective electron temperatures increasing from 0.5 to 0.8 eV with field. While these results demonstrate the existence of band gap electrons, and the possibility of impact ionization in SiO_2, the effect of the Au electrode on the distribution prevents, at present, a quantitive comparison of these results with various dielectric breakdown theories .

INTRODUCTION

The present investigation concerns the determination of the hot electron energy distribution in SiO_2 under high field stressing conditions. This data is necessary in order to understand high field electron transport, impact ionization, and dielectric breakdown in SiO_2, for which some models have recently been proposed [1-6] as well as to account for experimental results of high field charge generation [7,8] and electron trapping [9,10].

The method consists of biassing the metal-oxide-silicon (MOS) capacitor in vacuum with the top metal electrode positive. Electrons tunnel from the Si into the SiO_2 conduction band where they are accellerated by the applied field. The electrons are transported through the insulator and the thin Au top electrode, into vacuum, where they are collected and their energy measured.

This technique was first used by Mead [11] and subsequently by many other investigators. Electron energy loss in insulators such as Al_2O_3 and SiO (but not SiO_2) have been studied. For reviews of this technique the reader is referred to papers by Handy [12] and Eckertova [13]. Energy angular distributions were measured for Al_2O_3 by Hrach et. al. [14].

The main difficulty with the previous investigations was that the insulators were poorly characterised and their prorerties varied greatly from sample to sample. The conduction mechanism, electron trapping and electric field distribution in the insulator were largely unknown, and often a partial breakdown or 'forming' process had to be used to increase or stabilize the electron emission.

This experimental uncertainty is largely eliminated when using SiO_2, which exhibits a well behaved Fowler Nordheim electronic conduction mechanism [15], and internal photoemission characteristics [16] closely follow those of an ideal insulator. Trapped charge can be kept at a minimum during the experiment so that electric fields may be considered to be uniform. High quality SiO_2 films can be obtained with the minimum of defects and partial breakdowns.

An experimental difficulty with SiO_2, which may have prevented it's investigation in the past, are the extremely low emission currents obtained, which necessitate special instrumentation. The main difficulty in interpreting the results of this type of experiment is to account for the effects of the top electrode, as will be discussed later.

SAMPLE PREPARATION

SiO_2 was grown etched and regrown on <100> silicon substrates of 2 Ω-cm resistivity to form thin oxide areas of 0.08 cm dia. surrounded by thicker 1000 Å oxide. Thin (120 Å) Au dots of 0.24 cm dia. were evaporated overlaying the thin oxide areas and were contacted by thick Au dots evaporated on the thick oxide. The Au was 99.999% pure and the evaporation carried out at a background pressure of 3×10^{-6} torr with the sample at room temperature (water cooled holder). The samples were subsequently diced and mounted on TO-5 headers. The thickness of the thin Au was measured using optical transmission. The capacitors were checked for ionic contamination using capacitance vs voltage techniques and mobile ion concentrations of $<5 \times 10^{11}$ cm^{-2} were obtained.

MEASUREMENT TECHNIQUE

Pilot experiments had shown the electron yield from SiO_2 to be extremely low so that an apparatus of great sensitivity was constructed as is shown in fig.1. An electron multiplier (Galileo model 4219 'Spiraltron') with a dark count of <0.5 electrons per sec (sometimes as low as 0.02 electrons/sec) was used as the detector. To effectively interface the sample to the the multiplier a special electron energy analyser was constructed (see fig.1). Computer simulations and calibration using LaB_6 thermionic source showed that the analyser efficiently collected low energy electrons and achieved an energy resolution of 0.25 ev. The analyser was Au plated to minimise work function differences and an aperture of 0.05 cm dia. was placed in close proximity to the sample, at the cathode potential, to create a field free region near the cathode. A magnetic shield surrounded the whole structure. Measurements were carried out under a vacuum of 1×10^{-7} torr using a Freon cold trapped oil diffusion system.

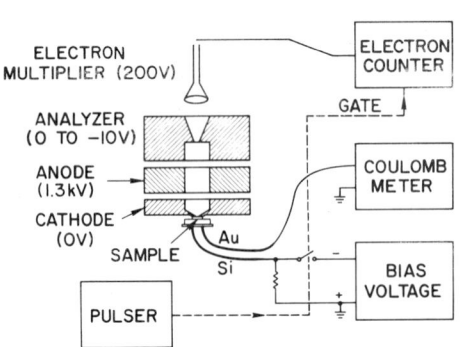

Fig.1 Schematic diagram of the apparatus used in the electron emission experiment.

For measuring the electron emission, the sample's substrate was pulsed positive at fields high enough (>6MV/cm) to cause currents to flow by the Fowler Nordheim conduction mechanism. The emitted electrons were counted during the pulse as a function of the retarding potential on the analyser electrode and the total charge passed through the sample was integrated using a Keithley model 610C electrometer. In order to minimize charge trapping in the oxide the pulse width was adjusted (from 1ms to 100s) to keep the total charge per pulse below 1×10^{-9} Cb.

RESULTS

Electron emission was seen at all fields at which appreciable Fowler-Nordheim conduction occurred. At a constant applied voltage the emission was stable and was proportional to the charge passed through the sample at all except the lowest voltages, where non-Fowler-Nordheim time varying currents were observed. A grave experimental difficulty was the occurrence of non-shorting breakdowns. Such breakdowns would not affect the current appreciably but could result in a large, unstable electron emission. The experiments had to be carried out in the absence of such breakdowns and this resulted in very poor sample yield. Reproducible results were obtained on a few samples with 300 Å thick oxides but not for 600 Å oxides, due to the poor yield.

Results for the 300 Å oxides with 120 Å Au electrodes are shown in fig.2. The transfer efficiency, T, is given by:

$$T = \frac{\text{ELECTRONS COUNTED / DISCRIMINATOR EFFICIENCY}}{e \times \text{CHARGE PASSED THROUGH SAMPLE}} \tag{1}$$

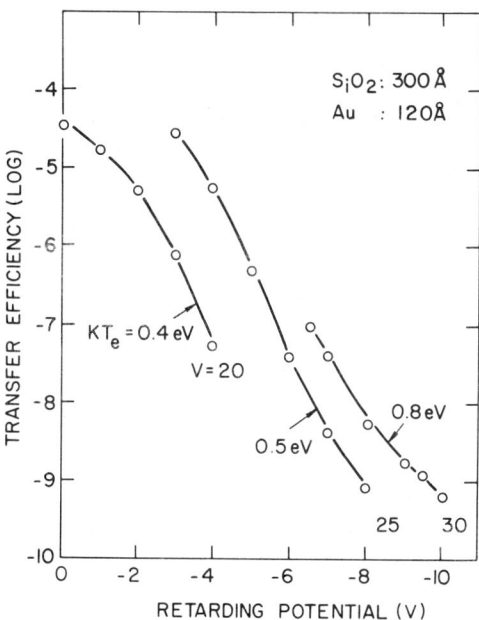

where e is the electronic charge. Fig. 2 corresponds to the integral distribution of electron kinetic energies up to the value of the retarding potential. The electron energy in the oxide is increased with respect to the vacuum energy by the electron affinity of the oxide [3] (about 1V for SiO_2). The measurements were limited at the low energy end by saturation of the electron multiplier or falloff of the collection efficiency, and at the high energy end by background noise or spurious electron pickup. The normalization of the 20V curve in fig.2 is uncertain due to difficulty in measuring the current, the other curves are all normalized according to eq.1.

The curves in fig.2 all show an exponential falloff in transfer efficiency characteristic of a Maxwellian distribution. The curves do tend to flatten out at high energies indicating higher electron temperatures in the distribution's tail. This efect may be caused by the Au electrode as will be dscussed later. The electron temperatures calculated from the slopes near the tail end are given in fig.2.

Fig.2 Transfer efficiency for electron emission for a Si-SiO₂-Au sample for applied voltages of 20,25 and 30V.

DISCUSSION

The paupacity of the results prevents any firm conclusions at this stage, however taking fig.2 as a true measure of the electron energy in SiO_2 we shall calculate electron mean free paths and impact ionization coeficients and see later how this may be modified by the Au electrode.

Simple relations given by Wolff [16] and Schockley [17] will suffice to obtain an estimate of the electron mean free path, λ, under the assumption of a Maxwellian distribution and scattering by LO optical phonons :

$$kT_e = \frac{(eF\lambda)^2}{4} \times E_r \qquad (2)$$

where kT_e is the electron temperature and E_r is the optical phonon energy (0.15 eV for SiO_2 [4]). Eq.2 predicts an electron temperature increasing with the square of the field, which is approximately true for the experimental results, and a value for λ of 7 Å is calculated.

Taken at their face value the results of fig.2 give an extremely small value for impact ionization in 300 Å thick SiO_2 at fields up to 10MV/cm. An upper limit (assuming an infinite cross section for impact ionization above the ionization threshold) to α, the impact ionization coefficient per unit length, is

$$\alpha = \frac{n(E_I)}{F} \qquad (3)$$

where $n(E_I)$ is the electron distribution integrated beyond the ionization energy E_I and F is the electric field. Assuming the transmission through the thick oxide to be >1% and a value for E_I of 9ev [6], fig. 2 puts an upper limit on α at 10 cm^{-1} at a field of 10 MV/cm. Reported values are much higher than this [4,7]. This discrepancy may be due to a deficiency in the previous measurements, which were rather indirect, or may be accounted for by the effect of the Au electrode.

Effect of the Au electrode
There is reason to believe that the electron distribution in the SiO_2 may be much hotter than the external one due to the selective filtering action of the Au electrode. In the metal the electrons are assumed to lose energy only in large increments due to electron-electron scattering [12] (phonon scattering is much weaker for these electron energies.). Thus an electron, once scattered, is lost from the distribution. Over the energy range of interest the mean free path for electron-electron scattering decreases rapidly with electron energy. Experimental results [19,20] give electron attenuation lengths in Au decreasing from 40 Å to about 8 Å for electron energies increasing from zero to 10eV above the Au vacuum barrier. This dependence means that the high energy electrons are probably transmitted through pinholes rather than the bulk Au.

The observed distribution therefore probably consists of two parts, the low energy part, mostly transmitted through the Au, and greatly influenced by the Au transmission characteristics, and a high energy part, mostly transmitted through pinholes [21], with an energy distribution characteristic of the insulator.

In fig.2 the flattening of the distribution at high energies could be the beginning of this high energy part. The actual electron energy distribution in the oxide may therefore be considerably hotter than the external one, with appropriate consequences for impact ionization.

CONCLUSION

In spite of difficulty of application, the electron emission structures have given the only direct method for measurig the high energy electron distribution in SiO_2 .The top metal electrode is the main drawback and carefull experimentation would be neccessary to eliminate this variable.

The experimental results clearly show the existence of above bandgap electrons in SiO_2 however it is not yet resolved whether there could be sufficient impact ionization as required by some breakdown theories [6,7]. Further experiments are neccessary with varying thicknesses of Au and SiO_2 to settle this.

ACKNOWLEDGEMENTS

The author wishes to thank E.D Alley, A. Cramer, and E. Petrillo for sample preparation, G.A. Wardley for the design of the electron analyser and E. Munro for computer simulations. Valuable discussions were had with P.J. Price and T. W. Hickmott.

REFERENCES

1.   K.K.Thornber and R.P. Feynman, Phys. Rev. B1, 4099 (1976).
2.   D.K. Ferry, Appl. Phys. Lett. 27, 689 (1975).
3.   S. Baidyaroy, M.A. Lampert and B. Zee, J. Appl. Phys, 47, 2103 (1976).
4.   R.C. Hughes, Solid State electron., 21, 251 (1978).
5.   N. Klein and P. Solomon, J. Appl. Phys., 47, 4634 (1976).
6.   T.H. DiStefano and M. Shatzkes, J. Vac.Sci. Technol., 13, 50 (1976).
7.   P. Solomon and N. Klein, Solid State Comm., 17, 1937 (1975)
8.   M. Shatzkes and M. Av-Ron, J. Appl. Phys., 47, 3192 (1976).
9.   P. Solomon, J. Appl. Phys., 48, 3843 (1977).
10. E. Harrari, Appl. Phys. Lett., 30, 601 (1977).
11. C.A. Mead, J. Appl. Phys., 32, 646 (1961).
12. R.M. Handy, J. Appl. Phys., 37, 4620 (1966).
13. L.E. Eckertova, Phys. Stat. Solidi, 18, 3 (1966).
14. R. Hrach, Czech. J. Phys., B23, 234 (1973).
15. M. Lenzlinger and E.H. Snow, J. Appl. Phys., 40, 278 (1969).
16. B.E. Deal, E.H. Snow and C.A.Mead, J. Phys. Chem. Solids, 27, 1873 (1966).
17. P.A. Wolff, Phys. Rev., 95, 1415 (1954).
18. W. Shockley, Solid State Electron., 2, 35 (1961).
19. H. Kanter, Phys. Rev., B1, 522 (1970).
20. J.C. Tracy and J.M. Burkstrand, Crit. Rev. Solid State Sci., 4, 381 (1974).
21. R.A. Collins and J.P.A. Williamson, Phys. stat. sol., 27, 85 (1975).

ELECTRON TRANSPORT IN SILICON OXYNITRIDE

A.V.Rzhanov, K.P.Mogilnikov and V.A.Gritsenko
Institute of Semiconductor Physics, Academy
of Science of the USSR, Siberian Branch,
Novosibirsk, USSR

The present work deals with the investigation of silicon oxynitride films the composition of which is considered to be close to silicon dioxyde. The films were obtained under interaction of SiH_4 and NH_3 in H_2O vapor at $830°C$ in a low pressure reactor. The results of structure investigations have shown that the films were amorphous. The thickness of the films was in the range of 200÷1500 Å. Silicon of p- and n-types with resistance of 10 om.cm was used as a substrate. Semi-transparent aluminium of $5·10^{-3}cm^2$ was used as an upper electrode.
The investigation of internal emission in MIS-structures allows to determine the potential barrier at the interface and also to get some information of the dielectric properties near the interface. The photoemission current is given as (ref. 1):

$$ j = A(h\nu)(h\nu - \varphi_0 + \beta\sqrt{F})^k \exp(-x_m/L) = j_0 \exp(-x_m/L) \qquad (1) $$

here $A(h\nu)$ is an apparatus function; $\beta = \sqrt{q^3/4\pi\epsilon\epsilon_0}$ is Schottky constant; ϵ is the optical dielectric constant; $x_m = \beta/2\sqrt{F}$ is the

distance from the interface up to the potential maximum in a film; k is the parameter which has a value from 1 up to 5/2 depending on mechanisms of excitation and scattering in an emitter (ref. 2). The results of the measurements of photocurrent dependence of a field are given in Fig. 1. The magnitude L for SiO_2 (curve 3) is equal to 35 Å that is in good agreement with the other authors results (ref. 1).

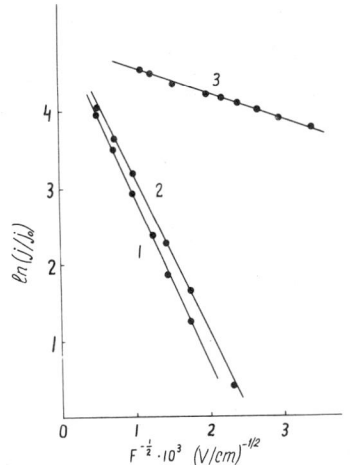

Fig. 1. The current-voltage dependencies
of photocurrent for silicon oxynitride
films, d=330 Å, n=1,55, curves 1,2 and for
silicon dioxide d=1100 Å, curve 3. For curve
1 $-h\nu$ =4,5 eV, 2,3 $-h\nu$ =5,0 eV.

The magnitude L for oxynitride turned out to be essentially less, near 6 Å, that indicates the high scattering in silicon oxynitride films.

The magnitude L is attributed the meaning of mean free
path for momentum exchange scattering in the work
(ref. 1), but later it was shown (ref. 3) that such
relaxation could not give strong field dependence of
photocurrent and meaning of mean free path for energy
relaxation should be given to the magnitude L.

The conception of mean free path can be introduced when
de Broglie wavelength for an electron is larger than
a mean free path. In silicon oxynitride for $h\nu$ =5 eV
and ϕ_o =4 eV we obtain:

$$\lambda_o = \frac{h}{m^*V} = \frac{h}{\sqrt{2m^*(h\nu - \phi_o)}} = 12,3 \text{ Å} \qquad (2)$$

that is twice as much as the obtained L value. Thus
the classical notions of the motion of photoinjected
electrons in silicon oxynitride films lose their sense

and the quantum notions should be used for the photoemission description. The high scattering in silicon oxynitride is connected to our opinion, with small-scaled fluctuations of a potential, which characteristic size is of interatomic distances order. The spectral dependence measurements of photoemission quantum yield have shown that the decreasing of photoemission efficiency is observed near 4.3 eV as it is shown in Fig. 2.

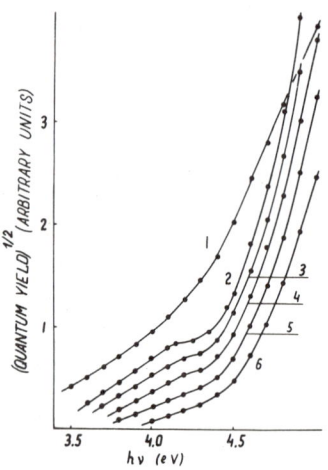

Fig. 2. The spectral dependencies of photoemission current at electronic emission from alluminium (1) and silicon (2-6) into silicon oxynitride n= 1,55 d= 330 Å. For curve 1 V_g= -5V, 2 - +20V, 3 - +17V, 4 - +13V, 5 - +9V, 6 - +5V.

The similar decreasing of the photoemission efficiency of the electrons from silicon is observed also at emission into silicon dioxyde and other dielectrics. It means, that this effect is connected with the electron excitation in silicon. At the energy of light quantum near 4.3 eV the direct transitions in Brillouin zone are observed in silicon (ref. 4). The finite electronic energy at these transitions is lower than the threshold energy, therefore these electrons give no contribution to the photoemission current. The same effect was previously ob-

served at external emission with caesium-covered pure surfaces
of silicon (ref. 5).
In MIS-structures not only the emission of electrons is possible but the emission of the holes as well. When the threshold
energy for the emission of the holes is less that the finite
energy of the holes at direct transitions, the intensive emission of the holes into a dielectric is possible. The results
of photoemission measurement in $Al-Si_3N_4$ - Si structure are
given in Fig. 3.

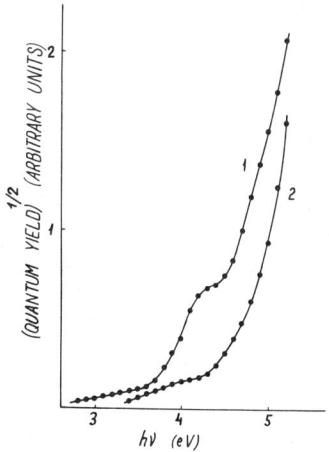

Fig. 3. The spectral dependence of photocurrent for silicon nitrade films, d=280 Å,
n=1,98 for curve 1 - V_g= -3V, for curve 2 -
+3V.

At negative voltage in alluminium the sharp increasing
of photoemission current at the quantum energies near
4 eV is observed. As the measurements of the flat
bands voltage have shown, at these quantum energies
the large positive charge near silicon is stored. This
fact shows the truth of the suggested explanation of
the photoemission pecularities from silicon and indicates the necessity of taking into account the conc-
rete mechanisms of excitation in semiconductors at
photoemission measurements, the magnitude of the poten-

tial barrier at the interface silicon-silicon oxynitride is close to the barrier magnitude at the interface silicon-silicon
dioxyde. However, oxynitride is high disordered in the comparison with thermal silicon dioxyde. Therefore the comparison of
current-voltage characteristics for these materials is of great
interest. The results of dark current measurements in the coordinates corresponding to Fowler-Nordheim law determining the
conductivity of silicon dioxyde are given in Fig. 4.

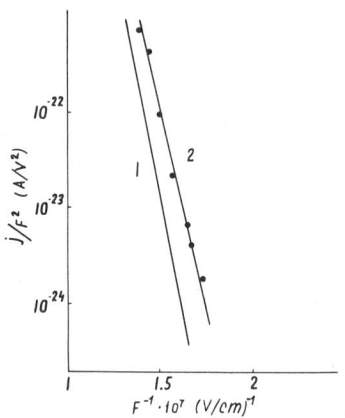

Fig. 4. The dark current-voltage characteristics of the films SiO_2 (1) and silicon
oxynitride (2), n=1,55.

As one can see from the figure the conductivity of silicon oxynitride also corresponds to Fowler-Nordheim law well. The magnitude of effective mass for electrons in silicon oxynitride which
is determined from Fig. 4 is near 0.4 m_o. Thus the effects connected with the disorder of silicon oxynitride structure do not
become apparent in tunneling.

A negative charge can be stored in silicon oxynitride while long-
time photocurrent passing through the film. The capture cross
section and the trappings concentration can be determined by
photocurrent kinetics experiment. The data obtained from these
experiments give for the trappings concentration $N_+ \sim 5.10^{17} cm^{-3}$
and for the capture cross section $\sigma \sim 10^{-13} cm^2$. Such capture
cross section corresponds to attractive Coulomb potential.

References:

1. C.N.Berglund, R.J.Powell, Photoemission into SiO_2: electron scattering in the image force potential well, J.Appl.Phys. 42, 573 (1971).

2. E.O.Kane, Theory of photoelectric emission from semiconductors, Phys.Rev. 127, 131 (1962).

3. M.Silver, P.Smejtek, J.Appl.Phys. 43, 2451 (1972) Comment on Electron Scattering in the Image Potential Well.

4. J.C.Phillips, The fundamental optical spectra of solids, Solid State Physics 18, 55 (1966).

5. F.G.Allen, G.W.Gobeli, Energy structure in Photoelectric emission from cs-covered silicon and germanium, Phys.Rev. 144, 558 (1966).

THE NATURE OF ELECTRON TUNNELING IN SiO$_2$

M. Av-Ron and M. Shatzkes
IBM System Products Division, East Fishkill
Hopewell Junction, New York

T. H. DiStefano
IBM Thomas J. Watson Research Center
Yorktown Heights, New York

I. B. Cadoff
Polytechnic Institute of New York
Brooklyn, New York

Experimental results for electron injection from aluminum into SiO$_2$, for $300°K \leq T \leq 575°K$, are explained in terms of a Murphy-Good emission model[1] generalized by the introduction of a Franz-type two-band dispersion relation.[2] For the study reported here, the I-V characteristics for high field are fitted with an electron effective mass ratio of 0.64, a SiO$_2$ bandgap energy of 8.64 eV, an Al-SiO$_2$ barrier height of 3.19 eV, and a relative dielectric constant of 2.31--all temperature-invariant. Sample-to-sample differences observed in intermediate fields are accounted for by assuming that at the Al-SiO$_2$ interface there are small areas in which the barrier height is lower. At low fields the current differs markedly from sample to sample, and depends on field and temperature in a manner that does not conform to any of the known emission models.

In comparisons of theory with experimental results, earlier workers encountered various difficulties. Lenzlinger and Snow[3] and Osburn and Weitzman[4] obtained the form of field dependence predicted for the field emission region, but the temperature dependence was so strong that they could make the model fit their experimental results only by assuming that some parameters depend on the temperature. Chou[5] and Av-Ron, Shatzkes, and Cadoff[6] encountered similar difficulties in the thermionic and intermediate regions of emission.[1] Recently, Weinberg and Hartstein[7] questioned whether the image potential applies to emission into SiO$_2$. Studying photo-assisted electron tunneling into SiO$_2$, they found that their experimental results agreed with Murphy-Good theory more closely when they neglected the image potential.

From results for tunneling in MOS with thin SiO$_2$, Maserjian[8] found that in part of the SiO$_2$ energy gap the dispersion relation is not parabolic, but rather is a Franz two-band relation.[2] In the study reported here, the Murphy-Good theory[1] was generalized by replacing the parabolic dispersion relation in the transmission function with Franz's form

$$\kappa[E(x)] = (2\pi/h)\{-2m^*E(x)[1 - E(x)E_g^{-1}]\}^{1/2}$$

where E(x) is the difference between the energy of the conduction bandedge and the energy of the tunneling electron, and E_g is the bandgap energy of the insulator. The result is that v(y), which appears in the Murphy-Good theory, is replaced by the function

$$u(y, r) = \{(-3y/4)(1 + y)^{1/2}[1 - (2y/r)][1 - (1 - y)r^{-1}]^{-1/2}\}K(m)$$

$$+ \{(-3/8)(r - 2)(1 + y)^{1/2}[1 - (1 - y)r^{-1}]^{1/2}\}E(m)$$

$$+ \{(3y/4r)(1 + y)^{-1/2}(r^2 - 4y^2)[1 - (1 - y)r^{-1}]^{-1/2}\}\Pi(n\backslash\alpha)$$

where $r = E_g/|w|$ (here w is the electron energy as defined in Reference 1); $y^2 = e^3F/4\pi\varepsilon_0\varepsilon w^2$; $n = (1 - y)/(1 + y)$; $m = \sin^2 \alpha = n(r - 1 - y)/(r - 1 + y)$; K(m), E(m), and $\Pi(n\backslash\alpha)$ are elliptic integrals of the first, second, and third kind, respectively.[9]

Currents were measured on Al-SiO$_2$ structures on which dry thermal SiO$_2$ had been grown at 1200°C to a thickness of 1000 Å and then chemically etched back to about 360 Å. Values of m*/m, E_g, ϕ, and ε were computed by fitting the high-field part of the experimental data to the generalized model. The barrier height was measured directly by internal photoemission for 300°K ≤ T ≤ 575°K; the results (Fig. 1) indicate a slight decrease with temperature. For ϕ, the values obtained from tunneling and from internal photoemission at 300°K agree with the value that Williams[10] reported, ϕ = 3.2 eV. Our value of E_g, 8.64 eV, agrees well with values derived earlier from optical measurements, 8 eV[11] and 9 eV.[12] Our value of ε is about 7.5% larger than Morey's optical dielectric constant for SiO$_2$.[13] Our value of m*/m is significantly larger than values given previously.[3,8] Figure 2 presents a comparison between high-field experimental J and calculated currents.

For F < 8 MV/cm, the quality of the fit was found to degrade with decreasing F, and the experimental J exceeded that calculated with the above parameters. This discrepancy differs appreciably from sample to sample. For ~6 MV/cm < F < ~8 MV/cm, depending on T, the discrepancy can be accounted for by assuming the existence of small regions having a lower barrier height, $\phi_1 \simeq 2.43$ eV, whose area differs from sample to sample about a mean of ~5 × 10^{-9} cm^2. Neutral traps near the interface could alter the current in such a way that its characteristic is described by a lower barrier.[14,15] Figure 3 shows (a) experimental results; (b) the current density J_0 associated with barrier height ϕ; and (c) $J_c = J_0 + \alpha J_1$, where J_1 is the current density for barrier height ϕ_1. Here $\alpha = 4.04 \times 10^{-6}$ is the fractional area of the perturbed region for this sample. For F < 6 MV/cm there is no agreement; the experimental data show a pattern that differs markedly from sample to sample and that present emission models to not account for.

Figure 4 presents some general trends, for F ≤ 8 MV/cm and 300°K ≤ T ≤ 575°K, that reflect the field and temperature dependence of the observed currents. Given here are the loci of $J_0(T, F)$ and $J_c(T, F)$, $J_c = J_0 + (\bar{\alpha} + 2\sigma)J_1$, where $\bar{\alpha} = 2.71 \times 10^{-6}$ and $\sigma = 1.11 \times 10^{-6}$ is the standard deviation. Also given are the experimental currents; the data are joined by dashed lines to indicate the trends. Some data at 5 and 6 MV/cm, which would mainly illustrate the variability of the current, were omitted from the figure to avoid clutter. At 8 MV/cm, the experimental data fall in the region bounded by J_0 and J_c for the entire temperature range. For fields of 7 and 6 MV/cm the temperature dependence is stronger than that of J_c, and for a lower field the temperature

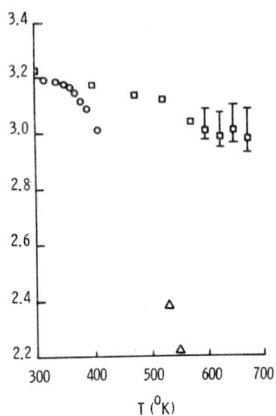

Fig. 1. Temperature dependence of Al-SiO$_2$ barrier height, as determined from photoemission (□) and tunneling data, from the data of Lenzlinger and Snow[2] (o), and from the data of Av-Ron et al.[5] (△)

at which the experimental J(T, F) exceeds J$_C$ is lower. At 5 MV/cm, the experimental currents are larger than J$_C$ at all the temperatures used. This behavior indicates an additional component of current that dominates at lower fields and has a temperature dependence stronger than the model predicts. This component is too strongly sample-dependent to be related to the intrinsic properties of SiO$_2$ or the interface. We can infer, therefore, that experiments to determine the temperature dependence of the current at constant field could give anomalous results, especially when the field is below 8 MV/cm, and could be erroneously interpreted as implying that the barrier height decreases with temperature.

In fitting experimental results to theory, the difference between the single-band model and the two-band model shows up chiefly in the temperature dependence. At a given temperature the two models can provide comparable fits to the experimental I-V characteristics. For a single-band model, however, the parameters must be adjusted in order to maintain the best fit when the temperature is changed; in the two-band model the temperature dependence is consistently accounted for with a single set of four parameters.

To determine whether the image potential applies, we fitted our high-field data to the one-band and two-band models, but neglected the image potential; i.e., we took $\varepsilon \to \infty$. Neither model produced as good a fit as the two-band model with $\varepsilon = 2.31$.

By fitting the Mg-SiO$_2$-Si data on Lenzlinger and Snow[3] to the present model, we obtained m*/m = 0.64, E$_g$ = 8.61 eV, ε = 2.32, and ϕ = 2.45 eV. The parameters pertaining to properties of the SiO$_2$ were the same for Mg as for Al; the value of ϕ was about the same as Deal et al.[16] obtained in photoemission and CV measurements. Our findings for m*/m differed from those of Lenzlinger and Snow, who report m*/m = 0.48 for Mg and m*/m = 0.39 for Al. Lenzlinger and Snow, and also Maserjian,[8] had suggested that the two-band model could eliminate this discrepancy.

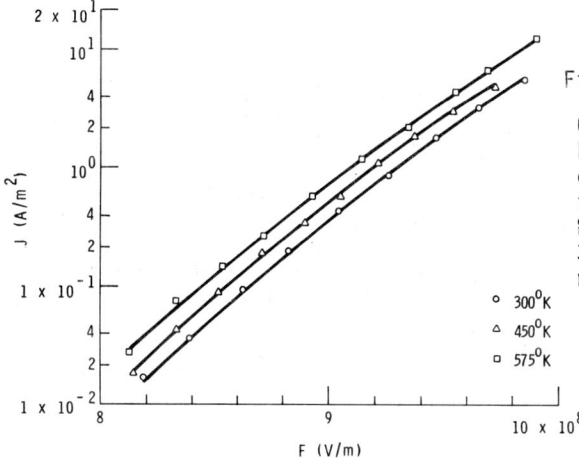

Fig. 2.

Current density vs. field for $F > 8 \times 10^8$ V/m. The solid curves are computed from the theory, with $m^*/m = 0.64$, $E_g = 8.64$ eV, $\varepsilon = 2.31$, and $\phi = 3.19$ eV. The discrete points represent experimental results.

∘ 300°K
△ 450°K
□ 575°K

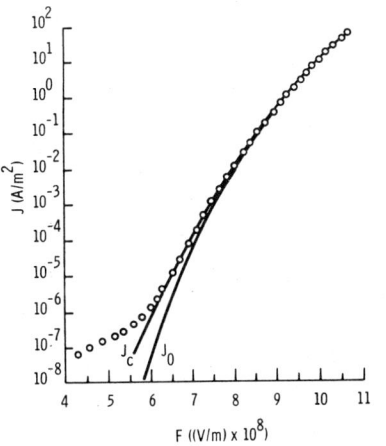

Fig. 3. Current density vs. field, for $T = 350^\circ$K. J_0 is the current computed from the theory. J_c is $J_0 + rJ_1$, where J_1 is computed from the theory for $\phi = \phi_1 = 2.43$ eV, the other parameters are as before, and r is the ratio of the area having this lower barrier to the total area; for this sample, $r = 4.04 \times 10^{-6}$.

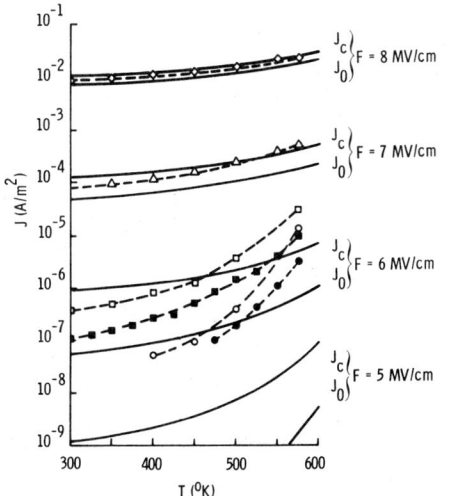

Fig. 4. Temperature density of the current. Solid lines are computed, J_0 is a lower bound where no area of lower barrier height is assumed at the interface, and J_c is an upper bound when an area of lower barrier height, two standard deviations larger than the mean, is assumed to exist. Shading indicates that a single sample was used for all temperatures; otherwise each data point was obtained with a different sample. □ , ■ and ∘ , ● correspond to 6 and 5 MV/cm, respectively.

In summary, we found that

1. The Murphy-Good theory, modified with a Franz-type dispersion relation, accurately represents emission into SiO_2 with parameters that are invariant with temperature and agree with independent measurements.

2. Parameters related to properties of the SiO_2--i.e., m^*, E_g, and ε--do not change when different metal electrodes are used.

3. Discrepancies noted in intermediate field ranges can be explained by assuming the existence of small areas, at the oxide interface, where only the barrier height is lower.

4. At low fields an extraneous component of current may be found, whose properties do not conform to theory.

ACKNOWLEDGMENT

We are grateful to F. Barson, E. L. Boyd, R. A. Gdula, P. E. Lokey, F. Stern, and R. H. Warren for contributions in various phases of this activity. We are particularly indebted to G. Goertzel, who wrote the algorithm for the two-band four-parameter fitting procedure.

REFERENCES

1. E. L. Murphy and R. H. Good, Jr., Phys. Rev. 102, 1446 (1956).
2. W. Franz, in Handbuch der Physik, ed. by S. Flugge, 17, p. 155 (Springer, Berlin, 1956). An English translation is available as TT64-16611 from the National Translation Center, Chicago, Illinois.
3. M. Lenzlinger and E. H. Snow, J. Appl. Phys. 40, 278 (1969).
4. C. M. Osburn and E. J. Weitzman, J. Electrochem. Soc. 119, 603 (1972).
5. N. J. Chou, J. Electron. Mat. 1, 344 (1972).
6. M. Av-Ron, M. Shatzkes, and I. B. Cadoff, Technical Report TR22.2052, IBM System Products Division, East Fishkill, Hopewell Junction, N.Y. (1976).
7. Z. A. Weinberg and A. Hartstein, Solid State Commun. 20, 179 (1976).
8. J. Maserjian, J. Vac. Sci. Technol. 11, 996 (1974).
9. M. Abramowitz and I. A. Stegun, Handbook of Mathematical Functions, National Bureau of Standards, Washington D.C. (1964).
10. R. Williams, in Semiconductors and Semimetals, Vol. 6, ed. R. K. Willardson and A. C. Beer, Academic Press, N.Y. pp. 93-139 (1970).
11. R. Williams, Phys. Rev. 140A, A569 (1965).
12. T. H. DiStefano and D. E. Eastman, Solid State Commun., 9, 2259 (1971).
13. G. W. Morey, The Properties of Glass, 2nd ed., Reinhold, N.Y. (1960); from this source we inferred average values of 1.465 for the refractive index and 2.15 for the optical dielectric constant.
14. C. Svensson and I. Lundstrom, J. Appl. Phys. 44, 4657 (1973).
15. H. Moes and R. Van Overstraeter, Electron. Lett. 9, 19 (1973).
16. B. E. Deal, E. H. Snow, and C. A. Mead, J. Phys. Chem. Solids, 27, 1873 (1966). The value of ϕ, based on CV, is 2.45 eV, whereas that resulting from photoemission is 2.25 eV--a value that ought to be adjusted to zero field. Lenzlinger and Snow[3] give 2.4 eV as derived from photoemission.

EVIDENCE FOR A BAND TAIL ON THE CONDUCTION BAND EDGE OF THERMAL SiO$_2$ FROM PHOTON ASSISTED TUNNELING MEASUREMENTS

A. Hartstein, Z. A. Weinberg, and D. J. DiMaria
IBM Thomas J. Watson Research Center, P.O. Box 218, Yorktown
Heights, New York 10598

ABSTRACT

Photon assisted tunneling measurements of electrons at the Al/SiO$_2$ interface have been extended into the range $0.4 \text{ eV} \geq \phi \geq 0.1 \text{ eV}$, where $\phi = \phi_B - h\nu$ is the effective tunneling barrier, ϕ_B is the Al/SiO$_2$ barrier height and $h\nu$ is the incident photon energy. When the effective barriers are smallest, the tunneling characteristics correspond to anomalously high barriers, i.e. ϕ_B seems to increase with photon energy. This apparent anomaly can be resolved with the assumption that a band tail of 0.1 eV exists on the SiO$_2$ conduction band edge.

Photon assisted tunneling measurements have been previously performed at the Al/SiO$_2$ interface of MOS devices using lasers.[1] It was found that the technique was very sensitive to the electronic structure of the interface. These experiments also showed that the classical image force interaction of an electron with a metal is not the proper quantum mechanical formulation of the interaction during a tunneling process. The theory of photon assisted tunneling without the image

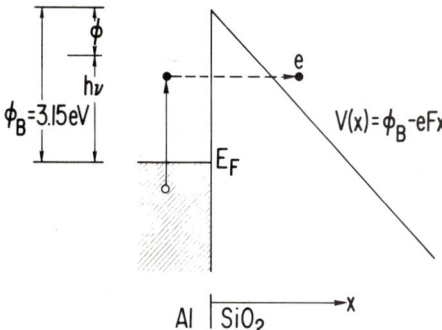

Fig. 1   A schematic illustration of the photon assisted tunneling experiment. E_F is the Al Fermi level, $h\nu$ is the incident photon energy, ϕ_B is the Al/SiO$_2$ barrier height and ϕ is the effective tunneling barrier. $V(x)$ represents the conduction band edge in the SiO$_2$ with an applied electric field (F). Electrons are photo-excited in the Al and some tunnel into the SiO$_2$ conduction band through the effective barrier ϕ.

force correction was found to give a good fit to experiment. These experiments were carried out using photon energies in the range such that the effective tunneling barriers were $0.4 \text{ eV} < \phi < 1.2 \text{ eV}$. In this paper we report the extension of these experiments into the range $0.10 \text{ eV} < \phi < 0.40 \text{ eV}$.

The photon assisted tunneling experiment is schematically illustrated in Fig. 1. This figure shows the interface between Al and SiO_2, showing the fermi level (E_F) in the aluminum and the conduction band edge (E_c) in the SiO_2. Electrons are photoexcited in the aluminum to energies $h\nu$ above E_F. They subsequently tunnel through the reduced barrier, $\phi = \phi_B - h\nu$, into the conduction band of the SiO_2. Since the effective tunneling barrier (ϕ) can be controlled by the incident photon energy and chosen to be small, this technique is very sensitive to the electronic structure of the interface.

The samples used in these experiments were MOS capacitor dots prepared by thermal oxidation of silicon to a thickness of 1000 Å, followed by evaporation of a thin semitransparent aluminum (100 Å) dot pattern. Various annealing procedures were employed with different results. These will be discussed later. The samples used for most of the data to be discussed were given a standard forming gas anneal at 400 C for 20 min.

The incident photons were provided by a grating monochromator system with an xenon arc lamp. Wavelengths in the range of 4500 Å to 4000 Å were used. A negative voltage bias was applied to the aluminum dot, and the subsequent current both with and without incident light was measured using an automated system. The difference between these two currents was calculated and stored in a computer based data logging system. Most of the data was obtained with a slit width allowing a bandpass of 50 Å in the incident photon wavelength. To check that this was indeed a narrow enough spread, some data was obtained with a bandpass of 25 Å. The curves were identical in the two cases, except for the intensity of the light incident on the sample, and the consequent magnitude of the photocurrents measured.

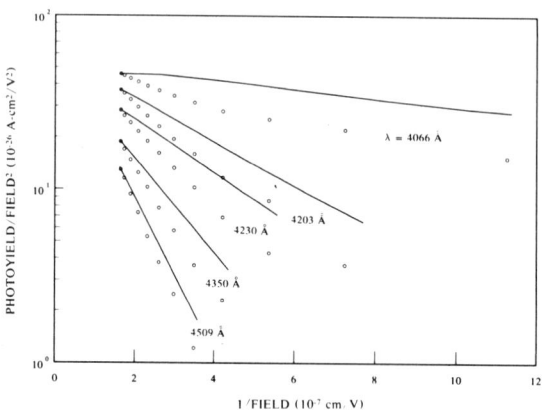

Fig. 2   Experimental data on the photon assisted tunneling of electrons at the Al/SiO_2 interface for 5 different photon wavelengths. Wavelengths correspond to effective barriers of 0.2, 0.3, 0.32, 0.4 and 0.5 eV. The solid curves are an exact calculation of the photon assisted tunneling process illustrated in Fig. 1 with $\phi_B = 3.15$ eV. Theory is scaled to data at highest field point.

The results of these measurements on one particular sample are shown in Fig. 2. The dots are the experimental data for five different incident photon energies. Data is plotted as log Y/F^2 vs $1/F$, where Y is the photocurrent and F is the electric field, since the process is essentially a photo-excited Fowler-Nordheim tunneling process, and this plot approximately yields a straight line.

The solid curves in figure 2 are a theoretical calculation of the photon assisted tunneling curves corresponding to the barrier (ϕ=3.15 eV) and effective mass (m* = 0.5) determined from previous measurements. The calculation used for this comparison is an exact Airy function calculation of the transmission of the tunneling barrier depicted in figure 1 using the above parameters and assuming a uniform distribution of photoexcited electrons.[1] This theory is seen to systematically deviate from the experimental results in the range where the effective tunneling barrier is the smallest. This deviation appears as if the Al/SiO_2 barrier increases with increasing photon energy.

A physically reasonable assumption which has been found to give a good fit to the experiment, is the assumption that a band tail exists on the conduction band edge of SiO_2. This idea has been modeled using the assumption that the Al/SiO_2 barrier height is not sharp, but rather is given by a gaussian distribution centered at ϕ_B and with a width Γ. The exact Airy function calculation was then numerically evaluated with this "fuzzy" barrier height. The results of this calculation with ϕ_B=3.25 eV and $\Gamma = 0.1$ eV are shown in figure 3 along with the experimental data. As can be seen the fit between theory and experiment is very good. We could not find another theory which gave as good a fit to even one of the curves in figure 3, let alone a fit to all of the curves using the same parameters. It should be noted that for large effective barriers this theory with ϕ_B and Γ or a theory without the gaussian broadening with a barrier $\phi_B' \approx \phi_B - \Gamma$ give the same results.

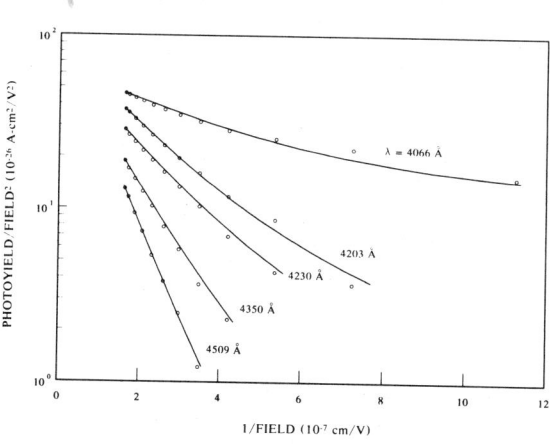

Fig. 3   Comparison of experimental photon assisted tunneling data with band tailing theory. Five wavelengths are shown. The theory is for $\phi_B = 3.25$ eV and $\Gamma = 0.1$ eV. Theory is scaled to data at highest field point.

54

Further evidence for the bandtail model is obtained from an analysis of the photon assisted tunneling curves from another sample in which no annealing steps were performed after the oxide was grown. When the data from this sample were analyzed in the same way, the gaussian broadening parameter Γ was found to be 0.25 eV. A broader band tail is qualitatively expected for unannealed samples when the band tail is a consequence of the disorder in the material.

Table 1 shows the results of this type of analysis on a series of samples which underwent various annealing treatment. The table shows that the barrier height changes with the annealing treatment, as has been found previously by other experimental techniques.[2] It also shows that the extent of the band tail is 0.25 eV for the unannealed sample, and that the band tail is only 0.1 eV wide following any type of annealing step.

TABLE 1

Sample	Final Annealing Treatment	ϕ_B(eV)	Γ(eV)
A	Post-Oxide N_2 1000°C 5 min..	3.10	0.1
B	Post-Oxide N_2 1000°C 30 min.	3.05	0.1
C	Pre-Metal Forming Gas 400°C 20 min.	3.05	0.1
E	Post-Metal Forming Gas 400°C 20 min.	3.25	0.1
F	No Anneal	3.05	0.25

We have utilized the photon assisted tunneling technique to study the details of the electronic structure of SiO_2 near the conduction band edge. It has been found that the conduction band edge is not sharp, but rather has a bandtail of 0.1 eV. Strictly speaking the experiment only probes the SiO_2 to a depth of 10 Å to 20 Å. The extent of the bandtailing found experimentally, even in the unannealed sample, is quite small compared to the bandgap of SiO_2.

It is not clear whether or not the band tailing described in this paper is consistent with the observation that in SiO_2 the electron mobility is not affected by shallow trapping.[3] In either case the band tailing we observe may be a property of the interface, or perhaps it may arise from a time dependent change in the energy bands due to lattice vibrations.[4]

REFERENCES

1. Z. A. Weinberg and A. Hartstein, Sol. St. Comm. 20, 179 (1976).
2. C. M. Osburn and E. J. Weitzman, J. Electrochem. Soc. 119, 603 (1972).
3. R. C. Hughes, Phys. Rev. Lett. 35, 449 (1975).
4. Z. A. Weinberg and G. W. Rubloff, these proceedings.

CHAPTER II
ELECTRONIC STRUCTURE AND SPECTRA

ELECTRONIC STRUCTURES OF CRYSTALLINE AND AMORPHOUS SiO_2

D. J. Chadi
Xerox Palo Alto Research Center, Palo Alto, CA 94304

and

R. B. Laughlin and J. D. Joannopoulos
Massachusetts Institute of Technology, Cambridge, MA 02139

The bulk electronic properties of crystalline and amorphous SiO_2 are examined by the tight-binding method.
For the crystalline case, we consider α-quartz and
β-cristobalite. The effects of topological disorder on
the electronic states are studied by using a Bethe-
lattice with a local atomic environment similar to that
of α-quartz. The effects of Si-O-Si bond-angle variations and topological disorder on the density of states
are discussed.

Crystalline SiO_2 occurs in many different allotropic forms (1). The primary
difference between the various crystalline forms is in the Si-O-Si angle between neighboring SiO_4 tetrahedra. This angle is 180° in β-cristobalite (2)
and very nearly 144° in α-quartz. Its mean value in amorphous SiO_2 is believed (3) to be near 144°.

Despite the diversity of its atomic structure, the electronic structures of
crystalline and amorphous SiO_2, particularly as revealed by optical absorption
measurements (4), are quite similar. This indicates that the local atomic
environment is the most important factor in determining the electronic structure of SiO_2. A simple tight-binding model with short-range interactions
should therefore be applicable to SiO_2.

The electronic structure of crystalline SiO_2 has been recently studied by
several authors using: the bond-orbital approach (5), the mixed-basis method
(6-7) involving localized functions and plane-waves, and the self-consistent
pseudopotential method (8). Previous to these molecular-orbital calculations
(9) on clusters were used in studies on SiO_2. Our main reason for constructing a parametrized tight-binding model for SiO_2 is to apply it to the $Si-SiO_2$
interface problem. The tight-binding method in conjunction with techniques
developed for treating disordered systems (10) provides an effective way for
dealing with lattice mismatch problems at the $Si-SiO_2$ interface. The interface studies are presented in a separate paper at this Conference (11). In
this paper we discuss our use of the tight-binding model in studying the electronic properties of SiO_2.

The Si-O nearest-neighbor interactions in SiO_2 are similar to those in a zinc-
blende crystal. They are of the type $V_{ss'\sigma}$, $V_{s'p\sigma}$, $V_{s'p\sigma}$, $V_{pp'\sigma}$, $V_{pp'\pi}$

where the primes reter to O orbitals. Let us consider α-quartz which has six
O atoms (per primitive cell) and 18 degenerate O p-states at the top of the
valence bands in the absence of interactions. The nearest-neighbor Si-O interactions cause a splitting to lower energies and a broadening of 12 of these
bands in which process they also acquire some Si character. The upper six
bands, however, remain degenerate and dispersionless. The dispersion in these
bands comes from interactions between orbitals on different oxygen atoms.
Since the O-O nearest-neighbor distance is about 1Å larger than the one for
Si-O and because of the localized nature of the oxygen orbitals, the O-O interactions are about an order of magnitude smaller than the Si-O interactions.
This results in a small dispersion and large effective masses for holes in the
upper valence bands. The O-O interactions have been included in our calculations. We have used photoemission data (12), optical absorption (4) and
photoconductivity (13) data to determine the tight-binding parameters (14).

The calculated band structure of α-quartz along the principal symmetry directions of the Brillouine zone is shown in Fig. 1. For β-cristobalite we find
the valence band maximum to be 0.3 eV higher in energy than in α-quartz but
the most noteworthy difference in the electronic structures of the two crystals is in the magnitude of the bandgap. α-quartz has an indirect (A to Γ)
gap of 9.2 eV whereas β-cristobalite has an appreciably smaller direct (Γ to Γ)
gap of 6.93 eV. We have calculated the dipole matrix elements (14) for optical
transitions in α-quartz and find the first allowed direct transition at Γ to
be at 11.65 eV in good agreement with structure at 11.7 eV in optical reflectivity data (4). Because of the different numbers of electrons and bands in
α-quartz and β-cristobalite, their electronic structures can be best compared
by looking at densities of state (shown in Fig. 2) in which these differences
are suppressed through an adjustment of the vertical scale. The density of
states for a topologically disordered SiO_2 as modeled by a Bethe-lattice (10)
is also shown in Fig. 2. The Bethe-lattice used in our calculations gives a
perfect tetrahedral environment to each Si atom and has a Si-O-Si bond angle
of $138°$, close to the $144°$ value in α-quartz (the $138°$ value was chosen for
computational purposes).

Figure 2 shows the behavior of the various valence states as a function of
both Si-O-Si bond angle variations and of topological disorder. The similarities between the α-quartz and Bethe-lattice densities of states for the oxygen
lone-pair states and the differences with the β-cristobalite results indicate
that these states are more sensitive to Si-O-Si bond angle variations than to
topological disorder. The peak near -7 eV which arises from pp'σ bonding between Si and O orbitals is influenced primarily by the tetrahedral environment
around the Si atoms and is not affected much by changes in structure that do
not alter the tetrahedral geometry. The largest changes in the density of
states occur for the sp'σ bonding states at -8 to -12 eV and in the corresponding sp'σ anti-bonding states at the conduction band edge. Figure 2 indicates that the structure in the filled states, when measured by high resolution photoemission spectroscopy, could provide a "fingerprint" of the type of
structure involved. It also demonstrates explicitly that the previously predicted (for the filled states) (15) and experimentally observed (12) smoothing
out of structure in the sp'σ bonding and anti-bonding regions in amorphous
SiO_2 is in fact disorder induced. Figure 2 also suggests the possibility of
"band-tailing" at an Si-SiO_2 interface resulting from bond-angle variations.
This is discussed in more detail in the paper by Laughlin, et al. (11).

The tight-binding method allows an easy and straightforward evaluation of Si
and O local densities of states in SiO_2 and a decomposition of these into

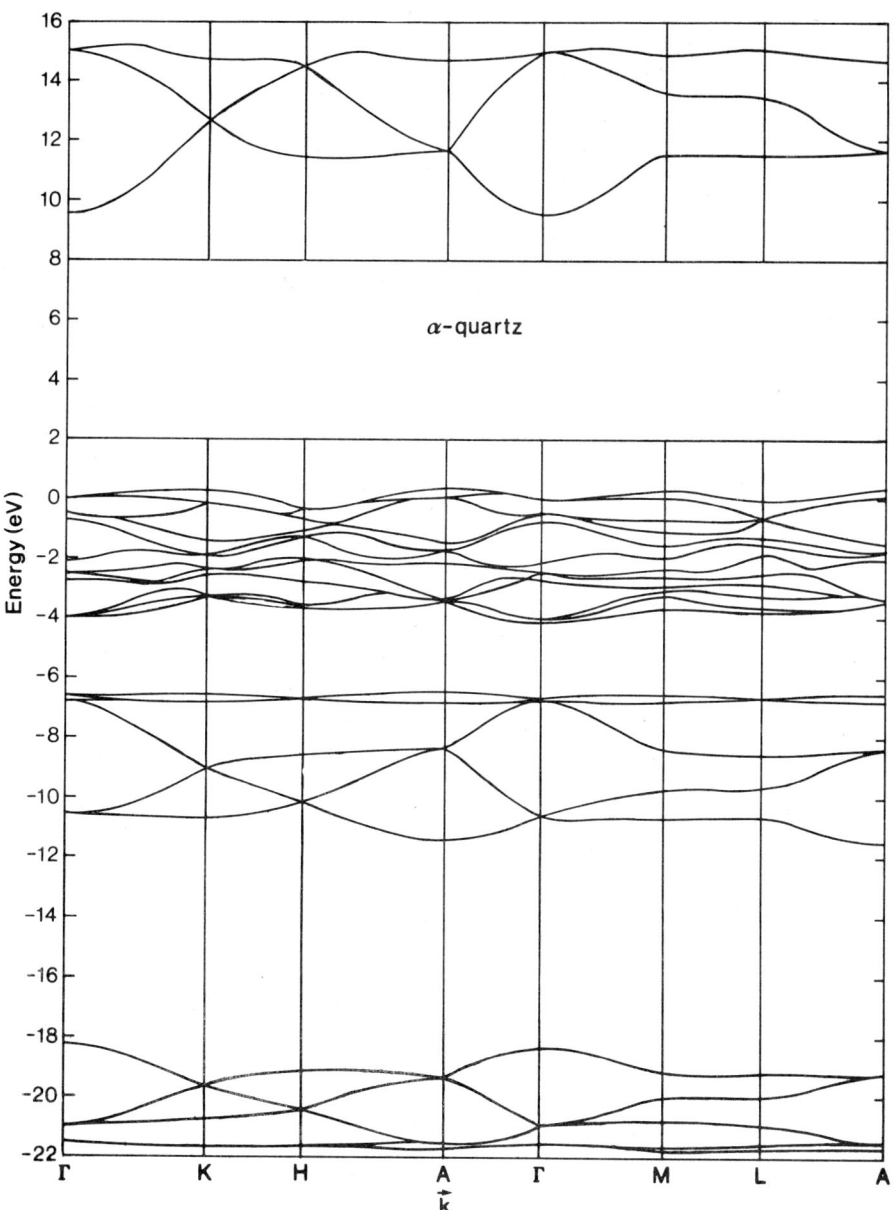

Fig. 1. Band structure of α-quartz. The valence band maximum
at A is 0.37 eV above zero (at Γ).

Table I.

$E_s = 4.42$ $E_p = 10.67$ $E_{s'} = -14.63$ $E_{p'} = -1.83$

$V_{ss'\sigma} = -2.85$ $V_{s's'\sigma} = -0.15$ $V_{sp'\sigma} = -5.4$ $V_{s'p\sigma} = -9.5$

$V_{pp'\sigma} = -5.4$ $V_{pp'\pi} = -1.4$ $V_{p'p'\sigma} = -0.45$ $V_{p'p'\pi} = -0.45$

Tight-binding interaction parameters (in eV) for SiO_2. The unprimed (primed) subscripts refer to Si(O) orbitals. A three-center interaction ($V_{x'y'} = -0.45$) was also used in the calculations.

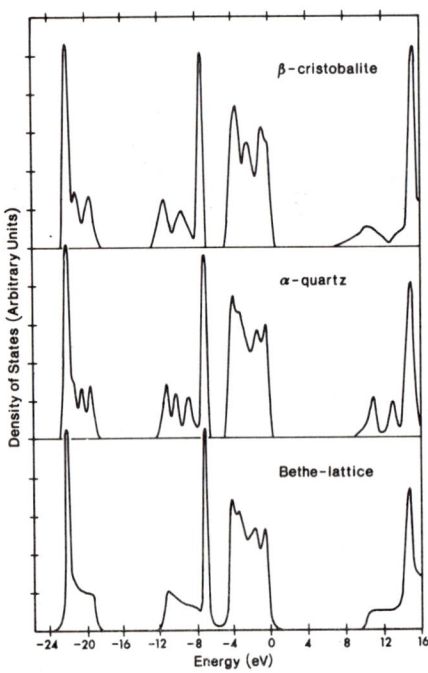

Fig. 2. Densities of state of β-cristobalite (top), α-quartz (middle), and Bethe-lattice SiO_2 (bottom). The valence band maximum of α-quartz has been chosen for the zero of energy. The densities of state have been slightly Gaussian broadened.

their s and p derived components which can then be compared with X-ray emission spectra (15). Our calculations give good agreement with O K_α and Si K_β X-ray emission spectra (15) in SiO_2. The Si $L_{2,3}$ emission spectra which gives information on the energy distribution of Si s and/or d derived orbitals shows two main structures. One is centered around -10 eV which is in the sp'σ bonding region (see Fig. 2) and is therefore expected. The other is centered around -4 eV in the O lone-pair region. Our calculations show no mixing of Si s-states with the O 2p-states in this region. We have not found it necessary to use Si d-states in our calculations and do not believe that there is appreciable mixing between these states and the O 2p-states. The suggestion (8) that excess Si produced during the experiment is responsible for the high energy peak of the Si $L_{2,3}$ emission spectra is the most plausible explanation.

In conclusion, we have shown that the tight-binding method is a useful technique in studying the electronic properties of crystalline and amorphous SiO_2. The method can be easily adapted to the study of the Si-SiO_2 interface (11).

REFERENCES

(1) R. W. G. Wyckoff (1965) Crystal Structures, Interscience, New York and references therein.

(2) The existence of SiO_2 in the β-cristobalite structure has not been conclusively established. The structure is however useful in theoretical studies of the effect of the Si-O-Si bond angle on electronic properties.

(3) R. L. Mozzi and B. E. Warren, J. Appl. Cryst. 2, 164 (1969).

(4) H. R. Philipp, Sol. State Commun. 4, 73 (1966); J. Phys. Chem. Sol. 32, 1935 (1971); E. Loh, Sol. State Commun. 2, 269 (1964).

(5) S. T. Pantelides and W. A. Harrison, Phys. Rev. B 13, 2667 (1976).

(6) P. M. Schneider and W. B. Fowler, Phys. Rev. Lett. 36, 425 (1976); E. Calabrese and W. B. Fowler (to be published).

(7) S. Ciraci and I. P. Batra, Phys. Rev. B 15, 4923 (1977).

(8) J. R. Chelikowsky and M. Schlüter, Phys. Rev. B 15, 4020 (1977); M. Schlüter and J. R. Chelikowsky, Sol. State Commun. 21, 381 (1977).

(9) A. R. Ruffa, Phys. Stat. Sol. 29, 605 (1968); Phys. Rev. Lett. 25, 650, (1970); J. Noncryst. Sol. 13, 37 (1973/1974); M. H. Reilly, J. Phys. Chem. Sol. 31, 1041 (1970); J. A. Bennet and L. M. Roth, Phys. Rev. B 4, 2686 (1971); J. Phys. Chem. Sol. 32, 1251 (1971); K. L. Yip and W. B. Fowler, Phys. Rev. B 10, 1391 (1974); ibid., 1400 (1974).

(10) J. D. Joannopoulos and F. Yndurain, Phys. Rev. B 10, 2764 (1977); J. D. Joannopoulos, Phys. Rev. B 16, 2764 (1977).

(11) R. B. Laughlin, J. D. Joannopoulos, and D. J. Chadi in the Proceedings of this Conference; and to be published.

(12) T. H. DiStefano and D. E. Eastman, Phys. Rev. Lett. 27, 1560 (1971); B. Fischer, et al., Phys. Rev. B 15, 3193 (1977).

(13) T. H. DiStefano and D. E. Eastman, Sol. State Commun. 9, 2259 (1971).

(14) D. J. Chadi, R. B. Laughlin, and J. D. Joannopoulos (to be published).

(15) See references in D. L. Griscom, J. Noncryst. Sol. 24, 155 (1977).

ELECTRONIC STRUCTURE OF α-QUARTZ AND THE
INFLUENCE OF SOME LOCAL DISORDER: A TIGHT BINDING STUDY

R. N. Nucho and A. Madhukar
Departments of Physics and Materials Science
University of Southern California
University Park
Los Angeles, CA   90007

ABSTRACT

A study of the electronic structure of α-quartz, the influence of short range departure from the α-quartz geometry, and the correlations between charge transfer from Si to Oxygen with variations in Si-O-Si bond angle is presented. For the first time, a framework is presented and utilized to shed light on the nature and origin of the several recently observed Si(2p) levels in XPS studies of bulk a-SiO$_2$ as well as the interfacial region near the Si/SiO$_2$ interface.

In this paper we report the results of (i) a tight-binding study of the electronic structure of α-quartz, including second-nearest neighbors, (ii) some changes in this electronic structure caused by short range departures from the α-quartz geometry, and (iii) an analysis of the Si(2p) core level chemical shifts in a-SiO$_2$ and near the Si/SiO$_2$ interface, from which some inferences regarding the chemical and structural nature of this interface are drawn for the first time. The calculations are motivated by ongoing ESCA studies[1,2] of the Si/SiO$_2$ system in which continuous monitoring of the valence and core levels, as a function of the oxide thickness, have revealed an interfacial region, about 7 Angstroms thick, and of chemical composition SiO$_x$, x < 2. The flexibility of the tight-binding method is exploited to simulate the influence of variations in the Si-O-Si as well as the O-Si-O angles on the band structure, density of states, and charge distribution. Calculated changes in the charge transfer from Si to Oxygen are, for the first time, correlated with the observed chemical shifts[2] of the Si(2p) core levels in bulk a-SiO$_2$ as well as in the interfacial SiO$_x$ layer.

The tight-binding integrals in our study of α-quartz are treated as parameters and determined by fitting to the recent pseudopotential based band structure calculation reported[3] by Chelikowsky and Schluter. An extended summary of the theoretical and experimental work on SiO$_2$ is to be found in that paper. The tight-binding basis consists of the four sp^3 hybrids on each of the three Silicon atoms and the three p

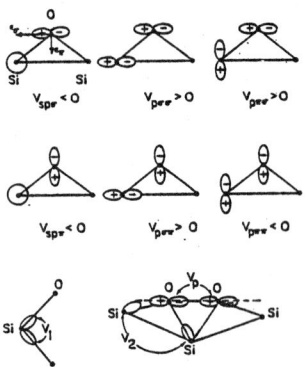

Fig. 1   Shows the various atomic matrix elements used in the text.

orbitals centered on each of the six Oxygen atoms in the hexagonal primitive cell of α-quartz. We neglect the Oxygen 2s state. The lowest eighteen of the thirty bands so obtained correspond to the valence bands. The system is viewed as alternating Si-O-Si and O-Si-O triads. Defining the unit vectors ε_σ and $\varepsilon_{\pi 1}$ as shown in Fig. 1, and $\varepsilon_{\pi 2}$ normal to the Si-O-Si plane, we note that the $p_{\pi 2}$ Oxygen orbital interacts weakly with the Si orbitals. The topmost

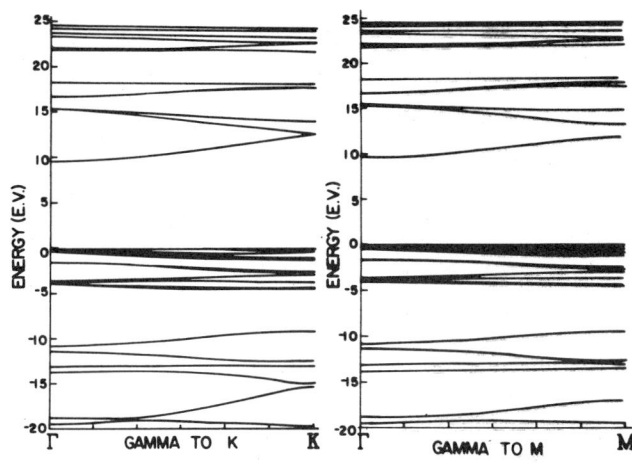

Fig. 2   Energy bands corresponding to the parameters (set 1) of Table I.

valence bands are largely derived from these orbitals and have consequently been referred to as "non-bonding." The matrix elements, in terms of atomic orbitals, are defined as in Fig. 1. In Table I we provide two sets of tight-binding parameters. The first set gives a good fit to the non-bonding valence bands and the low-lying conduction bands of the pseudopotential calculations. The corresponding band structure and density of states (DOS) are shown in Fig. 2 and Fig. 3 respectively. The second set yields a good fit to all the valence bands at some expense to the conduction bands and to the indirect gap. The valence band density of states for this case is shown in Fig. 4. A Gaussian width of .5 eV was used in the DOS calculations. The non-bonding bands are found to have largely Oxygen p character with only a little Si s and p mixing, while the lower bonding valence bands have mixed Si-*s*, Si-p, and Oxygen-p character. A more detailed description will be published elsewhere. In brief, the diagonal energies control the basic separation between conduction and valence bands, $V_{sp\sigma}$ and $V_{p\sigma\sigma}$ control the widths of the upper conduction and lower valence bands, and $V_{sp\pi}$ is a "fine tune" for the gap at the Γ point due to the pure s-character of the conduction band and p-character of the valence band at that point. The other nearest neighbor parameters yield only secondary effects. As for second neighbor parameters, we find that the band structure is not very sensitive to V_2 but that V_p is crucial in determining the width of the non-bonding bands. It is of interest to note that the indirect gap arises from a second order effect involving the O-Si interaction, $V_{p\sigma\sigma}$. It was only when this parameter was made large, and with $V_p \neq 0$, that an indirect gap was obtained. The price to pay was a lowering of the lower-lying valence bands.

The valence band DOS weighted by the appropriate[5] s and p photoelectric cross-sections at 1487 eV is shown in Fig. 5 (curve a) for comparison with XPS spectra for α-quartz[4] (curve b) and a-SiO_2[2] (curve c). The relative intensities of the bonding and non-bonding valence band regions can be brought into better agreement with experiment by increasing the ratio of s to p photoelectric cross-sections. The larger width of the bonding valence bands and the sharp peak at the non-bonding valence band edge are a consequence of fitting to the pseudopotential calculations which also show the same discrepancy. The calculated charge transfer from Si to Oxygen for α-quartz is found to be

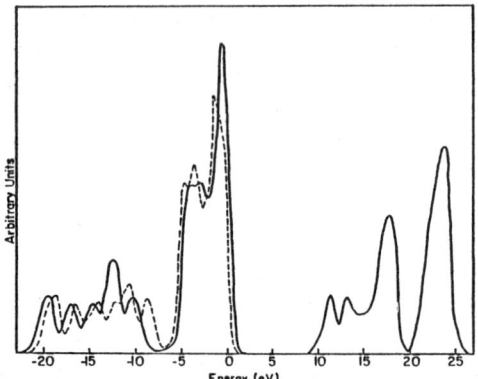

Fig. 3 Density of states corresponding
to the parameters (set 1) of
Table I.

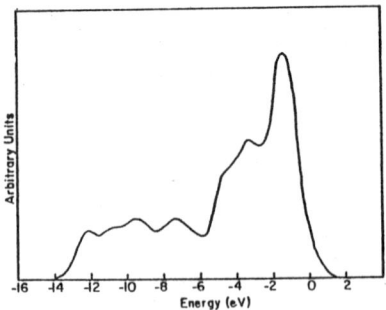

Fig. 4 Valence band density
of states corresponding
to the parameters (set 2) of
Table I.

about 1.17 electron.

The basic mechanisms of our model having
been exposed, we wish to exploit the
built-in flexibility of such a scheme to
examine the effect of various deforma-
tions.

(a) Si-O-Si Bond Angle Changes

The changes in the DOS as a function of
the Si-O-Si bond angle were examined.
The DOS for a Si-O-Si bond angle of 130°
is shown by the dashed line of Fig. 3.
We note the appearance of states in the
"gap" between the two sets of valence
bands, as well as the displacement of
the non-bonding valence band edge to
lower energy, thereby increasing the
fundamental gap. The opposite effect is
found when the angle is made larger than
the α-quartz value of 144°.

(b) O-Si-O Tetrahedral Angle Changes

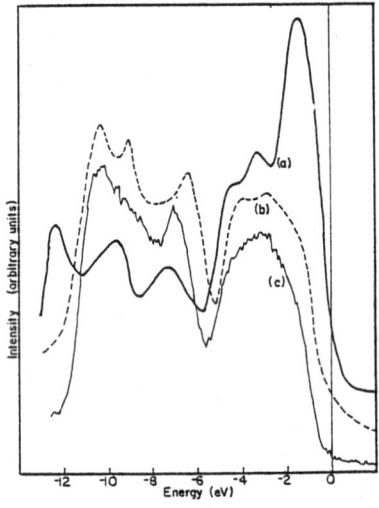

Fig. 5 Photoelectric cross-section weighted density of
states as described in text.

The major changes in the electronic structure due to variations in the tetrahedral angle arise from changes in the O-O interactions. This fact is illustrated in Fig. 6 where we have compared the following three curves: the non-
bonding valence band DOS corresponding to set 1 of Table I (Fig. 6a), the DOS
for a smaller tetrahedral angle when only the changes in the nearest neighbor
parameters are included (Fig. 6b), and finally (Fig. 6c) the DOS when changes
in all the parameters are kept. It is seen that the smaller tetrahedral angle,
which corresponds to a smaller O-O distance and therefore a larger interaction
matrix element, V_p, gives rise to a broadening of the non-bonding valence band.
The opposite effect is observed for larger tetrahedral angles.

We now turn to an analysis which, for the first time, correlates the valence band charge densities, as a function of the Si-O-Si bond angle, to the several different Si(2p) core level chemical shifts[2] observed in a-SiO$_2$ as well as in the interfacial region. This analysis has enabled us to suggest the possible distribution of the Si-O-Si bond angles and the ring sizes that may be associated with them. In addition, within this conceptual framework we are able to shed light on the presence of possible suboxides, ranging from Si$_2$O$_3$ to Si$_2$O, in the interfacial region.

Recent XPS measurements[2] have revealed the presence of several Si(2p) core levels in a-SiO$_2$, as well as in the neighborhood of the interface (see Table II). The presence of these peaks has been taken to indicate differing local structural and chemical environments, in both cases.[2] Considerable evidence for fluctuations in the Si-O-Si bond angle (θ) in glasses has been ac-

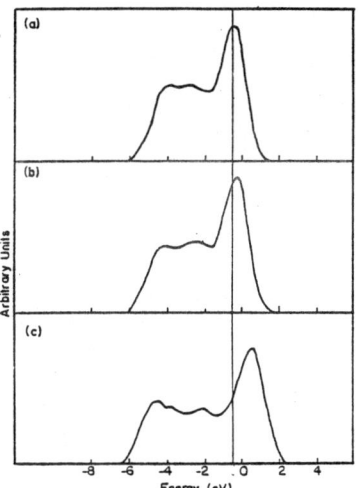

Fig. 6 Shows the influence of the change in the O-Si-O tetrahedral angle. Valence band edge of α-quartz (curve a) marks the energy zero.

cummulated over the years.[6] This suggests the possibility that the different chemical shifts observed reflect the influence of loss of charge from Si to Oxygen for different most likely values of the Si-O-Si bond angle, possibly corresponding to a distribution of ring sizes ranging from 4 to 8. We therefore calculated the Si to Oxygen charge transfer for various values of θ of which $\theta = 120°$ and $\theta = 144°$ are known[6] to correspond to 4 and 6 membered rings. From the observed difference of Si(2p) levels in Si element and SiO$_2$, and the calculated value of 2.34 for the total charge loss per Silicon for α-quartz, we obtain an estimate of the chemical shift per unit charge, within the accepted linear approximation. Employing this value, we calculate the Si(2p) core level shift as a function of θ. Comparison with experimental peak values of bulk a-SiO$_2$ shows that these peaks occur at $\theta = 120°$, $144°$, and approximately $180°$, the last value suggesting regions with possibly linear Si-O-Si bonds or large sized (8,9 and higher) rings. Comparison with the observed peaks in the interfacial region shows that the most dominant values of θ lie near $133°$ and $151°$ which may largely arise from 5 and 7 membered rings. In addition, in the interfacial region, peaks are observed at 101.60 eV, 100.4 eV, and 99.3 eV, which have been tentatively identified[2] as peaks due to Si^{+3}, Si^{+2}, and Si^{+1} states arising from Si$_2$O$_3$, SiO, and Si$_2$O structures. It is worthy of note that the chemical shifts of all these peaks scale identically with the binding energy shifts per unit charge calculated for bulk a-SiO$_2$ and SiO$_2$ in the interfacial region, the corresponding theoretically calculated peaks lying at 101.45 eV, 100.4 eV, and 99.35 eV.

We are presently investigating various aspects of the interplay between theory and experiment. The important message, nevertheless, is the conceptual framework presented here for the first time, within which interpretations of the valence and core levels of bulk and interfacial regions promises to be a fruitful avenue of investigation.

TABLE 1   Tight Binding Parameters Corresponding to
Figs. 3 and 4 (Notation is as Shown in Fig. 1).

Matrix Elements	$E(sp^3)$	$E(P_{ox})$	V_1	$V_{sp\sigma}$	$V_{p\sigma\sigma}$	$V_{p\sigma\pi}$	$V_{p\pi\sigma}$	$V_{sp\pi}$	$V_{p\pi\pi}$	V_2	V_p
Set 1	12.40	−1.45	−1.6	−7.0	9.00	4.00	4.00	−0.8	−0.7	1.5	1.76
Set 2	12.59	−1.26	−1.1	−4.0	4.35	4.00	4.00	−0.8	−0.7	1.5	1.45

TABLE 2   Observed and Calculated Si(2p) Levels in Bulk a-SiO$_2$
and the Interfacial Region of Si/SiO$_2$ Interface. θ
Denotes the Si-O-Si Bond Angle Corresponding to the Calculated Si(2p) Core Level Position.

BULK			INTERFACE		
Observed E_B [Si(2p)]	Calculated E_B [Si(2p)]	θ	Observed E_B [Si(2p)]	Calculated E_B [Si(2p)]	θ
103.0 eV	103.0 eV	~180°	102.7 eV	102.7 eV	150°
102.5 eV	102.5 eV	144°	102.3 eV	102.3 eV	135
102.1 eV	102.1 eV	120°			

We are deeply indebted to our colleagues at the Jet Propulsion Laboratory,
Drs. F. Grunthaner, J. Wurzbach, and J. Maserjian for numerous enlightening
discussions and making available their unpublished data. Partial support by
ONR and the A. P. Sloan Foundation is gratefully acknowledged.

REFERENCES

1. F. J. Grunthaner and J. Maserjian, "Experimental Observation of the Chem-
    istry of the SiO$_2$/Si Interface," IEEE Transactions Nuc. Sc. NS-24, 2801
    (1977).

2. F. Grunthaner and J. Maserjian, (Proc. This Conf.).

3. J. R. Chelikowsky and M. Schluter, "Electron States in α-Quartz: A Self-
    consistent Psuedopotential Calculation ," Phys. Rev. B15, 4020 (1977),
    and references to earlier theoretical and experimental works contained
    therein.

4. B. Fischer, R. A. Pollak, T. H. Distefano and W. D. Grobman, "Electronic
    Structure of SiO$_2$, Si$_x$Ge$_{1-x}$, and GeO$_2$ from Photoemission Spectroscopy,"
    Phys. Rev. B15, 3193 (1977), and references to earlier works therein.

5. J. H. Scofield, "Hartree-Slater Subshell Photoionization Cross Sections
    at 1254 and 1485 eV," J. Elec. Spec. Related Pheno. 8, 129 (1976).

6. See, for example, A. G. Revesz, "Pressure Induced Conformational Changes
    in Vitreous Silica," J. Non Cryst. Solids, 7, 77 (1972) and references
    therein.

ELECTRONIC STRUCTURE INVESTIGATIONS OF TWO ALLOTROPIC
FORMS OF SiO_2: α-QUARTZ AND β-CRISTOBALITE

I. P. Batra
IBM Research Laboratory, 5600 Cottle Road
San Jose, California 95193, U.S.A.

ABSTRACT

The extended tight binding method, with Gaussian basis sets and the
Hartree-Fock-Slater model for potential, is employed to investigate the
electronic structure of silicon dioxide in the α-quartz and β-cristobalite
structures. Results are compared with photoemission experiments and
previous theoretical calculations.

INTRODUCTION

During the last few years band theoretic techniques have been applied (1-5)
for gaining insight into the electronic properties of silicon dioxide
(SiO_2). Since SiO_2 is found in a variety of forms (6), it is important to
investigate the electronic properties of at least two allotropic forms
within the same theoretical framework. In the present paper, we report
the results of electronic structure calculations for α-quartz and
β-cristobalite. The unit cell of α-quartz contains three Si atoms and six
oxygen atoms in a hexagonal arrangement. In β-cristobalite, the SiO_4
tetrahedra occupy the positions of the diamond lattice and are connected
by straight Si-O-Si chains. The Si-O-Si bond angle is about 144° in
α-quartz. Further details of the crystal structure may be found in Ref. 6.

For both these systems we provide detailed band structure information,
total and orbital density of states and compare this with other theoretical
and experimental results available in the literature. There have been
conflicting reports about the nature of energy gap (direct or indirect)
for α-quartz. Furthermore, the effect of exchange parameter (α) in the
statistical exchange approximation (7) on the electronic structure of SiO_2
has not been explored previously. We address both these issues in the
following sections.

All calculations being reported here were performed non-self-consistently
within the framework of the extended tight-binding method (8,9). We used
single Gaussians for Si (the exponents were obtained from the (11s7p)
primitive set of Huzinaga (10)) and a double zeta [4s2p] set for oxygen
obtained from Bagus (10). The crystal potential, obtained by superimposing
atomic potentials, is handled in two parts. The divergent,$-2Z/r$, core part
is expressed in terms of linear combination of Guassians located at each
atomic site and is treated (4,9) in direct space. The remaining smooth

potential is represented by a truncated Fourier series in the reciprocal
space. Analytical evaluation (8,11) of the various matrix elements can
then be carried out. An advantage of ETB method for treating systems with
many atoms in the unit cell lies in the fact that one can readily describe
the nature of various bands in terms of their orbital and atomic origins
through Mulliken (12) population analysis.

Band Structure

Figure 1 shows the band structure for α-quartz along various directions in
the hexagonal Brillouin zone (the exchange parameter $\alpha=0.67$ was used).
Following Chelikowsky and Schlüter (3) (CS) we shall simply label the bands
in ascending order with $\Gamma(1)$ referring to the lowest valence band at Γ and
$\Gamma(24)$ referring to the top of the valence band at Γ. These twenty-four
bands fall into three nonoverlapping groups (13) containing 6, 6 and 12
bands respectively. The lowest group is primarily oxygen 2s, the
intermediate one is oxygen 2p bonded to silicon 3s and 3p, and the highest
one is due to nonbonding oxygen 2p orbitals.

The computed energy band gap is indirect, $K(24) \rightarrow \Gamma(25)$, and is equal to
7.6 eV. However, the indirect gap, $M(24) \rightarrow \Gamma(25)$, is only 0.05 eV larger in
our calculation. Thus within the accuracy of our calculation the indirect
gap may be located at either one of these positions. The direct gap,
$\Gamma(24) \rightarrow \Gamma(25)$, is 7.8 eV. CS have computed using pseudopotential method an
indirect gap $(M(24) \rightarrow \Gamma(25))$ 9.2 eV wide (for $\alpha=0.8$). Calabrese and Fowler
(5) (CW), on the other hand, using a mixed basis set obtain a direct gap
at Γ with magnitude 6.3 eV for $\alpha=1.0$. This value as well as our value of
7.6 eV is considerably lower than the experimental value of 8.9($\pm$0.2) eV
obtained by photoinjection experiments (14). However, as will be seen
below, our results for $\alpha=0.8$ are in excellent agreement with experiments.

Figure 2 shows the computed band structure for α-quartz for three different
values of the exchange parameter along a few particular directions in
Brillouin zone. It should be remarked that results for a general value of
α were obtained in an approximate fashion starting with $\alpha=0.67$ results.
The overall effect of increasing α is to reduce width of valence bands and
to increase the value of energy gap. Thus for $\alpha=0.8$ the indirect gap,
$K(24) \rightarrow \Gamma(25)$, is 8.7 eV as opposed to 7.6 eV for $\alpha=0.67$. This value is in
much better agreement with CS and is in excellent agreement with the
experimental value (14) of 8.9$\pm$0.2 eV.

In Fig. 3 we have displayed similar results for β-cristobalite. Our
calculation shows that this material has a direct gap of 9.8 eV ($\alpha=0.67$) at
Γ. Thus the suggestion by CS that an idealized form of SiO_2 might yield
too low a value for energy gap is not realized. Complete band structure
for β-cristobalite may be found in ref. 4.

Total and Orbital Densities of States

Figure 4(a) shows our calculated TDOS for α-quartz ($\alpha=2/3$). In the same
panel, results of pseudopotential calculations (3) and various photoemission
experiments (14,15,16) are reproduced. The remaining panels 4(b)-4(g) give
calculated orbital densities of states. The lowest valence bands (-17 to
-20 eV) are primarily derived from the oxygen 2s orbitals with relatively

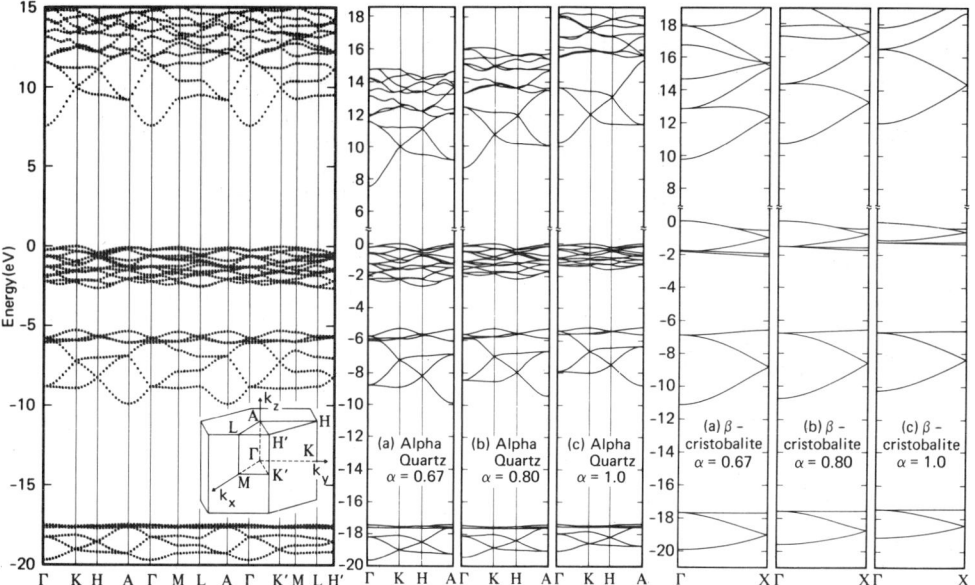

Fig. 1. Energy bands of   Fig. 2. Energy bands of   Fig. 3. Energy bands
α-quartz (the exchange   α-quartz for different   of β-cristobalite for
parameter α=0.67).   values of the exchange   different values of
                               parameter.   the exchange parameters.

small amounts of oxygen 2p, silicon 3s and 2p characters. Our calculated
width of these bands agrees very well with CW, but CS obtain somewhat
narrower bands. Also the position of oxygen 2s bands obtained by CS are
~5 eV too low with respect to experiment.

Moving up in energy, we encounter oxygen 2p bonding bands extending from
-10 to -5 eV. Most of the structure in these bands arises due to M-point
singularities. The lower and upper parts of these bands contain silicon
3s and 3p respectively. This assignment is in good agreement with the
findings of CW. The width of our bands, which is in good agreement with
the experimental (UPS) data, is somewhat narrower than CS results.
Furthermore CS predict a peak in TDOS at about -12 eV which is absent in
our calculations as well as in the UPS data.

The nonbonding oxygen 2p bands are about 3 eV wide in our calculation are
somewhat narrower than the corresponding bands obtained by CS and in UPS
experiments. There is about 2 eV wide region of zero density of states
which separates our bonding and nonbonding bands. The lower lying
conduction bands are predominantly silicon s and p with relatively small
admixture of oxygen 2p. This is to be contrasted with CW who suggested

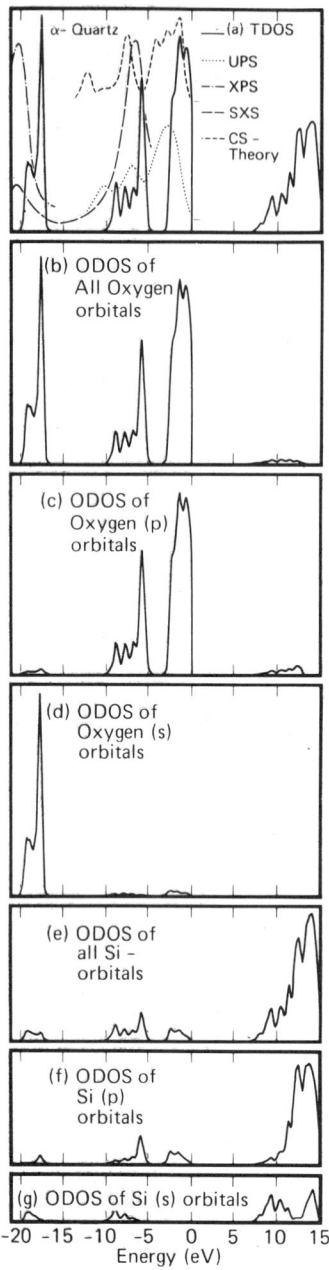

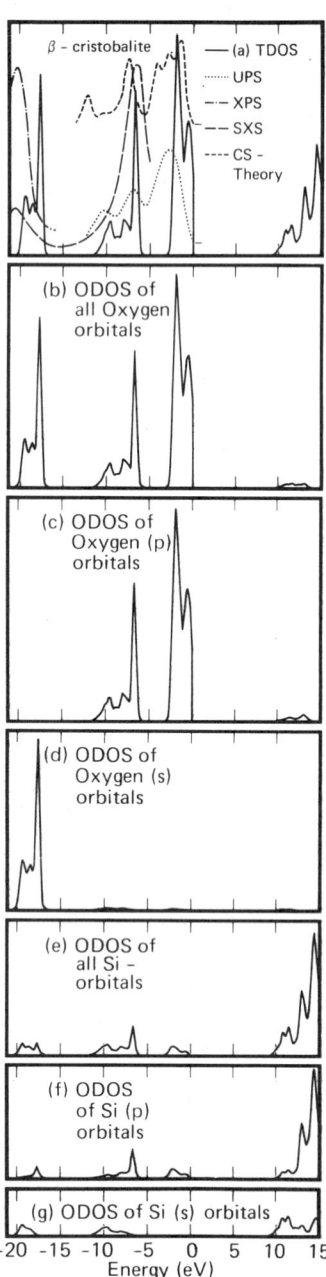

Fig. 4. Total and orbital densities of states for α-quartz. CS theory results are from ref. 3. Experimental results are reproduced from refs. 14-16.

Fig. 5. Total and orbital densities of state for β-cristobalite. For other explanations see Fig. 4.

that oxygen s components play a significant role in the lowest conduction bands.

In Fig. 5 total and orbital densities of states for β-cristobalite are shown and compared with experiments. Most of the peaks in TDOS arise due to L point singularities. One major difference between α-quartz and β-cristobalite is that Si-O bonding bands in β-cristobalite lie deeper by about 1 eV. This results in a 3 eV wide region of zero density of states between bonding and nonbonding bands. Orbital character of various bands are similar in both structures. It appears that α-quartz gives a somewhat better representation of experimental data but for qualitative purposes either one should suffice.

REFERENCES

1. S. T. Pantelides, Phys. Lett. A54, 401 (1975); S. T. Pantelides and W. A. Harrison, Phys. Rev. B13, 2267 (1976). Also see M. A. Reily, J. Phys. Chem. Solids 31, 1041 (1976); S. T. Pantelides, Comm. Solid State Phys. 8, 55 (1977).

2. P. M. Schneider and W. B. Fowler, Phys. Rev. Lett. 36, 425 (1976).

3. M. Schluter and J. R. Chelikowsky, Solid State Commun. 21, 381 (1977) and Phys. Rev. B15, 4020 (1977).

4. S. Ciraci and I. P. Batra, Phys. Rev. B15, 4923 (1977).

5. E. Calabrese and W. B. Fowler, to be published.

6. R. W. G. Wyckoff (1965) Crystal Structures, Interscience, New York.

7. J. C. Slater, Phys. Rev. 81, 385 (1951).

8. E. E. Lafon and C. C. Lin, Phys. Rev. 152, 579 (1966); R. C. Chaney, T. K. Tung, C. C. Lin, and E. E. Lafon, J. Chem. Phys. 52, 361 (1970).

9. I. P. Batra, S. Ciraci, and W. E. Rudge, Phys. Rev. B15, 5858 (1977); and references cited therein.

10. S. Huzinaga, Approximate Atomic Wavefunctions II, University of Alberta, Edmonton, Alberta, Canada; P. S. Bagus (private communication).

11. I. P. Batra, IBM Research Report, RJ2096, October (1977).

12. R. S. Mulliken, J. Chem. Phys. 23, 1833 (1955).

13. T. H. DiStefano and D. E. Eastman, Phys. Rev. Letters 27, 1560 (1971).

14. T. H. DiStefano and D. E. Eastman, Solid State Commun. 9, 2259 (1971).

15. H. Ibach and J. E. Rowe, Phys. Rev. B10, 710 (1974). Also see L. F. Wagner and W. E. Spicer, Phys. Rev. B4, 1512 (1974).

16. D. W. Fischer, Advn. X-Ray Analysis, 13, 159 (1970).

BAND STRUCTURES AND ELECTRONIC PROPERTIES OF SiO_2

W. Beall Fowler, Philip M. Schneider* and Eduardo Calabrese**
Department of Physics and Sherman Fairchild Laboratory
Lehigh University, Bethlehem, Pa. 18015

ABSTRACT

The results of mixed-basis calculations of the electronic energy bands of SiO_2 in both the idealized beta cristobalite and alpha quartz structures are discussed. Simple tight-binding fits to the lowest conduction bands are presented and effective masses and densities of states are obtained. The valence and lowest conduction bands of idealized tridymite are obtained by the simple tight-binding technique; the tridymite results are then compared with results on cristobalite and quartz.

INTRODUCTION

We have recently reported the results of electronic band-structure calculations on silicon dioxide in both the idealized beta cristobalite (1) and alpha quartz (2) structures. These calculations were carried out by the mixed-basis method, and the results were discussed in some detail and were compared with experiment and with the results of other calculations.

In this paper we shall summarize some of the more important results of (1) and (2) and shall present some extensions of this work. The reader should note the existence of several other papers in these Proceedings on SiO_2 band structures. In addition, the review article by Griscom (3) contains detailed discussions of most of the work in this field up to the present time. The present paper should thus be considered in the context of this wider literature.

The results of (1) and (2) may be highlighted as follows (except where noted, these results apply to both beta cristobalite and alpha quartz):
1) The upper valence bands are relatively flat, with a maximum at $\vec{k} = 0$. Hole effective masses are correspondingly large, of order 3 to 6 electron masses. The wave functions are composed primarily of oxygen $2p\pi$ orbitals, with some participation by Si 3d.
2) The lower valence bands involve silicon-oxygen σ bonds. Silicon wave function admixture is relatively small.
3) The conduction-band minimum is at $\vec{k} = 0$, and the effective electron mass is estimated to be ∿0.5 electron masses.

---

*Present address: Freie Universität Berlin, Germany.
**Permanent address: Università Degli Studi di Parma, Italy.

4) The optical absorption edge is direct forbidden. This is consistent with the absence of an n=1 exciton below the absorption edge (4) (∿9 eV).
5) Some details of optical absorption and photoemission spectra as well as x-ray emission spectra can be explained by these results.
6) The valence bands may be fitted with some success by a very simple tight-binding model involving only oxygen p orbitals and two adjustable potential-energy parameters.
7) The mixed-basis calculations underestimate the band gap and overestimate the oxygen 2s-2p energy difference. It is argued in (2) that this difficulty probably arises from our treatment of exchange.
8) Other published theoretical results on alpha quartz by Chelikowsky and Schluter (5) predict an indirect band gap and conduction bands considerably wider than those which we have obtained.

TIGHT-BINDING FIT TO CONDUCTION BANDS

Although the lowest conduction bands in both cristobalite and quartz were found to be free-electron-like, the lowest energy point at H in quartz was found to deviate from this picture, and it was suggested (2) that this band resembled a broadened oxygen 2s band. Acting upon this idea, we have fitted the lowest conduction band with a simple tight-binding scheme involving only oxygen s functions and one adjustable potential-energy parameter. As expected, this fit works very poorly for higher conduction bands; however, it works well for the lowest conduction bands in both cristobalite and quartz, using the same value of the parameter, and because of its simplicity we can obtain densities of states and effective masses.

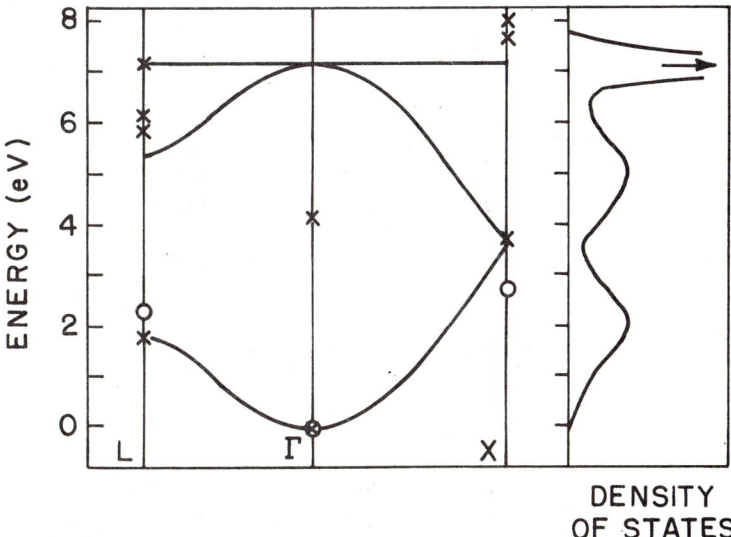

Fig. 1. Tight-binding fit to the idealized beta-cristobalite conduction bands (1). Circles represent free-electron energies, x's represent computed energies. The density-of-states curve was obtained from the tight-binding bands by a root sampling technique with Gaussian broadening.

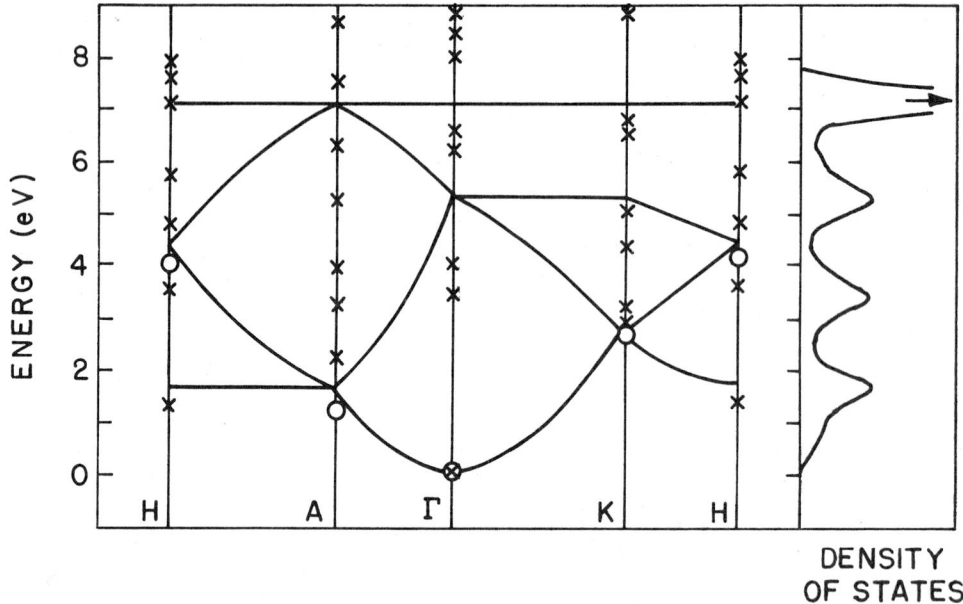

Fig. 2. Tight-binding fit to the alpha-quartz conduction bands (2). See
caption of Fig. 1 for further details.

The results are shown in Figs. 1 and 2. We consider only results associated
with states up to ∿3.5 to 4 eV; beyond this, the effects of states from
silicon cannot be ignored. It is interesting to note that in both cases
there is a density of states peak ∿2 eV above the bottom of the band.

From the curvature of these bands we find the effective mass to be .67 m_e for
cristobalite; for quartz the mass is anisotropic, having a value .65 m_e along
the k_z direction [Γ→A] and .70 m_e within the k_x-k_y plane [Γ→M]. These figures
are approximately twice those obtained by Chelikowsky and Schluter (5) and
reflect the fact that their bands are considerably wider than ours. Thus,
despite the simplicity of our model, these figures seem within reason.

TIGHT-BINDING BANDS OF IDEALIZED TRIDYMITE

It is of interest to know the band structures for as many modifications of
SiO_2 as possible, not only because of interest in each particular version but
also for the opportunity to study similarities and differences and thus perhaps obtain further insight into the properties of amorphous SiO_2. Of the
modifications which have not yet been considered, tridymite is of particular
interest because one structural model of amorphous SiO_2 involves tridymite-
like regions which are connected to each other in ways which preserve the
Si-O coordination but which destroy long-range order (6). Furthermore,
idealized tridymite is closely related to idealized beta cristobalite (7).
The latter may be generated by placing an oxygen midway between each pair of
silicons in the diamond lattice, while the former consists of the same
operation in the wurtzite lattice. There are 4 silicons and 8 oxygens in the

coordination unit of idealized tridymite, as opposed to 2 and 4 for cristobalite, 3 and 6 for quartz.

Because tight-binding fits have proved of some value for cristobalite and
quartz, we have begun our study of tridymite by constructing tight-binding
valence and conduction bands, using the same parameters as were used for
cristobalite and quartz. The conduction bands are shown in Fig. 3. Again,
the minimum is at k = 0. The density of states turns out to be identical to
that for cristobalite, and again the reader is reminded not to take seriously
the results above ⌄4 eV because of the simplicity of the model. The computed
effective mass is .67 m_e, and is found to be nearly isotropic; this occurs
because the c/a ratio of the hexagonal lattice of tridymite is close to the
ideal close-packing value.

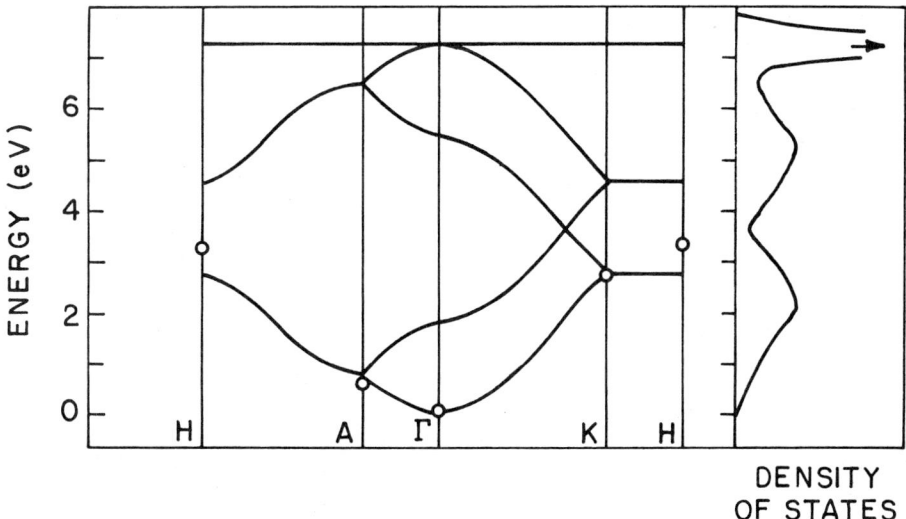

Fig. 3. Tight-binding estimate of the idealized tridymite conduction bands.
See caption of Fig. 1 for further details.

We do not present the valence bands here, but rather in Fig. 4 illustrate the
density of valence states obtained from tight-binding bands for tridymite,
along with those for cristobalite and quartz. The same parameters were used
in all three cases. It should be noted that the tridymite and cristobalite
densities of valence states curves are identical.

SUMMARY AND CONCLUSIONS

It thus appears that tridymite and cristobalite may be expected to have quite
similar optical and electronic properties, inasmuch as their densities of
states as computed by simple models are identical. When quartz is included
in the comparison some interesting differences emerge, including extra density-

of-states peaks in conduction and valence bands. The latter effect has also been suggested by Pantelides and Harrison (8) on the basis of tight-binding fits for cristobalite in which the Si-O-Si angle was varied.

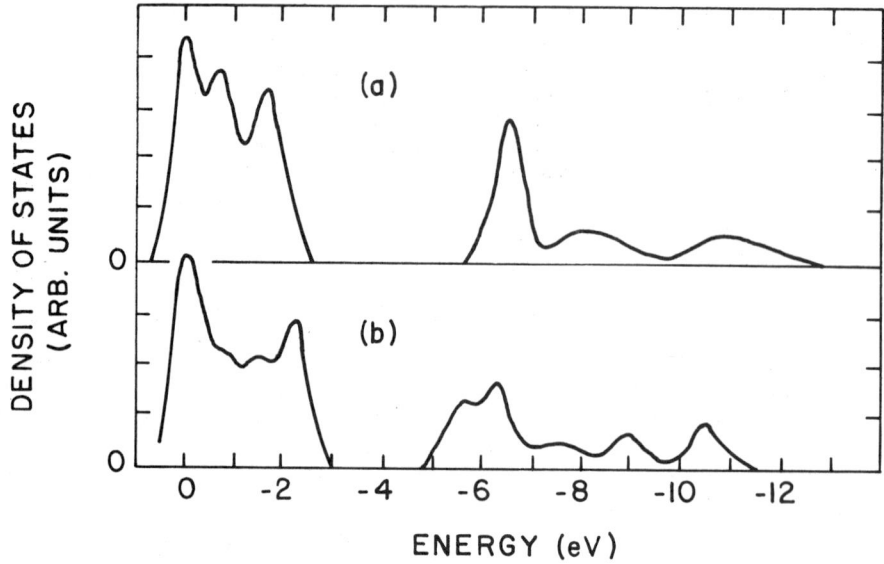

Fig. 4. Densities of states associated with tight-binding valence bands, obtained by a root sampling technique with Gaussian broadening: (a) idealized beta cristobalite and idealized tridymite; (b) alpha quartz.

In all 3 modifications a density-of-states peak occurs ∿2 eV above the bottom of the conduction band. As suggested previously (1), (2), this may be related to structure in the lower-energy part of the uv absorption spectrum.

ACKNOWLEDGMENT

This research was partially supported by the NSF, Grant No. DMR 71-01798 A02.

REFERENCES

(1) Philip M. Schneider and W. Beall Fowler, Phys. Rev. Lett. 36, 425 (1976); Philip M. Schneider and W. Beall Fowler, Phys. Rev. B17 (1978).
(2) Eduardo Calabrese and W. Beall Fowler, Phys. Rev. B17 (1978).
(3) D. L. Griscom, J. Non-Cryst. Solids 24, 155 (1977).
(4) T. H. DiStefano and D. E. Eastman, Solid State Commun. 9, 2259 (1971).
(5) J. R. Chelikowsky and M. Schluter, Phys. Rev. B15, 4020 (1977).
(6) J. H. Konnert, J. Karle and G. A. Ferguson, Science 179, 177 (1973).
(7) Helen D. Megaw, Crystal Structures: A Working Approach, W. B. Saunders, Philadelphia (1973).
(8) S. T. Pantelides and W. A. Harrison, Phys. Rev. B13, 2667 (1976).

K X-RAY SPECTRA OF AMORPHOUS AND CRYSTALLINE SiO$_2$

C.Sénémaud - M.T. Costa Lima
Laboratoire de Chimie Physique - 11 rue Pierre et Marie Curie -
75 005 PARIS - FRANCE -

INTRODUCTION

Various spectroscopical methods have been applied to obtain information about the electronic structure of silicon dioxide. A review of recent experimental and theoretical results on its electronic structure has been made recently by D.L.Griscom (1). In X-ray spectroscopy, which takes place among these experimental methods, electronic transitions between a core level and the valence or conduction states are studied. The energetic distribution of the SiK and L$_{2,3}$ inner level being narrow, the spectra reflect closely the band structure of the substance. The role of transition probabilities allows moreover to have information on the symmetry of the electronic states involved. Thus a K spectrum provides principally the distribution of p states while a L spectrum involves s or d states. By this method, studies of insulators or conductors can be performed with the same accuracy.

During our investigation of X-ray spectra of silicon in the pure element (2) and in some compounds (3), we have studied the spectra of dioxide SiO$_2$ in amorphous and crystalline form. The purpose of this paper is to present these spectra and discuss them with comparison to other experimental data and to recent theoretical works available on this compound.

Spectra were analysed with bent crystal spectrographs adapted to electronic or photonic excitation. The samples were, for emission spectra, either vitreous silica or α-quartz. For absorption spectra, thin screens (a few thousand Å thick) are needed. They were prepared either by evaporation with an electron gun of silica powder, or by deposit of α-quartz powder on a 2 μm makrofol foil.

RESULTS

K Emission and absorption spectra of amorphous and crystalline SiO$_2$ (a-SiO$_2$, c-SiO$_2$) are shown in Fig. 1 . (curves 1 and 2). These spectra are not corrected for broadening effects which consist of the width of the K level (0.4 eV) and of the instrumental effect (about the same order). They are in agreement with previous results (4).

The general shape of K$_\beta$ is similar in a-SiO$_2$ and c-SiO$_2$. Only a small broadening of the curve is observed from crystalline to amorphous phase. The total widths are respectively 11.7 and 12.5 eV; the shoulder (labelled C) situated on the high energy side of the peak B , and the small structure (labelled A) on the low energy side of B , are better resolved on c-SiO$_2$ than on a-SiO$_2$ emission. In both spectra, the K$_\beta$' emission (labelled D) is situated at $^-$14 eV from K$_\beta$ maximum and its relative intensity is about 20% of K$_\beta$.

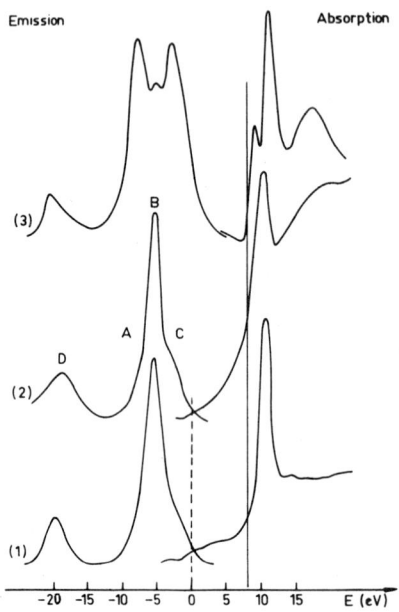

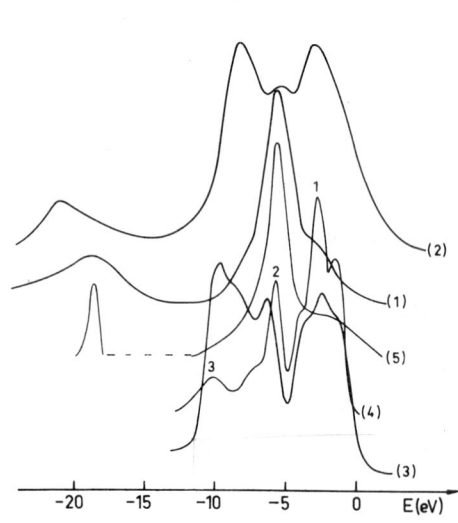

Fig.1 X-ray emission and absorption
spectra of SiO_2

    (1) K spectra of amorphous SiO_2

    (2) K spectra of α -quartz

    (3) L spectra of α -quartz from
ref.(5).

Fig.2 Spectral and theoretical data
for SiO_2

    (1) X-ray emission $K\beta$

    (2) X-ray emission $L_{2,3}$ from
ref.(5)

    (3) XPS from ref.(7b). The high-
energy edge is adjusted to E=0

    (4) DOS from ref.(7b)

    (5) Calculated $SiK\beta$ from ref.(9)
Peak 2 of curve (4) and maxi-
mum of curve (5) are adjusted
to $K\beta$ peak.

In order to locate the top of the valence band, the lorentzian distribution of the inner level and the instrumental function must be taken into account. In practice, we can consider that these broadening effects give rise to the tails of the emission and the position of the top of the band can be obtained as shown Fig. 1 (zero of the energy scale).

A strong absorption maximum occurs on both spectra at the same energy. Its low energy edge is more abrupt for a-SiO$_2$ than for α-quartz spectrum which shows moreover a noticeable extension of the discontinuity bottom. The absorption maximum is higher for a-SiO$_2$ than for c-SiO$_2$, but this may be due to an effect of the thickness of the screen which was higher for the crystalline form than for the amorphous for technical reasons.

The energy range which appears between the top of the valence band and the absorption edge gives the value of the energy gap. In fact a precise value is difficult to deduce from the spectra on which the limits of the gap are not strictly defined. Owing to the apparent extension of the absorption edge of α-quartz, the energy gap is smaller in this case than for a-SiO$_2$. This will be discussed further.

The L$_{2,3}$ emission and absorption spectra of α-quartz obtained by O.A.Ershov et al[5] are shown comparatively to K-spectra in Fig. 1. No difference has been observed between L spectra of glassy and crystalline SiO$_2$ (5).

To adjust K and L spectra in the energy scale, we have taken into account our energetic value of the SiO$_2$ Kα_1 transition. The Kβ and L$_{2,3}$ spectra describe respectively p and s or d states; they should be complementary, which is in fact observed.

DISCUSSION

Crystalline SiO$_2$
Emission (Fig.2). Various calculations on electronic distribution of SiO$_2$ have been reported : molecular cluster calculations or band calculations of crystallographic structures.

Most of the molecular calculations concern SiO$_4$ cluster. According to these theoretical works, the high energy part C of the valence band is principally due to non-bonding oxygen 2p like orbitals (1t$_1$, 5t$_2$, 1e). Parts A and B can be related to bonding orbitals involving oxygen 2p states and silicon 3s or, respectively, 3p states (5a$_1$ and 4t$_2$ orbitals). The low energy emission D is attributed to bonding orbitals involving oxygen 2s states and 3p or 3s silicon states (3t$_2$, 4a$_1$). The data of a molecular orbital calculation LCLO by Yip and Fowler (6) are reported in table 1 comparatively to the experimental data of K and L spectra.

The XPS spectrum of the valence band region of α-quartz from ref.(7b) is shown in Fig.2 . There is a good agreement between the position of XPS and SXS structures.

In order to obtain a more realistic description of the electronic structure of SiO$_2$, band structure calculations have been recently performed. Two calculations have been made for β-cristobalite. One,by Pantelides and Harrison (7a),is semi-empirical, based on tight-binding method and concerns the valence band. The other,by Schneider and Fowler (8),gives both valence and conduction states : the crystal potential is obtained from atomic Si and O potentials in Hartree-Fock-Slater model and the wave functions are constructed from Bloch sums of

TABLE 1   Values in eV are given with respect to $K\beta$ maximum and
$4t_2$ orbital energies

Orbitals		Calculated values (6)	Experimental values	
			K Spectrum	L Spectrum(5)
	$1t_1$	4.95		
non-bonding 0 2p	$5t_2$ $1e$	3.44 1.22	3.3	3.1
bonding 0 2p	$4t_2$ $5a_1$	0 − 1.26	0 − 2.6	0.9 − 2.4
bonding 0 2s	$3t_2$ $4a_1$	− 14.84 − 17.71	− 14.0	15.4

atomic-like functions and combinations of plane waves. A more recent calculation by Schluter and Chelikowsky is made for α -quartz in the OPW scheme (9).

The non-empirical calculation (8) gives a total band width of 11.5 eV, in good
agreement with our experimental data. In Fig.2, curve (4) from reference (7b)
corresponds to a 0.5 eV gaussian broadening of the density of states (DOS)
curve in order to obtain the theoretical XPS spectrum. Curve (4) reproduces
well the overall features of K and L spectra. Curve (5) from reference (9)
takes into account the dipole matrix elements for the 1s transition. This curve
is in excellent agreement with our $K\beta$ spectrum. The theoretical curve obtained
in the same reference for a 2p transition is in disagreement with the $L_{2,3}$
emission in the lower binding energy region. The 3d contribution which is not
taken into account in this calculation could explain the experimental result.

Pantelides et al (7b) show that the nearest neighbor oxygen-oxygen p orbitals
interactions are important and that the electronic distribution cannot be
described by the simplified molecular representation of separated bonding and
non bonding orbitals : there is in fact an hybridization of these orbitals.

Absorption From the band calculation of Schneider and Fowler for β-cristoba-
lite (8), the lowest conduction band minimum Γ_1 is composed principally of
Si 3s states with an 0 2s contribution. Transitions to these states are allowed
from Si 2p but forbidden from Si 1s. This result is well supported by X-ray absorption data : the first L absorption maximum energy is smaller than the first
K maximum energy which may correspond in the same theoretical scheme to the
transitions to Si 3p states located at $L_{2'}$, about 1 eV beyond.

From this calculation the value of the energy gap is 8 eV; it corresponds to
a direct but optically forbidden transition. Experimental values of the gap
vary between about 8 and 9 eV. In Fig.1 we have indicated the limit of the
band gap by a vertical line located at 8 eV from the top of the valence band.
All the absorption structures are situated beyond this limit. In particular

the first L absorption maximum lies at 9 eV from the top of the valence band. The presence of excitons has been invoked to explain optical and electron energy-loss spectra of SiO_2. It appears from our results (Fig.1) that if the first L absorption maximum was an exciton line, the gap would be higher than 9 eV which is unlikely.

Amorphous SiO_2 As already seen, Pantelides and Harrison established that the band structure of SiO_2 is mainly determined by the short-range order (7b) ; they predicted only a filling of the dip between peak (2) and (3) in the DOS curve of SiO_2 (Fig.2). On our spectra, an overall broadening of $K\beta$ is observed from crystalline to amorphous phase. Similar results have been deduced from XPS spectra of $c-SiO_2$ and $a-SiO_2$ (10). This suggests that the high energy part of the DOS could also be influenced by the atomic arrangement.

Owing to the shape of the absorption edge, our spectra indicate that the energy gap is higher in $a-SiO_2$ than in α-quartz. This result is in disagreement with other measurements of the energy gap; it suggests that the extension of α-quartz absorption edge could be due to an impurity state. In order to conclude on this point, experiments with other samples must be carried out.

REFERENCES

(1) D.L. GRISCOM   J. non. cryst. solids 24, 155, 1977.

(2) C. SENEMAUD M.T. COSTA LIMA   to be published.

(3) C. SENEMAUD   M.T. COSTA LIMA   J.A. ROGER   A. CACHARD
     Chem. Phys. Lett 26, 431, 1974.

     M.T. COSTA LIMA   C. SENEMAUD   Chem. Phys. Lett 40, 157, 1976.

(4) K. LAÜGER   Dissertation   München 1968.

(5) O.A. ERSHOV   A.P. LUKIRSKII   Sov. Phys. Sol. State 8, 1699, 196/.

(6) K.L. YIP   W.B. FOWLER   Phys. Rev. B10, 1400, 1974.

(7) a _ S.T. PANTELIDES   S.A. HARRISON   Phys. Rev. B13, 2667, 1976.

     b _ S.T. PANTELIDES   B. FISHER   R.A. POLLAK   T.H. DISTEFANO
         Solid. St. Comm. 21, 1003, 1977.

(8) P.M. SCHNEIDER   W.B. FOWLER   Phys. Rev. Lett. 36, 425, 1976.

(9) M. SCHLUTER   J.R. CHELIKOWSKI   Sol. St. Comm 21, 1123, 1977.

(10) D.A. STEPHENSON   N.J. BINKOWSKI   J. non Cryst. Solids 22, 399, 1976.

THE OPTICAL ABSORPTION SPECTRUM OF SiO_2[‡]

Sokrates T. Pantelides
IBM T. J. Watson Research Center, Yorktown Heights, NY 10598

ABSTRACT

We show that inclusion of electron-hole interactions not only produces an exciton at 10.4 eV, but also modifies the interband continuum in a substantial way. The result is a spectrum consisting of three excitonic peaks, in good agreement with the observed spectrum. These peaks are identified in terms of the bond-orbital model as members of a crystal-field split Frenkel-type exciton corresponding to the free oxygen atom $2p \rightarrow 3s$ transitions.

INTRODUCTION

The optical absorption spectrum of SiO_2, shown[1] in Figs. 1a and 2, has been the subject of considerable interest and controversy[2]. Following early attempts to interpret the spectrum,[2,3] DiStefano and Eastman[4] concluded from a series of photoconductivity and internal photoemission measurements on Si-SiO_2 structures that the band gap is about 9 eV. These results raised the question why no strong excitons appear in the absorption spectrum below the band edge as is the case in other ionic wide-gap materials, such as the alkali halides. The observed peaks thus became the subject of conflicting interpretations in terms of excitons, interband transitions and various combinations thereof (see e.g. Fig. 6 in Ref 2 for summary). More recently, Schneider and Fowler[5] calculated a band structure for cubic SiO_2 and suggested that excitons are *not* formed because of a symmetry-forbidden edge. Chelikowsky and Schlüter (CS)[6] actually calculated an interband spectrum (Fig. 1c) and found an indirect bandgap of 9.2 eV, which is consistent with the measurements of Ref. 4. They also found that matrix elements suppress absorption below ~10.4 eV. They therefore suggested that the first observed peak corresponds to an exciton, as previously proposed by others,[2,3] while the rest of the spectrum is due to inter-band transitions. In support of the latter assignment, they displayed the experimental spectrum shifted by 0.7 eV and noted the resulting good agreement between theory

[‡]Supported in part by the Office of Naval Research under Contract No. N00014-76-C-0934.

and experiment (Fig. 1b and 1c). More recently, Mott[7] investigated the subject and suggested that there exists an allowed exciton at 9 eV which has a dispersion of ~2.4 eV due to hopping, resulting in the 10.4 eV peak (see also the paper by Mott in these Proceedings).

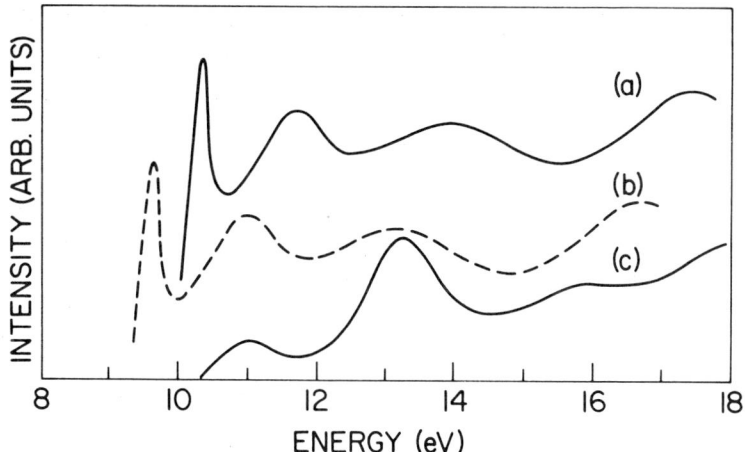

Fig. 1   (a) The optical absorption spectrum of SiO_2 measured by Klein and Chun [Ref. 1; the spectrum shown is actually $E^2 \varepsilon_2(E)$]. (b) The same spectrum shifted by 0.7 eV in Ref. 4 in order to align it with the theoretical interband spectrum, curve (c).

In this paper we present two completely independent treatments of electron-hole (e-h) interactions and conclude that the observed peaks are a series of excitonic resonances. This interpretation is consistent with but goes beyond a band-theoretic approach and is also consistent with the known data.

BAND-THEORETIC APPROACH AND THE EFFECT OF E-H INTERACTIONS

In most semiconductors which have large dielectric constants, e-h interactions tend to be weak and produce excitons with very small binding energies and oscillator strengths. As a result, the interband spectrum remains essentially unmodified and is in general well reproduced by band-structure calculations. In the case of wide-gap insulators, however, dielectric constants are small, making e-h interactions very strong. The resulting excitons are then localized and carry considerable oscillator strength. This oscillator strength is pulled out of the interband continuum (the total oscillator strength remains constant) in such a way that the final spectrum is substantially different from the independent-particle spectrum. This assertion has been supported by theoretical calculations[8,9] and by a study of 39 x-ray absorption spectra of alkali halides.[10]

In this section we start with the interband spectrum of SiO_2 calculated by CS^6, which indicated that the 10.4-eV peak should not be attributed to interband transitions. We will then incorporate e-h interactions by means of a model calculation, which indicates that e-h interactions strong enough to produce an excitonic peak at 10.4 eV also modify the interband spectrum very strongly, so that all the observed peaks are excitonic in nature.

The model calculation we intend to carry out is similar to that done previously[9] for LiF. We start with the interband spectrum calculated by CS and assume it can be simulated by a two-band model and momentum-independent transition matrix elements. In such a model, the absorption coefficient for interband transitions is given by the imaginary part of a two-particle Green's function G^o describing independently propagating electrons and holes. We therefore construct an analytical form for $G^o(E)$ as in Ref. 9, chosen so that its imaginary part, when broadened by 0.5 eV, reproduces CS's interband spectrum. E-h interactions are then introduced in the contact approximation,[8,9] i.e. as a single on-site matrix element V in the Wannier representation. The new absorption spectrum is given by the imaginary part of the new Green's function G, which satisfies Dyson's equation:

$$G = G^o + G^o V G. \tag{1}$$

Eq. (1) is solved directly to yield

$$G(E) = G^o(E)/[1 - G^o(E)V] \tag{2}$$

so that a bound state occurs when $G^o(E) = 1/V$. The value of V was chosen to yield a bound state at 10.4 eV.

The results of the calculation are shown in Fig. 2. The dashed curve is $-\text{Im } G^o(E)$, with E having a 0.5 eV imaginary part for broadening, and corresponds to CS's interband spectrum (cf. with Fig. 1c). The solid curve marked theory is $-\text{Im } G(E)$, also broadened by 0.5 eV. These results clearly show that if the 10.4-eV peak is indeed an exciton, e-h interactions also modify the interband spectrum strongly, so that the final spectrum is excitonic in nature, i.e. consists of a series of excitonic resonances (sometimes called metastable excitons). Note that whereas the two peaks in the interband spectrum are separated by ~2.3 eV, the two peaks in the final spectrum are separated by 1.4 eV, in excellent agreement with experiment. The spectrum at higher energies is not reproduced very well as one might expect, since oscillator strength at those energies would have to be brought down by e-h interactions from even higher energies, which are left out by the present model. Nevertheless, the calculated peak at 15 eV compares rather well with the observed peak at 14 eV.

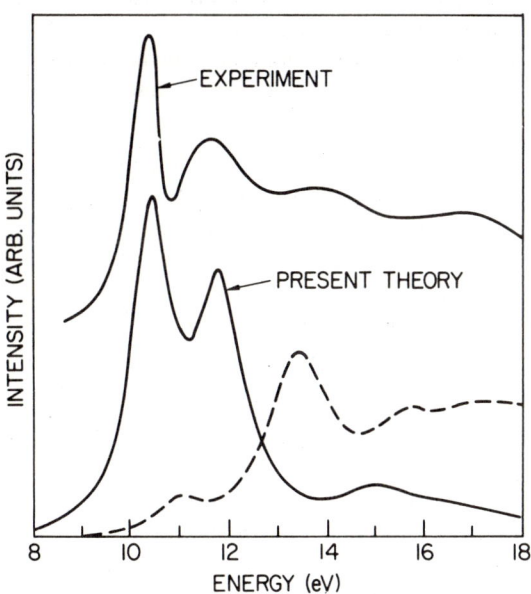

Fig. 2   Theoretical absorption spectrum compared with the optical absorption spectrum $[\varepsilon_2(E)]$ measured by Philipp (Ref. 1).   The dashed line is interband spectrum before e-h interactions.   See text.

Apart from the success of the above model calculation in predicting the energy separations and relative heights of the peaks in the optical spectrum, the main conclusion is that in wide-gap insulators one cannot ignore the effect of e-h interactions on interband transitions, as done in semiconductors.   In contrast, e-h interactions have a rather pervasive effect, so that excitonic peaks dominate the entire spectrum. One should not, therefore, hope to extract a band gap from such a spectrum.

BOND-ORBITAL APPROACH

The above calculation indicates that the absorption spectrum of SiO_2 can be interpreted as a series of excitonic peaks arising from localized excitations, but does not provide information about the nature of states contributing to the two observed peaks.   As was the case for alkali halides,[10] an atomistic point of view can be complementary and useful.   For this purpose, we make use of the tight-binding bond-orbital model[11] for SiO_2, according to which the valence band states arise from three different bond orbitals, associated with a Si-O-Si bonding unit: B_y, which is a pure O $2p$ orbital perpendicular to the Si-O-Si plane (lone-pair or non-bonding p orbital); B_x, which is ~85% an O $2p$ orbital in the Si-O-Si plane perpendicular to the Si-Si axis and ~15% Si sp^3 hybrid (partially bonding orbital); and B_z, which is ~65% an O $2p$ orbital along the Si-Si axis and ~35% Si sp^3 hybrid (bonding orbital). These three orbitals at each bonding unit have been found adequate for a calculation of the valence bands and the related photoemission and x-ray-emission spectra.[11]

According to the same bond-orbital scheme, the lowest unoccupied band states arise from a bond orbital denoted in Ref. 10 by A_+, consisting mainly of O $3s$ and Si hybrid admixtures. We suggest that during photoabsorption the electron and the hole do not occupy propagating band states. Instead, e-h interactions tend to localize the electron and the hole in the same bonding unit, so that excitonic peaks are expected at the $B_y \to A_+$, $B_x \to A_+$, and $B_z \to A_+$ energies, in that order. Note that these transitions are essentially localized on an O atom and correspond to the atomic transition $2p \to 3s$. In the free atom the transition energy is about 9.2 eV. In the solid, B_y, B_x, and B_z may be viewed as the "crystal-field" split members of the O $2p$ orbital, resulting in a "crystal-field" split Frenkel-type exciton with peaks at 10.4 eV, 11.7 eV, and ~14 eV. These energies cannot be predicted by some simple quantitative calculation, but the interpretation may be further supported by a variety of arguments. First, CS's calculation showed that the 10.4-eV should not be attributed to interband transitions. Second, the calculation of the previous section demonstrated that if the 10.4-eV peak is indeed an exciton, the other peaks are also predominantly excitonic in nature. Third, using the identification presented above, one can infer that the three peaks ought to be successively shorter, because of the smaller weight of O $2p$ in each successive B, and also broader, because of the shorter lifetime of a hole in each succesive B. Both these observations are in agreement with experiment and consistent with the calculation of the previous section.

REFERENCES

1. H. R. Philipp, **Solid State Comm. 44**, 73 (1966); G. Klein and H. U. Chun, **Phys. Stat. Sol. B 49**, 167 (1972).
2. See recent review by D. L. Griscom, **J. Non-Cryst. Sol. 24**, 155 (1977).
3. A. R. Ruffa, **J. Non-Cryst. Solids 13**, 37 (1973/74).
4. T. H. DiStefano and D. E. Eastman, **Solid State Comm. 99**, 2259 (1971).
5. P. M. Schneider and W. B. Fowler, **Phys. Rev. Lett. 36**, 425 (1976)
6. J. R. Chelikowsky and M. Schlüter, **Phys. Rev. B 15**, 4020 (1977).
7. N. F. Mott, **Adv. in Phys. 26**, 363 (1977).
8. J. Hermanson, **Phys. Rev. 166** 893 (1968).
9. S. T. Pantelides, R. M. Martin and P. N. Sen, Proc. IV Intern. Conf. on VUV Rad. Phys., E. E. Koch, ed., (Pergamon/Vieweg, 1974).
10. S. T. Pantelides, **Phys. Rev. B 11**, 2391 (1975).
11. S. T. Pantelides and W. A. Harrison, **Phys. Rev. B 13**, 2667 (1976); S. T. Pantelides, **J. Vac. Sci. Technol. 14**, 965 (1977).

INELASTIC ELECTRON SCATTERING IN SiO$_2$

A. E. Meixner, P. M. Platzman, and M. Schlüter
Bell Laboratories, Murray Hill, New Jersey 07974

ABSTRACT

We have measured the cross section for 200 keV electrons scattered from thin amorphous SiO$_2$ foils. Energy losses from 0 to 600 eV were studied. This includes features at the oxygen K-edge, the silicon L$_{2,3}$-edge, and the low energy valence to conduction band transitions. The low energy spectra are also measured for finite momentum transfers out to q = 1.0 Å^{-1} The results are discussed in the light of one electron band structure theories and possible excitonic effects.

The cross section of inelastically scattered high-energy electrons with momentum transfer, q and energy transfer ω may be written as (1)

$$\frac{d\sigma}{d\Omega d\omega} = \frac{r_0^2}{\pi(q\hbar/mc)^4} \quad S(\vec{q},\omega) \tag{1}$$

where

$$S(\vec{q},\omega) = \sum_{if} |<f| \sum_j e^{i\vec{q}\cdot\vec{r}_j}|i>|^2 \delta(E_f - E_i - \hbar\omega). \tag{2}$$

$S(\vec{q},\omega)$ reflects the spectrum of electronic excitations in the system. The cross section may also be written in terms of the complex dielectric response function $\varepsilon(\omega,q)$ as

$$\frac{d\sigma}{d\Omega d\omega} \sim Im\left(\frac{1}{\varepsilon(\omega,\vec{q})}\right). \tag{3}$$

This form allows us to compare electron energy loss spectra to optical reflectivity or absorption spectra which measure other functions of $\varepsilon(\omega,0)$. In particular, loss spectra near q = 0 compare directly to spectra obtained by Kramers-Kronig transformation from optical data. Beyond that, loss spectra for q ≠ 0 yield information about the electronic transitions that are dipole forbidden and thus inaccessible to optical experiments.

We report here scattering experiments of 200 keV electrons on amorphous 1000 Å SiO$_2$ films. These films were grown in a wet atmosphere on a single crystal Si wafer as part of a MOS device structure. The Si was then selectively etched away. Experimental details are described elsewhere (1). In Fig. 1, spectra in the energy range of the Silicon L$_{2,3}$-edge and the oxygen K-edge

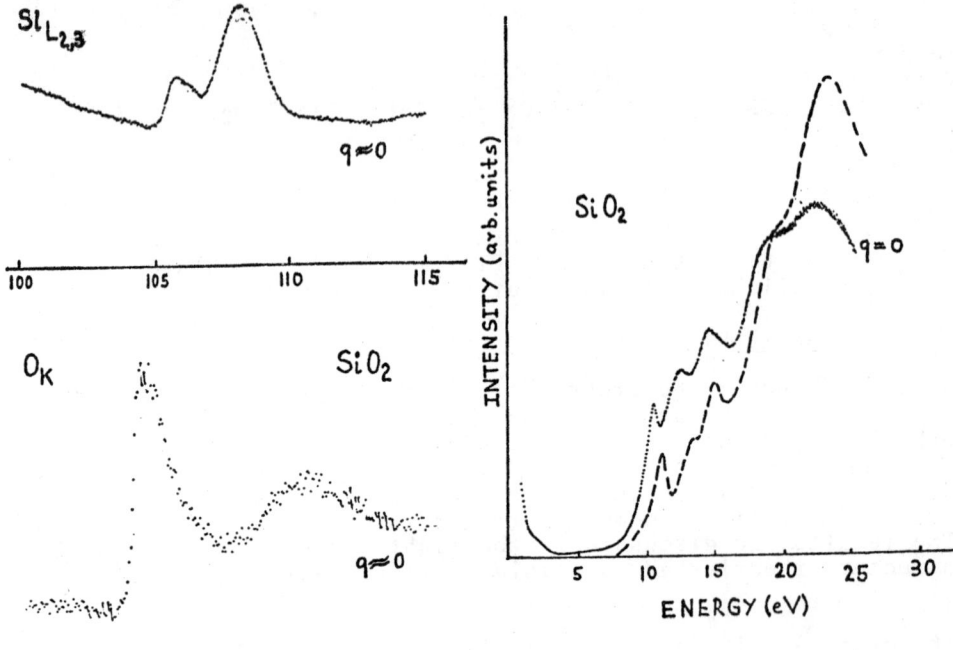

Fig. 1 Silicon $L_{2,3}$- and oxygen K-edge electron energy loss spectra obtained for q = 0 momentum transfer on an amorphous SiO_2 film.

Fig. 2 Comparison of a q = 0 loss spectrum in the low-energy range of SiO_2 (data points) with the $Im(1/\varepsilon)$ function computed from optical UV data reported in ref. 4 (dashed line)

are shown for q ≠ 0. The spectra compare roughly in their energetic positions to earlier measurements by Koma and Ludeke (2) The spectra are strongly affected by excitonic effects as earlier noted by comparing the threshold energies to XPS photoemission data (3). In Fig. 2, the low energy valence to conduction band loss spectrum for q = 0 is compared to the $Im (1/\varepsilon)$ function computed from optical UV reflectivity data (4). Excellent agreement is found for all major peak positions. Both experiments measure volume properties in contrast to low-energy electron loss spectroscopy which can be quite surface-sensitive Data obtained by the latter technique (2,5) show some differences in peak positions as compared to the present results. In addition, there is an extra peak at about 17 eV in the low-energy electron loss data which is missing in our volume data. This peak may be due to the surface plasmon.

In Fig. 3, we show the low-
energy spectra for several finite
momentum transfers. Rather
marked spectral changes occur
which can be partially explained
in terms of one-electron transitions based on bandstruc-
ture models and in terms of excitonic effects. The UV reflectivity spectrum which, as we saw
in Fig. 2, corresponds well to
the q = 0 loss spectrum, was
recently analyzed on the basis
of a pseudopotential band calculation of α-quartz (6). This
analysis yielded that the sharp
structure at about 10.5 to 10.8
eV corresponds to a direct,
allowed exciton, overlapping
with some indirect and dipole
forbidden transitions. Structures above 11 eV could be identified with structures in the
one-electron joint density of
states. Exciton as well as band-
to-band transitions were identified to correspond to oxygen 2p
to 3s-like transitions. This
picture seems to be consistent
with the q-dependence of the

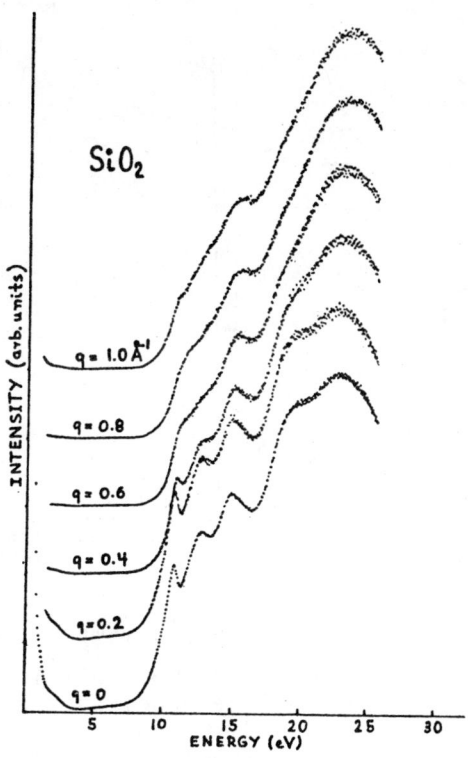

present loss spectra. The strong
excitonic peak and the associated band edge decrease rapidly
in intensity with finite momentum transfer. While these, at
q = 0 dipole allowed transitions
become weaker, resonant but di-
pole forbidden transitions become stronger for finite q-values
and "smear out" the spectrum.

Fig. 3 Comparison of loss
spectra of amorphous SiO_2 in
the low-energy range for several momentum transfers ranging
from q = 0 to q = 1.0 Å^{-1}.

The oxygen 3s character of the lowlying conduction states can
further be checked by the q-dependence of transitions from the
oxygen 2s quasi core-like states. In Fig. 4, loss spectra for
q = 0 and q = 1.0 Å^{-1} up to about 35 eV are compared. While for
q = 0, no structure is observed at the high-energy side of the
main peak, for q = 1, an additional loss occurs around 31 eV.
This energy corresponds approximately to the sum of the 2s
core level binding energy, measured by XPS and the optical gap
(6). Small excitonic shifts may be present, but are not resolved
in the current data.

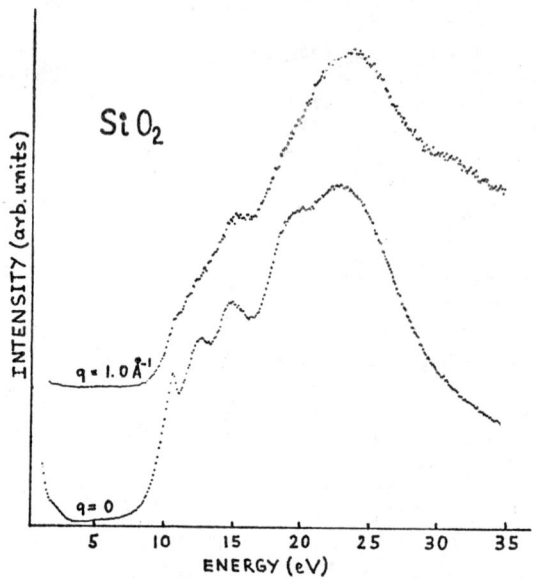

Fig. 4 Comparison of the q = 0 and
q = 1 Å^{-1} loss spectra of amorphous
SiO$_2$ for energies up to 35eV. Note
the appearance of a structure around
31 eV in the q = 1 Å^{-1} spectrum due
to dipole forbidden oxygen 2s-3s
transitions.

REFERENCES

1. A. E. Meixner, M. Schlüter, P. M. Platzman, and G. S. Brown,
Phys. Rev. B17, 686 (1978); G. S. Brown and A. E. Meixner, to
be published.

2. A. Koma, R. Ludeke, Phys. Rev. Lett. 35, 107 (1975).

3. S. T. Pantelides, W. A. Harrison, Phys. Rev. B13, 2667 (1976).

4. H. R. Philipp, J. Phys. Chem. Solids 32, 1932 (1971).

5. H. Ibach, J. E. Rowe, Phys. Rev. B10, 710 (1974).

6. J. R. Chelikowsky, M. Schlüter, Phys. Rev. B15, 4020 (1977).

ELECTRONIC STRUCTURE OF SiO_2

FROM ELECTRON ENERGY LOSS SPECTROSCOPY

J. OLIVIER, P. FAULCONNIER and R. POIRIER

THOMSON-CSF, Laboratoire Central de Recher-

ches - Domaine de Corbeville, 91401 ORSAY, FRANCE

ABSTRACT

The electronic structure of SiO_2 has been studied by electron energy loss spectroscopy, and an energy level model of both filled and empty states is constructed from the ELS and available optical data. For the thermally grown SiO_2, the transitions are assumed to originate on the three principal peaks in the valence band density of states and the O (2s) core state, respectively, to terminate on a single peak within the conduction band density of state. We also report E.L. spectra due to excitations out of the deeper Si (2p), Si (2s), and O (1s) core levels, with 250, 500, 1400 electron-beams. The excitations originating on the Si (2p), Si (2s), terminate on levels in the conduction band and on an exciton lying 2 ev below the conduction band edge. We have also excitations out of the O (2s) and O (1s) core level terminating on an exciton lying 1 ev below the conduction band edge.

INTRODUCTION

Low-energy electron loss spectroscopy (ELS) may be used as a tool to study excited states of solid materials, involving excitations from valence-band or from relatively shallow core states (such as d-core states, Ref. 1) into both bulk and surface-related final states.

Moreover, we report here detailed excitation spectra of deep core electrons of binding energies > 100 ev.

In E.L.S., initial occupied states as well as final empty states are involved. Therefore an assignment of an observed transition to certain accupied energy levels might be difficult unless additionnal information. Therefore, we have used available optical data, and an energy level diagram for SiO_2 is constructed.

EXPERIMENTAL TECHNIQUE

A single-pass VARIAN cylindrical-mirror analyzer with an integral coaxial electron gun are used for AES and ELS.
The primary beam has an angle of incidence normal to the surface while the scattered electrons are detected about a conical surface of apex angle of about 85°.

The loss spectra are measured in the reflexion mode and are taken as the second derivative $\frac{d^2 N}{dE^2}$ of the energy distribution to enhance detail. With a peak to peak modulation voltage of 0.4 V, the energy resolution as measured by the full width at half-maximum of the second derivative of the elastic peak is 0.7 ev at a primary electron energy of 150 ev. The average time required for scanning a 50 ev wide loss spectrum is ~ 15 min. By using the analyzer in the conventional mode, Auger analysis can be carried out on the same surface.

The amorphous silicon dioxide layers were thermally grown and was 500 Å thick.

RESULTS AND DISCUSSION

Second derivative energy-loss spectrum of silicon dioxide surface is shown in Fig. 1 for 150 ev electrons.

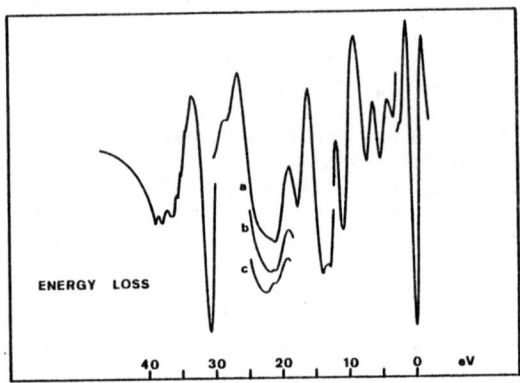

Fig. 1 Second derivative of the energy loss
spectrum of SiO_2. Primary energy is
150 ev.

The energies of the single-particle electronic transitions for SiO_2 are 10.5, 12.7, 13.7, 17.8, 20.9 ev.

At lower energies close to the values of loss peaks for the pure Si substrate (Ref.4-5), a metastable oxide phase $Si\,O_x$ is observed (3.3, 5.1 and 7.2 ev).

The Si (2p) and Si (2s) core excitation spectra, shown in Fig. 2, like the O (2s) and O (1s) core excitation spectra shown in Fig. 3 exhibit main peaks which are similar in relative energy position and line shape. But for higher values of the energy losses, the agreement is better for O (1s) and O (2s) than for Si (2p) and Si (2s).

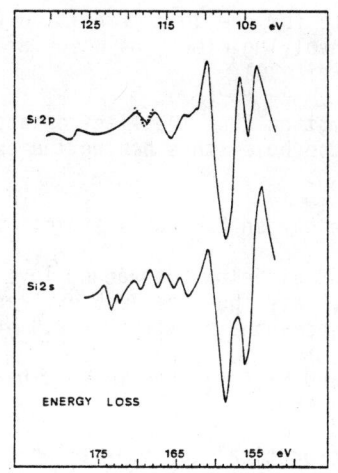

Fig. 2 Comparison of the electron
excitation spectra of the
Si (2p) and Si (2s) core
levels, the former taken
with Ep = 250 ev, the later
with 500 ev.

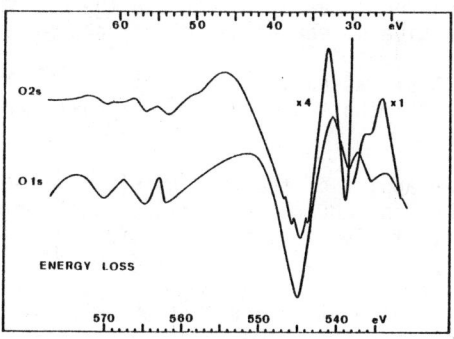

Fig. 3 Comparison of the electron
excitation spectra of the
O (2s) and O (1s) core
levels, the former taken
with Ep = 250 ev, the
later with Ep = 1400 ev.

In Fig. 1 a primary electron energy of 150 ev exites transitions from the
valence band and the O (2s) core state. The energy values are slightly different from those obtained by IBACH and ROWE (Ref.6) and KOMA and LUDEKE (Ref 7).

The UPS and XPS valence-band spectra have three main peaks and a shoulder on
the low binding energy side of the non-bonding O (2p) band at high concentrations of silicon (Réf. 8-9). The UPS valence-band spectrum is shown in Fig. 4
with energy positions according ot the review of GRISCOM (Ref. 11).

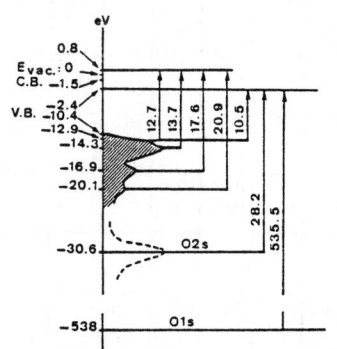

Fig.4 U.P.S. valence-band spectrum
and transition energies obtained from E.L.S.

REILLY has pointed out that the final state for the 10.2 transition (10.5 ev in the loss spectrum) may be an exciton involving electrons which are tightly bound in atomic-like oxygen 3s orbitals (Ref. 10).

Our data seem to confirm the RUFFA's assumption that the electron should be in a rather large hydrogenic orbit about the hole, this beeing the exiton model due to WANNIER (Ref. 12).

The 28.2 and 535.5 ev transitions from the oxygen 2s and 1s states (see table 1) fit in this picture (Fig.4).
The positions of these levels, 30.6 and 538 ev below the vacuum level, give an excitonic final state of - 2.4 and - 2.5 ev. Thus the 10.5 ev transition starts from the first peak in the UPS valence-band spectrum -(12.9 ev).

The first excitations out of the Si (2p) and Si (2s) core levels, have the same final state, 3.5 ev below the vacuum level.

These two excitonic levels seem confirm theoretical considerations of PANTELIDES (Ref. 13) : an anion-centered exciton is expected to have a smaller binding energy than an exciton centered on a cation because the electron-hole interaction is weaker due to effective screening by the valence electrons.

The remaining 12.7, 13.7, 17.8 and 20.9 are assumed to originate on the four peaks in the V.B. density of states to terminate on aconduction level lying $\sim$ 0.8 ev above the vacuum level. (Fig. 14)

We have not found the nearly exact overlapping of the core (Si (2p) and 0 (1s) and valence E.L.S. spectra that led KOMA and LUDEKE (Ref. 6) to conclude that the final states which terminate the first five optical transitions are all different.

The 20.9 ev transition is a weak peak located at the edge of the wide volume plasmon structure that peaks at 22.3 ev when the primary beam energy is increased to 500 ev (Fig. 1 b-c).

Table 1 collects the energy values of the excitations out of the V.B., the 0 (2s), 0 (1s), Si (2p) and Si (2s) core levels.

The energetic value of the Si (2s) core level is reported from SIEGBAHN et al (Ref. 14).

TABLE 1

INITIAL STATE													
B.V.	Transitions		10.5			4 T.							
	Final State		- 2.4			+ 0.8							
0 (2s)	T.		28.2		31		36.6			49.4	52.9	56.3	+61.9
	F.S.		- 2.4		+ 0.4		+ 6			+18.8	+ 22.3	25.7	30.3
0 (1s)	T.		535.5		538.4		545				561.9	564.1	569.9
	F.S.		- 2.5		+ 0.4		+ 7				-23.9	+26.1	+31.9
Si (2p)	T.	104.8		107.4			112.3	114.4	118.3	127.6			
	F.S.	- 3.5		- 0.9			+ 4	+ 6.1	+10	19,3			
Si (2s)	T.	156.1		158.7			163.4						
	F.S.	- 3.5		- 0.9			+ 3.8						
Average Final State		- 3.5	- 2.4	- 0.9	+ 0.4	+ 0.8	+ 3.9	+ 6.4	+10	+19	+23.1	+25.9	+31.1

All these results, reported on Fig. 5 fit together to forme a picture of the band structure of SiO_2. We obtain a set of empty conduction states levels, the three first beeing close to the edge of the C.B. and their energies in the range of the final states of the transitions seen in X-ray absorption.

One recognize too the 6.4. ev final states level seen by the same technique.

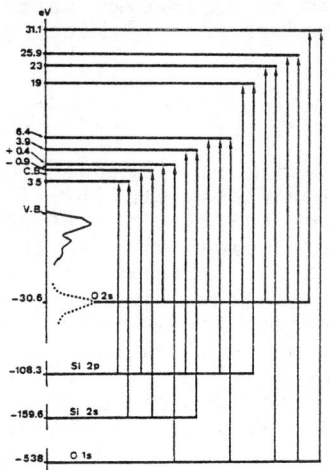

Fig. 5   Empirical energy-band diagram for SiO_2.

Except for our 10.5 ev excitonic transition, this transition scheme for SiO_2 is near to the PANTELIDES and HARRISON picture (Réf.2) involving a single final state slightly above the vacuum level.

REFERENCES

1. J. OLIVIER, P. FAULCONNIER, Séminaire sur les méthodes d'analyse et de caractérisation des couches minces. Les Arcs-Bourg St-Maurice FRANCE 23/27.1.78.
2. S.T. PANTELIDES, W.A. HARRISON, Phys. Rev. B 13, 2667 (1976)
3. H. IBACH, J.E. ROWE, Phys. Rev.B.9, 1951 (1974)
4. J.E. ROWE, A.P.L., 25, 576 (1974)
5. H.R. PHILLIP, J. Phys. Chem. Solids, 32, 1935 (1971)
6. H. IBACH, J.E. ROWE, Phys. Rev. B., 10, 710 (1974)
7. A. KOMA, R. LUDEKE, Phys. Rev. Lett, 35, 107 (1975)
8. T.H. DI STEFANO, D.E. EASTMAN, Phys. Rev. Lett., 27, 1560 (1971)
9. B. FISHER, R.A. POLLAK, TH. DI STEFANO, W.D. GROBMAN, Phys. Rev. B, 15, 3193 (1977)
10. M.H. REILLY, J. Phys. Chem. Solids, 31, 1041 (1970)
11. D.L. GRISCOM, J. Non-Crystal. Solids, 24, 155 (1977)
12. A.R. RUFFA, Phys. Stat. Sol., 29, 605 (1968)
13. S.T. PANTELIDES, Phys. Rev. B, 11, 2391 (1975)
14. K. SIEGBAHN, D. HAMMOND, H. FELLNER-FELDEGG, E.F. BARNETT Science, 176, 245 (1972)

THE ABSORPTION AND PHOTOCONDUCTIVITY SPECTRA OF VITREOUS SiO$_2$

A. Appleton, T. Chiranjiví and M. Jafaripour-Ghazvini
Birkbeck College (University of London) Malet Street London WC1

INTRODUCTION

The far ultraviolet reflectivity spectra of vitreous SiO$_2$ and α-quartz are well known from work by Loh (1964), Philipp (1966,1971) Platzöder and Platzöder and Steinmann (1968). DiStefano and Eastman (1971) and Zakis *et al*. (1973) have measured the photo-conductivity spectrum of amorphous SiO$_2$. At lower photon energies we have absorption measurements in the glass by Edwards (1966) and many observations of impurity and defect bands – mainly those induced by various forms of radiation – (review by Lell *et al*. (1966)). However, remarkably little information has been published on the fundamental absorption edge which lies between about 7.9 and 8.4 eV in the glass and about 0.5 eV higher in α-quartz.

For the purpose of exploring these regions we have used polished platelets 5 mm and 0.2 mm thick of synthetic z-cut α-quartz and water-free silica, the former generously donated by Professor C.H.L. Goodman and Mr R.W.T. Rabbetts of S.T.L. and the latter by Dr G. Hetherington of Thermal Syndicate Ltd. In addition we have prepared films of thickness down to 0.7 μm by blowing Spectrosil and Vitreosil tubing in an oxyacetylene flame.

RESULTS

The absorption spectra are shown in Figs. 1 and 2. Also plotted in Fig. 1 are absorption coefficient curves above 9.8 eV deduced from reflectivity values supplied by H.R. Philipp (private communication). In addition is plotted a photoconductivity spectrum obtained on Vitreosil films about 11 μm thick using "sandwich" geometry and semi-transparent Aℓ electrodes.

In both the glass and the crystal, the "tail" beneath the exponential edge was found to vary from one specimen to another (as noted also by Edwards, 1966) but was insensitive to temperature. Temperature-dependence of the tail has, in fact, been reported by Edwards and by Bates (1976) who used very thick specimens better suited to observing small changes in this region of low absorption. Photoconduction associated with the tail in the glass showed a similar variation between specimens. Photo-current in this region was not observed by DiStefano and Eastman possibly because their specimens thermally grown on silicon were only 0.5 μm thick.

The exponential edge across which the absorption coefficient increases by 2 to 3 orders of magnitude is slightly temperature-dependent in the glass (Fig. 2) and considerably more so in

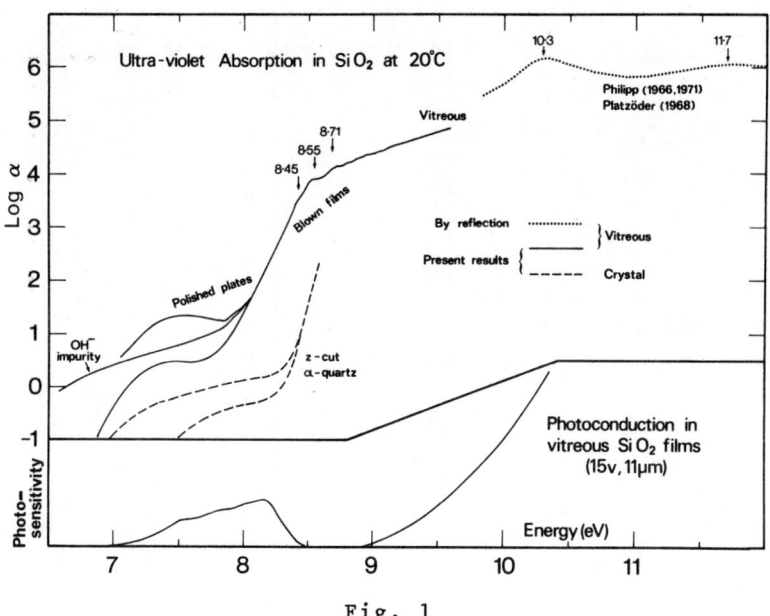

Fig. 1

α-quartz. A shoulder at about 8.45 eV (more clearly visible
on the linear plot of Fig. 3) suggests that the edge is due
to a barely-resolved peak broadened more at high than at low
temperatures. More peaks are just resolved at higher energies.
Their positions do not change noticeably with change of
temperature. Unfortunately the upper region of the absorption
edge is inaccessible in the case of α-quartz due to our
inability to prepare sufficiently thin specimens.

Above 9 eV there is no evidence of further peaks to the limit
(9.6 eV) of our observations. In this region the photocurrent,
having been absent from about 8.4 to 8.9 eV, resumes as expected
from the results of DiStefano and Eastman and Zakis *et al*. In
our films of thickness 11 μm with a steady bias voltage of 15 V,
the current under constant illumination is found to fall with a
time-constant of typically 20 minutes. The spectrum shown is
the average of several scans in opposite directions.

INTERPRETATION

Mott (1977) has considered the interpretation of the reflectivity
spectra of amorphous and crystalline SiO_2. In common with many
earlier authors (e.g. Philipp 1971, and Platzöder 1968) he
considers the 10.3 eV reflectivity peaks to be excitonic but
puts the band gap higher (10.6 eV) than most previous estimates.
It happens that our results also suggest a gap of about 10.6 eV
using Mott's estimate of the exciton binding energy as 2.2 eV.
However, on the basis of the present results, we feel that
this is more likely to be the direct gap. An indirect gap
of lower energy may also exist.

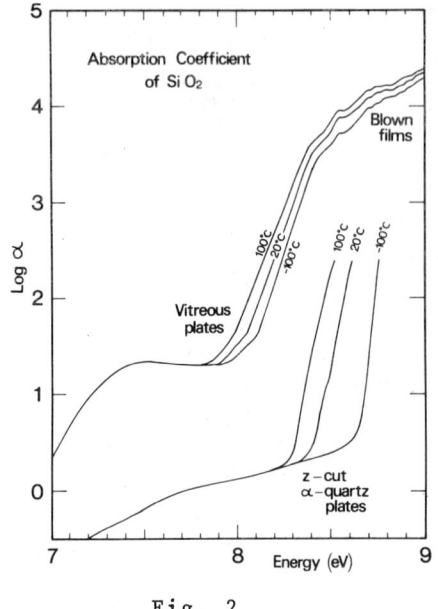

Fig. 2

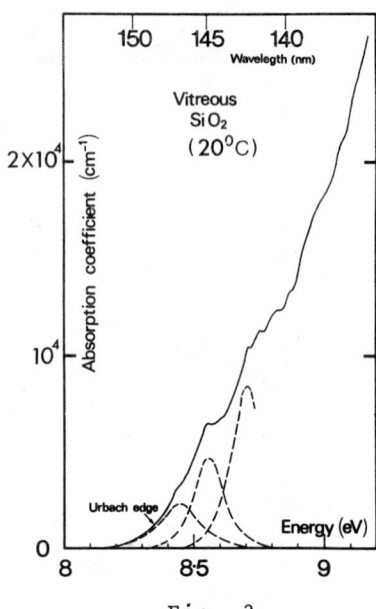

Fig. 3

Exponential (Urbach) absorption edges are usually attributed to broadening of exciton lines by random internal electric fields associated with structural and thermal disorder (Dow and Redfield 1970) and this interpretation seems appropriate here. Certainly, despite the rather low absorption coefficients, our peaks observed above 8.4 eV in the glass must be associated with processes characteristic of a fully-connected bonding network since, unlike the weaker and broader peak centred at 7.5 eV, they are essentially constant from one specimen to another. We therefore suggest that the peaks above 8.4 eV in the glass are due to direct excitons weakly allowed as compared with those giving the prominent peak at 10.3 eV. Presumably the weaker excitons are associated with interband transitions of lower energy than that to which the 10.3 eV peak belongs. Mott's estimate of an exciton binding energy of 2.2 eV thus implies a direct band gap of about 8.45 + 2.2 = 10.65 eV. Very low transition probabilities are calculated by Chelikowsky and Schlütter (1977) for the direct interband transitions of lowest energy in α-quartz. Similarly Schneider and Fowler (1976) find that the lowest energy optical transitions in β-cristobalite are forbidden. It is therefore not unreasonable to expect low transition probabilities for the lowest-energy interband transitions and their associated excitons in amorphous silica.

The simplest interpretation of the onset of intrinsic photo-conduction in the glass at about 9 eV is that this energy corresponds to the lowest interband transitions (not necessarily direct). Any excitons of higher energy would be unstable owing to a high probability of autoionization to produce free carrier-pairs. It should, however, be borne in mind that excitons which are of too low an excitation energy to be degenerate with free electron-hole pair states may, in principle, also contribute to the photocurrent by ionizing at the electrodes or at defects which they reach by diffusion. Our photoconductivity spectrum, plotted only to 10.4 eV in Fig. 1, in fact rises monotonically to at least 13 eV and is such that the square root of the photosensitivity (current divided by light intensity) is roughly linear in energy above 9 eV. As with the results of DiStefano and Eastman there is no recognisable feature at 10.3 eV confirming a high probability of ionization of the exciton.

Because of its variability our absorption tail must clearly be related to defects. In the glass these are most likely to be non-bridging oxygen centres. This identification is supported by the fact that the presence of water modifies the tail as shown in Fig. 1. So also does the addition of sodium (Sigel 1973/4). Both these impurities are known to bond to non-bridging oxygen atoms (e.g. Browell and Hetherington, 1965).

The sharp drop of photoconductivity between 8.2 and 8.4 eV corresponds to the rapidly rising attenuation of the light in the 11 μm thick samples due to the 8.45 eV exciton which evidently must have a high probability of decay by phonon emission (Dexter, Klick and Russell, 1955) giving neither photocurrent in this region nor photoluminescence.

Finally, we remark on the widths of the absorption bands. The absorption line for the production of an exciton localised at a defect is characteristically greater than that of a free exciton for which self-trapping can occur only after a delay (Mott 1977). Such a difference in breadths (half-value widths 0.8 eV as against 0.2 eV) is indeed observed between the tail centred at 7.5 eV and the peaks of higher energy. The tailing of the 10.3 eV exciton to lower energies was explained by Mott as due to sampling of a density of exciton band states over a range of wavevectors K (moving excitons) due to breakdown of the K selection rule in the glass and crystal alike. There are many attractive features of this explanation which would imply that for our excitons of lower energy the bands are much flatter. One wonders, however, whether the 10.3 eV peak might alternatively be attributed to excitons formed with high transition probability by more-or-less direct (vertical) electronic transitions from a group of closely-spaced lone-pair electronic states about 2 eV below the top of the valence band. The short lifetime of the excitons combined with the effects of disorder may then lead to broadening giving the single feature observed.

REFERENCES:

(1) Bates, C.W., *Applied Optics*, 15, 2976 (1976).

(2) Browell, T.P. and Hetherington, G., *Hilger Journal*, 9, 41 (1965).

(3) Chelikowsky, J.R. and Schlüter, M., *Phys. Rev. B.*, 15, 4020 (1977).

(4) Dexter, D.L., Klick, C.C. and Russell, G.A., *Phys. Rev.* 100, 603 (1955).

(5) DiStefano, T.H. and Eastman, D.E., *Solid St. Commun.*, 9, 2259 (1971).

(6) Dow, J.D. and Redfield, D., *Phys. Rev. B.*, 1, 3358 (1970), *Ibid.*, 594.

(7) Edwards, O.J., *J. Opt. Soc. Am.*, 56, 1314 (1966).

(8) Lell, E., Kreidl, N.J. and Heusler, J.R., *Prog. in Ceramic Sci.*, 3, 1 (1966) *Pergamon Press, N.Y.*

(9) Loh, E., *Solid St. Commun.*, 2, 269 (1964).

(10) Mott, N.F., *Adv. Phys.*, 26, 363 (1977).

(11) Philipp, H.R., *Solid St. Commun.*, 4, 73 (1966), *J. Phys. Chem. Solids*, 32, 1935 (1971).

(12) Platzöder, K., *Phys. Stat. Sol.*, 29, K.63 (1968).

(13) Platzöder, K. and Steinmann, W., *J. Opt. Soc. Am.*, 58, 588 (1968).

(14) Schneider, P.M. and Fowler, W.B., *Phys. Rev. Lett.*, 36, 425 (1976).

(15) Sigel, G.H., *J. Non-cryst. Solids*, 13, 372 (1973/4).

(16) Zakis, Y.P., Trukhin, A.N. and Zhimov, V.P., *Soviet Physics - Solid State*, 15, 216 (1973).

CALCULATED AND MEASURED AUGER LINESHAPES IN SiO_2

D. E. Ramaker, J. S. Murday and N. H. Turner,
Naval Research Laboratory; G. Moore and M. G. Lagally,
University of Wisconsin, Madison; and J. Houston,
Sandia Laboratory

ABSTRACT

The Si $L_1L_{23}V$, Si $L_{23}VV$ and O KVV Auger lineshapes in SiO_2 have been measured and calculated. The Auger transition energies and linewidths are estimated from one electron calculations; the intensities are estimated from electron densities local to the core hole multiplied times atomic Auger matrix elements. The calculated energies and measured values agree within the uncertainties. The calculated intensities are in general agreement with experiment, however there is some disagreement with K_β X-ray emission observations of the local electron density on silicon. The calculated linewidth is too small when there are two localized holes in the final state.

INTRODUCTION

The interest in the electronic structure of SiO_2 has resulted in numerous theoretical and experimental investigations of the valence electron density of states. A reasonably consistent model of the valence electron states has been distilled from this work (1,2,3) and is in close correspondence to a similar model for the silicates as deduced from the corresponding spectra (4). Good agreement has been achieved with calculated and measured lineshapes for solid Si (5,6) and between Auger lineshapes calculated via MO theory and gas phase data (7). Molecular orbital approaches have been successful in interpreting photoemission and x-ray emission in solid SiO_2(1,3). We will therefore use this approach to quantitatively interpret the Si $L_1L_{23}V$, Si $L_{23}VV$, and O KVV Auger spectra as measured from a thermally grown oxide on a silicon wafer.

CALCULATION OF THE AUGER TRANSITIONS

The Auger transitions will be calculated by assuming the valence electron DOS in SiO_2 can be described in terms of the LCAO-MO model. The transitions will involve the appropriate core hole, Si L_1, L_{23} and O K, coupling with the local electron DOS as represented by the occupation of the atomic orbitals centered on that atom (i.e. atom with the core hole).

The transition energies will be calculated from the formulation of Shirley (8)

$$E_{KXY}(Z) = E_{KXY}{}^o + R - F(X,Y) \tag{1}$$

TABLE 1

One Electron Binding Energies[+] (SiO_2)

Orbital Group	C_{2V} (Si_2O)	T_d (SiO_4)	Orbital Energy (eV)	Orbital Linewidth (eV)	Orbital Composition (%) Si		O	
					3s	3p	2s	2p
O_{2p} non-bonding	$2b_1$ $7a_1$	$1t_1$ $1e$ $5t_2$	6	3.3	–	–	–	100
"O_{2p}" bonding	$5b_2$	$4t_2$	9	1.5	–	50	–	50
		$5a_1$	12	1	49	–	1	50
"O_{2s}" bonding	$6a_1$	$3t_2$	23	2	–	9	91	–
		$4a_1$	26	2	15	–	85	–
Si_{2p}			102					
Si_{2s}			151					
O_{1s}			532					

[+] Energies referenced to Fermi level

where $E_{KXY}{}^o$ is the transition energy deduced from the one electron binding energies referenced to the Fermi level, R is the intra orbital and outer atomic orbital term, and F(X,Y) is the two electron interaction energy describing the coupling of the final state holes. Our adaptation of this formula from atoms to molecules, in particular the R and F(X,Y) terms, is described in more detail elsewhere (9).

Fischer (3) have correlated the photoemission and x-ray emission measurements of the filled DOS so that they can be compared to each other on the same absolute energy scale. For this paper we will use the value of 102 eV as the Si_{2p} binding energy (4,10). From the work of Bernett (4) compared to that of Fischer this would lead to a value of 4 eV between the top of the valence band and the Fermi level. The one electron binding energies of the various electron levels to be used in this paper are summarized in Table 1. Table 2 presents the Auger transition energies as well as the values of E^o, F(X,Y) and R for the experimental choice of local electron densities. The choice of these densities is the subject of the next paragraph.

The transition intensities are calculated from the expression

$$E_{KXY} \propto c_{Xn}^2 \, c_{Ym}^2 \, P_{Knm} \tag{2}$$

where c_{Xn}^2 and c_{Ym}^2 are the populations of the n,m central atom orbitals in the X,Y molecular orbitals respectively and P_{Knm} is the appropriate atomic Auger matrix element obtained from the results of McGuire (11) and Walters and Bhalla (12). The populations have been deduced by assuming the O_{2p} non-bonding orbitals are pure O_{2p} (13,14). The OK_α, SiK_β and SiL_{23} x-ray emission intensities, coupled with state counting, determine everything else except the amount of Si 3p versus Si 3s in the "O_{2p}" bonding group. We will choose

TABLE 2

Calculated Auger Transitions

			E^0_{KXY}	R	$F(X,Y)^+$	$E_{KXY}^\dagger$	I*	K	Γ
L_1	L_{23}	V							
		$4a_1$	23	9.6	− 8.3	25	0.5	−	5.2
		$3t_2^1$	26	9.6	− 7.6	28	0.3	−	5.2
		$5a_1^2$	37	9.2	−11.1	35	1.7	−	4.2
		$4t_2^1$	40	9.1	− 9.6	39	1.8	−	4.7
L_{23}	V	V							
	$4a_1$	$3t_2$	52	5.5	− 8.1	49	0.1	3.3	11.8
	$3t_2^1$	$3t_2^2$	53	5.6	− 8.2	51	0.2	3.6	12.4
	$3t_2^2$	$5a_1^2$	64	4.8	− 7.3	62	0.4	−	4.3
	$4a_1^1$	$4t_2^1$	65	4.4	− 7.2	63	0.8	−	4.7
	$3t_2^2$	$4t_2^2$	67	4.5	− 7.1	65	2.8	0.4	5.5
	$5a_1^1$	$4t_2^2$	78	3.8	− 7.4	74	2.4	1.0	5.7
	$4t_2^1$	$4t_2^2$	81	3.5	− 7.1	78	7.6	1.5	7.2
K	V	V							
	$6a_1$	$6a_1$	480	5.4	−17.3	468	1.0	−	4.2
	$6a_1$	$7a_1,2b_1$ (S)	500	5.7	−18.7	483	2.0	3.6	5.5
	$6a_1$	$5b_2$ (S)	497	4.4	−12.2	487	.5	1.8	3.7
	$6a_1$	$7a_1,2b_1$ (T)	500	5.7	−18.7	491	.5	3.6	5.5
	$7a_1$ $2b_1$	$7a_1$ $2b_1$	520	6	−21.9	504	3.1	−	6.8
	$7a_1$	$2b_1$	520	6	−19.7	505	2.0	1.1	6.8
	$5b_2$	$5b_2$	514	3.5	− 9.8	508	.4	−	3.2
	$5b_2$	$7a_1,2b_1$	517	4.8	−12.6	509	2.0	0.5	5.0

+ Unless the transition is designated singlet (S) or triplet (T), $F(X,Y)$ is calculated from $F^0(n,m)$ terms only for Si, from F^0 and F^2 terms for O.

† Energies measured relative to Fermi level and in eV.

* The intensities are related only for the common core hole; no attempt has been made to relate the $SiL_1L_{23}V$, $L_{23}VV$ and O KVV total intensities.

a 3p/3s ratio of 3 for this group. The MO calculations of Tossell (13) would augment the p to s ratio while the calculations of Yip and Fowler (14) would decrease it. The electronic charge configuration is Si $3s^{0.65} 3p^{1.77}$ and O $2s^{1.8} 2p^{5.0}$. The populations are presented in Table 1 and the transition intensities in Table 2.

Each transition will be represented by a Gaussian centered about the transition energy and whose area is proportional to the calculated intensity. The linewidths of the Gaussians are calculated from

$$\Gamma_{KXY} = \Gamma_K + \Delta_X + \Delta_Y + 2K \tag{3}$$

where Γ_K is the core level width, $\Delta_{X,Y}$ are the valence level widths and 2K is the singlet-triplet splitting. A simple summation of the contributions is used because the various sources of linewidth are not all Gaussian or Lorentzian or any other common functional form. The core level widths are taken to be

O_{1s} Γ_{1s} = 0.2 and Si_{2s} Γ_{2s} = 2 eV estimated from the lifetime broadening (11); Si_{2p} Γ_{2p} = 1.2 eV estimated from the spin orbit splitting. The valence level widths are deduced from x-ray emission and photoemission data and are presented in Table 1. The singlet-triplet energy differences 2K for the various two hole final states were calculated from one center Slater exchange and coulomb integrals. Where this energy was on the order of the linewidth contributions above, the 2K values were added into the linewidth. In the O KVV transitions the magnitude of 2K was large enough to require separate transitions for the singlet and triplet final states.

EXPERIMENTAL MEASUREMENTS

The Auger spectra were taken on a ∿ 500 Å thick thermal oxide grown at 950°C in pure oxygen in a diffusion furnace on a clean silicon wafer. The sample was transferred to a UHV system for analysis. The Auger process was stimulated by a 1 keV electron beam incident normal to the sample; the signal was detected by a single pass CMA with a modulation voltage of 1 eV (peak to peak). The experimental data was integrated to an N(E) form, the background subtracted and inelastic electron contributions deconvolved according to procedures presented by Houston (15).

RESULTS AND DISCUSSION

When comparing the calculated and experimental lineshapes, three parameters have been allowed to vary. The first is a multiplicative constant which simply adjusts the total intensity of the calculated lineshape. This parameter is necessary because there has been no attempt at absolute quantification. The second parameter, δE, permits the full calculated lineshape to be shifted up or down in energy. The δE could reflect sample charging shifts, extra atomic relaxation energies, or other processes which would not change the lineshape, but would affect the Auger kinetic energies. The third parameter, $\delta\Gamma$, is a constant which is added to each of the individual transition linewidths. This would account for any source of line broadening, such as is observed with sample charging, which uniformly affects all transitions.

The comparison of the measured and calculated $L_1L_{23}V$ lineshape is shown in Figure 1. This feature is only that part of the full lineshape which originates from the "O_{2s}" bonding group. There are two transitions which contribute to the $L_1L_{23}V$ lineshape presented in Figure 1; the higher energy peak corresponds to the $4t_2$ orbital, the lower energy peak to the $5a_1$ orbital. The discrepancy between the calculated and measured intensities shown in Figure 1 is a reflection of the relative populations of the Si 3s and 3p in the $5a_1$ and $4t_2$ orbitals respectively. A 10% change in the relative s-p populations in the $5a_1$-$4t_2$ orbitals would be sufficient to bring the relative peak heights into better agreement.

The calculated and measured Si $L_{23}VV$ lineshapes are presented in Figure 2. The calculated $L_{23}VV$ lineshape is dominated by the ratio of the Si populations in the $3t_2$ and $4t_2$ orbitals. The major difference between the experimental and calculated lineshapes is the failure of the calculated line to reproduce the lowest energy peak. This peak corresponds to the transitions $L_{23}3t_23t_2$ and $L_{23}3t_24a_1$, i.e. to transition involving only the "O_{2s}" bonding group. There are at least two reasons which might explain the discrepancy. From Table 2 one can observe that transitions are in fact predicted to occur at the proper energy, but their intensity is much too low. This may be due to an insufficient amount of Si 2p electron density in the "O_{2s}" bonding group.

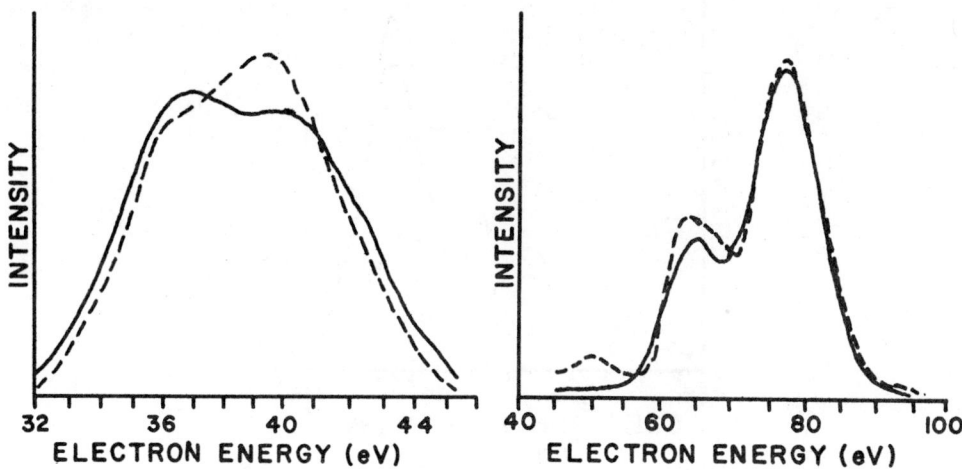

Fig. 1: The calculated (-) and experimental (--) Si $L_1L_{23}V$ Auger transitions. The fitting parameters are δE = 1 eV and $\delta \Gamma$ = 0.

Fig. 2: The calculated (-) and experimental (--) Si LVV Auger transitions. The fitting parameters are δE = 1 eV and $\delta \Gamma$ = 2 eV.

It would be necessary to increase the intensity of the calculated $L_{23}3t_23t_2$ transition by a factor of 5 to reproduce the observed 50 eV peak. In order to do that the Si 3p contribution to the $3t_2$ orbital would increase by a factor of 2; this would cause unacceptable increases in the $L_{23}4t_23t_2$ Auger and the Si K_β' x-ray emission intensities. Another possibility is that the 50 eV peak does not originate from the Si $L_{23}VV$ Auger in SiO_2 but comes from deconvolution problems or some other source.

The calculated and experimental O KVV lineshapes are presented in Figure 3. The major peak at 503 eV is reasonably matched, but there is too much intensity in the two lower peaks of the calculated transitions. The major peak is determined principally by the O_{2p} nonbonding electrons, the peak at 482 eV by the product of O_{2s} bonding and O_{2p} nonbonding populations and the peak at 465 eV by the square of the O_{2s} bonding populations. Since the O_{2p} nonbonding population is essentially 16 electrons in all calculations, to lower the 482 eV peak intensity by about 25% would require transferring 25% of the O_{2s} bonding electron density onto the Si. Such a transfer would increase the Si population in the "O_{2s}" orbital group by more than a factor of two; this transfer was also suggested by the Si LVV intensities but it is contradicted by the x-ray emission data. It may be that the x-ray emission matrix element samples different parts of the electron wavefunction than does the Auger matrix element. Another possible origin for the intensity disparities is the use of atomic Auger matrix elements in a system whose wavefunctions are not truly atomic in nature.

The values of δE are all considered to be within the errors inherent in the calculation. The extra linewidth contributions $\delta \Gamma$ to the linewidth was essentially determined by the highest energy peaks in the three Auger lines. The value of $\delta \Gamma$ got progressively larger as one progressed from the Si $L_1L_{23}V$

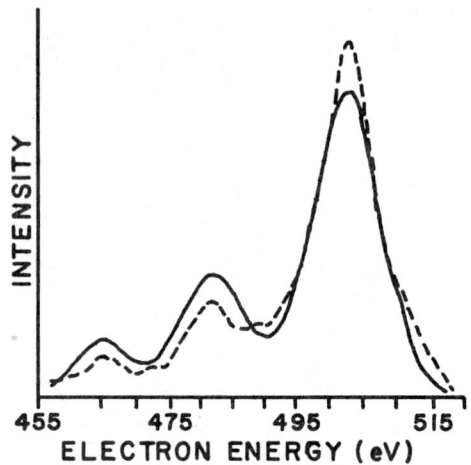

Fig. 3: The calculated (–) and experimental (--) O KVV Auger transitions. The fitting parameters are $\delta E = 3$ eV and $\delta\Gamma = 5$ eV.

line where the final hole states are far removed from each other to the O K 2p2p transition where the final hole states are more localized. This suggests that we have not fully accounted for the line broadening mechanisms for the localized two hole final state.

REFERENCES

1. S. T. Pantelides and W. A. Harrison, Phys. Rev. B 13, 2667-2691 (1976).
2. J. R. Chelikowsky and M. Schluter, Phys. Rev. B 15, 4020-4029 (1977).
3. B. Fischer, R. A. Pollak, T. H. DiStefano and W. D. Grobman, Phys. Rev. B15, 3193-3199 (1977).
4. M. K. Bernett, J. S. Murday and N. H. Turner, J. Elec. Spectrosc. and Related Phenom. 12, 375-393 (1977).
5. P. J. Feibelman, E. J. McGuire and K. C. Pandey, Phys. Rev. B15, 2202 (1977); P. J. Feibelman and E. J. McGuire, Phys. Rev. B17, 690 (1978).
6. D. R. Jennison, Phys. Rev. Lett. 40, 807-809 (1978).
7. R. W. Shaw, J. S. Jen and T. D. Thomas, J. Elec. Spectrosc. and Related Phenom. 11, 91-100 (1977).
8. D. A. Shirley, Phys. Rev. A7, 1520-1528 (1973).
9. D. E. Ramaker and J. S. Murday, in preparation.
10. B. Carriere and B. Lang, Surf. Sci. 64, 209-223 (1977).
11. E. J. McGuire, Phys. Rev. 185, 1-6 (1969); Phys. Rev. A3, 587-594 (1971).
12. D. L. Walters and C. P. Bhalla, Phys. Rev. A4, 2164-2170 (1971).
13. J. A. Tossell, J. Phys. Chem. Solids 34, 307-319 (1973).
14. K. L. Yip and W. B. Fowler, Phys. Rev. B 10, 1400-1408 (1974).
15. J. Houston, J. Vac. Sci. Technol. 12, 255 (1975).

IS SILICON DIOXIDE COVALENT OR IONIC?*

Walter A. Harrison
Applied Physics Dept., Stanford University, Stanford, CA, 94305

ABSTRACT

The electronic structure of SiO_2 can be rather accurately obtained from
LCAO theory using universal parameters which have recently become available.
The results may be characterized by a covalent energy and a polar energy.
The former is the matrix element between a silicon sp^3-hybrid and an oxygen
p-state; the latter is half the difference between the energy expectation
value of the two. Since the covalent energy is four times as large as the
polar energy, approximate calculations should be made using a covalent basis
with corrections for polarity, and never the other way around. Using this
approach and the universal parameters (no input from SiO_2 except for
structure), we estimate the band energies and widths, equilibrium bond angles
and rigidity, optical dielectric constant which determines the refractive
index, and the effective transverse charge tensor which determines the
infrared absorption and the static dielectric constant.

INTRODUCTION

This question might be regarded as a bad one since the term "covalency" has
been applied to such a variety of essentially unrelated attributes that it
has become almost without meaning. Thus the first point of the discussion
is to give a meaning to the question.

Everyone will agree that diamond is a covalent solid. Its electronic
structure can be described rigorously (1) in terms of bond Wannier
functions, doubly occupied, and conceptually very close to the original two-
electron bond proposed in 1916 by G. N. Lewis (2). Each atom is bonded by
two electrons to each of its four nearest neighbors. We may then shift the
relative energy on the two carbon atoms in the primitive cell as in going
from C to BN to BeO . It is ordinarily said that this increases the
"ionicity" of the solid. I prefer the term "polarity" to distinguish
electronic structure from that of ionic crystals such as NaCl with more
than four nearest neighbors and no possibility of a description in terms
of two-electron bonds. The two terms are interchangable. If we stick to
tetrahedral solids we can analytically span the entire range from two-
electron bonds to atomic states on the nonmetallic atom – the polar or ionic
limit. Either the covalent or the ionic concept is allowed for intermediate
compounds, and if we do a machine calculation of the energy bands, the
machine will not care which we have in mind. If on the other hand we make
an approximate calculation of the bands or some other property the accuracy
of the result will depend upon which parameter we treat as small, or if we

guess an answer without calculation one concept may lead us to guess the right answer.

SiO$_2$ also has a structure which allows an electronic structure in either limit. We need to consider that electronic structure to see which concept will best serve our intuition. A very good way to do this is to formulate an LCAO calculation of the electronic structure, which itself can give a good representation of the true energy bands, and compare the results with those obtained in the polar and in the covalent limit. There have been a number of LCAO treatments of the bands (3). However, recently developed universal LCAO parameters have reduced approximate electronic structure calculations to the back-of-the-envelope level (4) and we may as well start from scratch.

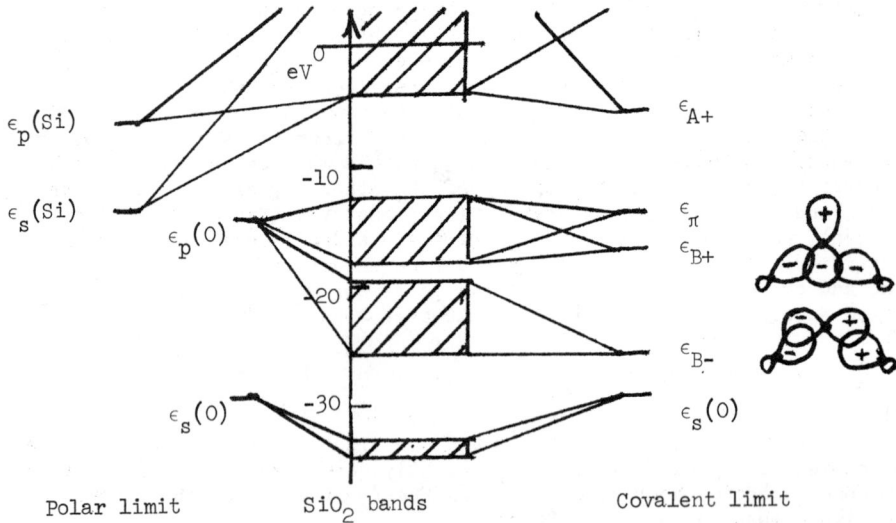

Fig. 1. The energy band structure of SiO$_2$ obtained by Chelikowsky and Schlüter (Ref. 5. The energy $\epsilon_s(0)$ has been raised 5 eV as they suggested and their zero of energy taken to match the π-state energies) is shown in the center for comparison with the band positions given by LCAO theory and universal parameters. To the left are the term values, the levels obtained by neglecting the covalent energy and all other interatomic matrix elements. To the right are the bonding and antibonding levels obtained by retaining the covalent energy but neglecting the polar energy, the silicon sp-splitting, and all other interatomic matrix elements; the corresponding bond orbitals are also sketched. It is concluded that the covalent bond provides the correct starting approximation.

THE ELECTRONIC STRUCTURE

The electronic states are to be represented as linear combinations of oxygen 2s- and 2p-states and silicon 3s- and 3p-states. Their energies are taken equal to the atomic term values (4), illustrated to the left in Fig. 1. These in fact give the electronic structure in the polar limit, neglecting all interatomic matrix elements. We see that the sixteen valence electrons in the molecular unit are all accommodated in the eight oxygen levels for the molecular unit but with a band gap of less than an electron volt. All nearest-neighbor interatomic matrix elements are obtained directly (4) in terms of the bond length, 1.61 Å, and a band calculation could be performed using them. For describing the covalent limit it is sufficient to isolate the "bonding unit" (3) consisting of an oxygen atom and the two silicon hybrids, of energy $(\varepsilon_s + 3\varepsilon_p)/4$, directed toward it. We neglect for the moment the coupling between the oxygen s-state and the hybrids since the s-state is so low in energy. The only remaining matrix element is that between the hybrid and a σ-oriented oxygen p-state. This matrix element is called the covalent energy and is given by (3,4).

$$W_2 = 1/2 \, (V_{sp} + \sqrt{3} \, V_{pp\sigma}) = \frac{3.37\hbar^2}{md^2} = 11.0 \text{ eV} . \quad (1)$$

We may also define a polar energy W_3 which is half the difference between the silicon hybrid energy and the oxygen p-state energy, 2.9 eV. The average of the two energies is written $\bar{\varepsilon} = -11.2$ eV. The energy of bonding orbitals (and antibonding orbitals) which are even and odd with respect to the reflection planes of the bonding unit are immediately written down. They are

$$\varepsilon_{B+} = \bar{\varepsilon} - \sqrt{2W_2^2 \sin^2 \theta + W_3^2}$$

$$\varepsilon_{B-} = \bar{\varepsilon} - \sqrt{2W_2^2 \cos^2 \theta + W_3^2} \qquad (2)$$

where θ is the angle, seen from one silicon, between the oxygen position and the other silicon in the bonding unit. The corresponding antibonding levels appear with a plus sign before the square root. These levels are plotted (neglecting W_3) to the right in Fig. 1 along with schematic sketches of the corresponding orbitals. There are enough electrons to occupy the oxygen 2s-state, all bonding levels, and the nonbonding π-levels, leaving a gap of about 9 eV.

We could incorporate the polar energy and the matrix elements between bonding units to obtain the full bands, but will instead sketch the more accurate bands obtained by Chelikowsky and Schlüter (5) by a self-consistent pseudo-potential calculation in the center of Fig. 1. It is seen by comparison that the levels in the covalent limit are in very good accord with the true band structure with the gap determined almost entirely by the covalent energy. In contrast, the polar limit to the left omits the principal source of the band gap and the bonding energy. There would seem little question but what we should regard SiO_2 as a covalent solid. This result is in tune with the rather open structure with two-fold and four-fold coordinated atoms, contrasting with the more closely packed ionic structures.

THE CONSEQUENCES

More importantly, we believe that essentially all of the properties of the material are determined by the electronic structure and this approach has given such a simple description of that electronic structure that we can, without additional parameters, make estimates of virtually any property of the system. The description is crude so the accuracy of the results is limited, but the calculations are extremely elementary. We will illustrate several.

We are less interested in the electronic structure per se than in other properties but we may explore it slightly beyond the narrow-band limit to the right in Fig. 1. An immediate contribution to the breadth of the individual bands comes from the intra-atomic matrix elements between adjacent bonding units due to the matrix element, of magnitude $(\varepsilon_p - \varepsilon_s)/4$, between the two hybrids sharing the intervening atom. This matrix element contributes a half of the valence-band width in elemental silicon. It is reduced here by the polarity of the bond, transferring charge off the silicon atoms and onto the oxygen. A numerical polarity β was defined by Pantelides and Harrison (3) for the even and odd bond orbitals, given in each case by the polar energy divided by half the splitting between bonding and antibonding states; these are obtained immediately as 0.20 and 0.56 , respectively, for our parameters. By decomposing the bond orbitals in our elementary calculation we see that the effect of this polarity is to reduce the band width by a factor $1/2(1 - \beta)$ from the contribution $\varepsilon_p - \varepsilon_s$ obtained for elemental silicon. This gives contributions to the band widths of 1.4 and 2.3 eV , for the B+ and B- bands respectively, seen from Fig. 1 to be about half the true width. As in elemental silicon the contributions from interatomic matrix elements are needed for a good quantitative result.

We note that in our discussion above we neglected the matrix elements between the hybrids and the oxygen s-state, as did Pantelides and Harrison (3). We may immediately estimate the effect of that matrix element, found to be 6.7 eV in analogy with Eq. 1, and obtain the shift in the s-band by second-order perturbation theory. It is -3.5 eV, making up much of the discrepancy with the true bands illustrated in Fig. 1; the remaining discrepancy may be partly from inaccuracy of the parameters and partly from the error in this perturbation-theoretic estimate. The same coupling also causes the breadth of the s-band indicated in Fig. 1. This shift is a big enough contribution to the bond formation energy that we do not believe the s-state can be treated as a core state in the treatment of bonding properties.

This may be illustrated by analysis of the angular rigidity of the structure, the bond angle at the oxygen in particular. The principle contribution to the angle-dependent energy is the sum of the gains in one-electron energy from the bonding states of Eq. 2. These are minimized at $\theta = 45°$, corresponding to an angle of $90°$ at the oxygen. The conclusion is qualitatively correct; this is the source of the bend at oxygen. However, the observed angle is $144°$. Some of the discrepancy arises from coupling with the oxygen s-state, though including it still leaves a considerable discrepancy.

The polarity of the bonding described above contributes to the effective charges of the atoms. By decomposing the bond orbitals we can add up the portions of the electronic charge to be associated with each atom and sub-tract them from the nuclear charge. This leads directly to an effective charge $Z*$ of the oxygen equal to the sum of the two polarities (3), or 0.72 . It is a static charge which is not directly observable. Of more interest is the transverse charge describing the dipole induced by displace-ment of the oxygen atom. This includes the static charge $Z*$ but also includes contributions from transfer of charge between atoms due to the variation of the matrix element in Eq. 1 with bond length. It is a simple and straightforward calculation (3) to obtain the effective charges of 0.72 for displacement perpendicular to the sketches to the right of Fig. 1 (to the plane of the bonding unit), 0.50 for vertical displacements, and 2.04 for horizontal displacements. These estimates (3) appear to be the first theoretical estimates of these charges and appear to be in agreement with the observed infrared spectra. They also determine the contribution to the static dielectric constant arising from atomic displacements.

The optical dielectric constant is also readily estimated. It is elementary in any case but becomes trivial if we set the bond angle at the oxygen at $90°$ so individual hybrid-p_σ-bonds can be constructed between each set of nearest neighbors. Decomposition of charge into atoms allows the construc-tion of bond dipoles along the bond vector $\vec{d}$. An applied field has the effect of modifying W_3 by $e\vec{d} \cdot \vec{\mathcal{E}}/2$ and therefore induces an additional dipole in each bond given by

$$\vec{\delta p} = e^2 (1 - \beta^2)(\beta/W_3)(\vec{d} \cdot \vec{\mathcal{E}}) \vec{d}/2 . \qquad (3)$$

Where β is the bond polarity, given by the earlier expressions for $\theta = 45°$. We may average over angle, multiply by the density N_{Si} of silicon atoms in SiO_2 and by the four bonds per silicon to obtain a susceptibility of

$$\chi = (2N_{Si}/3)d^2 e^2 (\beta/W_3)(1 - \beta^2) = 0.06 \qquad (4)$$

giving the main part of the observed 0.11 (corresponding to the observed dielectric constant of 2.4); the rest is presumably from the π-states.

We should emphasize that the calculation of the electronic structure and the variety of properties has not used any experimental input from SiO_2 except the structure itself, nor has it required a computer. It required the universal LCAO parameters and was made elementary by a proper assessment of the nature of the electronic structure.

REFERENCES

(1) E. O. Kane and A. B. Kane, to be published.

(2) See, for example, C. A. Coulson, Physical Chemistry, An Advanced Treatise, Vol. V, edited by H. Eyring, Academic Press, New York (1970).

(3) S. T. Pantelides and W. A. Harrison, Phys. Rev. B13, 2667 (1976), and references therein.

(4) W. A. Harrison, Festkörperprobleme (Advances in Solid State Physics, XVII) 135, J. Treusch (ed.) Vieweg, Braunschweig (1977). A more complete description will appear in W. A. Harrison, <u>The Physics of the Chemical Bond</u>, W. H. Freeman (San Francisco, in press).

(5) J. R. Chelikowsky and M. Schlüter, Phys. Rev. <u>B15</u>, 4020 (1977).

*Research sponsored by NSF Grant DMR77-21384.

CHEMICAL BOND AND RELATED PROPERTIES OF SiO_2

K.Hübner
Physics Section,Wilhelm Pieck University,Rostock,GDR

INTRODUCTION

We present in this paper semiempirical models for the chemical bond and related properties of SiO_2.These models are based on the semiempirical dielectric theory and the empirical pseudopotential method, introduced by Phillips [1] and other investigators for semiconductors and extended by the present author in several aspects The purpose of our models is to describe different electronic and structural properties of some SiO_2 modification by few parameters and to investigate with the help of the same parameters the trends of these properties in the various allotropic forms of SiO_2.

CHEMICAL BOND

It is commonly accepted that SiO_2 has a mixed ionic-covalent chemical bond.The ionicity parameter f_i of different SiO_2 forms is determined from the relation

$$\varepsilon = 1 + \frac{(\hbar\omega_p)^2}{E_g^2} = 1 + 1.26\, d_{Si-O}^2\, (1-f_i) \qquad (1)$$

for the electronic dielectric constant,where d_{Si-O} is the nearest-neighbour distance (in Å).This relation was derived by an investigation of the bond-charge polarization and implies that the shift of the bond charge from the centre of the chemical bond toward the anion has to be considered as proportional to $f_i^{1/2}$,at least for small ionicities [2].With the assumption that the covalent part E_h and the ionic part C of the average bonding-antibonding gap $E_g = (E_h^2 + C^2)^{1/2}$ in relation (1) act as average symmetric and antisymmetric potential,respectively,responsible for the location of the bond charge,we get its position $(d_{Si-O}/\pi)\arcsin(f_i^{1/2})$ measured from the centre of the chemical bond. The resulting distances between the bond charges and the oxygen cores are very small (~ 0.1 Å) for all SiO_2 forms and seem to favour structural transitions from one SiO_2 form to another as well as modifications of the short-range order in amorphous SiO_2 [2].

Although it is the Fermi surface where the most important chemical bonding effects take place, the ionicity parameter can be determined in agreement with the data derived above also from the ratio of the antisymmetric density-of-states gap E_{as} and the total width E_{VB} of the valence band, observable in photoemission experiments (Ref.3).

$$f_i = \frac{1}{2} (1 - \cos \pi \frac{E_{as}}{E_{VB}}) \qquad (2)$$

IONIZATION ENERGIES AND EFFECTIVE ATOMIC CHARGES

Within the method of relative sharing probabilities of the valence electrons at the atomic constituents of a compound, developed by the author some years ago, the factors $(1 \mp f_i)/2$ are used to weight renormalized atomic values of special properties P of silicon and oxygen to get the corresponding value of SiO_2.

$$P(SiO_2) = (\frac{1 - f_i}{2}) P(Si) + (\frac{1 + f_i}{2}) P(O) \qquad (3)$$

This method yields for different SiO_2 forms ionization energies I which are in agreement with corresponding experimental data (Ref.3). Its extension to SiO_x leads to a slight downward bowing of I(x), and together with the downward bowing of the optical gap (Ref.4) to the conclusion that the electronic affinity has a weak upward bowing in dependence on x.

The application of the relative sharing probabilities to the calculation of effective atomic charges is connected with some problems due to the nonbonding electrons at the oxygen bridge between two silicon atoms (Ref.3).

REFRACTIVE INDICES AND THEIR PRESSURE AND TEMPERATURE DEPENDENCE

Taking into account the special role of nonbonding electrons by a two-oscillator model for the refractive index $n = \epsilon^{1/2}$, it can be shown that the nonbonding electrons are delocalized with bond compression and that the total one-oscillator gap E_g in (1) is approximatively independent of the nearest-neighbour distance (Ref.2). Therefore, we get from the scaling of the plasma frequency ω_p in (1) with the macroscopic density ϱ the simple relation

$$n = (1 + (n_\alpha^2 - 1) \varrho / \varrho_\alpha)^{1/2} . , \qquad (4)$$

where α-quartz was used for the fit. The values of n obtained from (4) are in good agreement with experimental ones of all crystalline and amorphous SiO_2 forms (Ref.2). Furthermore, from relation (4) useful expressions follow for the pressure and temperature coefficients of the refractive indices of different

SiO_2 modifications in terms of the corresponding coefficients
for α-quartz (Ref.2).

EMPIRICAL PSEUDOPOTENTIAL MODEL

The electronic structure and the stability of the idealized β-
cristobalite form of SiO_2 were investigated with the help of the
empirical pseudopotential method (Ref.5).Local empirical pseudopotential form factors of oxygen and silicon were transfered from
MgO and Si,respectively,to SiO_2.This transfer involves the determination of an appropriate electronic dielectric screening and interpo
lations of the atomic form factors with respect to the required
reciprocal lattice vectors.Following the perturbation expansion
of March et al.(6) for the Dirac density matrix it can be shown
that the higher expansion terms yield exactly calculable corrections
of the free-electron gas,which depend only on the effective
potential and which lead to the Thomas-Fermi approximation for
the valence-electron density (Ref.7 and 8)

$$n(\vec{r}) = \frac{1}{3\pi^2} (k_F^2 - V(\vec{r}))^{3/2} , \tag{5}$$

where k_F is the Fermi wave number of the disturbed free-electron
gas.The difference of the results obtained from (5) (using empirical pseudopotentials) with respect to complete charge-density
calculations by solving the eigenvalue problem (with the same
potentials) and summation over all occupied states were proved
to be less than 10%,even for ionic semiconductors (Ref.7 and 8).

The valence-charge density distribution in β-cristobalite calculated with the help of (5) from the real-space potential $V(\vec{r})$ agrees
qualitatively with corresponding results of more complicated
methods (Ref.9 and 10) and confirms the mixed ionic-covalent
character of the chemical bond of SiO_2.These results indicate that
the empirical pseudopotential method and the Thomas-Fermi-approxi-
mation,which are basically nearly-free-electron approaches,may
be useful even for wide-gap insulators like SiO_2.

The long-range disorder effect due to different oxygen positions
in real β-cristobalite (Ref.11),which has a bent Si-O-Si axis,was
simulated by a corresponding modification of the pseudopotential.
The resulting redistribution of valence charge from oxygen into
the chemical bond leads to the conclusion that a β-cristobalite
form with bent Si-O-Si bonds is structurally more stable than the
idealized form.

STRUCTURE OF AMORPHOUS SiO_2

The bonding angle Φ at the oxygen bridge is a fundamental structural order parameter for SiO_2.Using the values of Φ compiled by
Pantelides and Harrison (12),we get qualitatively the same dependence of the ionicity parameter on Φ as derived by these authors
for the effective atomic charge in dependence on Φ .Using more
recent experimental data for Φ we get for the fourfold coordi-

nated SiO_2 polymorphs a new empirical relation between f_i and $\bar{\Phi}$ (Ref.13).The $\bar{\Phi}$ distribution is almost symmetrical around the mean value of $\bar{\Phi}$ for amorphous SiO_2 ($\sim 150^\circ$) and has for the average ionicity of different amorphous SiO_2 forms ($f_i = 0.57$) an appreciably smaller width than commonly assumed for amorphous SiO_2.

From a detailed analysis of this relation between the polymorphism of SiO_2 and its amorphous structure,which takes also into account energetic data,we propose the following model for the structure of amorphous SiO_2 (Ref.13).The structure of amorphous SiO_2 in a quasi-equilibrium state consists of randomly orientated metastable submicrocrystallites with mainly α-quartz, α-cristobalite,and α-tridymite structure of limited long-range order,which give together with their disordered network-like environment no long-range order. This intermediate-range-order model with nonplanar 6-membered rings and fewer 5- and 7-membered as well as broken rings of almost identical SiO_4 tetrahedra leads to a natural connection of the rival network and microcrystallite hypotheses.The submicrocrystallites are assumed to be ordered domains within a network which must be disordered and strained,since the submicrocrystallites have different polymorphic structures.
This model explains various transport phenomena along prefered directions in amorphous SiO_2 as well as irregularities of some material properties,when corresponding parameters pass the values of the crystalline polymorphs (Ref.13).Furthermore,it is in agreement with corresponding conclusions which we have drawn from an analysis of phonon spectra and elastic properties of amorphous SiO_2 (Ref.14).

CHEMICAL DISORDER IN SiO_x

To test the validity of the current SiO_x models,we have applied the random-bonding model and the mixture model to the quantitative interpretation of known core-level shifts measured with XPS (Ref.15) The total shift of Si core levels in SiO_x was calculated in terms of chemical and Fermi level shifts.For relaxation effects it was shown that they play no decisive role.In the electrostatic bond-charge model for the chemical shift the potentials of the effective atomic charges of an Si atom and its four nearest neighbours as well as of the bond charges within a tetrahedron and the average influence due to the statistical neighbourhood of the tetrahedron considered were taken into account.From the excellent agreement of the Si core-level shift,calculated within the random-bonding model without free parameters,and the experimental results we get further evidence for the random-bonding nature of SiO_x and conclude that SiO and other intermediate compositions of SiO_x between Si and SiO_2 exist as stable chemical compounds.

Work is in progress to extend the random-bonding statistics to chains and rings.

REFERENCES

(1) Phillips,J.C. (1973) <u>Bonds and Bands in Semiconductors</u>, Academic Press, New York/London.

(2) K.Hübner, Chemical bond and related properties of SiO_2.II. Structural trends, <u>phys.stat.sol.(a)</u> 40, 487 (1977).

(3) K.Hübner, Chemical bond and related properties of SiO_2.I. Character of the chemical bond, <u>phys.stat.sol.(a)</u> 40, 133 (1977).

(4) H.R.Philipp, Optical properties of non-crystalline Si,SiO,SiO_x and SiO_2, <u>J.Phys.Chem.Solids</u> 32, 1935 (1971).

(5) K.Hübner, Chemical bond and related properties of SiO_2.V. Pseudopotential study of bonding and structure, to be published.

(6) March,N.H.,Young,W.H.,and Sampathar,S. (1967) <u>The Many-Body Problem in Quantum Mechanics</u>, University Press, Cambridge.

(7) R.Pickenhain and K.Hübner, Refined application of pseudo-potentials in mixed crystals, <u>Proc.Conf.Mixed Crystals '75</u>, Reinhardsbrunn 1975 (Ed.Academy of Sciences,GDR), p.139.

(8) R.Pickenhain and A.Milchev, The quadratic response of a Fermi gas, <u>phys.stat.sol.(b)</u> 77, 571 (1976); 79, 549 (1977).

(9) J.R.Chelikowsky and M.Schlüter, Electron states in α-quartz: A self-consistent pseudopotential calculation, <u>Phys.Rev.B</u> 15, 4020 (1977).

(10) S.Ciraci and I.P.Batra, Electronic-energy-structure calculations of silicon and silicon dioxide using the extended tight-binding method, <u>Phys.Rev.B</u> 15, 4923 (1977).

(11) A.F.Wright and A.J. Leadbetter, The structures of the β-cristobalite phases of SiO_2 and $AlPO_4$, <u>Phil.Mag.</u> 31, 1391 (1975).

(12) S.T.Pantelides and W.A.Harrison, Electronic structure,spectra, and properties of 4:2-coordinated materials.I.Crystalline and amorphous SiO_2 and GeO_2, <u>Phys.Rev.B</u> 13, 2667 (1976).

(13) K.Hübner and A.Lehmann, Chemical bond and related properties of SiO_2.IV.Structure of amorphous SiO_2, <u>phys.stat.sol.(a)</u> in the press.

(14) A.Lehmann and K.Hübner, Phonons in disordered SiO_2, to be published.

(15) K.Hübner, Chemical bond and related properties of SiO_2.III. Core-level shifts in SiO_x, <u>phys.stat.sol.(a)</u> 42, 501 (1977).

TOPOLOGICAL EFFECTS ON THE BAND STRUCTURE OF SILICA*

M. F. Thorpe
Dept. of Physics, Michigan State Univ., E. Lansing, Michigan

ABSTRACT

Under some circumstances, the electronic properties (using a tight binding model) and the vibrational properties (using a rigid ion model) of silica networks can be expressed in terms of the Connectivity Matrix. This matrix contains information about the connectivity of the network and its properties are fairly well understood as a result of extensive studies on silicon networks. If the range of the overlap parameters becomes too large, this description breaks down and structural parameters such as the dihedral angle enter in an unavoidable way.

In crystalline forms of silica, traditional band structure methods are far superior to this approach. However in comparing crystalline and vitreous silica, it is useful to have a common language when discussing the main features in the density of states. A detailed discussion is given for the oxygen 2s band in silica.

INTRODUCTION

Silica exists in many forms; almost all of which (e.g. cristobalite, tridymite, quartz and vitreous silica) can be regarded as being made up of nearly perfect SiO_4 tetrahedra linked together via the oxygen ions so that each oxygen has two silicon neighbors. Most of these forms also occur in a high and low form, in both of which the Si-O-Si bond angle is ~145°. In this paper we shall refer to the high form to mean an idealized high form in which the Si-O-Si bond is straight.

Some of these structures are very complicated arrangements of silicon and oxygen ions. However for some purposes it is sufficient to understand only the topology of silica network and to ignore complicated geometrical aspects. If the oxygens ions are disregarded for the moment, then the topology of the silica network is described by the silicon network in which each silicon atom has exactly four neighbors. The connectivity matrix for this network $\underset{\sim}{M}$ is defined by

$$M_{ij} = 1 \quad \text{if } i,j \text{ are nearest neighbors}$$
$$= 0 \quad \text{otherwise} \tag{1}$$

Thus $\underset{\sim}{M}$ is a symmetric matrix with four non zero elements of in each row and

*Work supported in part by N.S.F.

column. Because of extensive work on silicon networks (2) a fair amount is known about the spectrum of $\underset{\sim}{M}$. For example its eigenvalues ε are bounded by

$$-4 \leq \varepsilon \leq 4 \tag{2}$$

where the lower bound is only reached if the network does not contain odd <u>rings</u> (2). Also it is known that there is some correlation between the peaks in the density of states of $\underset{\sim}{M}$ and the occurence of <u>rings</u> of bonds in the network.

The Vibrational Spectrum

It has been shown by Sen and Thorpe (3) that if one uses only nearest neighbor central forces along the Si-O bonds in a silica network, then the phonon density of states can be expressed in terms of the eignevalues ε of the conductivity matrix (of course, it also depends on the masses, the central force constant and the Si-O-Si bond angle θ). Due to space limitation we shall not pursue these results here.

The Electronic Properties

One of the earliest calculations of the band structure of silica in the high cristobalite form was by Weaire and Thorpe (2) in a section of a paper concerned with amorphous semiconductors in general. High cristobalite has the silicon ions on a diamond cubic structure with the oxygen ions placed mid-way between nearest neighbor silicon ions so that each Si-O-Si bond is straight.

A tight binding model can be set up as follows. Hybridized "sp^3" orbitals are formed on the silicon ion from the silicon 3s and 3p states. The matrix element between these four orbitals is V_1 so that the splitting between the 3s and 3p states is $4V_1$. On the oxygen ion "sp" hybridized orbitals can be set up involving the oxygen 2s state and the 2p state along the Si-O-Si bond. These two states are connected by a matrix element V_1' and have a diagonal energy ΔE so that the 2s state is at $\Delta E + V_1'$ and the 2p state is at $\Delta E - V_1'$. The other two oxygen 2p states are perpendicular to the bond and are described as "non-bonding" or "lone pair" and are at an energy $\Delta E - V_1'$. The parameters V_1, V_1' and ΔE are essentially atomic in nature and we choose them to be $-1.75eV$, $-7eV$ and $-14eV$ respectively (2,4). The hybridized "sp^3" silicon orbitals have a matrix element V_2 with the neigbhoring oxygen "sp" orbital. This is essentially a bonding/ antibonding parameter and is chosen to be $-14eV$ in order to get the band gap approximately correct. No attempt has been made to choose these parameters very accurately as we are only after a schematic band structure. Much more sophisticated calculations have been done recently (5,6) and we could obtain an improved set of tight binding parameters by trying to fit the computed band structures.

The eigenvalues of the model described above can be related to the connectivity matrix for the diamond structure and the energy eigenvalues E are given in terms of ε

$$[E-4V_1][(E-\Delta E)(E^2 - E\Delta E - V_2^2) - EV_1'^2] - V_2^2[E^2 - E\Delta E - V_2^2] = V_1 V_1'V_2^2 \varepsilon \tag{3}$$

The single band of the connectivity matrix generates four bands in cristobalite
via the quartic equation 3. In addition there are flat bands that lead to
delta functions in the density of states at energies

$$\frac{1}{2} [(\Delta E + V_1') \pm \sqrt{(\Delta E + V_1')^2 + 4V_2^2}]$$

$$\frac{1}{2} [(\Delta E + V_1') \pm \sqrt{(\Delta E - V_1')^2 + 4V_2^2}]$$

(4)

This band structure is shown in Fig. 1. The density of states for any other
high form of silica (i.e. with straight Si-O-Si) bonds could·be obtained from
equations (3) and (4) by putting in the appropriate spectrum of eigenvalues ε

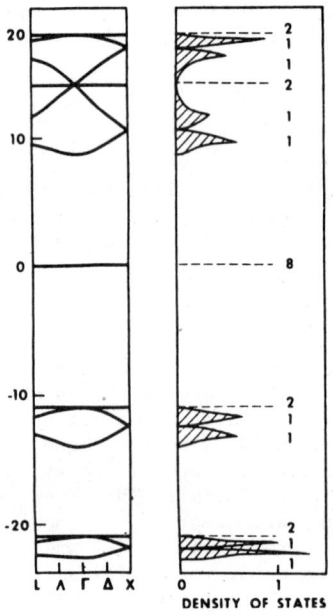

Fig. 1. The band structure and density of
states for the high form of cristobalite
using the tight binding Hamiltonian described
in the text. The energies are in eV and the
numbers beside the bands give the total number of states in that band (not counting spin).
A constant (7eV) has been added to the energy
to put the lone pair at zero energy. The
band gap is between the flat lone pair band
and the bottom of the conduction band at 8.7eV.
The conduction bands in this model are not
reliable because of the neglect of the silicon
3d states.

The range of the overlap parameters could be extended considerably without
requiring any other information about the structure other than its connectivity
matrix. This has been done in the case of silicon networks (2) where one can
include all parameters that only involve nearest neighbor Si-Si bonds. When
one goes to overlaps between orbitals on more distant Si-Si bonds, the geometry
of the structure enters in an unavoidable way and more information is needed
than. just the connectivity matrix.

The main deficiency of this model is that the lone pair band has no width.
(Some of the other bands are also too narrow but this could be fixed by adjusting
some of the parameters). To give these bands a width, parameters must be introduced that also require a knowlege of the geometry of the structure (7).

Comparing this work to other calculations of the density of states, the characteristic shape of the diamond cubic density of states shown in the lowest two bands in Fig. 1 is always apparent. The upper of these bands is a bonding band made up mainly of oxygen 2p and silicon 3s states whereas the lower band is almost entirely oxygen 2s. This correspondence is particularly clear in the work of Schneider and W. Beall Fowler (6) and Ciraci and Batra (8). The very narrow bands that correspond to our delta functions can clearly be seen in these calculations.

For other silica structures in the high form, the delta functions will persist, but the characteristic two peaked density of states for the connectivity matrix of diamond will be replaced by some other density of states. For example keatite has the same topology as GeIII (9,10). This is particularly interesting because it contains five fold rings of bonds so that the eigenstates of the connectivity matrix near −4 cannot be formed and there will be a small gap in the density of states between the delta function and the rest of the band in the lowest two bands in Fig. 1. Tridymite has the same topology as the wurtzite structure (1) and so has the same density of states as cristobalite (11). The spectrum of the connectivity matrix of quartz has not previously been studied and we examine it in the next section.

Quartz

Quartz contains three molecular units per unit cell. Removing the oxygen ions we can generate the eigenvalues of the connectivity matrix by Fourier transforming to a 3x3 matrix

$$\varepsilon(\varepsilon^2-4) - 2\gamma_k(\varepsilon-2\cos k_z c) = 0 \qquad (5)$$

where
$$\gamma_k = 2\cos k_x a \, [\cos k_x a + \cos \sqrt{3}k_y a] \qquad (6)$$

The parameters a, c are related to the hexagonal unit cell of quartz. The density of states is found by solving the cubic equation at each point in $\underline{k}$ space and integrating over the Brillouin zone. It can be seen from Fig. 2 to have three peaks rather than two in the case of cristobalite. The spectrum in even about the center because the structure can be divided into two sublattices so that there are only even rings of bonds present. This characteristic structure is clearly seen in the work of Calabrese and Fowler (12). It is difficult to compare with the work in reference (5) as a detailed density of states is not calculated.

The Oxygen 2s Band

This result can be most directly applied to the oxygen 2s band (the lowest band shown in Fig. 1). We take the overlap between 2s states on adjacent oxygen ions to be V. This should be independent of the structure to a good approximation as the 0-0 distance is always around 2.60Å in silica. Each oxygen has a matrix element V to six other neighboring oxygens and we can write down a matrix for the band that is independent of the Si-0-Si bond angle θ and depends only on the topology. By writing down the equations for the amplitude of the wave function (after the style of Ref. 2), it can be shown that there are N states

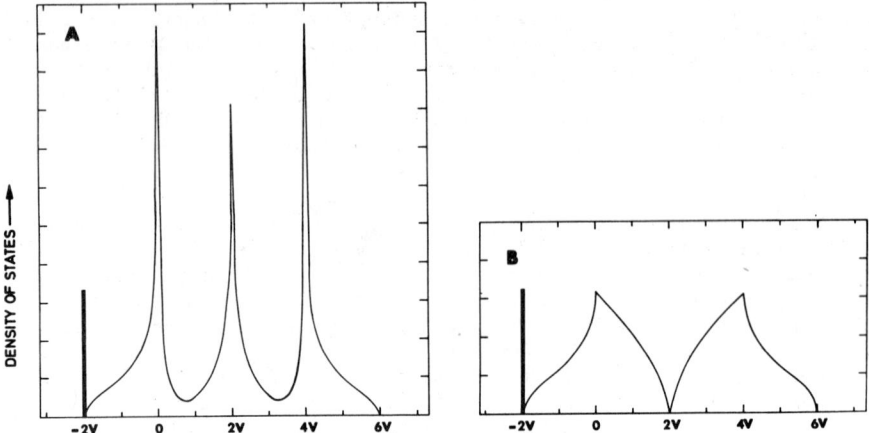

Fig. 2.   The oxygen 2s bands in (A) quartz and (B) cristobalite and tridymite. The vertical scale is arbitrary but the same in both cases.   The horizontal scale is in units of V where V ≃ -.25eV.   One half of the weight is the delta function.

in a band with

$$E = V(2 + \varepsilon) \tag{7}$$

and another N states at an energy

$$E = -2V \tag{8}$$

where N is the number of SiO_2 molecules.   This is shown in Fig. 2.   The delta functions arise because one can select chains or rings of oxygen ions in the structure such that the amplitude alternates +1, -1, etc.   There are exactly N such independent states.   The peak at E=2V (corresponding to ε=0) in the quartz structure is primarily due to the large number of eight fold rings in quartz where the amplitude of the states around a ring goes 1, i, -1, 1, i, -1, -i. There is also a peak in the density of states of the connectivity matrix at ε=0 on square lattice for a similar reason.   The peaks at E=0 and 4V (corresponding to ε=-2 and +2) occur because of states that can be formed on the hexagonal rings of bonds.   There are only half as many hexagons in quartz as in cristobalite (see Table 1) so that although the peaks are higher for quartz in Fig. 2 there is less weight in the vicinity.

CONCLUSION

We have shown that the effect of structure on the density of states of silica can be determined by expressing everything in terms of connectivity matrix.

Moments and Rings

	Bethe Lattice	Quartz	Cristobalite Tridymite
2	4	4	4
4	28	28	28
6	232	244	256
8	2092	2492	2716
n_6	0	6	12
n_8	0	44	0

Table 1. The moments of the connectivity matrix for different lattices. These are just the number of returns to the origin after a walks of n steps; shown here for n=2,4,6,8. These can be obtained from the tree diagrams, (i.e. the Bethe lattice) and the additonal contributions from the rings. The number of six and eight fold rings are given by n_6 and n_8.

While this leads to a rather oversimplified model for the vibrational and electronic properties, it is probably rather accurate for the oxygen 2s band. Unfortunately experimental techniques (photoemission and soft x-ray emission) do not as yet have the resolution ($\sim$.25eV) to see these structural effects.

REFERENCES

(1) Sosman R. B. (1965) The Phases of Silica, Rutgers University Press, NJ.

(2) D. Weaire and M. F. Thorpe in Computational Methods for Large Molecules and Localized States in Solids edited by F. Herman, A. D. McLean and R. K. Nesbet (Plenum Press, New York 1973) p.295.

(3) P. N. Sen and M. F. Thorpe, Phys. Rev. B15, 4030 (1977).

(4) These can be obtained from for example E. Clementi "Tables of Atomic Functions", I.B.M. J. of Res. and Devel. 9, 2 (1965). In Ref. (2) the values of V_1 and V_2 were misprinted in the text.

(5) J. R. Chelikowsky and M. Schlüter, Phys. Rev. B15, 4020 (1977).

(6) P. M. Schneider and W. Beall Fowler, Phys. Rev. Lett. 36, 425 (1976) and to be published.

(7) S. T. Pantelides, Phys. Lett. A54, 401 (1975); S. T. Pantelides and W. A. Harrison, Phys. Rev. B73, 2667 (1976).

(8) S. Ciraci and J. P. Batra, Phys. Rev. B15, 4923 (1977).

(9) R. Alben, S. Goldstein, M. F. Thorpe and D. Weaire, Phys. Stat. Sol (b) 53, 545 (1972).

(10) J. D. Joannopolous and M. L. Cohen, Phys. Rev. B7, 2644 (1973).

(11) M. F. Thorpe, J. Math. Phys. 13, 294 (1972).

(12) E. Calabrese and W. Beall Fowler, Phys. Rev. B (April, 1978).

CHAPTER III

THERMAL AND STRUCTURAL PROPERTIES

HEAT PULSE EXPERIMENTS ON VITREOUS SiO$_2$ IN THE TEMPERATURE RANGE 2.5 - 300 K

W. Block, M. Meissner and K. Spitzmann
Institut für Festkörperphysik
Technische Universität Berlin, West Germany

ABSTRACT

The thermal conductivity measured with a steady-state method and both the specific heat and the thermal diffusivity obtained from a transitory method have been compared for crystalline and vitreous SiO$_2$ in the temperature range 2.5 - 300 K. The data for crystalline SiO$_2$ are consistent for both methods; an unexpected difference, however, is observed for the thermal diffusivity data of vitreous SiO$_2$ below 15 K. Depending on the sample length, the thermal diffusivity obtained from a heat pulse experiment is smaller than the corresponding stationary value. These results, derived from thermal conductivity experiments with different time scales, are discussed with respect to a theory which includes the tunneling states model into the heat diffusion problem.

INTRODUCTION

It has been proposed[1] that the thermal anomalies in glasses[2] arise from a distribution of localized two-level tunneling states characteristic for the amorphous state. Strong evidence for resonant scattering of phonons by the two-level systems is given by ultrasonic experiments below 1 K.[3] The physical nature of these tunneling states is not yet established. Above 1 K the tunneling model cannot explain all the experimental data. The ultrasonic attenuation has given evidence for a non-resonant scattering involving a relaxation process.[3] Recently, Zaitlin and Anderson[4] and Jäckle[5] have calculated the thermal conductivity curve for vitreous SiO$_2$ between 0.1 K and 10 K using the non-resonant scattering relaxation times. On the other hand, the "excess" specific heat of glasses has been attributed to the existence of low-frequency modes in addition to the Debye contribution. Comparing the results from neutron experiments with the corresponding "excess" specific heat,[6] it is argued that this "excess" above the Debye density of states is due to a strong broadening of low-lying transverse acoustic branches.

We started this work to study the influence of the two-level systems on the thermal conductivity in the non-resonant scattering region. This has been done with regard to one of the consequences of the tunneling states model: the interaction of the phonon system with the two-level systems is expected to be a function of the time duration of the experiment. This effect is due to a relaxation time τ which is required for the two-level systems to come into thermal equilibrium with the phonons. If the duration time of the experiment is t $\lesssim \tau$, the influence of the tunneling states on the heat-carrying phonons will vanish. Goubau and Tait[7] tried to study this problem with 10^{-4} sec time-scale measurements of the specific heat of thin SiO_2 and PMMA samples in the temperature range 0.1 - 2.5 K. Their results did not agree with the prediction based on the tunneling states model. However, for SiO_2, differences of up to 40% were actually present between their long-time and short-time measurements in the resonant scattering region (below 1 K).

We report simultaneous measurements of the thermal conductivity and the thermal diffusivity in the temperature range 2.5 - 300 K. Results obtained from thermal conductivity measurements with steady-state methods are typical of those for long-time experiments; thermal diffusivity data obtained from heat pulse experiments are typical of those for short-time measurements. Considering the anomalous thermal properties of glasses we would like to emphasize the difference between experiments on crystalline and vitreous SiO_2. From the Debye theory long-time and short-time experiments are expected to result in identical data for crystalline solids. For vitreous SiO_2 an unexpected time dependence has been observed in our experiments up to a 10^{-10} sec time-scale.

RESULTS AND DISCUSSION

In Fig. 1 our experimental arrangement is shown.[8] Two cylinders of the same specimen are mounted with a thermocouple between the lower face (x = L) and the point of measurement (x = x_o) inside the sample. This problem has been solved with the following arrangement: an Au(Fe) *vs*. Chromel-thermocouple and an indium foil have been pressed under presence of GE 7031 varnish to a thickness of about 50 μm between the two polished surfaces. On the top of the sample (x = 0) a constantan heater has been evaporated. In case of stationary heat flow the input thermal power Q will cause a temperature gradient $\Delta T/(L-x_o)$ from which the thermal conductivity K_{stat} can be calculated. In the case of transient heating the constantan film acts as pulse heater. The thermal diffusivity α can be calculated from the temperature-time profile within the time t_{max}, and the specific heat C_p is given by the maximum temperature rise ΔT.

In Fig. 2 we compare our experimental results for the thermal diffusivity of crystalline and vitreous SiO_2. Considering the data for crystalline quartz ($\perp$c, Valpey Fisher) there is a very good agreement between results from the stationary and the heat pulse method in the whole temperature range. Below 20 K our values are smaller than the diffusivities α_{stat} which can be calculated from the thermal conductivity of a perfect crystal.[9] This is mainly due to the fact that at diffusivities $\alpha_{stat} \gtrsim$ 500 cm^2/sec the indium foil will influence the temperature-time profile of the heat pulse experiment.

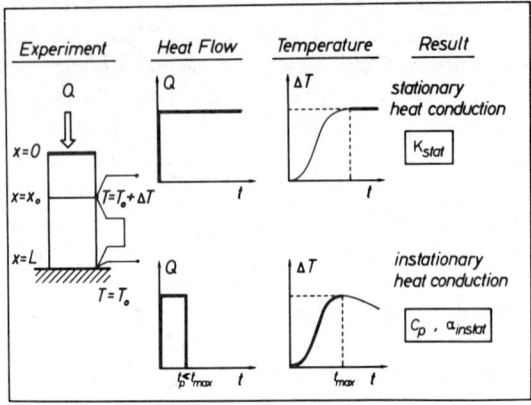

Fig. 1. Experimental arrangement

For vitreous SiO_2 (Suprasil I and W, Schott-Heraeus) both the stationary and the transient method lead to identical data only in the temperature range 15 - 300 K. At lower temperatures we observe a reduction in α, which is $\alpha/\alpha_{stat} \approx 0.5$ at the lowest temperature of about 2.5 K. In contrast to the heat pulse values our data obtained from the stationary method agree very well with former thermal conductivity measurements.[10]

In a more detailed analysis we varied the time duration of our heat pulse experiments. We measured α of vitreous SiO_2 for different sample lengths x_o = 0.1 ... 1 cm, which give rise to characteristic temperature-time profiles within t_{max} = 10^{-3} ... 10^2 sec between 2.5 - 300 K. The important result is seen in Fig. 3, where the ratio α/α_{stat} versus 0.1 t_{max} has been plotted for 3 different sample lengths. From this analysis one may obtain a characteristic time $t' \approx 0.1$ sec which should be interpreted as a lower limit to obtain $\alpha/\alpha_{stat} \equiv 1$ in a heat pulse experiment. For example, with a long sample (x_o = 1 cm) the heat pulse experiment yields $\alpha \approx \alpha_{stat}$ at T = 5 K; with a short sample (x_o = 0.1 cm) the heat pulse experiment yields identical data at temperatures above 15 K.

With regard to the tunneling states model it is obvious that a relaxation time t' = 0.1 sec is well outside the theoretical predictions. On the other hand, it is not yet clear whether the tunneling states are the dominant scattering centers in the plateau region of the thermal conductivity. We suggest that this problem should be treated by involving the heat diffusion problem more properly. The linear flow of heat inside the sample is governed by the Fourier-Ansatz

$$Q = - K \cdot \frac{\partial T}{\partial x} .$$
(1)

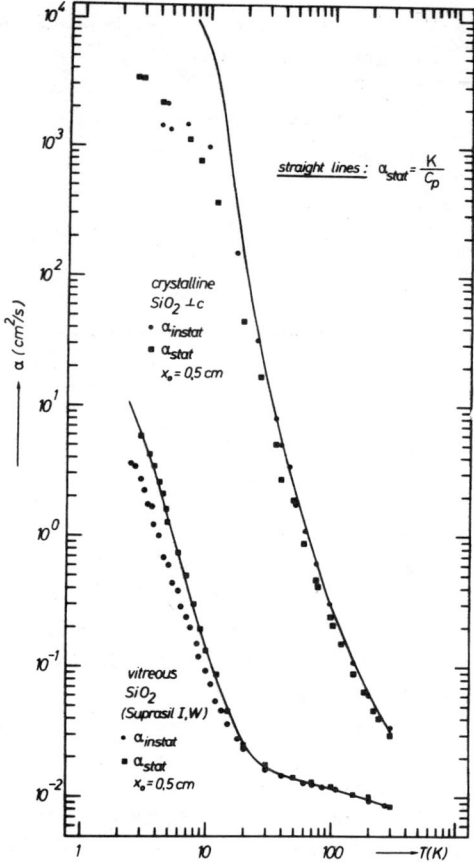

Fig. 2 (See text)

and the equation for conservation of the heat flow

$$\frac{\partial Q}{\partial x} + C_p \cdot \frac{\partial T}{\partial t} = 0 \ . \tag{2}$$

In a recent paper Heinrichs and Kumar[11] treated the heat diffusion problem by including the tunneling model. The new aspect is given by the following Ansatz for Eq. (2):

$$\frac{\partial Q}{\partial x} + \left(\frac{\partial E}{\partial t}\right)_{\text{phonon}} + \left(\frac{\partial E}{\partial t}\right)_{\text{2-level}} = 0 \tag{3}$$

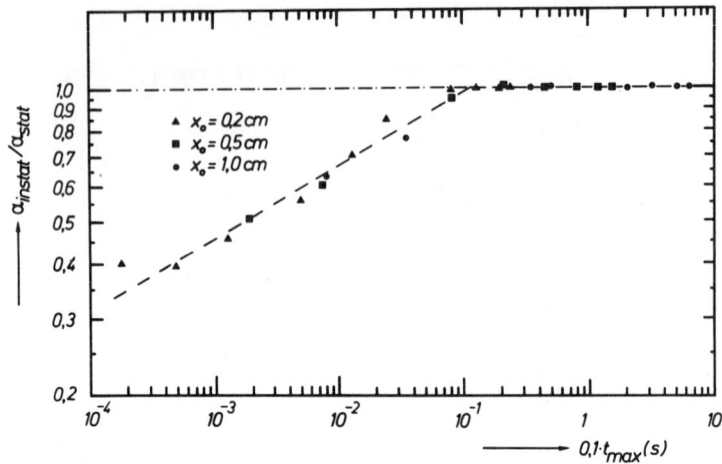

Fig. 3 (See text)

where the second term is given by the heat-carrying phonons, and the third term represents the transfer of energy from the phonon system to the tunneling states. Assuming resonant interaction, the authors derived a more detailed solution, where different terms (K/C_t), where C_t is a time-dependent specific heat, appear. At very short times $C_t = C_{phonon}$ is dominated by the phonon system, and at long times it reaches the value $C_t = C_{phonon} + C_{2\text{-level}}$. We suggest that the heat diffusion in glasses should be related to a modified equation of energy conservation as given in Eq. (3). On the other hand, a physical model for energy transfer with relaxation times of about 10^{-1} sec in glasses at low temperatures is still unknown. For a more detailed theoretical analysis further experimental investigations on this time-dependent effect are necessary.

REFERENCES

1. P. W. Anderson, B. I. Halperin, C. M. Varma, *Phil. Mag.* **25**, 1 (1972); W. A. Phillips, *J. Low Temp. Phys.* **7**, 351 (1972).
2. A recent review is given by R. O. Pohl, G. L. Salinger, *Annals N.Y. Acad. Sci.* **279**, 150 (1976).
3. J. Jäckle, L.Pichè, W. Arnold, S. Hunklinger, *J. Non-Cryst. Solids* **20**, 365 (1976).
4. M. P. Zaitlin, A. C. Anderson, *Phys. Rev. B* **12**, 4475 (1975).
5. J. Jäckle, in *The Physics of Non-Cryst. Solids* (G. H. Frischat, Trans. Tech. Publ. ed.) 568 (1976).
6. N. Bilir, W. A. Phillips, *Phil. Mag.* **32**, 113 (1975).
7. W. M. Goubou, R. H. Tait, *Phys. Rev. Lett.* **34**, 1220 (1975).

8. R. Krüger, M. Meissner, J. Mimkes, A. Tausend *Phys. Stat. Sold. (a)* **17**, 471 (1973); M. Meissner, thesis (1974), TU Berlin.

9. $\alpha_{stat} = K_{stat}/C_p$ for crystalline SiO_2 calculated from smoothed data of

   K: recommended values given by Y. S. Touloukian *et al.* (1970), *Thermophysical Properties of Matter, Vol. 2*, The TPRC Data Series, IFI/ Plenum, New York; R. C. Zeller, thesis (1971), Cornell University;

   C_p: G. K. White, J. A. Birch, *Phys. Chem. Glasses* **6**, 3 (1965); E. F. Westrum, unpublished, cited in R. C. Lord, J. C. Morrow, *J. Chem. Phys.* **26**, 230 (1957).

10. $\alpha_{stat} = K_{stat}/C_p$ for SiO_2 calculated from smoothed data of

    K: G. A. Slack, *Cryogenics* **9**, 384 (1969); R. A. Fischer, G. E. Brogdale, E. W. Hornung, W. F. Giauque, *Rev. Sci. Instr.* **40**, 365 (1969); R. C. Zeller, see Ref. (9).

    C_p: E. F. Westrum, see Ref. (9); P. Flubacher, A. J. Leadbetter, J. A. Morrison, B. P. Stoicheff, *J. Phys. Chem. Sol.* **12**, 53 (1959); R. C. Zeller, see Ref. (9).

11. J. Heinrichs, N. Kumar, *Phys. Rev. Lett.* **34**, 1406 (1976).

THERMAL CONDUCTIVITY OF SiO$_2$

Baxter H. Armstrong

IBM Scientific Center, Palo Alto, California 94304

ABSTRACT

A new theory of U-process relaxation, along with the acoustic phonon Boltzmann
equation, is applied to the thermal conductivity of SiO$_2$. Relaxation due to
dislocations and the contribution of optical modes are explored heuristically.
The results compare favorably with experiment from 1 - 500°K using Grüneisen
gammas, the U-process threshold constant, and boundary scattering length as
adjustable parameters, with possible adjustment of the Debye temperature to
reflect variable participation of optical modes.

INTRODUCTION

It has been proposed (1) that the acoustically perturbed phonon Boltzmann equation in the relaxation-time approximation be used to determine lattice thermal
conductivity. The attenuation length provided by the Boltzmann equation in the
Woodruff-Ehrenreich (WE) formulation (2) is taken as the thermal phonon mean
free path (mfp). This mfp is ascribed to a "propagation group" of phonons, and
the absorbing or "reservoir" group in the proposed theory are those phonons
engaging in U-processes. Thus, the sum over reservoir modes is restricted to
modes lying above the U-process threshold. Although the WE theory is conventionally expected (3) to be valid only if $\hbar\omega \ll kT$ when $\omega\tau > 1$, arguments
are advanced in (1) that it retains validity when $\hbar\omega \sim kT$ (ω is phonon angular frequency, τ is reservoir phonon relaxation time). In this application,
the propagating phonon takes the place of the acoustic phonon of sound absorption theory, and reservoir phonons take the place of thermal phonons. The
reservoir is taken to include optical modes, but, for a first approximation,
these are excluded from the propagation role.

It is further proposed that the U-process relaxation time is determined by the
decay rate for a lattice vibration whose wave vector falls outside the unit
cell in reciprocal space. Such a vibration is conventionally assumed to be
dynamically equivalent to one whose wave vector lies within the cell, in the
U-process interpretation of replacement of the larger wave vector by one modulo
a reciprocal lattice vector $|\vec{Q}_0| = 2\pi/a_0$ (a_0 is a lattice constant). The
theory proposes that this replacement be viewed as a transition requiring a
finite but short time, which is to be identified with the U-process relaxation
time. It is assumed that a U-process has associated with it a displacement
wave $u = u_0 \exp[-t/(2\tau) - iQ_0 z]$ which satisfies a continuum wave equation
such as

$$\rho \ddot{u} = G(\partial^2 u/\partial z^2) + \eta(\partial^2 \dot{u}/\partial z^2) \tag{1}$$

conventionally associated with viscoelastic dissipation. G is an elastic constant, ρ the density and η the viscosity. The dynamical equivalence of states modulo $\vec{Q}_0$ within and without the unit cell is invoked to postulate that this displacement wave satisfies Eq. 1 for a viscosity extrapolated to zero frequency. The "dispersion relation" for this equation and wave is

$$\tau = \eta(1 \pm \sqrt{1 - \rho \, c \, a_0/(\pi\eta)})/(4G) \tag{2}$$

(c is the sound velocity) which permits the purely decaying displacement wave to exist if

$$\eta \geq \rho \, c \, a_0/\pi \tag{3}$$

Available data for six cubic crystals show longitudinal but not transverse viscosities obeying Eq. 3, consistent with isotropic U-process selection rules which permit only $T + T \rightarrow L$ and $L + T \rightarrow L$ transitions. This finding leads to the approximation employed in (1) that only one polarization branch occurs in the final state of the U-process. Since η is independent of frequency and temperature over large regions, τ can be taken as a constant adjustable parameter in the thermal conductivity formulas. For the six cubic crystals mentioned above for which η is available, $\rho \, c \, a_0/(\pi\eta) \ll 1$, so that the magnitude of $\tau \approx \eta/(4G)$.

The WE theory as interpreted above yields

$$\Lambda_{\omega j} = 2 \, \rho \, c_j^3/(\gamma_j^2 \, T \, \Delta C \, \omega \, F(\omega\tau)) \tag{4}$$

for the mfp of thermal phonons of polarization j, or $\Lambda_{\omega j}/c_j$ for the thermal conductivity relaxation time. $F(\omega\tau)$ is an angular integral expression with the limits $2\,\omega\tau/3$ and $\tan^{-1}(2\,\omega\tau)$ for small and large $\omega\tau$, respectively. ΔC is the contribution to the specific heat per unit volume of modes above the U-process cutoff $\omega_c = k \, T_D/(\hbar b)$ in the Debye approximation. T_D and T are Debye and absolute temperatures, b is the usual U-process threshold parameter, and γ_j^2 is a Grüneisen constant which couples propagating modes of polarization j with reservoir modes (1,3). Equation 4 for the thermal mfp was used in (1), along with boundary and Rayleigh scattering expressions to evaluate the thermal conductivity from the gas-kinetic theory expression, using the Grüneisen gammas, τ, b, and the Rayleigh scattering strength as adjustable parameters. Favorable results were demonstrated for LiF, Ge, [4]He, KH_2PO_4, Al_2O_3, $ZnSO_4$, and, to a lesser extent, NaCl. The b-values found for LiF and solid He agreed with previously established results, and agreement with experiment was generally as good as for relaxation-time fits of the Callaway type, with comparable number of parameters.

Numerical experimentation revealed crystals in the same symmetry classes as those for which the above-noted comparisons were made, but for which apparent unrealistic values of τ and/or b were required to fit the data. The NaCl comparison of (1), made for both unannealed and annealed samples, as well as that for solid He, suggests that these anomalous parameter values result from the influence of dislocations on U-process relaxation, and the contribution, possibly influenced by motion of dislocations, of optical modes to the conductivity, as well as the possible failure of optical modes to participate in the reservoir action. SiO_2 is one of the crystals for which the fundamental

U-process model outlined above provides a fit only with parameter values which are unrealistic in terms of their physical definitions. It is also a species for which considerable structural knowledge, thermal conductivity, and acoustic absorption data exist. Hence, we undertake herein to study the application of the theory to this crystal, supported by a heuristic model for the influence of dislocations on U-process phonon relaxation.

APPLICATION TO SiO_2

Reduction to Computation

Addition of the reciprocal mfp as given by Eq. 4, to Λ_0^{-1} and Λ_{RS}^{-1} to include boundary and Rayleigh scattering, provides a resultant mfp $\Lambda_{\omega j}^{(R)}$ which is used in the formula $K = \frac{1}{3} \sum_{\omega j} c_{\omega j} \Lambda_{\omega j}^{(R)} c_j$ to compute the thermal conductivity K. Λ_0 is taken as a phenomenological length parameter upon neglect of the details of Casimir scattering; $\Lambda_{RS}^{-1} = d \omega^4 / c$ where d is an analogous parameter. One readily obtains in the usual acoustic approximation of elementary Debye theory the result (1):

$$K = \frac{c_M \Lambda_1 \rho (x_D')^2}{W} \int_0^1 \frac{t^4 \exp(x_D't) dt}{[\exp(x_D't) - 1]^2} \left\{ \frac{1}{1 + B_L t + B_{RS}^{(L)} t^4} + \frac{2}{1 + B_T t + B_{RS}^{(T)} t^4} \right\} \quad (5)$$

c_M is the Debye mean velocity, W the molecular weight

$$B_j \equiv 9(k/\hbar) R \bar{n}^{2/3} \Lambda_1 \gamma_j^2 \Delta J_4 (x_D) F(At) / (2 c_j^3 W T_D^2) \quad (6)$$

$$B_{RS}^{(j)} \equiv \Lambda_1 d_1 (\omega_D')^4 / c_j \quad (7)$$

$\Lambda_1 = \Lambda_0 s$, the number of unit cells per unit volume N defines s according to $N = \rho L s/W$, and R is the gas constant. $A \equiv \eta \omega_D / (2G)$ where ω_D is the Debye frequency and G the elastic constant appropriate to η. $\bar{n}$ is the number of atoms per unit cell, $d_1 = sd$, and $x' = T_D'/T$, where the prime is used to denote Debye temperature/frequency excluding optical modes. $J_4(x)$ is a member of the family of Debye theory integrals defined by Ziman (4), and $\Delta J_4(x_D)$ signifies the integral between the limits $T_D/(bT)$ and T_D/T. $F(\omega\tau)$ is the imaginary part of the bracketed postfactor in Eq. 4.8 of (2). Equation 5 can be straightforwardly integrated by numerical means; 48-order Gaussian integration was employed in (1).

Dislocation Relaxation and Optical Mode Contribution

We assume dislocations will oscillate harmonically with amplitude x_0 in response to the strain field of the propagation wave. Normal relaxation will be disrupted since dislocations carry irregular local temperature distributions, and mix properties of adjacent regions over the distance x_0. Such mixing may produce relaxation over this distance rather than $c\tau$ as for U-processes (2). We postulate that in pure materials, in response to the comparatively large strain of a thermal wave, x_0 will be a sizable fraction of the wave length of the propagating wave. Thus, it will be generally larger than $c\tau$ and able to dominate the U-process relaxation in the presence of sufficient dislocation density. Furthermore, the time τ_D for this relaxation to occur must be ~ the period of the wave. Thus we estimate $\tau_D \cong 2\pi/\omega$ which yields a frequency-independent function $F(\omega \tau_D) \simeq \tan^{-1}(4\pi)$ for Eq. 4. The consequences of this model are phenomenologically explored herein without determination of the density or other properties of dislocations. We expect the magnitude of b inferred from experiment to be larger in the presence of dislocations than in their absence, since they can simulate U-processes by dissipating momentum out of the phonon field at any frequency. Both these effects appear to be independent of the static scattering by dislocations which is well known.

Optical modes, which were assumed in (1) not to contribute to propagation, may do so in crystals with very low optical branches. Interaction of optical phonons with dislocations may occur, along with conversion to acoustic modes such that they contribute to thermal conductivity. This effect is potentially immensely complex, so we select herein a simple phenomenological approach of Debye temperature adjustment of three types. I, termed "partial inclusion of optical modes" consists of ignoring the mode sum over optical modes in the basic expression for K as in (1) while using the conventional Debye temperature in the integration limits. This has the effect of excluding optical modes from explicit propagation while spreading out the acoustic spectrum to cover the region normally occupied by optical modes. Improved agreement with experiment due to this procedure may indicate interaction between optical and acoustic modes affecting the conductivity, or simulate low-lying optical modes along with modified values of the adjustable gammas and boundary length parameter. The second method, II, is to calculate the conductivity upon removal of optical modes from the reservoir. This is accomplished by omitting the optical mode sum from the specific heat appearing in Eq. 4 and replacing T_D by T_D' in J_4 as well as the other factors of Eq. 5. Finally, the third way, III, treats T_D in Eq. 5 as an adjustable parameter. This will reflect a different modal contribution balance than I and may ultimately lend itself to physical interpretation. Manipulation of T_D may also accommodate some of the substantial anisotropy such as occurs in SiO_2, but which is ignored in the theory.

Comparison with Experiment for SiO_2

Immediate agreement of the U-process relaxation theory given in (1) and outlined above cannot be expected for SiO_2 since $\rho c a_0/(\pi \eta)$ is not < 1 as required by Eq. 3. Calculation of r for X, Y, and Z wave-propagation studied by Lamb and Richter (5) for L, FS, and SS waves in SiO_2 yields values ranging from 1.71 up to 8.0. Hence, it is unlikely that $r < 1$ obtains in other directions, and U-processes involving a shift of two reciprocal lattice vectors will be required, or slower relaxation processes may occur. Figure 1a compares methods I and II described above with experimental data tabulated by Touloukian, et al. (6). The agreement is quite good over limited portions of each curve; however, points from a given experiment migrate between the curves. This appears to confirm the effect of optical modes as dependent upon nonintrinsic properties related to purity and condition of the specimen. Curve A in Fig. 1a for heat flow parallel to the c-axis was computed for $\gamma_L = 1.2$, $\gamma_T = 0.881$, $b = 8$ and $\Lambda_1 = 6.4$ mm. The Rayleigh scattering coefficient $d = 1.5 \times 10^{-45}$ sec^3 was obtained from the isotopic scattering data of Slack (7), and used for all

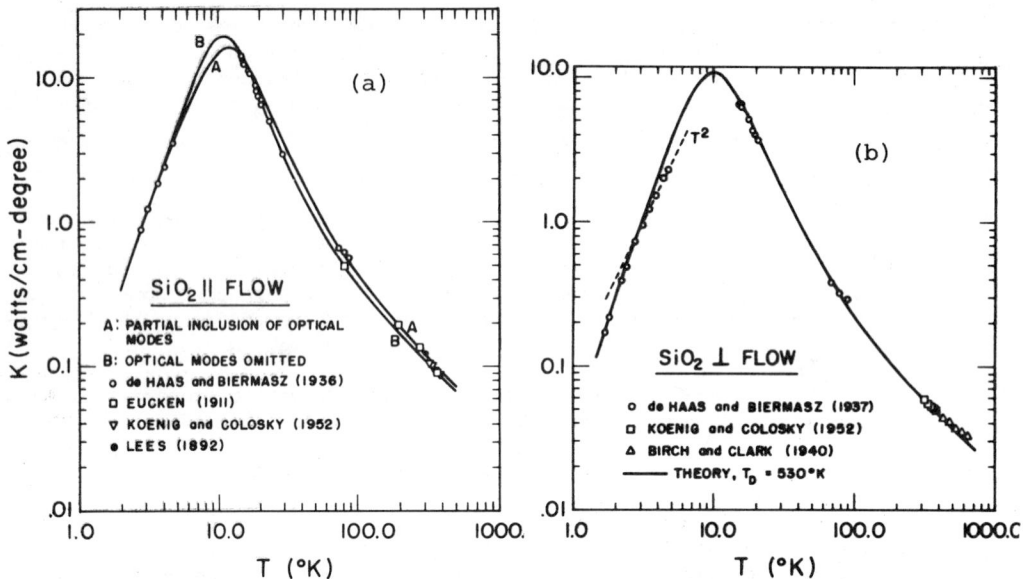

Fig. 1. Comparison of theory and experiment, thermal conductivity of SiO_2.

curves. As demonstrated in (1), γ_L and γ_T are not completely independent parameters. Except in the immediate vicinity of the maximum, they collapse into a
single gamma ($\equiv \gamma_M$) equal in this case to 0.93. The effect of $\omega\tau_D$ = constant
was accomplished numerically by setting $A = 10^4$ in all curves of Figs. la, b.
This produces the asymptotic constant value $F(\omega\tau_D) = \pi/2$, but does not distinguish between true frequency indepenedenc of F and a very large relaxation time.
Curve B computed for complete exclusion of optical modes fits the low-temperature
data very well with parameters γ_L = 2.30, γ_T = 2.232, b = 5.75 (γ_M = 2.03). In
view of our use of the Debye approximation, the molecule is taken as "unit cell"
so that $\bar{n}$ = 3, and Curve B corresponds to $T_D' = T_D/(3)^{1/3}$ replacing T_D in Eq.
6 with T_D = 585°K.

Figure lb for heat flow perpendicular to the c-axis derives from method III
for the parameters γ_L = 1.800, γ_T = 1.316 (γ_M = 1.392), b = 12.0, and a reduced
T_D = 530. Conventional low-temperature dislocation scattering is evident from
the region of T^2 dependence as noted on the figure.

The computational methods of (1) were employed; parameter values were obtained
by interactive numerical experimentation with the FORTRAN program under a VM/
370 (CMS) operating system on an IBM 145 computer.

REFERENCES

(1) B. H. Armstrong, "Transport of Thermal Phonons in Dielectric Solids. I.
Thermal Conductivity," submitted for publication. (2) T. O. Woodruff and H.
Ehrenreich, *Phys. Rev.* 123, 1553 (1961). (3) H. J. Maris, in *Physical Acous
tics*, VIII, ed. by W. P. Mason and R. N. Thurston, Academic Press, New York
(1971). (4) J. M. Ziman (1963) *Electrons and Phonons*, Oxford Univ. Press,
Oxford. (5) J. Lamb and J. Richter, *Proc. Roy. Soc.*, A293, 479 (1966).
(6) Y. S. Touloukian, R. W. Powell, C. Y. Ho, and P. G. Klemens (1970) *Thermophysical Properties of Matter*, 2, IFI Plenum, New York. (7) G. A. Slack,
Phys. Rev., 105, 829 (1957).

NEUTRON DIFFRACTION BY VITREOUS SILICA

Adrian C Wright
J. J. Thomson Physical Laboratory, Whiteknights,
Reading, RG6 2AF, U.K.

and

Roger N Sinclair
Materials Physics Division, A.E.R.E., Harwell,
Didcot, Oxon. OX11 ORA, U.K.

ABSTRACT

A series of neutron diffraction experiments have been performed on vitreous
silica in order to investigate its structure and also the inherent
limitations of the theoretical methods employed in neutron data analysis.
Measurements include the total scattering cross-section, total diffraction
and elastic diffraction. Total diffraction experiments have been carried
out, using both conventional steady state reactor facilities and pulsed
source time-of-flight techniques, to a maximum scattering vector of 40Å^{-1},
yielding much higher real space resolution than has been obtained to date.
The results are compared with the X-ray data of Warren (1) and Konnert and
Karle (2).

INTRODUCTION

Recent advances in neutron scattering techniques have made it possible to
determine the diffraction patterns from amorphous solids with much greater
accuracy than hitherto and, with the advent of accelerator-based time-of-
flight diffraction, to make measurements to much higher values of momentum
transfer. As a result it is worthwhile to repeat the earlier work of
Lorch (3) on vitreous silica in order both to provide a high quality, high
resolution correlation function to compare with the X-ray work of Warren (1)
and Konnert and Karle (2) and to test the validity of the methods used in
neutron data analysis to correct for departures from the static approximation.

The neutron correlation function differs from its X-ray counterpart in that it
refers to nuclei rather than electrons and the relative weighting of the
individual component correlation functions (Si-O, O-O and Si-Si) is different.
The weighting for each component is proportional to the product of the
scattering amplitudes of the two atoms involved. For X-rays, the form factor
ratio $f_{Si}(Q):f_O(Q)$ is approximately 14:8 whereas for neutrons the ratio of
the scattering lengths $b_{Si}:b_O$ is 0.41491:0.5804.

One of the difficulties in neutron diffraction experiments on amorphous
solids, especially those containing elements of low atomic number, is in
correcting for departures from the static approximation. For this reason a
range of measurements has been carried out on vitreous silica using both
conventional and time-of-flight techniques for which the static approximation

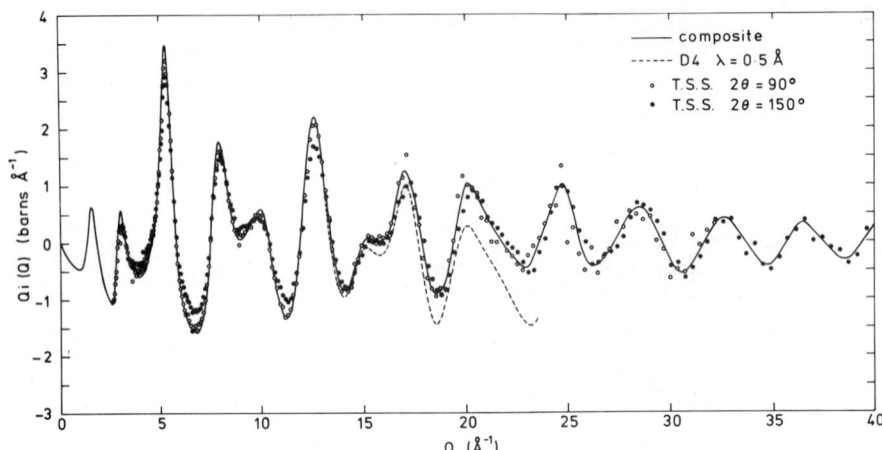

Fig. 1. The interference function Qi(Q)

distortions are different. The present paper briefly summarises these measurements which will be reported in detail elsewhere (4).

OUTLINE THEORY

The neutron correlation function

$$T(r) = 4\pi r \rho^0 \left(\sum_j \bar{b}_j\right)^2 + \frac{2}{\pi} \int_0^\infty Qi(Q)M(Q)\sin rQ \, dQ \qquad (1)$$

is obtained by Fourier transformation of the interference function

$$Qi(Q) = Q\left[I(Q) - \sum_j \overline{b_j^2} \, p_j \, (\lambda, 2\theta)\right] \qquad (2)$$

where Q is the scattering vector $(4\pi/\lambda)\sin\theta$, λ being the incident wavelength and 2θ the scattering angle. I(Q) is the corrected diffraction pattern in absolute units, M(Q) is the modification function due to Lorch (3) and ρ^0 the average number of composition units (SiO_2) per unit volume. The j summation is taken over the atoms in one composition unit, $\bar{b}$ and $\overline{b^2}$ are isotopically averaged neutron scattering lengths and $p(\lambda, 2\theta)$ is the so-called Placzek correction for departures from the static approximation, the exact form of $p(\lambda, 2\theta)$ depending on the type of measurement employed (5). A full account of the above theory has been given elsewhere (6,7) together with the method of extracting T(r) from the total scattering cross-section.

EXPERIMENTAL TECHNIQUE

In order to ensure consistency between the different diffraction experiments the same "Spectrosil" sample and vanadium standard were used in each case. Reactor measurements were performed at a series of incident wavelengths between 0.5 and 2.62Å using the D4 liquids diffractometer at I.L.L., Grenoble, and the CURRAN twin-axis machine at Harwell. The data were corrected for

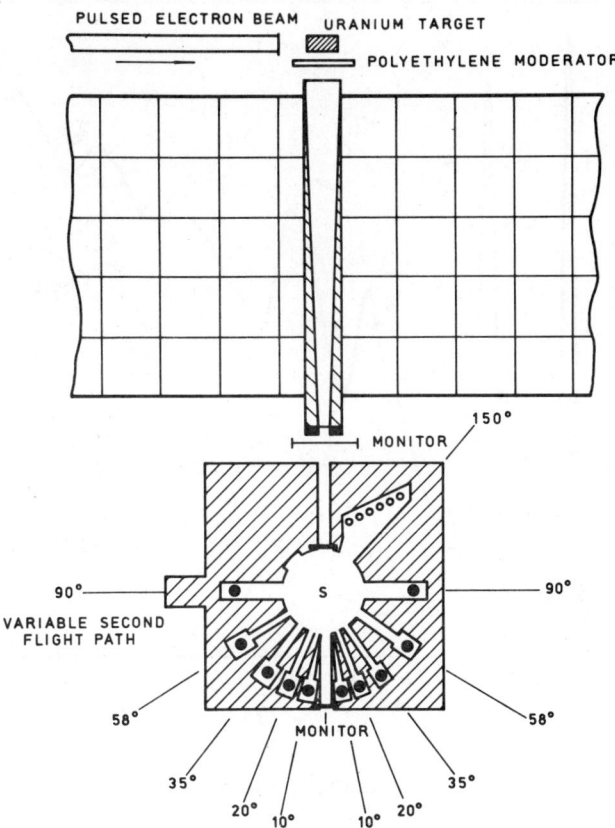

Fig. 2. The Harwell linac total scattering spectrometer

background, absorption, multiple scattering and self shielding and the self
scattering was calculated assuming a detector efficiency proportional to
neutron wavelength. The interference function obtained at 0.5Å is given in
Fig. 1.

An alternative method of measuring I(Q) is to vary the incident wavelength at
fixed 2θ and this is the basis of the time-of-flight technique. Diffraction
patterns were obtained at six scattering angles with the total scattering
spectrometer (T.S.S.) at the Harwell electron linac which is illustrated
schematically in Fig. 2. A pulsed polychromatic beam of neutrons, produced
by the target/moderator assembly, passed through a tapered collimator in the
shielding wall onto the sample placed in the centre of the spectrometer well.
The incident beam was monitored by a fission chamber and the scattered
neutrons were detected by ^{3}He counters placed at the scattering angles shown.
Each neutron time-of-flight was measured with a multishot time-of-flight
analyser and the accumulated data stored in a Digital Equipment Corporation
GT40 computer display system as a 2048 x 2μs channel time-of-flight spectrum.
Additional corrections, to those described in the previous paragraph, were
made for counter deadtime and the incident spectrum shape before calculating

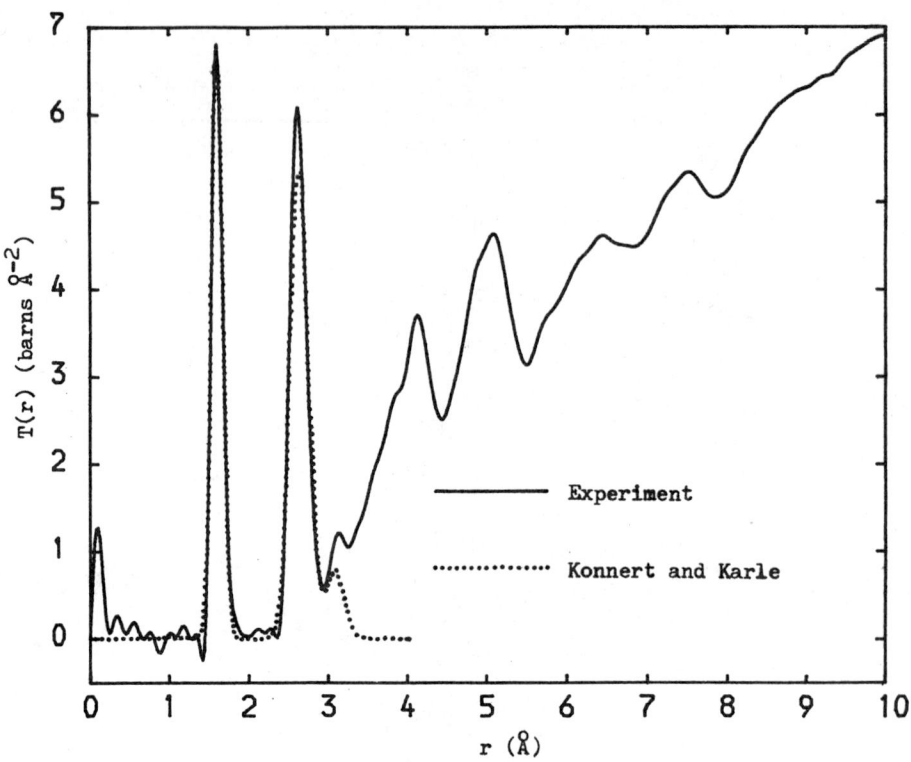

Fig. 3.   The correlation function T(r)

Qi(Q). The results for the two highest scattering angles are compared with
the reactor data in Fig. 1.

A measurement was also made of the total scattering cross-section $\sigma(E)$ as a
function of neutron energy, E, between 5 meV and 1.6 eV. This was very
similar to the T.S.S. experiment except that a ^{3}He detector was placed in the
straight-through beam and the sample comprised a set of 5 cm diameter
"Spectrosil" discs with a total thickness of 4 cm. The data show that the
incoherent scattering cross-sections of Si and O are extremely small and the
limiting behaviour of $\sigma(E)$ at high energies was used to obtain the effective
value of the average kinetic energy per atom used in calculating $p(\lambda,2\theta)$. A
Fourier transform of the interference contribution to $\sigma(E)$ yielded a real space
correlation function very similar to that in Fig. 3 except that the resolution
is poorer and the error ripples larger.

DISCUSSION

It is immediately apparent that the interference functions from the different
experiments are not in agreement. The 0.5Å reactor data drops below the
composite curve at high scattering angles due to the fact that the computed
self scattering is too high. A similar effect is present in the longer
wavelength measurements where there is an additional accompanying reduction in

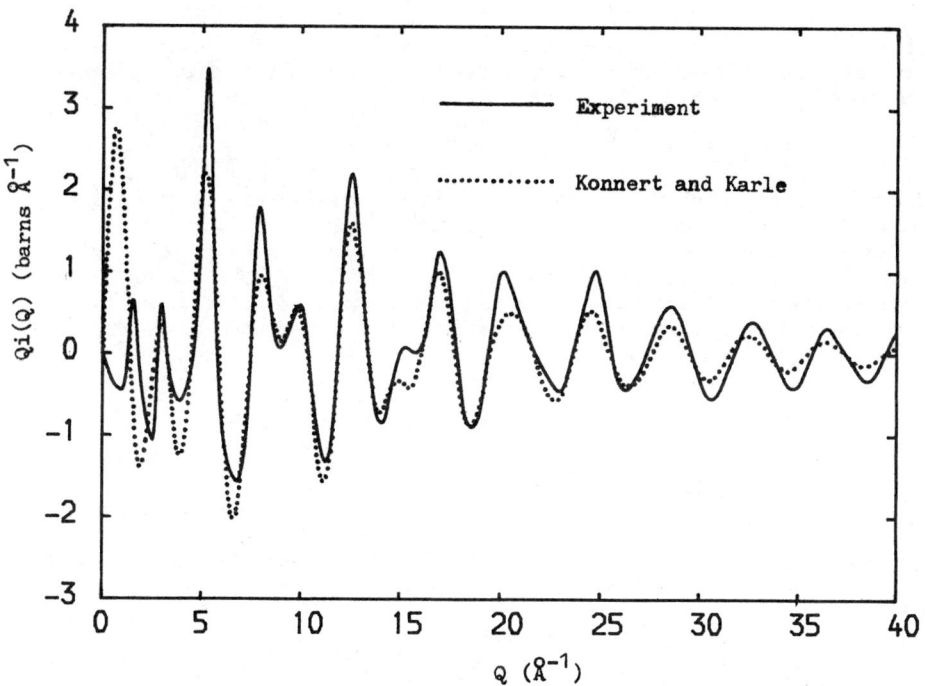

Fig. 4. A comparison of the experimental interference
function with the X-ray fit of Konnert and Karle (2)

the amplitude of oscillation. A decrease in amplitude is also seen in the
linac data, this time at low Q corresponding to low incident energies. It can
thus be concluded that the most reliable data at any given value of Q are those
obtained with the highest incident energy/lowest scattering angle combination
and this criterion was adopted in deciding the "best" line for the composite.

The Fourier transform of the composite interference function is shown in
Fig. 3. The first peak (Si-O) at 1.60Å is completely isolated while the
second (O-O) at 2.62Å is very much more prominent than in the corresponding
X-ray correlation function due to the different relative scattering powers of
Si and O. The peak widths at half height are respectively 0.16Å and 0.24Å
compared with the experimental resolution of 0.14Å. Owing to the relatively
small scattering length of Si the Si-Si peak at 3.13Å is very small and so the
best method of extracting the Si-Si distribution from a combination of X-ray
and neutron data will be to use the latter to obtain the O-O distribution
which can then be subtracted from the X-ray correlation function leaving the
Si-Si peak free for more detailed analysis.

A calculation of the neutron interference and correlation functions,
corresponding to the peak shape function fit of Konnert and Karle (2) to the
first three interatomic distances, is compared with experiment in Figs. 3 and 4.
The peak widths are slightly broader than the experimental values and this is
substantiated by the fact that the calculated interference function decays more
rapidly. A similar calculation was made for the Si-O + O-O distribution of
Mozzi and Warren (1) but these authors' use of a set of δ-functions instead

of a continuous distribution led to spurious effects in $Qi(Q)$ at very high Q. In real space however their first (Si-O) peak is in very good agreement with the neutron data while the second (O-O) is too broad.

CONCLUSION

The experiments described provide a neutron interference function and correlation function of comparable quality to the X-ray work of Warren and Konnert and Karle, but of higher resolution, and will be of great value in deciding between the various models of the structure of vitreous silica. The main limitation on accuracy is in the application of corrections for departures from the static approximation and further work is needed in this area.

REFERENCES

(1) R. L. Mozzi and B. E. Warren, The structure of vitreous silica, J. Appl. Cryst. 2, 164 (1969).
(2) J. H. Konnert and J. Karle, The computation of radial distribution functions for glassy materials, Acta Cryst. A29, 702 (1973).
(3) E. A. Lorch, Neutron diffraction by germania, silica and radiation-damaged silica glasses, J. Phys. C 2, 229 (1969).
(4) R. N. Sinclair and A. C. Wright, to be submitted to J. Non-Cryst. Solids.
(5) R. N. Sinclair and A. C. Wright, Static approximation distortions and neutron time-of-flight diffraction using the Harwell linac, Nucl. Instr. Meth. 114, 451 (1974).
(6) A. C. Wright, The structure of amorphous solids by X-ray and neutron diffraction, Adv. Struct. Res. Diffr. Meth. 5, 1 (1974).
(7) A. C. Wright and A. J. Leadbetter, Diffraction studies of glass structure, Phys. Chem. Glasses 17, 122 (1976).

RAMAN SPECTRA AND ATOMIC CONFIGURATIONS IN VITREOUS SILICA

Stephen W. Barber (retired)
Owens-Illinois, Inc., Toledo, Ohio.

ABSTRACT

Raman spectra of Type III vitreous silica specimens S1 and S2 stabilized at 1000°C and 1300°C, were compared. Two spectra of S1 stabilized first at 1000°C, then at 1300°C, were compared with those of S2 stabilized first at 1300°C, then at 1000°C. All spectra contain a prominent 600cm^{-1}-band, which corresponds to normal modes in cristobalite and tridymite but not in quartz, as well as other broader bands at ~400cm^{-1} and ~800cm^{-1} where strong quartz bands occur. Comparisons indicate that significantly different equilibrium mixtures of crystal-like configurations occur in the glass at 1000°C and 1300°C and that approaches to these equilibria are reversible with respect to temperature.

These observations and many with respect to other structure sensitive properties reported in the literature support a general theory for glass-forming systems of which silica is considered the best and most studied example.

INTRODUCTION

The primitive cells of quartz, tridymite and cristobalite crystals of silicon dioxide are instances of three geometric arrangements of Si and O atoms stable under atmospheric pressure at temperatures between 0 and 2000°K. The reconstructive transitions, quartz $\rightleftharpoons$ tridymite $\rightleftharpoons$ cristobalite $\rightleftharpoons$ melt, are very slow and all three melting temperatures [T(q) ~1700°K, T(t) ~1945°K, T(c) ~2000°K] are time-dependent requiring many hours to approach the reproducible values reported in the literature (1). Entropies of melting per mole SiO_2 [S(q) ~0.546R, S(t) ~0.56R (estimate), S(c) ~ 0.577R] are smaller than R(ln 3) = 1.099R, the entropy of mixing for an equimolar solution of three species (2), where R is the gas constant. It is plausible, therefore, to hypothesize that the stabilized melts consist, at least in part, of mixtures of primitive crystal-like configurations (cells) stabilized by the entropy of mixing [-N(q) ln N(q) - N(t)ln N(t) - N(c)ln N(c)]R where N(q), N(t), N(c) are mole fractions and that the reported equilibrium melting points, T(q), T(t) and T(c) are, therefore, the time dependent liquidi of mixtures which require long times for the reconstructive transformations to achieve new equilibrium compositions after any sudden change in T.

This hypothesis implies a certain sequence of reconstructive processes during the formation of a stable glass from any one of its crystalline polymorphs and consequently certain features of the final arrangements of atoms in the glass. The Raman spectra reported here were intended to verify these implications independently and, in this way, test the hypothesis.

HYPOTHETICAL MODEL OF GLASS-FORMING PROCESSES IN SILICA

For convenience, let the stable arrangements of atoms (i.e. primitive cells) be called mers and let those species identified with a crystalline polymorph be called crystamers to distinguish them from other species, called liquimers, which conceivably may exist only in the melts and be incompatible with any crystal lattice [Ubbelohde's (3) anticrystalline clusters]. Glass formation can then be described as the copolymerization of two or more species of crystamer and, if present, liquimers while crystal formation can be described as polymerization of a single species. Formation of a stable silica glass from any one of its crystalline polymorphs would then consist of the following sequence of transformations:
1. During melting at T, crystamers of any one species transform slowly to other species until a mixture of equilibrium composition at T is approached. Rate of melting and composition of melt depend on T.
2. During cooling of the melt, the entropy of mixing compensates the transition entropies so that the mixture remains stable at T < T(melt) except at certain interfaces.
3. During cooling through the liquidus, a eutectic structure evolves and freezes in.
4. During stabilization of the glass at constant T(s), the mole fractions of different crystamer species slowly approach equilibrium values which are smooth functions of T(s) in the range $1200^\circ K$ to $\sim 2100^\circ K$ (4). Small irreversible changes in eutectic structure may also occur during such stabilization.

Raman spectra of stable silica glasses formed in this way would exhibit the mode frequencies characteristic of each crystal represented in the mixture where some frequencies might be shifted slightly. Spectra reported in the literature indicate that this is indeed the case (5). Furthermore, the relative intensities of different species bands should follow relative concentrations of the different mer species as functions of stabilizing temperature, T(s).

EXPERIMENTAL PROCEDURE

Raman spectra from three specimens of Type III silica as described by Bruckner (4) were studied. All specimens were supplied by Amersil, Incorporated, and are specified as Suprasil 2, made by pyrolysis of $SiCl_4$, containing ~ 1000ppm OH, ~ 100ppm Cl, $\sim$ zero other impurities, annealed at $1120^\circ C$ for 4 hours by Amersil. This type was chosen instead of Type I (<8ppm OH) because the starting material contains no crystals, it stabilizes faster and the OH and Cl impurities were assumed to increase the rates of transitions but have only secondary effects on relative values of N(c), N(t) and N(c).

Specimens S1, S2, S3 were plates $1"x\frac{1}{2}"x$ 1/8" having edges squared and polished to admit a laser beam without distortion or change of polarization. Specimens S1 and S2 were stabilized at 1000°C for 120 hours and 1300°C for 2 hours, respectively. The S3 specimen was a control stabilized at T(s) midway between 1000°C and 1300°C. After the Raman spectra of S1 and S2 were recorded, S1 was stabilized at 1300°C for 2 hours and S2 at 1000°C for 120 hours. Second spectra of each were then compared with first spectra to verify the assumed reversibility of stabilizing processes at each temperature.

RESULTS

Spectra of S1, S2, and S3 are superimposed in Fig. 1. The narrow well-separated bands centered at $600cm^{-1}$ are clearly between two isosbestic points (6) as indicated by arrows. This band is unambigously associated with prominent ir-active modes at slightly higher frequencies ($620cm^{-1}$) in cristobalite (7) and tridymite (8) both of which have very similar high melting temperatures. Quartz, which melts about 300° lower, contributes no intensities to this band but contributes strongly to the broad bands around $400cm^{-1}$ and around $800cm^{-1}$. Comparison of intensities at these three frequencies indicate two effects of the two stabilizations on each specimen:
i. Stabilizations at 1000°C and 1300°C lead to two different equilibrium distributions of modes. That at 1000° is the richer in quartz-like modes while that at 1300°C is the richer in modes characteristic of the higher melting tridymite and (or) cristobalite. The transitions between these two states are reversible functions of temperature.
ii. Relatively small irreversible increases in mode density at both $600cm^{-1}$ and $800cm^{-1}$ occurred during stabilization at both 1000°C and 1300°C while changes at $400cm^{-1}$ were more complicated. These effects are associated with the irreversible loss of OH from both specimens; ~0.03% from S1, ~0.04% from S2.

DISCUSSION

The stabilizing times, 2 hours and 120 hours, are less than required for complete stabilization. They were compromised to avoid surface devitrification and minimize other exceedingly slow irreversible processes such as conceivable changes in "texture" of the eutectic melt. As was assumed, the irreversible processes detected had only small secondary effects on equilibrium distributions of crystamer modes and did not obscure them.

The observed loss of OH during 120 hours at 1000°C plus 2 hours at 1300°C may have been the only significant irreversible process, but this cannot be assumed. A parallel experiment is in progress with specimens of Infrasil 2 having OH < 8ppm (from Amersil, Inc.) using stabilization times 2 to 3 times longer (9) to verify the presence of other irreversible processes in dry atmosphere.

The association of the $600cm^{-1}$ band with cristobalite-like atomic architecture is justified by Gaskell and Johnson (7). As they note, others have assigned this band to nonbridging oxygen defects of the kind assumed in neutron irradiated silica, but this assignment is put in serious doubt by their analysis showing that "It is natural, therefore, to associate the mode near $620cm^{-1}$ in vitreous

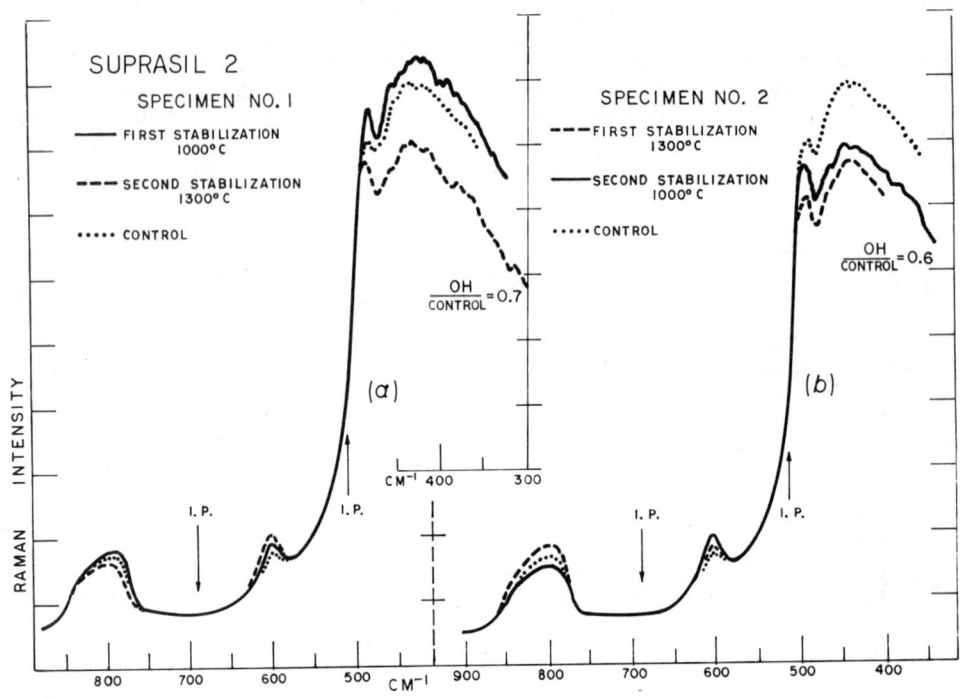

Fig. 1. Raman spectra of Suprasil 2 stabilized
at 1000°C and at 1300°C. Isosbestic points are
indicated by arrows labelled I.P. OH-content of
the control was ~1000ppm.

silica with the corresponding band in α-cristobalite ....." (p 188).
Increased intensity of this band and its closer approach to 620cm⁻¹
caused by neutron irradiation of vitreous silica are to be expected
due to the drastic quenching of thermal spikes from temperatures
which strongly favor cristobalite crystamers in the presence of
other species in lesser but sufficient amounts to assure the conditions for Raman activity described in (7).

Comparisons of the Raman spectra of vitreous silica with those
of its crystalline polymorphs (5a, 7) show that vitrification
shifts many modes from the depolarized bands at frequencies above
700cm⁻¹ into the many times more intense broad polarized continuum
below 500cm⁻¹. This shift can be understood as a consequence of
the eutectic structure assumed to evolve during melting and stabilization. Primitive elements of such a structure must be much
larger than any of the crystamers of which they are composed. For
this reason the modes localized in them are expected to have the
observed lower frequencies in broader everlapping bands belonging
more generally to A-classes among which A₁ contributes intense
polarized Raman bands like those observed. The well known shift
of modes out of this continuum to higher frequencies, expecially
into the next higher frequency band at 600cm⁻¹, caused by neutron
irradiation can be understood as due to the effects of thermal
spikes on the structure and composition of the eutectic material.

ACKNOWLEDGMENTS

Important contributions to this study by the following persons is
gratefully acknowledged: A. C. Kreutzer of Amersil, Incorporated,
for glass samples; colleagues E. J. Hornyak, R. R. Bebee, T. W.
Roberts and W. F. Reiter of Owens-Illinois, Incorporated, for
assistance in planning, executing and monitoring specimen preparation; and especially Professor Duane Burow and his assistants,
M. Diem and R. Hohmann of the University of Toledo Department of
Chemistry for taking the Raman spectra.

REFERENCES

(1) Sosman, R. B. (1965) The Phases of Silica, Rutgers Univ. Press.
(2) JANAF Thermochemical Tables (1971) NSRDS-NBS 37.
(3) Ubbelohde, A. R. (1965) Melting and Crystal Structure,
     Clarendon Press, Oxford.
(4) R. Bruckner, Properties and Structure of vitreous silica. I,
     II, J. Non-crystalline Solids 5, 123, 177 (1970).

(5) a. Wong, J. and Angell, C. A. (1976) Glass Structure by
        Spectroscopy, Marcel Dekker, Inc., New York.
     b. Barber, S. W. (1974) Frozen Configurations, Phonons and
        Anomalous Properties of Vitreous Solids, LD00028, Univ.
        Microfilms, Ann Arbor, Mich. A review.
(6) Brode, W. R. (1943) Chemical Spectroscopy, Wiley, New York.
(7) Gaskell, P. H. and Johnson, D. W. The optical constants of
     quartz, vitreous silica and neutron irradiated vitreous
     silica II, J. Non-crystalline Solids 20, 153-169 (1976).
     See p 188, especially.
(8) Plendl, J. N., Mansur, L. C., Hadni, A., Brehat, F., Henry,
     P., Morlot, G., Naudin, F., and Strimer, P., Low temperature
     far infrared spectra of SiO_2 polymorphs, J. Phys. Chem.
     Solids 28, 1589 (1967).
(9) Hetherington, G., Jack, K. H., and Kennedy, J. C., The vis-
     cosity of vitreous silica, Phys. and Chem. Glasses 5, 130
     (1964).
(10) Mallinder, F. P. and Procter, B. A., Elastic constants of
     fused silica as a function of large tensile strain. Phys.
     and Chem. Glasses 5, 91 (1964).

Electrostriction and Piezoelectricity of Thermally Grown SiO_2 Films

K. Misawa, A. Moritani, and J. Nakai

Department of Electronics, Faculty of Engineering,
Osaka University, Suita, Osaka 565, Japan

Abstract

It has been found in this experiments with the use of a new and simple optical technique that vibrating phenomena in Si MOS samples take place in a similar manner with a piezoelectric bimorph when an alternating electric field is applied to the samples and that this stress field is induced by both the electrostrictive and piezoelectric effect. The electrostrictive and piezoelectric constants of the SiO_2 film have been determined by the present technique.

Introduction

Thermally grown SiO_2 films are generally considered to be in amorphous state with small crystallites which are formed at localized nucleation centers such as impurities or imperfections on the Si surface. Piezoelectricity and/or electrostriction of these SiO_2 films have not been extensively studied, which may affect the electrical characteristics of MOS transistors. In this paper, We report a new and simple technique to study electromechanical effect which has been observed in Si MOS (ref.1 and 2) and determine the values of stress which results from the piezoelectricity and electrostriction.

Experimentals

Sample Preparation

The MOS samples were fabricated on the (100) surface of p- and n-type Si substrates. Carrier concentrations were 1.5×10^{15} cm^{-3} for the p-type and 3×10^{14} cm^{-3} for the n-type sample, respectively. SiO_2 films with the thickness of 2300 Å were grown by thermal oxidation at 1100°C in wet O_2 atmosphere. The metal electrodes were formed by evaporation of Al onto the SiO_2 films. One end of the sample was clamped on a brass plate and the other end was set free and the $\langle 0\bar{1}0 \rangle$ crystalline axis of Si was set in the x direction as schematically shown in Fig.1, where a medium (1) is the SiO_2 film with Young's module E_1, the piezoelectric constant d_{31} and the electrostrictive constant γ_{13}, and a medium (2) is the Si substrate with Young's module E_2. Thickness of the medium (1) and (2) are t_1 and $2t_2$, respectively. Length and width of the Si-SiO_2 system and width of the electrode on the SiO_2 film are 1, b, and c, respectively.

Measurements of Mechanical Vibrations

A He-Ne laser with the output power of 1 mW was used as a light source.

The laser beam was incident on the sample and the reflected light was focussed on to a photomultiplier after passing a slit which cuts half of the laser beam as shown in Fig.2. When the sample is vibrated by the alternative electric field, the direction of the reflected laser beam is vibrated. As a result, the intensity of the light after the slit is modulated. The output of a lock-in amplifier which detects the vibration of light intensity phase-sensitively was recorded as a function of frequency f_m of the ac modulation voltage ΔV_g. In this experimental arrangement, we measure $\Delta I_o/I_o$, where ΔI_o and I_o are the dc and ac component of the output of the photomultiplier, respectively. $\Delta I_o/I_o$ is expressed as follows(ref.2 and3):

$$\frac{\Delta I_o}{I_o} = \frac{24cLt_1 1\Delta\sigma_v}{\pi rbE_2(2t_2)^2}\left|\frac{\partial Y_1(\alpha_n,\nu,X)}{\partial X}\right| \quad (1)$$

where, L, r and $\Delta\sigma_v$ are a distance between the sample and the slit, a radius of the laser beam at the slit and the change in stress at the medium (1), respectively, and

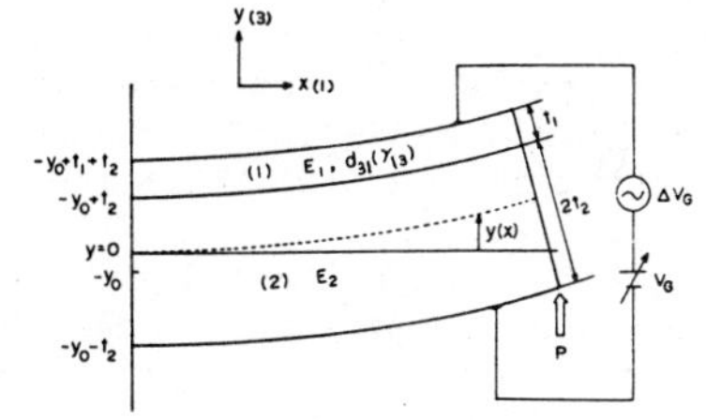

$$Y_1(\alpha_n,\nu,X)$$

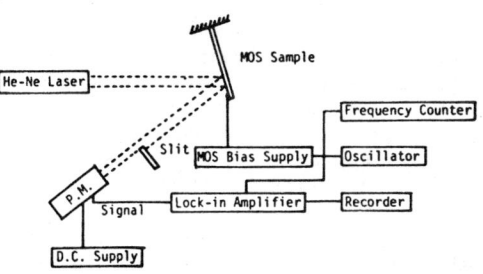

Fig.1. Schematic diagram of a clamped-free beam composed of SiO_2 and Si with rectangular shape.

Fig.2. Schematic diagram of the electronic and optical components of the system to measure the vibration phenomena.

$$= \frac{1}{\alpha_n^2\nu(1 + \cos\alpha_n\sqrt{\nu}\cosh\alpha_n\sqrt{\nu})}\left\{(\cos\alpha_n\sqrt{\nu} + \cosh\alpha_n\sqrt{\nu})(\cos\alpha_n\sqrt{\nu}X - \cosh\alpha_n\sqrt{\nu}X)\right.$$

$$\left. + (\sin\alpha_n\sqrt{\nu} - \sinh\alpha_n\sqrt{\nu})(\sin\alpha_n\sqrt{\nu}X - \sinh\alpha_n\sqrt{\nu}X)\right\} \quad (2)$$

here, $X = x/1$, $\alpha_n = (\frac{6\rho_2 S\pi^2 f_m^2}{E_2 t_2^3 b})^{\frac{1}{4}}1$, $\nu = f_m/f_n$.

The resonant frequency f_n is given by (ref.3),

$$f_n = \frac{\alpha_n^2 t_2}{2\pi 1^2}\sqrt{\frac{E_2}{3\rho_2}} \quad (3)$$

where ρ_2 is the density of Si.

Result and Discussion

Experimental Observations

Figure 3 shows an experimental result of $\Delta I_o/I_o$-f_m characteristics for the bias voltage V_g = 0.0 V, −1.0 V and −3.0 V at room temperature, where ΔV_g = 4.0 V_{ptp}. They have a narrow and sharp maximum at 585 ± 0.5 Hz. The lineshape is very similar to a resonant curve observed in the piezoelectric bimorph. The maximum point agrees with the fundamental natural frequency of the MOS sample which has been calculated to be 596 ± 30 Hz from eq.(3).

Figure 4 shows an experimental result of $\Delta I_o/I_o$-f_m characteristics in the region of the second resonant frequecy for V_g = 0.0 V and −1.45 V at room temperature in the p-Si MOS sample, where ΔV_g = 0.5 V_{ptp}.

Figure 5 shows experimental results of $\Delta I_o/I_o$-V_g characteristics. The f_m-values used were 585 Hz for the n-Si MOS and 1689 Hz for the p-type Si MOS. ΔV_g of 0.5 V_{ptp} was applied. In Fig.5 $V_c(n)$ = −1.55 ± 0.05 V and $V_c(p)$ = −2.85 ± 0.05 V was obtained, where the notations are described in the figure caption.

The vibrating phenomena observed in the present experiment indicate that stress field is generated along the surface of the Si MOS samples. When we experimentad on Au-Si Schottky-barrier system in place of the MOS sample, no vibrating phenomena could be observed. This gives an evidence that those stress field is induced in the SiO_2 films at the Si-SiO_2 interface.

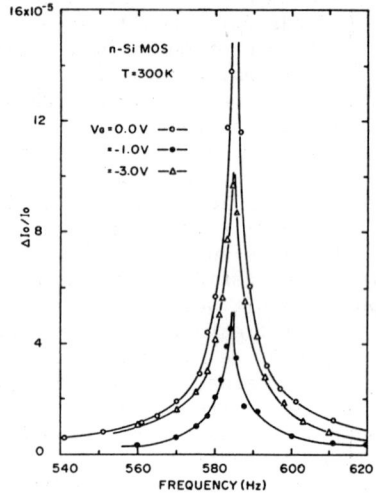

Fig.3. Experimental curves of $\Delta I_o/I_o$ as a function of frequency of the modulation voltage for V_g = 0.0 V, −1.0 V and −3.0 V in the n-Si MOS. $\Delta I_o/I_o$ for V_g = −3.0 V is negative, because it was in antiphase with respect to those for V_g = 0.0 V and −1.0 V.

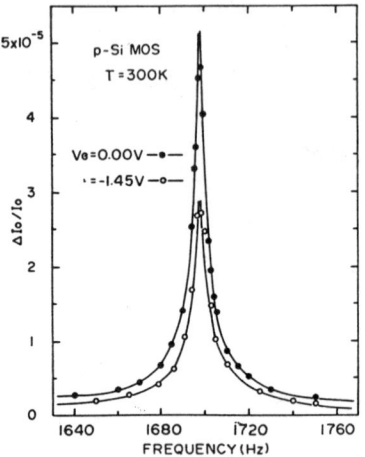

Fig.4. Experimental curves of $\Delta I_o/I_o$ as a function of frequency of the modulation voltage for V_g = 0.0 V and −1.45 V in the p-Si MOS.

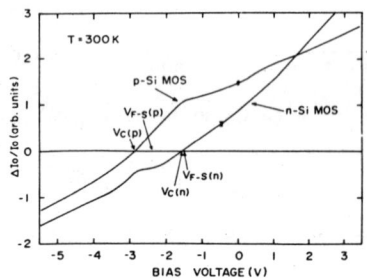

Fig.5. Measured curves of $\Delta I_o/I_o$ as a function of the bias voltage in the n-Si MOS sample, and the p-Si MOS sample. The points marked as V_{F-S}(n) and V_{F-S}(p) are the flat-band voltages of the n- and p-Si MOS sample obtained from the quasi-static technique, respectively. The points marked as V_c(n) and V_c(p) are the bias voltages where the curves cross the abscissa for the n- and p-type samples, respectively.

Determination of the Electrostrictive and Piezoelectric Constants

The equation for the piezoelectric effect in the present system modulated by ΔF_3 may be written in the quadratic approximation in the form

$$S_1' = s_{11}T_1' + d_{31}\Delta F_3 + 2\gamma_{13}F_{30}\Delta F_3 \quad (4)$$

$$D_3' = \varepsilon_{33}\Delta F_3 + d_{31}T_1' + 2\gamma_{13}F_{30}T_1' \quad (5)$$

where S_1', D_3' and T_1' represent the fundamental component of, strain, electric displacement and stress, respectively, s_{11}, d_{31}, γ_{13} and ε_{33} are the compliance, piezoelectric, electrostrictive, dielectric constant, respectively, and F_{30} and ΔF_3 are the static electric field and the modulation electric field. The stresses induced by piezoelectricity and electrostriction are given by

$$\Delta\sigma_{vp} = \frac{d_{31}\Delta V_I}{s_{11}t_1} \quad (6)$$

$$\Delta\sigma_{ve} = \frac{2\gamma_{13}V_I\Delta V_I}{s_{11}t_1^2} \quad (7)$$

respectively, where V_I is the voltage drop across the SiO_2 film.

For the purpose of determining the values of d_{31} and γ_{13}, transformation from Vg (or ΔVg) to V_I (or ΔV_I) has been made, using the experimental results of capacitance-voltage characteristics obtained from the quasi-static technique(ref.4). The obtained V_I-Vg curves are shown in Fig.6. It is evident from eq.(1) and (7) that in the case of the electrostrictive effect, $\Delta Io/Io$ must be zero at V_{F-I} (see the figure caption of Fig.6 for the notation.) and change in the same manner with V_I-Vg curves. It is found that the difference between the calculated curve shown in Fig.6 and the experimental curves in Fig.5 lies in a longitudinal shift. It is reasonable to consider that there exists piezoelectricity in SiO_2 films and it may contribute to $\Delta Io/Io$ by an constant amount and cause the longitudinal shift in $\Delta Io/Io$, as is seen from eq.(6).

The change in stress $\Delta\sigma_v$ can be obtained by performing the best-fit to the experimental curves in Fig.3 and 4 with the use of eq.(1). A best-fitted curve is shown in Fig.7 where L = 100 cm and r = 1.4 ± 0.1 mm. A correction

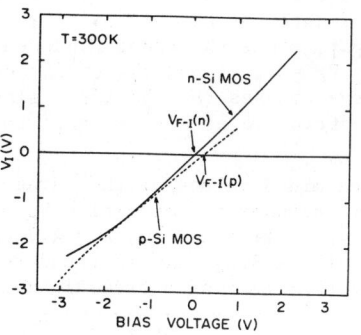

Fig.6. Calculated curves of the voltage drop across the SiO_2 film V_I as a function of Vg in the n- and p-type samples. The points marked as $V_{F-I}(n)$ and $V_{F-I}(p)$ are the bias voltages which produce a vanishing electric field at the SiO_2 for the n- and p-type MOS sample, respectively.

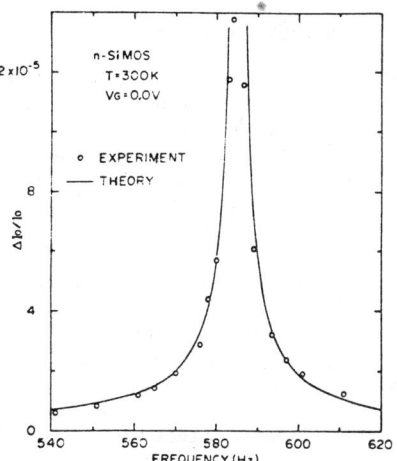

Fig.7. The best-fi to the $\Delta Io/Io$-fm characteristic at Vg = 0.0 V for the n-type MOS sample.

148

on $\Delta\sigma_V$ due to the difference of effective length of the electrode from the sample length has been made. d_{31} can be determined from the value of $\Delta\sigma_V$ at $V_g = V_{F-I}$, since the change induced by the electrostrictive effect is equal to zero at $V_g = V_{F-I}$, and γ_{13} can be determined from the point $V_c(n)$ where $\Delta I_o/I_o$ $-V_g$ curve crosses the abscissa, since the change in stress induced by the electrostrictive effect is equal to that induced by the piezoelectric effect at V_c.

The determined values of the parameters are listed in Table I, where $\Delta\sigma_V$ was determined under the condition $V_g = 0.5\ V_{ptp}$, $t_1 = 2300$ A for n-type, 2800 A for p-type, $2t_2 = 250\ \mu m$, b = 4.7 mm and l = 22.5 mm. The obtained values of d_{31} of the SiO_2 film is found to be about three orders of magnitude smaller than that of a crystallized quartz.

Sample	n-Si MOS	p-Si MOS
$\Delta\sigma_V$ (dyn/cm^2)	$4.8\pm1.0\times10^3$	$1.0\pm0.2\times10^3$
d_{31} (cm/V)	$9.2\pm1.9\times10^{-14}$	$2.1\pm0.4\times10^{-13}$
γ_{13} (cm^2/V^2)	$7.6\pm1.5\times10^{-19}$	$1.1\pm0.2\times10^{-18}$

Table I. Experimentally determined values of the change in stress $\Delta\sigma_V$, the piezoelectric constant d_{31}, the electrostrictive constant γ_{13} of the SiO_2 film where the subscripts 1 and 2 indicate the direction x and y, respectively as shown in Fig.1.

Acknowledgements

The authors would like to thank Professor C.Hamaguchi and Dr.M.Yamada for helpful discussions during the course of this work. They would also like to thank Research Laboratory of Matsushita Electronics Corp. for providing the Si MOS samples. This work was partially supported by the Grant-in-Aid for Scientific Research on "Surface Electronics" from the Ministry of Education of Japan.

References

1. K.Misawa, A.Moritani and J.Nakai, Japan.J.Appl.Phys. 15 (1976) 1309.
2. K.Misawa, A.Moritani and J.Nakai, Japan.J.Appl.Phys. 15 (1976) 2103.
3. H.Nukiyama and T.Suzuki, J.Inst.Elect.Commun.Engrs.Japan No.231 (1942) 367.
4. M.Kuhn, Solid-St.Electron. 13 (1970) 873.

CRITICAL NEED FOR $S(k,\omega)$ DETERMINATIONS IN AMORPHOUS
SiO_2: CALCULATION OF PHYSICAL PROPERTIES VIA FROZEN
LIQUID PHONONS

*

Edward Siegel
*International Atomic Energy Agency,Vienna,Austria
*Present Address: Molecular Energy Research Institute, &
Ashway Ltd., New York, New York

ABSTRACT

The Siegel(1)-Percus-Yevick(2) theory of liquid phonons in liquids is extended to the concept of frozen liquid phonons (FLP) in amorphous solids following Takeno and Goda(3), and to powdered crystalline solids following Ohshima
et.al.(4), Nakanishi and Matsubara(4) and Matsubara and Siegel(4). The connection of the SPY theory's liquid phonons to those of the independently derived generalized R.P.A.theory of Hubbard and Beeby(5) and the scattering theory of Egelstaff(6), as well as to corroberating experimental data of glass
and liquid phonon dispersion relations(esp. SiO_2 and BeF_2) (6)(8-13) is put
on firmer ground. Connection is also made with the recent work of Tanttila(21).
New characteristic features of amorphous solids in terms of their FLP dispersion relations: isotropic non-monaticity, an effective FLP temperature, scaling of these parameters and derived properties with respect to temperature,
pressure and composition(allowing for configurational and substitutional disorder), a law of corresponding states between glasses, liquids and powders,a
distinctive(set) of disorder parameter(s) and a new model with predictive
capacity from $S(k,\omega)$ determinations of Anderson localization in glasses are
for the first time presented. Application is attempted for amorphous SiO_2 and
BeF_2.

INTRODUCTION

Leadbetter et. al.(12), for amorphous SiO_2 and BeF_2, has shown via inelastic
neutron scattering, that the phonon spectra, dispersion relations and density
of states (d.o.s.) of amorphous solids closely approximate those of the crystalline state except for certain, yet unnoticed, qualitative differences.
Analogously (8-13)many authors have observed this striking, yet qualitatively
different, similarity between liquids and crystalline solids, though the work
of Lovesey and Coply(14) cautions against uncritical universal acceptance of
this finding.
Takeno and Goda(3) have developed a FLP theory of glasses remniscent of the
liquid phonon theories of deBoer(13), Hubbard and Beeby(5),Egelstaff(6),
Singwi(15), Egelstaff(16) and Rahman(17),   Bose and Goertze(17) and de
Lieuw(17). In fact Hubbard and Beeby claim that their liquid phonon dispersion relation is equally valid for glasses, liquids, hot solids and plasmas.
Similarly Ohshima er. al.(4) and Matsubara et. al.(4) have conjecturally developed a FLP theory of crystalline powders and applied it successfully to
melting temperature depression and superconducting transition temperature enhancement in powders using a harmonic and anharmonic Einstein model.

Siegel(1), using the Percus-Yevick(2) theory of liquid colletive coordinates,

has developed a simple one-parameter liquid phonon theory of liquids. This is
a simple minded model compared to the R.P.A. approach of Hubbard and Beeby,
or the alternate scattering and molecular dynamics theories (6,16,17).It
depends upon a simplified hard sphere model and S(0), and so contains no atom
dynamics nor atom coordination. It also contains no transverse polarization
liquid phonons, both observed via neutron scattering(8-13) and predicted(15,
16) in liquids and glasses. Yet this simple theory predicts successfully
sound velocity, specific heats, optical properties, conductivity, melting entropy and S(0) in liquid metals and alloys.

Well known in glasses(and powders(19)) are the striking(and similar) anomalies in specific heat(3,18,21), thermal conductivity, acoustic properties(22),
etc. Takeno's successful explanation of the specific heat anomaly in glasses,
follower more recently by Tanttila, in which all types of composition and
bonding glasses behave identically indicating that the anomaly is an inherent property of the amorphous state, based on the FLP model indicate that to
zeroth order "a glass is a glass is a glass". This will be reinforced here
and modified to " a glass is a glass is a glass is a liquid is a powder" ie.
they are each special cases of a FLP disordered phase model of matter.

EXISTENCE OF A DISORDER PARAMETER AND FLP TEMPERATURE

One should stress at the outset, even soberly appreciating the Lovesey-Cop-
ley cautionary note, that the similarity in theoretical and experimental
models and data in glasses, liquids and polycrtstalline powders implies that
the configurational disorder(and substitutional disorder if alloy glasses are
considered) can be described by a similar model.

Firstly, we must note(8) that polycrystalline solids must be $\underline{k}$ or $\underline{r}$ space averaged to give the same isotropic properties(including FLP spectra) as liq-
uids. We here assume that this is approximately true for glasses. Then there
will exist one isotropic FLP dispersion relation for a glass, given alternately in terms of $S(k,\omega)$ in the SPY collective coordinate, Hubbard-Beeby generalized R.P.A., Egelstaff scattering or Singwi-Rahman molecular dynamics theories as

$$\omega^{SPY}(k) = k \left[k_B T / m S(0) \right]^{1/2} \tag{1}$$

$$\omega^{HB}(k) = \hbar^2 k^2 / 2 m S(k) \tag{2}$$

$$\omega^{E}(k) = \left[k^2 (k_B T / m S(k)) - i k^2 V(k,\omega) \right]^{1/2} \tag{3}$$

where $V(k,\omega)$ is a damping function determining the FLP anharmonicity and
these FLP dispersion relations change as temperature, pressure and composition vary. It will determine all properties of the glass involving collective
oscillations.

How then does this disordered system, the glass, differ from an ordered crystalline solid? Firstly, its FLP dispersion relation is isotropically non-
monatonic. For glasses this has only been experimentally explicitly illustrated by Leadbetter,Apling and Wright (12) for a-BeF$_2$, though strongly implied
for a-GeO$_2$ by them and for a-SiO$_2$ by Leadbetter and Stringfellow(12). From
such a FLP dispersion relation one can now define qualitatively unique disorder parameter(s).Isotropically, each glass is characterized by a particular
wavevector cut-off $|k_c|$, its functionally corresponding frequency maximum cut-
off ω_c (seen as an effective FLP temperature), a curvature $(\partial \omega / \partial |\underline{k}|)_{k=k_c}$ and

and a negative slope $(\partial\omega/\partial|\underline{k}|)_{k>k} < 0$. These are three independent dis-or-
der parameters for the longitudinal FLP mode. If transverse FLP modes exist,
(12,24) each will also be characterized by three more independent dis-order
parameters.

In the Hubbard-Beeby, Egelstaff, Singwi-Rahman, Takeno-Goda and Tanttila models there also exist a roton-like minimum at higher wavevectors $k > k_c$. The
existence of this is still a moot point due to the large anharmonicity of the
FLP modes expected, and seen (6,12),ie. the large value of $Im\,\omega(k > k_c)$. But
if these roton-like modes are considered or found to be well defined in an
amorphous system, they can easily be included in this FLP theory be defining
further qualitatively distinctive dis-order parameters around the roton min-
imum at k_{c2}; $|\underline{k}_{c2}|$ and its associated ω_{c2}, the roton minimum curvature
$(\partial^2\omega/\partial|\underline{k}|^2)_{k=k_{c2}}$ and the positive upward slope beyond the roton minimum
$(\partial\omega/\partial|\underline{k}|)_{k > k_{c2}}$. These two slopes, of course, are not constant, and so must
be suitably defined near their respective FLP maximum and roton minimum. It
is similarly not clear whether the resolution of neutron scattering measurements beyond the FLP maximum is sufficient to resolve roton branches of different polarizations (though Leadbetter et,al. and Egelstaff find no such
good resolution), so here we purely concentrate on the FLP maximum peak only.
However at a later stage the roton minimum, with or without branch polarization, can be included and will not alter this theory qualitatively.

Associated with the FLP maximum at k_c is the vanishing of the FLP group velocity at k_c, where the FLP dispersion relation slope becomes zero(this would
also happen at the roton minimum k_{c2} if it exists and can be experimentally
resolved). This has been observed in liquid Rb(25) experimentally for FLP modes.

The meaning of the FLP maximum frequency ω_c is of great importance. Assuming
$\hbar\omega_c = k_B\theta$ allows us to define an effective FLP temperature(1), a one- parameter characterization of the FLP modes. This philosophy is identical with
that of the Debye theory. Our rationale is simply that since the Debye theory
works so well for such a wide range of crystalline solids, independent of
bonding type or crystal structure, an analog theory might be a valid first
order approximation, independent of glass structure or bonding. Since $S(k,\omega)$,
$S(k)$ or $S(0)$ will be parametrically functions of temperature, pressure and
composition, this effective FLP temperature will be also. This is not an
artificial compromise of the concept of the effective FLP temperature as a
one-parameter theory of glasses, but a natural consequence of its definition
via experimental liquid structure factor measurements. We have chosen the
simplest definition of a glass effective FLP temperature and not optimized
the definition to enhance calculational ability of the model. This could be
done if desired at a later stage, since it is not claimed that this definition is unique nor even that a one-parameter FLP theory is the best approach;
it is merely the simplest first step.

The effective FLP temperature can now be used in any standard crystalline
solid calculation of any property in which a Debye temperature appears. We
must caution that because of the qualitative difference between FLP modes
and crystalline solid phonons very different predictions of properties may
result in such an approach for glasses. What we have done is to map a glass
(or liquid or powder) into a FLP gas as one does in the Debye theory of crystalline solids; once this is done one can proceed just as in the Debye
theory. The FLP modes are assumed harmonic,to a first approximation, because
we concentrate on their properties for k around k_c, and not too far above.
Singwi(13,15),in liquids, has introduced finite liquid phonon lifetimes and

gotten the same results as Egelstaff[13], who introduced liquid phonon mode polarization explicitly. This may mean that in glasses we have the option of considering FLP anharmonicity or FLP mode polarization branches; the inclusion of both may be redundant. But here we avoid such as yet unanswered, albeit important, questions. In SPY theory neither is included for liquids, while in the Hubbard-Beeby, Egelstaff, Singwi and Raman theories anharmonicity is either included or found.

SCALING OF FROZEN LIQUID PHONON PROPERTIES IN GLASSES

From equations (1)-(3) we see, since $S(k,\omega) = S(k,\omega; T,P,x_j)$ experimentally, (where x_j is the concentration of the jth impurity in an alloy glass), that they should more correctly be rewritten as explicit functions of temperature, pressure and composition. For the Egelstaff form (eqn.3) we must additionally assume a knowledge of $V(k,\omega)$, the damping function, as a function of T,P and x_j. Thus, in any of these FLP possible forms, variation of T,P or x_j produce a continuous family of congruent FLP dispersion relations. This in turn will lead to a scaled set of whichever disorder parameters one allows into the model one constructs of a glass. Thus one can have, derived from the experimental T,P or x_j dependence of liquid structure factors experimentally observed via neutron scattering, the following possible scaled set of disorder parameters: $k_c = k_c(T,P,x_j)$, $\omega_c = \omega_c(T,P,x_j)$, $\theta = \theta(T,P,x_j)$ for the FLP maximum at k_c with additional similar scaling of roton minimum parameters possible. We term this congruence, or near congruence, of FLP dispersion relations and their derived dis-order parameters scaling. This has been observed for a series of metallic glass alloys as functions of temperature and composition by Guntherodt et. al.(25) of the Basil group, but is a commonly accepted property of all glasses. What distinguishes this approach to scaling in glasses is that the same FLP model with experimentally derived (via $S(k)$) scaling of dis-order parameter(s) holds at all temperatures, pressures and compositions.

LAW OF CORRESPONDING STATES WITH LIQUIDS AND POWDERS

The FLP model will lead quite naturally to a law of corresponding states between FLP modes, between their associated dis-order parameters, and hence between glasses, liquids and powders. The deviation of a Debye phonon will take place gradually, first to a FLP as the crystal is powdered, then at higher temperature to a glass FLP, and finally at still higher temperature to a liquid phonon(1). Scaling will take place within each phase. Scaling between the phases is the law of corresponding (FLP) states. We can say nothing at present about the transition region between these phases. Presumably this will involve evolution of FLP modes or between FLP and liquid phonon modes. This is reminiscent of the Kawasaki critical slowing down of modes used in some critical phenomena theories, and is presently being studied.

ANALOG PHENOMENOLOGICAL CALCULATION OF GLASS PROPERTIES

We have carefully avoided the important theoretical question of how to get from the pair potential, local bonding model, through the pair distribution function to the liquid structure factor, and hence to the FLP dispersion relation. This was done on purpose. While this consideration is extremely important, the alternate theories of glass structure and bonding cannot be generally applied, yet what is clearly needed is a general model valid for all glasses. In this model we utilize the (experimental) analog computer of neutron scattering $S(k,\omega)$ determinations to derive the FLP mode dispersion relations. The local bonding and coordination number are implicit in these measurements, but explicitly avoided. We cannot claim that this is a first principles theory, but merely a generally applicable proceedural model of glasses. It allows wide applicability to many systems at the expense of first principles rigor.

FERMI GLASS FORMATION AND ANDERSON LOCALIZATION

We briefly treat this application(23). Anderson(20) developed a theory of the onset of electron(hole) localization via configurational(lateral) and substitutional(vertical) disorder. Inherent is this theory is the coordination number occurring in the minimum metallic conductivity and localization distance. But the coordination number is the first neighbor peak integral of the pair distribution function, which itself is the Fourier transform of $S(k,\omega)$. We thus can use (1)-(3), depending upon which form is adopted, to express the coordination number, and hence the minimum metallic conductivity and localization distance, in terms of the FLP dispersion relation. Physically this means that the electron-FLP backscattering for $k > k_c$ can tend to localize the electron by trapping it in a cloud of backscattering FLP modes. This resembles polaron trapping, but with qualitatively distinct FLP modes comprising the polaron.

APPLICATION TO a-SiO_2, GeO_2 and BeF_2

We here merely outline the derivation of the FLP modes in these amorphous solids. Space limitations preclude any actual calculations of properties. Using the Leadbetter et. al(12) neutron scattering data on a-SiO_2, GeO_2 and BeF_2 we derive their FLP modes. Those for BeF_2 are explicitly experimentally derived (12). For this glass two average acoustic branches are derived from neutron scattering data, with parameters $k_c = 7.5$ A^{-1}, $\omega \cong 25$-45 cm^{-1} for the lower branch, $\omega_c \cong 55$-65 cm^{-1} for the upper branch. Optical branches are not derived and the exact perscription for the derivation is not given. The alternate equivalent derivation would be via (1)-(3) with an explicit $S(k)$ from neutron data. This lack of explicit perscription precludes here a derivation of FLP dispersion relation for SiO_2 and GeO_2. However, were an explicit liquid structure factor given, our FLP perscription could be applied. Hence the great need for $S(k)$ determinations in glasses. With thus derived FLP modes many properties of the glass can then be calculated via an analog to the Debye theory, using θ. We should note that the expected anharmonicity, ie. large value of Im $\omega(k)$, does appear beyond k_c so that the question of a roton minimum is a moot point. The FLP mode representation is merely a mapping of the configurational disorder into a qualitatively distinct group of coupled collective coordinate modes. We then neglect the FLP modes beyond k_c and proceed with a harmonic FLP model. Were the anharmonicity small, we could include the roton minimum region, but it hardly seems worthwhile in view of the ill defined FLP dispersion relation at high wavevectors.

REFERENCES

(1) E.Siegel, Jnl. Phys. Chem. Liq. 4,205,211,217,233,259(1975);9,(1976).
(2) J.K.Percus and G.T.Yevick, Phys. Rev. 110, 1 (1958).
(3) S.Takeno and M.Goda, Prog. Theo. Phys. 48, 5, 1468 (1972).
(4) K.Ohshima,T.Fujita and T.Kuroishi, preprint,Nagoya Univ.(1976); Jnl. Phys. Soc.Japan 40,1,90(1976); 41, 4, 1234 (1976).
(5) J.Hubbard and J.Beeby, Jnl. Phys.C,2,2,556 (1969).
(6) P.Egelstaff,An Introduction to the Liquid State,Academic Press,N.Y.(1967)
(7) P.W.Anderson, S.F.Edwards, M.Lax and T.Lukes-private communications.
(8) U.Dahlborg and K.Larsson, Ark.Fys. 33,271 (1965).
(9) S.J.Cocking, Dissertation, University of London (1966).
(10) K.Skold and K. Larsson, Phys. Rev. 161,102 (1967).
(11) S.H.Chen,C.J.Eder,P.Egelstaff,B.Haywood and F.Webb,Phys.Letters 19,269(19
(12) A.J.Leadbetter and M.W.Stringfellow, Inelastic Neutron Scattering,I.A.E.A

154

Vienna (1972); A.J.Leadbetter, A.C.Wright and A.J.Apling, Amorphous Materials,
    Ed. by R.H.Douglas and B.Ellis, Wiley, N.Y. (1972).
(13) J.deBoer, Phonons and Phonon Interactions, Ed. by T.Bak, Benjamin,N.Y.,
      (1966).
(14) J.R.D.Copley and S.W.Lovesey, Repts. Prog. Phys. 38,4,461(1975).
(15) K.S.Singwi, Phys.Rev. 136,A969(1967); Physica 31 (1965).
(16) P.A.Egelstaff, Rept. R4101, A.E.R.E. Harwell (1962).
(17) A.Rahman, Inelastic Neutron Scattering ,I.A.E.A., Vienna (1972).
(18) R.C.Zeller and R.O.Pohl, Phys.Rev.B4,2029(1971).
(19) L.Slutsky and L.Halsey, Physical Chemistry,An Advanced Treatise, Ed. by
      H.Eyring, Academic Press,N.Y.(1969); T.Matsubara and M. Nakanishi,-
      to be published; T. Matsubara and E.Siegel, Statphys-13 (to be publi-
      shed in Proc.Israel Acad.Sci)and International Conference on Lattice
      Dynamics(to be published in Jnl. de Physique).
(20) P.W.Anderson, Phys.Rev.109,1492(1958).
(21) W.Tanttila, Phys.Rev.Letters39,9,554(1977).
(22) H.S.Hunklinger,Habilatationschrift,Max Planck Institut fur Festkorper-
      forschung,Stuttgart; K.Dransfeld, W.Arnold, H.S.Hunklinger, U.Strom-
      private communications.
(23) E.Siegel, International Conference on Lattice Dynamics (1977); Statphys-
      -13 (1977); to be published in Phys.Rev.
(24) P.Marcus and T.Schneider-private communication.
(25) M.Tosi-private communication.
(26) H.Guntherodt-private communication.

PROPERTIES OF LOCALIZED SILICON-DIOXIDE CLUSTERS IN LAYERS OF DISORDERED SILICON ON SILVER

Cheol Jung Kim, K. Shu, H. Oona, and S. O. Sari
Optical Sciences Center, University of Arizona, Tucson, AZ 85721

ABSTRACT

Reflectivity spectra have been studied of silane-vapor-deposited silicon on metallic silver in the wavelength range between 0.7 and 15 μm. The infrared properties of local centers consisting of silicon dioxide clusters and silicon-oxygen complexes in silicon have been examined using an effective medium approximation. This approach can be justified in films containing relatively large concentrations of oxygen between approximately 0.5 to 15 volume percent as measured by proton-induced x-ray emission analysis. Oscillator strengths are estimated for prominent oxygen-induced vibrational transitions in films and are compared to the corresponding parameters measured in silicon dioxide for vibrational modes near 8.3 and 10 μm in the near infrared. The form of the complex dielectric function ε for electronic transitions near the fundamental edge is also given and its dependence on oxygen content is discussed.

***

When oxygen is present in silicon in relatively large fractional volume ratios, the optical spectrum of the semiconductor is modified in both the near-infrared and near the semiconducting band edge. It is the object of this paper to examine some systematic features of these changes which we have observed and which we believe can be explained from relatively simple considerations. The production of silicon by reactive means at high temperature is likely to lead to direct precipitation of any oxygen present to produce silicon-dioxide as the semiconductor lattice is formed. Thus, the result could be an inhomogeneous composite consisting of small glass droplets condensed in a randomly disordered matrix of silicon. Alternatively, a rather more homogeneous medium could be formed with a composition SiO_x where $x \leq 2$. Although it does not seem straightforward to distinguish silicon dioxide in silicon from oxygen in silicon, since their infrared absorption bands are similar, the behavior of silicon dielectric properties near its fundamental gap could in fact allow such a distinction to be made.

The reflectance spectra to be discussed were obtained using methods similar to other recent studies of silicon absorption (1,2). Though perhaps provisional, it seems reasonable to proceed in our analysis by assuming that our silicon films contain silicon dioxide precipitates and consequently to use a suitable method to analyze the resulting mixed solid. For this purpose, we have chosen to use an effective medium approximation (3,4) which can be written for a two-medium composite as

$$X \frac{\varepsilon_1 - \varepsilon}{\varepsilon_1 + 2\varepsilon} + (1 - X) \frac{\varepsilon_2 - \varepsilon}{\varepsilon_2 + 2\varepsilon} = 0 \tag{1}$$

Here ε_1 and ε_2 are the complex dielectric constants of the two media in question and X is the concentration of the first medium. This result holds at optical frequencies for random inclusions that are small compared to a wavelength (4) and yields an effective dielectric function ε. Within the limits in which we utilize this result, no percolation effects are expected.

In this study we examined a number of silane vapor-deposited films of disordered silicon on silver which had been formed at temperatures near 620°-650°C and which contained varying amounts of oxygen. Since the layers investigated had a thickness small compared to the absorption length of light incident on them, account was taken of film interference using a method previously given (5,6) to extract information on oxygen absorption. From location and depth of fringe structure both real and imaginary parts respectively of the layer dielectric function could be extracted from our reflectance data. The structure observed in the wavelength range examined between 0.6 and 15 μm is dominated by two oxygen absorption resonances near 8.3 and 9.7 μm and by the electronic energy gap transitions in silicon.

Fits to reflectance line shapes near the resonance transition frequencies yield room-temperature optical parameters as given in Table 1. These are combined with oxygen concentrations measured by proton-induced x-ray emission to give approximate oscillator strengths. These parameters can be compared to those in SiO_2 (fused quartz) in its opaque region by measuring its normal-incidence reflectance between 7 and 11 μm and fitting the result to a model composed of two Lorentz oscillators and a background dielectric constant of 2.25. This analysis is admittedly simplistic but can be motivated by assuming that the vibrational components studied are composed of symmetric and antisymmetric vibrations of bridge oxygen connecting tetrahedral SiO_4 clusters in glass. This model is thus analogous to a water-like molecular picture of localized oxygen centers in silicon consisting of Si_2O and in fact gives comparable oscillator strengths. Moreover, it is consistent with the idea that optical structure in a random network may be caused by configurations which are invariant in such a network. This criterion could well be satisfied by localized structure in a random lattice.

TABLE 1   Near-Infrared Oxygen Absorption Parameters in Silicon Containing Relatively Large Volume Fractions of Oxygen

Silicon Layer	N/V (cm^{-3})	$\lambda_0(\mu)$	$4\pi N/V$ $e^{*2}/\mu(sec^{-2})$	$\gamma(sec^{-1})$	$(e^*/e)^2$
1	3×10^{20}	9.7	4.5×10^{26}	2×10^{13}	11
		8.2	1.2×10^{26}	1×10^{13}	2
2	6×10^{20} (estimated optically)	9.7	8×10^{26}	2×10^{13}	
		8.2	1.9×10^{26}	1×10^{13}	
3	6×10^{21}	9.7	4.8×10^{27}	2×10^{13}	5
		8.2	2.3×10^{26}	1×10^{13}	0.3
SiO_2 (quartz)	5×10^{22}	9.14	2.4×10^{28}	2×10^{12}	3
		8.45	3.5×10^{27}	2×10^{13}	0.5-1

Notes: N/V is the atomic oxygen concentration in each case above. For layers 1 and 3 oxygen concentration was measured using x-ray emission analysis. In case 2 it is taken from optical data. Resonance wavelengths given for silicon layers are approximate. Line strength and concentration values are estimated to hold within a factor of 1.5. Oscillator strength determination is estimated to be accurate within a factor of 2. Values given here are numerically consistent with Table I of Ref. 6.

The line strengths of Table 1 are given in terms of one-particle parameters without reference to complications due to clustering. However, we may include such effects by examining Eq. (1) in more detail. Assuming ε_{1I}, $\varepsilon_{2I} \ll \varepsilon_{1R}, \varepsilon_{2R}$ and $X \ll 1$ we may write

$$\varepsilon_I = \varepsilon_{1I}\delta_1{}^2 X + \varepsilon_{2I}(1 - \delta_2{}^2 X) \tag{2}$$

where $\delta_1 = 3\varepsilon_{2R}/(2\varepsilon_{2R} + \varepsilon_{1R})$ and where $\delta_2{}^2 = \delta_1{}^2 - 3(\varepsilon_{1R} - \varepsilon_{2R})^2/(2\varepsilon_{2R} + \varepsilon_{1R})^2$. These depolarization factors δ are of order unity. Using the result that the average oxygen concentration is $N/V)_{avg} = 2 N/V)_{SiO_2} X + N/V)_{Si_2O} (1 - X)$, we find from Eq. (2) that the oxygen oscillator strengths are readily determined in the cases that solely clusters (ε_1) or localized centers in a silicon background (ε_2) dominate the absorption. These numbers are of the same order of magnitude as mentioned above. However, if both phases are present, the fraction of each phase contributing to optical structure cannot easily be determined from the vibrational transitions discussed.

Concurrently, additional information on clusters can nonetheless be obtained by careful examination of the silicon fundamental edge by determining both ε_R and ε_I from reflectance measurements. An example of such a fit is shown in Fig. 1.

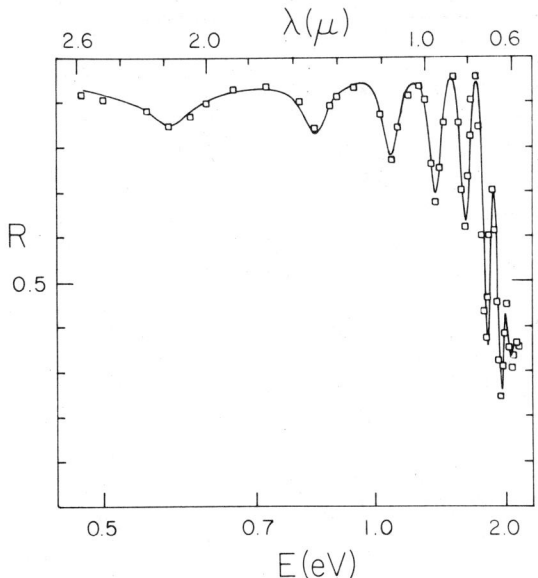

Fig. 1 Comparison of experimental (square points) and calculated (solid line) reflectance curves for a thin layer of silicon on silver near the lowest electronic gap. For wavelengths less than 0.8 μm a Gaussian phase average of the reflectance has been included in the numerical fit to account for film surface roughness. The roughness parameter obtained in this case is ∿200 Å.

In Fig. 2 we have plotted the imaginary dielectric function in this interband region for silicon containing increasingly larger fractions of oxygen as given by the three cases listed in Table 1. The absorption onset is observed to shift to higher energy. This is in qualitative agreement with Eq. (2) assuming that the absorption is dominated by silicon excluded volume between clusters, although the magnitude of the absorption decrease given in this model is too small. Moreover, if we examine the real part of the dielectric function (Fig. 3), a minimum is observed to develop in semiconductor films with increased oxygen concentration. Again, in the spirit of Eq. (2) we may write

$$\varepsilon_R = (\varepsilon_{1R} - \delta_3 \varepsilon_{2I}^2) \, \delta_1 X + \varepsilon_{2R}(1 - \delta_1 X) \tag{3}$$

where $\delta_3 = (2\delta_2^2 - 3)/3\varepsilon_{2R} > 0$. This model predicts a minimum as the SiO_2 cluster fraction increases as appears to be observed, although the magnitude change obtained experimentally is again larger than predicted. The observed behavior in this latter instance is thus seen to result from a mixing of real and imaginary dielectric components of the constituent phases of the resulting composite medium. More detailed examination of such scattering effects in multiple-phase structures will be discussed elsewhere.

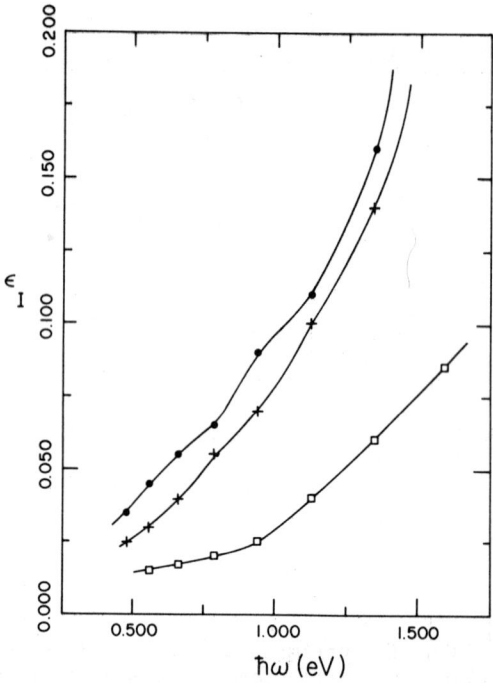

Fig. 2   Imaginary part of the dielectric function ε_I versus photon energy for films of increasing SiO_2 cluster content. Cases 1,2, and 3 of Table 1 are •, +, and □, respectively. The numerical values of ε_I obtained for low oxygen content suggest that our layers are composed of disordered silicon (cf. Ref. 1).

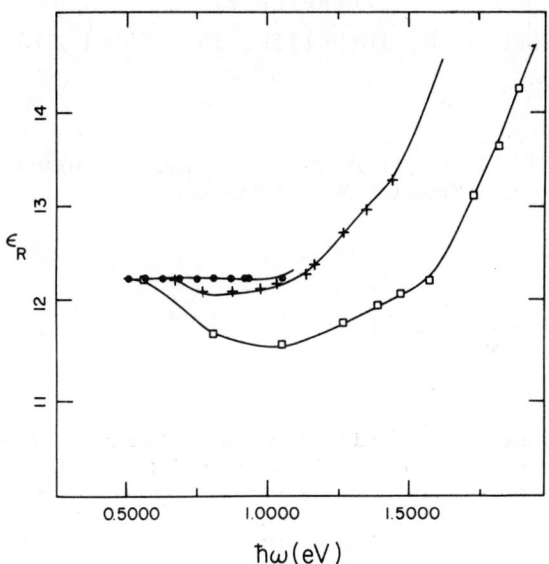

Fig. 3 Real part of the dielectric function ϵ_R versus photon energy for the same cases as in Fig. 2 showing a minimum observed in ϵ_R near the silicon first electronic edge as a function of increased oxygen concentration. The absolute value of ϵ_R for $\omega \ll \omega g$ is only approximately determined.

This work has been supported by the Joint Services Optical Program through the University of Arizona and in part by the National Science Foundation.

REFERENCES

1. M.H. Brodsky and A. Lurio, Phys. Rev. B9, 1646 (1974).
    M.H. Brodsky, R.S. Title, K. Weiser, and G.D. Pettit, Phys. Rev. B1, 2632 (1970).

2. G.A.N. Connell and J.R. Pawlik, Phys. Rev. B13 787 (1976).
    G.A.N. Connell, R.J. Temkin, and W. Paul, Adv. Phys. 22 (1973).

3. R. Landauer, J. Appl. Phys. 23, 779 (1952).

4. I. Webman, J. Jortner, and M.H. Cohen, Phys. Rev. B15, 5712 (1977).

5. S.O. Sari, P. Hollingsworth Smith, and H.S. Gurev, Phys. Rev. B15, 4817 (1977).

6. S.O. Sari, P. Hollingsworth Smith, and H. Oona, J. Phys. Chem. Solids (in press).

160

CHAPTER IV
DEFECTS AND IMPURITIES IN THERMAL SiO$_2$

THE PROPERTIES OF ELECTRON AND HOLE TRAPS IN THERMAL
SILICON DIOXIDE LAYERS GROWN ON SILICON*

D. J. DiMaria
IBM Thomas J. Watson Research Center
Yorktown Heights, NY 10598

ABSTRACT

Capture, optical and thermal discharge, and location of trapped electrons and holes in thermal silicon dioxide layers grown on silicon are reviewed. Charge trapping is discussed in oxides grown under various conditions (for instance, steam versus dry oxidation) and in oxides where impurites have been added by ion implantation or grown into the film at elevated temperatures. Characteristics of the trapping sites in terms of their spatial location, capture cross section, photoionization cross section, trap density, and energy depth in the forbidden gap are summarized for known traps. Techniques such as capacitance-voltage (C-V(1)) and photocurrent-voltage (photo I-V(2)) used in trap characterization are reviewed.

I.   INTRODUCTION

Thermal silicon dioxide SiO$_2$ is a very important material in the electronics industry where it is extensively used as a high quality insulator. An important example is the gate oxide of an insulated gate field effect transistor (IGFET) formed by the oxidation of single crystal silicon. SiO$_2$ films formed by thermal oxidation of Si are amorphous and contain lower number of impurties than other forms of SiO$_2$. For example, mobile Na$^+$ levels lower than 10^{10} ions/cm^2 ($<10^{15}$cm^{-3}) can be obtained with current standard semiconductor technology.

SiO$_2$ has a large band gap of approximately 9 eV (3-5) which forms blocking contacts at interfaces with metals and semiconductors (6-11). This is illustrated in Fig. 1 where the energy band diagram for zero electric field in the insulator is shown. The sandwich configuration of the metal-SiO$_2$-Si (MOS) structure is one that is used almost exclusively to contact electrically the SiO$_2$ layer.

Once an electron is injected into the SiO$_2$ conduction band, it will move rapidly with a mobility of 20-40 cm^2/V-sec (6,12,13) towards the contact which has been biased positively. As the electron traverses the oxide layer, it will scatter off phonons which can limit the energy that the electron will acquire in the applied electric field. Measured scattering lengths from 10-60 Å can be found in the literature (14-16). Electron injection from a contact over or through the large interfacial barrier can be accomplished by using light (internal photoemission (6-11)) or by using large electric fields (Fowler-Nordheim tunneling (17-21)). Other means use the generation of hot carriers in the Si layer (22-27).

As a free conduction band electron moves across the oxide layer, it can be captured by sites in the SiO$_2$ band gap formed by impurities or defects which are called traps (28,29). This is shown

---

*This work was supported in part by the Defense Advanced Research Projects Agency, the Department of Defense, and monitored by the Deputy for Electronic Technology (RADC) under contract No. F19628-76-C-0249.

schematically in Fig. 2. Thermal SiO_2 films will capture less than one out of every 10^5 electrons that flow through a film of approximately 1000 Å (capture probability of 10^{-5} (30-32)). However, impurities such as Al (33-36), Na^+ (32), H_2O (22, 30-32, 37-39), As (40), P (40), W (2,41-43), etc., can be added which could increase the capture probability to as high as unity. Radiation damage or effects caused by high energy photons (x-rays or vacuum ultraviolet light (VUV)) or electrons appear in the SiO_2 layer as the generation and subsequent partial trapping of holes in sites that existed prior to irradiation (44-46) or as the generation of neutral centers (31, 47, 48) both of which can trap electrons. Energetic ions cause atomic displacement damage (49) which can also act as trapping sites (33).

Holes can be injected and trapped in SiO_2 (22-25, 50-53). However, hole injection from the contacts into the SiO_2 valence band is more difficult because of the larger energy barriers they see compared to electrons (see Fig. 1). Usually holes are generated in the SiO_2 layer itself by using photons with energies greater than the band gap (44,45,54). Holes move relatively slowly compared to electrons with a mobility determined from transit time measurements that is $\leq 2 \times 10^{-5}$ cm^2/V-sec at room temperature (55). This slow motion of holes is believed to be limited by the presence of shallow traps in the band gap near the top of the SiO_2 valence band (55-60). The microscopic hole mobility could be much larger. Deeper traps are found near the interfaces which capture some of the holes indefinitely at room temperature (44,46,51,61-64) and lead to failure of many solid-state devices in radiation environments.

Trapping and detrapping of electrons and holes in thermal amorphous SiO_2 films with or without the presence of impurities or defects will be discussed here. Most of the experimental work to date has been concerned with electron capture on sites in the SiO_2 band gap. Some experiments exploring optical and thermal detrapping properties have also been performed.

This review will be divided into distinct sections covering briefly all aspects of trapping and detrapping in SiO_2. Section II discusses the physics of carrier capture into a trapping site and discharge from a trapping site to the conduction or valence band. Section III reviews experimental techniques used to measure capture cross sections, photoionization cross sections, trapped charge location, number of traps, energy depth of traps, etc. Section IV lists and discusses all trapping centers in amorphous thermal SiO_2 known to date.

II.   PHYSICS OF TRAPPING AND DETRAPPING

A.   Capture (Trapping)

Trapping sites are characterized by a capture cross section σ_c which is a measure of the ability of the potential of the trap to attract a free carrier and capture it into a bound state. Traps can be coulombic attractive, neutral, or coulombic repulsive with the various capture cross sections having ranges varying from $10^{-12} - 10^{-14}$ cm^2, $10^{-14} - 10^{-18}$ cm^2, and $10^{-18} - 10^{-21}$ cm^2, respectively (6,26,28,30-32,34,35,38-40,42,46-48,52,53,61,65-82). Energy is given off during capture usually by phonon emission (non-radiative capture), although photons can be given off with a much smaller probability (radiative capture (28,67,69)). Radiative capture cross sections are small ($10^{-18} - 10^{-21}$ cm^2) for any trap potential and similar in magnitude to photoionization cross sections as expected from detailed balance (28,67,69).

(i)   Coulombic Attractive Capture

When a thermal free carrier gets within the capture radius r_c of a coulombic attractive center which is about 100 Å at room temperature (28,72), it can be captured by losing energy in steps (for non-radiative capture) through hydrogenic like states of the trapping site emitting a single phonon at each step during the first stages of the cascade process (65-68). Figure 3a shows the potential energy distribution for a coulombic attractive center and compares it to a neutral center (Fig. 3b) and a coulombic repulsive center (Fig. 3c). The capture radius for the attractive center is that radius at which the coulomb potential is $2kT/q$ from the bottom of the conduction band edge where k is Boltzman's constant, T is the absolute temperature, and q the electron charge (28). At room temperature the capture cross section σ_c should be appproximately equal to πr_c^2 or $\approx 10^{-12}$ cm^2. The carrier also has a finite probability of absorbing phonons and being re-emitted to the conduction band as it steps down between the excited levels (65-68). As the temperature is lowered, the capture cross section would increase since the capture radius increases and the probability for re-emission becomes less because of decreasing numbers of phonons (28,65-68).

The capture cross section of coulombic attractive sites depends on the applied field through the shrinkage of the capture volume, and this has been shown to be proportional to the $-3/2$ power of electric field (70,71,78). This dependence has been observed experimentally (78,79) as shown in Fig. 4.

If the carriers are not at thermal equilibrium but are "hot", they will be more difficult to capture (68,69,70). Carrier heating can occur from the energy gained from the electric field being larger than that lost to phonon scattering in the SiO$_2$ layer, or carriers can initially be injected from a contact with excess energy. Examples of coulombic attractive traps are the 2.4 eV trap (6,79), electron capture by or annihilation of trapped holes (46,51,61), and positively charged centers found by Ning et al. (26,77,78) which may be one of the two centers formerly mentioned.

(ii)   Neutral Center Capture

The majority of trapping centers in SiO$_2$ fall into this class of sites which have capture cross sections between 10^{-14} cm^2 and 10^{-18} cm^2 (2,30-32,34-36,39,40,42,43,47,48,66,74,76). These traps are neutral initially and are usually deeper than 4 eV, at least for electrons, as measured from the bottom of the conduction band edge (2,30,36,40,42). Their capture radius is on the order of atomic dimensions (10^{-8} cm), and therefore σ_c would be about 10^{-16} cm^2 (66). The capture cross sections should exhibit weaker temperature and electric field dependence than coulombic attractive centers since the capture radius should be a weaker function of both (66).

These traps could capture either electrons or holes possibly with multiphonon emission (76). Multiphonon emission is a less probable process than the single phonon cascade emission used for coulombic attractive capture (65,66). The capture cross section would decrease with temperature as the lattice motion decreases (68,76) which reduces the probability of the energy level of the trap from moving to and crossing a band edge where a free carrier can be captured. Subsequently, the energy level would move to a new energy position in the band gap relaxing by multiphonon emission (68). The potential energy distribution for this type of center is shown in Fig. 3b. Examples of neutral centers would be W related sites (2,42,43); ion implanted Al, As, and P related sites (34-36,40); and water related sites (30-32,39).

(iii)   Coulombic Repulsive Capture

Little is known about these sites. They can be formed, for example in the case of free electron capture, by a site which has already trapped one or more electrons and, therefore, has a net negative charge as has been observed in semiconductors (28,65,68). An incoming electron (hole) would have to first surpass the repulsive energy barrier of a negatively (positively) charged site before it would be within the region of short-range attractive forces (28,65,68) as shown in Fig. 3c. Capture then probably proceeds by multiphonon emission because the repulsive sites will not have excited states of large orbit like coulombic attractive capture (65).

These sites could also be dipole type centers (which might overlap with the neutral trap classification) where the site consists of both negatively and positively charged centers in close proximity. A dipole

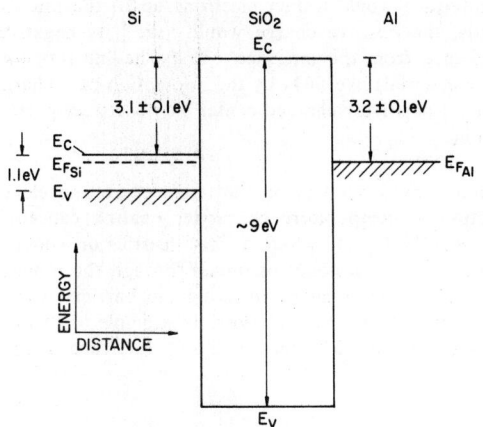

Fig. 1. Zero field energy band diagram of MOS structure constructed from references 6-11.

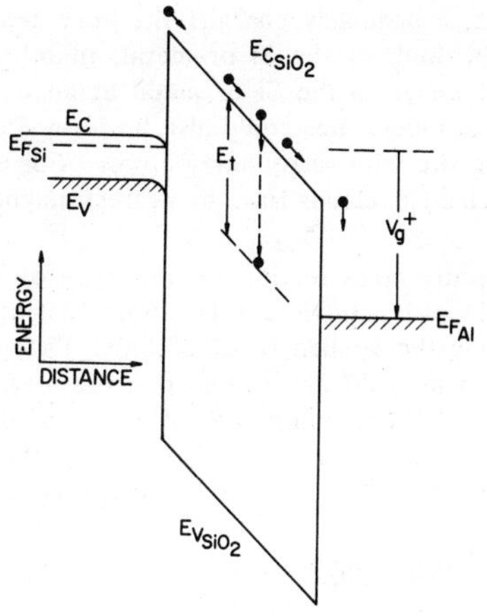

Fig. 2. Energy band diagram of MOS structure under positive voltage bias V_g^+ illustrating schematically electron capture by a trapping state located energetically in the SiO_2 bandgap.

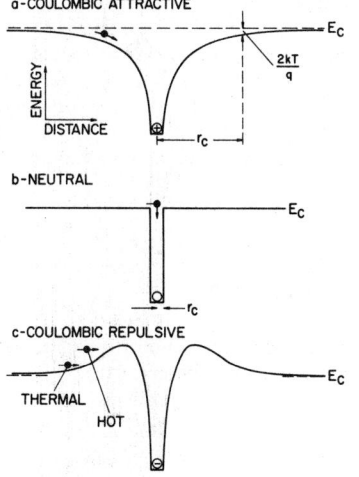

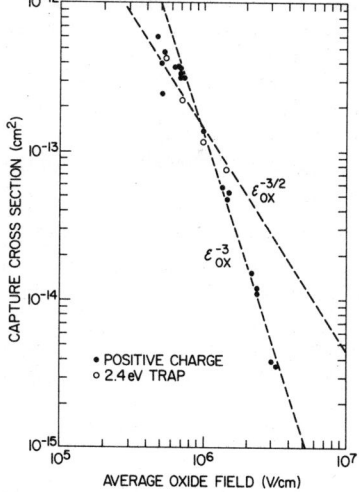

Fig. 3. Schematic representation of potential energy distribution for (a) coulombic attractive, (b) neutral, and (c) coulombic repulsive capture of SiO_2 conduction band electrons.

Fig. 4. Capture cross section for coulombic attractive traps as a function of the average oxide field (voltage bias ÷ oxide thickness). Data for 2.4 eV trap are from reference 79 and data for positively charged centers are from reference 78.

trap would have directionality, being more attractive in one direction than another (74). Dipole traps also could be formed by nearness of charged trapping sites to the metal-SiO_2 or Si-SiO_2 interface. For example, a positively charged site very near an interface would attract electrons up to the interface from the bulk of the Si or metal; in other words, the positive charge would "see" its negatively charged image in the Si or metal at an equal distance from the interface. Traps having properties similar to dipole sites could also be formed in polar materials like SiO_2 by the interaction of a charged trapping site with surrounding atoms. For example, a positively charged center would attract portions of the electron clouds from its nearest neighbor atoms.

The capture cross section for this type of trap should be dependent on the temperature and electric field (28,68). It should be an increasing function of temperature or carrier heating caused by increasing the applied field (28,68). This can be seen in Fig. 3c where a "hot" carrier or a thermal carrier in a lattice at elevated temperatures can more easily surmount or tunnel through the repulsive barrier of this trapping site. Distortion of the repulsive potential by lowering the barriers with an applied electric field would also increase the capture probability (68). A possible example of this type of center is electron capture by positively charged sodium ions Na^+ near the Si-SiO_2 interface (32).

B.   Detrapping

Once an electron is captured by a trapping site, it can be discharged by phonons, photons, field emission (tunneling), or hot carrier impact ionization (83). These various processes are illustrated in Fig. 5. From photodetrapping or thermal detrapping experiments, photoionization cross sections σ_p, thermal ionization rates, and trap energy depths E_t can in principle be determined (83). Trap energy depths determined by these two means never seem to follow the simple relation (29) of $E_{t_{opt}}/E_{t_{therm}}$ = K/n^2 where $E_{t_{opt}}$ ($E_{t_{therm}}$) is the energy depth of a trap from an energy band determined by photodetrapping (by thermal detrapping), n is the index of refraction, and K the static dielectric constant where the quantity K/n^2 = 1.7 for SiO_2. Regardless of the trap depth measured by photodetrapping, all traps tend to discharge at temperatures between 100-400°C as if thermal discharge were a characteristic of the SiO_2 layer and not the trapping sites. Fig. 6 shows the measured values of electron photoionization cross sections as a function of photon energy for the presently known optically accessible traps in SiO_2 which are the 2.4 eV trap (84) and traps related to ion-implanted As or P (40,85). The energy thresholds in this figure are indicative of the electron trap depth from the bottom of the conduction band edge. The photoionization cross section should not be strongly dependent on the applied electric field or lattice temperature at energies well above threshold (86).

At high enough applied fields, electric field emission of trapped carriers from traps will occur if the dielectric or breakdown strength of SiO_2 is not exceeded. Examples of high field emission have been seen on traps related to W (42) and Na^+ near the Si-SiO_2 interface (32).

Trapped charge ionization by hot carrier impact has been used as an explanation for charge detrapping in two cases: the 2.4 eV trap (79,87) and the radiation induced neutral centers (48).

C.   First Order Rate Processes

The trapping-detrapping of a thermal carrier on a trapping site can be described by a first order kinetic rate equation. This equation for electron trapping and detrapping is given by

$$\frac{dn_t}{dt} = k_c (N-n_t) - k_p n_t - k_t n_t \tag{1}$$

where $k_c = n_c v_{th} \sigma_c$,

$$k_p = F_p \, \sigma_p \quad ,$$

$$k_t = N_c \, v_{th} \, \sigma_c \, \exp(-E_t/kT) \quad ,$$

n_t is the trapped charge density, N is the total trap density, n_c is the conduction band electron density, v_{th} is the thermal velocity, σ_c is the capture cross section, F_p is the local photon flux, σ_p is the photoionization cross section, N_c is the effective density of states in the conduction band, and E_t is the trap energy depth from the conduction band edge. A similar equation can be written for hole trapping and detrapping. The first term on the righthand side of the equation represents the capture rate of free thermal conduction band electrons while the last two terms represent trap discharge rates by photons and phonons, respectively. Of course, not all phenomena are operative for a given trapping center in a given experiment, and this equation is merely used to give the overall picture. If there is more than one type of trapping center present, then a series of equations like Eq. 1 are needed to describe the trapping-detrapping kinetics of each center.

Most experiments have been concerned with electron capture. With energetically deep traps with respect to kT and carrier injection techniques (particularly avalanche injection from the Si which will be described in the next section) which require no light, Eq. 1 reduces to

$$\frac{dn_t}{dt} = n_c v_{th} \sigma_c (N - n_t) \tag{2}$$

In order to relate this equation to the macroscopic current per unit area J measured in the external circuit, the assumption made in almost all cases is

$$\frac{J}{q} = n_c v_d \approx n_c v_{th} \tag{3}$$

where v_d is the drift velocity of an electron. The drift velocity magnitude for the applied fields of interest from .2 to 3 MV/cm has been shown by Hughes to be approximately equal to the value of the thermal velocity which is $\approx$ ·10^7 cm/sec (13). At these fields, an electron should essentially move in straight lines possibly gaining some significant energy with respect to kT from the electric field before having a collision with a phonon. At average fields greater than 1 MV/cm, Ning has argued from experimental evidence that electron heating by the field affects the value determined in this manner for a capture cross section of a coulombic attractive center (78). If the electron is significantly heated, $\sigma_c v_{th}$ in Eqs. 1 and 2 should be replaced by the average of the product of the microscopic cross section and carrier velocity over the actual hot electron energy distribution. However, in any case a macroscopic capture cross-section can always be defined (78) using $\sigma = C/v_d$ where $C = \sigma_c v_{th}$ is the capture rate constant (66). For hole capture in SiO_2, the low mobility of holes is believed to be trap limited (55-60); therefore, care must be taken in relating the measured current to microscopic hole capture. Also care must be used for trapping very near an injecting interface where the supply of impinging electrons on the interfacial region, some of which can be trapped, can be greater than those that can energetically surmount the interface barrier and be measured as a current in the external circuit.

As will be shown in the next section, the time evolution of trapped charge per unit area is the experimental quantity which is measured in all experiments. Therefore, Eq. 2 with the subsitution of Eq. 3 must be integrated over the SiO_2 thickness L to give

$$\frac{dQ}{dt} = \frac{J\sigma}{q}(qN_t - Q) \tag{4}$$

which has the solution for constant J

$$Q(t) = qN_t(1 - \exp(-t/\tau_c)) \tag{5}$$

where $Q = q \int_0^L n_t \, dx$, $N_t = \int_0^L N \, dx$, and $\tau_c = (J\sigma/q)^{-1}$ is the capture time constant. In Eq. 4, it is assumed that J and σ are not spatially dependent functions. This can be tested experimentally (32). J will be constant spatially provided the capture probability (ratio of the number of trapped charges to the number of injected carriers) is low (less than at least 10 to 20% (2,32,88)).

Detrapping phenomena are usually tested separately after the traps have been filled. This can be done by shutting off the carrier flow injected from the contacts and applying light, heat, or electric fields to the SiO$_2$ layer. As an example, consider photoionization of trapped charges for which Eq. 1 reduces to

$$\frac{dn_t}{dt} = -F_p \sigma_p n_t \tag{6}$$

or integrating over oxide thickness

$$\frac{dQ}{dt} = -q \sigma_p \int_0^L F_p n_t \, dx \tag{7}$$

where it is assumed that σ_p is not spatially dependent. F_p in SiO$_2$ layers thicker than at least 500 Å is spatially dependent because of optical interference in the multilayer MOS structures used in most experiments (89-92). On very thin films (<500 Å) or very thick films (>10,000 Å) with traps in the bulk of the oxide film, optical interference effects are not as critical (89-92) and Eq. 7 can be approximated by

$$\frac{dQ}{dt} \approx -\sigma_p F_p Q \tag{8}$$

which has the solution

$$Q(t) = Q(0) \exp(-t/\tau_p) \tag{9}$$

where $\tau_p = (\sigma_p F_p)^{-1}$ is the photoionization time constant. No significant retrapping of carriers after photoionization was assumed in the previous analysis which is a good approximation for electron traps in SiO$_2$ (capture probabilities are usually less than 10%).

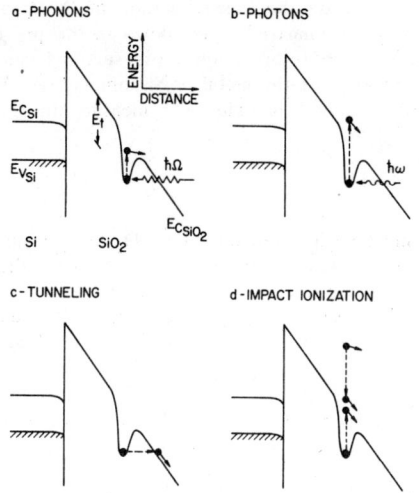

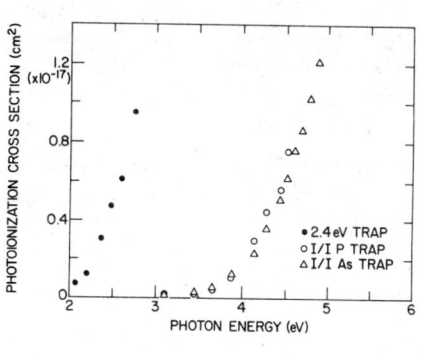

Fig. 6. Photoionization cross section as a function of photon energy for the 2.4 eV trap taken from reference 84 and for trapping sites related to ion implanted P and As taken from reference 85.

Fig. 5. Energy band diagram illustrating schematically several mechanisms for electron detrapping from a trap depth E_t by (a) phonons of energy $\hbar\Omega$, (b) photons of energy $\hbar\omega$, (c) tunneling, and (d) electron impact ionization.

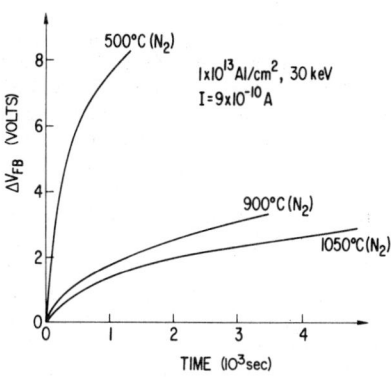

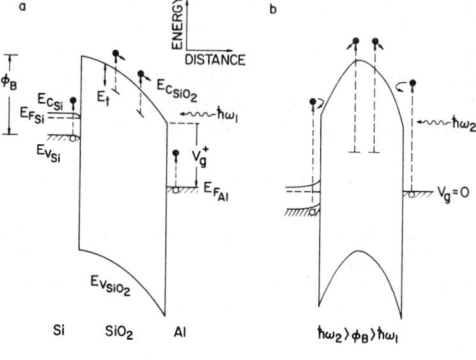

Fig. 7. Flat-band voltage shift ΔV_{FB} under a constant avalanche injection current I as a function of time for an SiO_2 layer ion implanted with 1×10^{13} Al ions/cm^2 at 30 keV. Various curves indicate effect of annealing on electron trapping. Taken from reference 34.

Fig. 8. Energy band diagram illustrating schematically photodetrapping of electrons (a) from shallow traps and (b) from deep traps where the built-in field due to the negative trapped charge is used to suppress internal photoemission from the contacts.

III.   EXPERIMENTAL TECHNIQUES

In this section, various experimental techniques used to measure the time evolution of the trapped charge density Q as trapping or detrapping occurs will be summarized. As shown in the previous section, microscopic quantities of interest like capture or photoionization cross sections can be determined from dQ/dt. Techniques to inject charge carriers from the metal or Si contacts into SiO_2 in order to study capture processes will also be summarized as will techniques to measure the spatial and energetic distributions of the traps.

A.   Carrier Injection

In order to study trapping kinetics, electrons or holes must be injected into the SiO_2 layer from the contact or internally generated in the SiO_2 layer. In most cases, electron injection without simultaneous hole injection can be accomplished because of the lower energy barriers for electrons at the interfaces as compared to holes (see Fig. 1). However, because of the high energy barriers (2 eV or greater depending on metal used (6-11)), special techniques must be used to inject carriers over or through the interface energy barriers.

For electron injection, internal photoemission (6-11) or avalanche injection (22-24) are typically used because of the simplicity of the MOS capacitor structures required. Internal photoemission uses light of energies between 2-5 eV to excite electrons from the contact at a negative voltage over the interfacial energy barrier into the oxide conduction band. The electrons then move towards and are collected by the contact at a positive voltage except for those electrons which may be trapped. Internal photoemission requires semitransparent metal electrodes to allow light penetration and can be used to inject carriers from either interface.

Avalanche injection can also be used to inject carriers over the Si-SiO_2 interface energy barrier. The Si substrate is driven into deep depletion by means of voltage pulses (22-24). In deep depletion, a large electric field from ionized donors or acceptors is developed between the bulk of the Si and the Si-SiO_2 interface by depleting this region of mobile charge carriers (electrons or holes). For electron injection, thermal electrons in bulk p-type Si drift or diffuse into the high field region where they can gain energy from the field until they impact ionize and lose energy by creating another electron and a hole across the Si bandgap of 1.1 eV. This impact ionization process is multiplicative and gives a hot electron distribution in the Si at the Si-SiO_2 interface. Some of the electrons can surmount the interface energy barrier and move into the SiO_2 conduction band. A similar injection of holes can be accomplished with n-type Si (22,24,47). Avalanche injection can give very high injected currents (10^{-3} A/cm^2 is easily obtainable) and does not require light and therefore semitransparent metal electrodes. However, only the Si can be used as an injecting contact.

Other techniques similar to avalanche injection which use the Si to generate hot carriers using visible light and/or electric fields have also been used to inject carriers (25-27). These structures are more complicated to fabricate since they require a MOS in a field effect transistor (MOSFET) configuration and not a simple capacitor as do the techniques previously described. The MOSFET configuration is required to allow the Si substrate to be biased independently of the SiO_2 layer. Verwey and then Young used epitaxial Si layers on bulk Si to form p-n junctions that could be independently forward biased to inject carriers into a high field region between the junction and the Si-SiO_2 interface (25,27). Ning and Yu used a MOSFET with a polycrystalline Si electrode in place of the metal electrode which could transmit visible light to the underlying bulk Si to generate minority carriers which would then gain energy in the high field region (26).

High electric fields (≥ 7 MV/cm) can also be used to inject electrons from the contacts via Fowler-Nordheim tunneling (17-21). However, electric fields in the SiO_2 layer of this magnitude tend to favor field ionization of trapped charges and SiO_2 breakdown which are not desirable for trapping studies.

Hole creation in the SiO_2 layer requires electron-hole generation across the SiO_2 bandgap. With VUV light of energies between 10-17 eV (44-46,54), electron-hole pairs are generated in the SiO_2 layer within $\lesssim$ 100 Å of the metal-SiO_2 interface. With the metal electrode biased positively, the more mobile electrons quickly flow to the metal contact leaving the less mobile holes moving towards the Si contact as a source of carriers for trapping studies.

If more than one trapping center is present, varing the injected current density J from low to high levels can be used as a means of separating out the different traps by their capture cross section (26,32). Traps with the largest capture cross sections would fill at the lowest current levels first. Then by increasing the injected current, the traps with the next largest cross section would fill. This process would be continued until at the highest current level, the traps with the smallest capture cross sections are filled.

B.   Trapping

In order to study the build-up of trapped charge Q, two main techniques are used both of which sense the internal fields created by this charge. The first and oldest is the capacitance-voltage (C-V) technique which uses changes in the electric field dependent capacitance of the Si layer near the Si-SiO_2 interface to measure the first moment of the charge distribution (1,93). The first moment is obtained from the measurement of the change in the voltage applied to the metal electrode necessary to return the total capacitance, after charge trapping, to its initial flat-band value (capacitance when the field in the Si near the Si-SiO_2 interface is zero). This voltage change referred to as the flat-band voltage shift ΔV_{FB} is related to the first moment of bulk trapped charge by

$$\Delta V_{FB} = - \frac{\overline{x} \, Q}{\varepsilon} \qquad (10)$$

where $\overline{x}$ is the centroid of the trapped charge distribution measured from the metal - SiO_2 interface and $\varepsilon = K\varepsilon_0$ is the static dielectric constant K of SiO_2 multiplied by the permittivity of free space ε_0 (1,93). By measuring the changes in ΔV_{FB} with time as trapping occurs, the capture cross section σ and total number of traps per unit area N_t can be determined from a data fit to Eq. 5 provided $\overline{x}$ is known and not changing significantly in time. Fig. 7 is an example of ΔV_{FB} as a function of time for trapping on sites related to ion implanted Al in SiO_2 (34). Even when $\overline{x}$ is not known, σ and an effective trap density $N_{t,eff}$

$$N_{t,eff} = \frac{\overline{x}}{L} N_t \qquad (11)$$

can still be obtained (26,31,32,34). Eq. 5 assumes that the injected current denity J is constant in time; however, Ning has shown that this is not a necessary condition to determine σ and $N_{t,eff}$ (26). If the trapping probability is nearly one (every injected electron gets trapped), the techniques described here cannot be used. However, techniques developed by Yun and Arnett (94,95) could be used for this special case. In MOSFETs shifts in the threshold voltage of the transistor which are equivalent to the flat-band voltage shifts for MOS capacitors for bulk oxide charging can also be used (1,93).

The second method of measuring trapped charge used quite recently is the photocurrent-voltage (photo I-V) technique (2,36,40,46,61,96-98). This technique, although similar in concept to C-V, has the advantage that it allows separate determination of $\overline{x}$ and Q as bulk trapping proceeds. The photo I-V technique uses voltage shifts of photocurrent-voltage characteristics after charging, referenced to an initial uncharged state. The photocurrent or internal photoemission current in these MOS structures is interface limited (14) and senses changes in internal fields caused by charge trapping. From simple electrostatics (2), the voltage shift for bulk trapped charge with the Si injecting is

$$\Delta V_g^+ = - \frac{\bar{x}\, Q}{\varepsilon} = \Delta V_{FB} \tag{12}$$

and with the metal injecting is

$$\Delta V_g^- = \frac{(L-\bar{x})\, Q}{\varepsilon} \tag{13}$$

Obviously, the solution of Eqs. 12 and 13 gives

$$\frac{\bar{x}}{L} = (1 - \frac{\Delta V_g^-}{\Delta V_g^+})^{-1} \tag{14}$$

$$Q = \frac{\varepsilon}{L} (\Delta V_g^- - \Delta V_g^+) \tag{15}$$

However, this technique does require a semitransparent metal electrode to allow light penetration. The light intensity can be reduced with screens or filters to allow only very small injected currents in the oxide layer to minimize additional trapping during the measurement (2,46). Reduced intensities also minimize perturbations of the trapped charge due to photodetrapping (2,46). Photo I-V measurements which require only very low fields (.2 to 2 MV/cm) and hence cause minimal field ionization of trapped charges have an advantage over Fowler-Nordheim current-voltage (dark I-V) measurements which give similar information but require very high electric fields (>6 MV/cm) (99,100). Dark I-V measurements are also sensitive to localized effects such as asperities on an interface (101) because of the strong electric field dependence of the current (17).

A third technique which has been used in some special cases is the photodepopulation technique (79,83,84,87,90,91,102,103). Here the trapped charge must be optically accessible since measurements of the total charge collected by one electrode or the other after photodetrapping are used to determine $\bar{x}$ and Q. By filling the traps for a certain amount of time, photodetrapping, refilling the traps for a different amount of time, and then continuing the sequence, data similar to Fig. 7 are obtained from which σ and N_t can be determined (79). This technique could also be used if the traps were discharged thermally.

C. Detrapping

Optical and thermal energy depths of trapping sites and photoionization cross-sections can be determined by measuring flat-band voltage shifts (30,42,85), photo I-V voltage shifts (85), or the current (83,84) during detrapping. Photoionization cross sections are obtained from data fits to Eq. 9 (84), or to Eq. 8 which is its time derivative (85). Energy depths or activation energies are obtained by studying detrapping rates as a function of photon (83-85) or thermal energy (30,42). As mentioned in previous sections, carrier injection from the contacts must be suppressed during photodetrapping. Fig. 8 shows two means of doing this. In Fig. 8a, the photon energy is kept below that required for photoinjection from the contact at a negative voltage. In Fig. 8b, the internal field of the trapped charge itself is used to suppress injection from the contacts while allowing photodetrapping of sites with energy depths greater than the energy barriers for photoinjection (2,36). For more than one type of trapping center, selective bleaching experiments could be used to determine individual σ_p and E_t for each trap.

D.   Spatial Distributions

The spatial location of trapped charge in the SiO_2 layer is important in probing the nature of the trapping sites either present in the as-grown oxide or purposely introduced. As seen in section III-B, the photo I-V technique or the photodepopulation technique can be used to determine the centroid $\bar{x}$ of the charge distribution. Combining these techniques with oxide etch-back experiments, where the charged oxide is etched back in steps, will give the spatial distribution of the trapped charge (91,104). Similarly, studying the charging in oxides of varying thickness will give information on the distribution (62,63).

IV.   TRAPS IN SiO_2

In this section, all known electron and hole traps in SiO_2 will be discussed in terms of their capture cross section, photoionization cross section, density, spatial distribution, optical threshold energy, and thermal activation energy. The traps are listed in groups based on the impurities or defects they are related to. Unless stated specifically, all trapping studies discussed were performed at room temperature.

A.   Sodium Related Traps

(i)   2.4 eV Trap

The 2.4 eV trap is a coulombic attractive site for electrons energetically situated in the SiO_2 band gap about 2.4 eV from the bottom on the conduction band. This trap was first observed by Williams in 1965 (6). Much work on the trapping-detrapping characteristics has been done in recent years by Feigl and his co-workers (79,83,84,87,91,104-107). The capture cross section is coulombic attractive as shown in Fig. 4 (79), and the photoionization cross section has values ranging from 10^{-18} to 10^{-17} cm^2 as shown in Fig. 6 (79,84) with a threshold energy at about 2.4-2.5 eV (91,104,107). This trapping site is possibly related to sodium grown into the SiO_2 lattice in an immobile configuration at elevated temperatures (1000-1200°C) particularly in steam oxides (106). The site is spatially distributed in a uniform manner in the SiO_2 layer (91,104). It is fairly monoenergic in energy having homogeneous dynamic (thermal) broadening with a homogeneous linewidth at 0°K of $.14\pm.01$ eV and inhomogeneous static broadening of $.34\pm.02$ eV due to local strains (83,107). The thermal broadening was proportional to $(\coth(T_0/2T))^{1/2}$ where $T_0 = 129°K$ (107). Trap densities from less than 10^{12} cm^{-3} to as high as 10^{15} cm^{-3} have been observed, and under ultra-clean growth conditions for SiO_2 these sites are not present in any measureable density (106). Significant thermal detrapping of captured electrons (up to 80%) has been observed at approximately 160°C (108). Recently these sites have been shown to be sensitive to variations in local strain in the oxide film with increasing temperature or oxide thickness (107). DiMaria et al. also have evidence of impact ionization of trapped electrons (79,87). The most extensive experimental work to date has been performed on this trap.

(ii) Na^+ Traps

These traps, first observed by DiMaria et al., are related to Na^+ ions immobilized at 77°K near the Si-SiO_2 interface (32,98). Their unique property is that there are several traps and that most of the trapping sites have very small capture cross sections of $10^{-19} - 10^{-20}$ cm^2 for the capture of electrons even though the ions are positively charged. A small number of the traps have a capture cross section of 2×10^{-15} cm^2. A one-to-one correspondence between the number of trapped electrons and the number of Na^+ ions (after correction for background oxide trapping) has been observed at least up to approximately 4×10^{12} Na^+ ions/cm^2 (32). Recent evidence shows that the trapping saturates for larger numbers of Na^+ ions (109). These sites also have the unusual property that when heated to room temperature after electron capture at 77°K, they do not detrap (109) even though the traps appear to be energetically too shallow for electron capture at room temperature (31,32). Electrons were not detrapped significantly at 77°K at applied electric fields up to 5 MV/cm (32). These sites are not to be confused with the 2.4 eV trap which readily captures electrons at room temperature. The Na^+ traps could be very shallow coulombic attractive sites which energetically become deeper with respect to the SiO_2 conduction band edge after electron capture, they could be dipole like traps due to the proximity of the Na^+ ions to the $Si-SiO_2$ interface, and/or they could be traps whose properties are influenced by the surrounding lattice.

B.   As-Grown Oxide Traps

(i)   Water Related Traps

These net neutral sites first observed by Nicollian et al. (22,30) are believed to be related to H_2O incorporated into the SiO_2 film (22,30). They also believe from some experimental evidence that the electron capture on a water related site is an electrochemical process and atomic hydrogen is released (30). However, Bulter et al. have argued recently that impurities such as sodium are also involved in forming this trapping site (106). These centers are found to some degree in all thermal SiO_2 films whether they are grown in steam or "dry" O_2 (22,30-32,37,39,47,91,106,110); however, steam oxides contain larger numbers of these traps (22,30,37,110). The electron capture cross sections have been reported in the range of approximately 10^{-17} to 10^{-18} cm^2 (30-32,39,47,110) with areal trap densities of about 10^{12} cm^{-2} or less (22,30-32,39,47,91,106,110). These traps cannot be detrapped with photons of energy up to 4 eV (22,30,91,96), but will discharge thermally at tempertures of 200°C or above in moist air with an activation energy of $\approx.35$ eV (22,30). Nicollian et al. believe that this activation energy is related to the diffusion of water from the ambient to the trapped electron sites where a chemical reaction then occurs to render the site neutral (30). The negative trapped charge in these sites is stable against detrapping up to average electrical fields of ≈ 5 MV/cm (30,37). They have a non-uniform spatial distribution in the SiO_2 film being fairly uniform in density in the bulk but with increasing density near the interfaces (22,30,39,91,96,104,106). Increased trapping near the $Si-SiO_2$ interface has been observed at 77°K into sites with similar capture cross sections (47,110). Most of these additional electrons trapped at 77°K were observed to detrap on heating to room temperature (110) unlike the behavior of Na^+ traps (see Section IV-A-ii). Recently, enhanced trapping in chemically vapor deposited (CVD) SiO_2 films is believed to be caused by water incorporated into the film during deposition (37).

(ii)   Repulsive Traps

Trapping sites with very small electron capture cross sections ($10^{-18}-10^{-21}$ cm^2) typical of coulombic repulsive capture have also been observed in as-grown SiO_2 films with areal densities from 10^{11} to 10^{13} cm^{-2} (26,31,32). These sites which are net neutral initially have been located near the $Si-SiO_2$ interface (81) and might be due to H_2O (81), excess Si (81), strained or broken bonds (111), and even possibly created by the high electric field conditions and/or current densities needed to detect them (111). They can be discharged thermally at temperatures between 200-400°C, and are not discharged with light of energy less than 2 eV (81). These sites are also fairly stable against high field detrapping (up to 7 MV/cm (32, 81)).

Solomon and Klein recently observed electroluminescense in SiO_2 at fields > 7 MV/cm with a quantum efficiency of $10^{-5}-10^{-4}$ photons/electron crossing 1000 Å of SiO_2 which was polarity independent (112). The light emission was proportional to current with a cut-off at 2.5 eV. This light could be associated with radiative capture into these traps or the water related traps where capture probabilities of $10^{-5}-10^{-4}$ are typical.

C.   Trapped Holes Sites and Hole Traps

(i)   Trapped Hole Sites

Electron trapping on or annihilation of positively charged holes generated or injected into SiO_2 films, some of which are subsequently trapped into energetically deep sites near the interfaces (24,46,51,61-63), is a coulombic attractive capture process at least for electron capture on holes trapped near the $Si-SiO_2$ interface (31,46,61). The electron capture cross section has a field dependence similar to that shown in Fig. 4 (46). The density of electron traps is equivalent to the density of trapped holes (48,61) which is dependent on SiO_2 processing conditions (growth temperature, annealing, etc. (53,113,114)) with areal densities between $10^{11}-10^{13}$ cm^{-2} (46,51-53,61-63). Electron capture on holes trapped near the $Al-SiO_2$ interface has been observed to proceed at a much slower rate (possibly due to a smaller capture cross section if all detrapping processes are ignored) than coulombic capture on holes trapped near the $Si-SiO_2$ interface (61). If the Al electrode is replaced by polycrystalline Si, more holes are trapped near this polycrystalline $Si-SiO_2$ interface as compared to the $Al-SiO_2$ interface, and electron capture by these trapped hole sites is very similar to that for the substrate single crystal $Si-SiO_2$ interface (61).

Positively charged centers introduced into SiO_2 films by various annealing treatments by Ning et al. (26,77,78) could possibly be trapped hole sites or the 2.4 eV trap both of which are coulombic attractive for electron capture with a cross section behaving like that in Fig. 4 (78). Ning also observed that this capture cross section was not strongly dependent on temperature between 77°K and room temperature (78) which was also seen by Aitken and Young for electron capture by trapped holes (31). Ning proposed that this temperature independence of the capture cross section was caused primarily by an effect, dependent on the applied electric field through Frenkel-Poole lowering of the energy barrier of the coulombic potential well of the trap in a regime where this lowering is much greater than kT (78), which tended to cancel the expected increase in the capture cross section with decreasing temperature (Section II-A-i). Ning also suggested that the departure of the measured capture cross section from an $E_{ox}^{-3/2}$ to an E_{ox}^{-3} dependence (E_{ox} is the average oxide field) as shown in Fig. 4 was due to electron heating in the electric field (78). As mentioned previously, the dependence on $E_{ox}^{-3/2}$ is characteristic of the shrinkage of the capture volume in the applied field (70,78).

(ii)   Hole Traps

Hole traps are the sites in an as-grown oxide which can trap holes introduced into the SiO_2 valence band by means of excitation across the SiO_2 bandgap with VUV light or x-rays (44-46,54), by high electric fields (20,21,51,64,99), or by injection usually from the Si substrate (22,24,52,53). At room temperature, deep hole traps which capture holes permanently are near the interfaces (22,46,61-63), particularly the $Si-SiO_2$ interface. These traps which have areal densities $\leq 1x10^{13}$ cm^{-2} (52,53) are net neutral before hole capture, and the measured capture cross section is coulombic attractive (10^{-13} to 10^{-14} cm^2 at low applied fields (52,53)). Ning explained this observation by postulating that the hole traps were compensated (52) similar to the conclusion of Yun and Arnett for attractive coulombic capture of holes or electrons in silicon nitride (94,95) which is also in a net neutral charge state initially (88). Hole traps are also very sensitive to processing conditions (53,113,114) and their density in "dry" O_2 grown oxides can be increased by high temperature annealing (above 950°C) for extended periods of time (53,113). They are believed to be caused by impurities introduced into the SiO_2 near the interfaces such as excess Si near the $Si-SiO_2$ interface (51) and/or caused by strained or broken bonds (115) possibly due to stress or stress relief resulting from viscous shear flow (113).

Attempts that have been made at hole detrapping with light or heat are usually subject to some controversy due to the possibility of annihilation of the holes by electrons introduced into the SiO_2 conduction band by the same light or heat (see section IV-C-i). Regardless of the exact physical mechanism, holes trapped near the $Si-SiO_2$ interface can not be discharged with light energies up to

4-5 eV (51), but they can be discharged thermally at temperatures between 150-400°C (24,51) with an activation energy of ≈.35 eV (22).

Energetically shallow hole traps which are at least partially responsible for the low mobility of holes in SiO_2 are observed to be populated at low temperatures (53,55-60,116). Spatially, these traps are thought to be uniformly distributed in the bulk of the SiO_2 film (55-60). Captured holes have been observed to be discharged from these traps on heating from 77°K to room temperature (53,55-60,116) and to be discharged at 77°K with light with energies less than 3 eV (116). Recently, DeKeersmaecker and DiMaria have shown that As or P implanted into SiO_2 films will give energetically deep stable hole trapping sites at room temperature in the bulk of the SiO_2 film (117). These bulk trapped holes can be easily annihilated with injected electrons in a manner similar to annihilation of holes trapped near the Si-SiO_2 interface (117).

D.   W Traps

Trapping sites related to W atoms deposited less than a monolayer thick ($\leq 10^{15}$ atoms/cm^2) between a thermal SiO_2 layer and a CVD SiO_2 or Al_2O_3 layer incorporated into an MOS structure have been studied recently (2,41-43). The electron capture cross section for these sites with a net neutral charge state initially is approximately $1-5 \times 10^{-14}$ cm^2 and depends on the number of deposited W atoms (42). The trap density is approximately constant regardless of small variations in the number of deposited W atoms and is about a factor of 10-40 less than the number of deposited W atoms (42). These observations have been used by Young et al. to argue for the existance of W clustering (42). Trapped electrons on these sites cannot be discharged with light energies up to 6 eV (2,42), but can be discharged thermally between 200°C and 400°C in N_2 with an activation energy of .9 eV (42). High electric fields $\gtrsim$ 6 MV/cm internally generated by large amounts of trapped electrons in the W layer or externally applied tend to discharge trapped electrons (42,43). Ion implanted W in SiO_2 layers was observed to have a smaller capture cross section of $\approx 1 \times 10^{-15}$ cm^2, consistent with the clustering model for the deposited W case (42).

E.   Radiation Induced Neutral Traps

Aitken and Young first observed increased electron trapping into neutral centers in SiO_2 films exposed to penetrating energetic electrons or x-rays (31). The capture cross sections for these neutral traps were in the range from 10^{-15} to 10^{-18} cm^2 with areal trap densities of 10^{10} to 10^{12} cm^{-2} (31,47,48). These sites are created in the bulk of the SiO_2 layer for penetrating radiation, but appear to be found predominantly near the exposed SiO_2 surface for low energy non-penetrating electrons (80,118). Ning has shown that the capture cross section is weakly dependent on the electric field (48). No enhanced trapping at 77°K was seen (47) implying temperature independent capture cross sections and trap densities which makes these traps different from traps in as-grown oxides (see section IV-B-i). These traps could be annealed out at approximately 500°C (47,48). Ning has also possibly observed hot carrier impact ionization of some trapped electrons on these sites (48). The origin of these neutral centers has been suggested by Aitken et at. to be due to broken or displaced bonds in the SiO_2 (47,48) tetrahedra. The surface traps observed by Fanet and Poirier which have similar capture cross sections and which were probably created by the low energy ($\approx$10 eV) electron beam they used are likely to be radiation induced neutral traps also. They have an optical depth of $\approx$5 eV from the SiO_2 conduction band edge (80).

F.   Traps Related to Implanted Ions

Ion implantation is a convenient way to get impurities into SiO_2. Chen et al. first studied the effects on carrier transport and charge storage in Au ion implanted SiO_2 films (119). Ions implanted into SiO_2 create numerous trapping sites due to the damage caused by the energetic ions (33). Usually this damage is so extensive that any carrier injected into the SiO_2 layer is trapped, and many of the techniques discussed in Section III cannot be used to analyze the trapping kinetics. If the oxide layer is subjected to high temperature annealing treatments (1000°C) in an inert gas such as N_2 after implantation, most of the traps due to damage are removed and trapping on sites related to the implanted ions themselves can be studied (34-36). Ion implanted Al, As, P, and B have been studied in this manner (40). B does not show any noticeable trapping above that seen in an as-grown oxide (40) and will not be discussed further. As and P, which are both in the fifth column of the periodic table, are very similar in their trapping-detrapping properties, although As has about twice the atomic mass of P. However, Al which is in the third column of the periodic table has much different properties compared to As and P, although it has approximately the same atomic mass as P.

(i)   Ion Implanted Al Related Traps

Traps related to implanted Al which are net neutral have electron capture cross sections in the range from 10^{-15} to 10^{-18} cm^2 with trap densities that are proportional to the implantation energy for a given ion fluence (number of implanted ions) and proportional to the fluence for a given ion energy (34,35,40). For an ion fluence of 10^{13} ions/cm^2, the total trap density varied between 10^{12} to 10^{13} cm^{-2} depending on ion energy (35,40). Captured electrons cannot be detrapped at light energies up to 6 eV (36,40), and a small amount of additional trapping was observed at 77°K (34). The trapped electrons and the implanted ions have been shown to have the same spatial distributions (36,40).

(ii)   Ion Implanted As or P Related Traps

Ion implanted As or P related trapping centers will be discussed together because they are very similar except for their capture cross sections and sensitivity to annealing conditions. P related traps are more sensitive because with their smaller capture cross sections it is necessary to reduce background trapping in the SiO_2 layer more. The electron capture cross sections of the dominant trap (site which has the highest trapping probability) for As and P related sites are approximately 1×10^{-15} cm^2 and 3×10^{-17} cm^2, respectively (40). The areal trap densities for these traps are not a function of the implantation energy as in the case of Al, are proportional to the ion fluence, and have a value of approximately half the value of the fluence (40). The trapped electrons have the same centroid as the implanted ions (40), and the electrons can be detrapped with light (40). The photoionization cross sections vary from 10^{-18} to 10^{-17} cm^2 with photon energy after threshold is reached which is approximately 4 eV (40,85) as is shown in Fig. 6. This threshold is consistent with theoretical predictions for substitutional P or As in an O site (85). No pronounced temperature dependence of σ_p between 77°K and 300°K was seen (117). For the As or P related sites, electrons could be thermally detrapped at temperatures between 100-350°C in N_2 with activation energies between .15 and .25 eV (117). No significant electron detrapping for either As or P related sites with applied electric fields up to at least 3 MV/cm was seen (117). As with the ion implanted Al case, the exact nature of these traps as related to the implanted impurity atoms is not known. As mentioned previously (Section IV-C-ii) holes can also be trapped on ion implanted As and P related sites (117).

V.   CONCLUSIONS

Electron and hole trapping-detrapping in thermal SiO_2 is extremely important in the electronics industry since the build-up of trapped charge changes operating characteristics of devices like IGFET transistors (38). An important example of this recently pointed out by Aitken et al. is the neutral centers caused by electron beams (see section IV-E) which only anneal out at high temperatures (31,47,48). Electron beam lithography has been used to shrink the dimensions of IGFET transistors to increase performance, but unless the SiO_2 layer is properly annealed those structures will degrade more rapidly (47,48). Charge trapping layers purposely incorporated into the insulating layers are currently used for non-volatile memory where information is stored as trapped charge for long periods of time. Examples of this are the dual-dielectric charge storage (DDC) cells which use less than a monolayer of W or WO_3 as a trapping layer interposed between thermal SiO_2 and CVD Al_2O_3 (41,43) (see section IV-D), and the floating gate avalanche injection MOS (FAMOS) structures which have a continuous polycrystalline Si layer for carrier trapping interposed between SiO_2 layers (120). As seen in this review the number of investigators in this area has increased rapidly over the past few years and probably will continue to increase in the future.

ACKNOWLEDGEMENTS

The author would like to acknowledge helpful discussions with A. B. Fowler, R. F. DeKeersmaecker, Z. A. Weinberg, D. R. Young, T. H. Ning, P. L. Solomon, and J. M. Aitken; and the critical reading of this manuscript by R. F. DeKeersmaecker, A. B. Fowler, D. R. Young and M. I. Nathan.

REFERENCES

1. A. S. Grove, (1967) Physics and Technology of Semiconductor Devices, Wiley, New York, Chap. 9.
2. D. J. DiMaria, J. Appl. Phys. 47, 4073 (1976).
3. E. Loh, Solid-St. Commun. 2, 269 (1976).
4. H. R. Philipp, Solid-St. Commun. 4, 73 (1966).
5. T. H. DiStefano and D. E. Eastman, Solid-St. Commun. 9, 2259 (1971).
6. R. Williams, Phys. Rev. 140, A569 (1965).
7. B. E. Deal, E. H. Snow, and C. A. Mead, J. Phys. Chem. Solids 27, 1873 (1966).
8. A. M. Goodman, Phys. Rev. 144, 588 (1966).
9. A. M. Goodman, Phys. Rev. 152, 785 (1966).
10. A. M. Goodman, J. Appl. Phys. 37, 3580 (1966).
11. A. M. Goodman, Phys. Rev. 152, 780 (1966).
12. R. C. Hughes, Phys. Rev. Lett. 30, 1333 (1973).
13. R. C. Hughes, Phys. Rev. Lett. 35, 449 (1975).
14. C. N. Berglund and R. J. Powell, J. Appl. Phys. 42, 573 (1971).
15. R. Poirier and J. Olivier, Appl. Phys. Lett. 21, 334 (1972).
16. J. Piesner, G. Aszodi, M. Nemeth-Sallay, and G. Forgacs, Thin Solid Films 36, 251 (1976).
17. M. Lenzlinger and E. H. Snow, J. Appl. Phys. 40, 278 (1969).
18. Z. A. Weinberg, Solid-St. Electron. 20, 11 (1977).
19. Z. A. Weinberg and A. Hartstein, Solid-St. Commun. 20, 179 (1976).
20. Z. A. Weinberg, W. C. Johnson, and M. A. Lampert, J. Appl. Phys. 47, 248 (1976).
21. R. Williams and M. H. Woods, J. Appl. Phys. 44, 1026 (1973).
22. E. H. Nicollian, A. Goetzberger, and C. N. Berglund, Appl. Phys. Lett. 15, 174 (1969).
23. E. H. Nicollian and C. N. Berglund, J. Appl. Phys. 41, 3052 (1970).
24. K. Nagai, Y. Hayashi, and Y. Tarui, Jap. J. Appl. Phys. 14, 1539 (1975).
25. J. F. Verwey, J. Appl. Phys. 44, 2681 (1973).
26. T. H. Ning and H. N. Yu, J. Appl. Phys. 45, 5373 (1974).
27. D. R. Young, J. Appl. Phys. 47, 2098 (1976).
28. A. Rose, (1963) Concepts in Photoconductivity and Allied Problems, Interscience, New York, Chap. 7.
29. N. F. Mott and R. W. Gurney, (1948) Electronic Processes in Ionic Crystals, Dover, New York, Chap. 4 and 5.
30. E. H. Nicollian, C. N. Berglund, P. F. Schmidt, and J. M. Andrews, J. Appl. Phys. 42, 5654 (1971).
31. J. M. Aitken and D. R. Young, J. Appl. Phys. 47, 1196 (1976).
32. D. J. DiMaria, J. M. Aitken, and D. R. Young, J. Appl. Phys. 47, 2740 (1976).
33. N. M. Johnson, W. C. Johnson, and M. A. Lampert, J. Appl. Phys. 46, 1216 (1975).
34. D. R. Young, D. J. DiMaria, and W. R. Hunter, J. Electron. Mat. 6, 569 (1977).
35. D. R. Young, D. J. DiMaria, W. R. Hunter, and C. M. Serrano, IBM J. Res. Develop. 22, 1978.

36. D. J. DiMaria, D. R. Young, W. R. Hunter, and C. M. Serrano, IBM J. Res. Develop. 22, 1978.
37. R. A. Gdula, J. Electrochem. Soc. 123, 42 (1976).
38. T. H. Ning, C. M. Osburn, and H. N. Yu, J. Electron. Mat. 6, 65 (1977).
39. A. Vshirokawa, E. Suzuki, and M. Warashina, Jap. J. Appl. Phys. 12, 398 (1973).
40. D. J. DiMaria, D. R. Young, R. F. DeKeersmaecker, W. R. Hunter, and C. M. Serrano, The Electrochemical Society Fall Meeting, Atlanta, Georgia, 1977, Abstract No. 212 (unpublished).
41. D. Kahng, W. J. Sundburg, D. M. Boulin, and J. R. Ligenza, Bell Syst. Techn. J. 53, 1723 (1974).
42. D. R. Young, D. J. DiMaria, and N. A. Bojarczuk, J. Appl. Phys. 48, 3425 (1977).
43. K. K. Thornber and D. Kahng, Appl. Phys. Lett. 32, 131 (1978).
44. R. J. Powell and G. F. Derbenwick, IEEE Trans. Nucl. Sci. NS-18, 99 (1971).
45. R. J. Powell, J. Appl. Phys. 46, 4557 (1975).
46. D. J. DiMaria, Z. A. Weinberg, and J. M. Aitken, J. Appl. Phys. 48, 898 (1977), and references therein.
47. J. M. Aitken, D. R. Young, and K. Pan, J. Appl. Phys., (1978).
48. T. H. Ning, unpublished.
49. E. P. EerNisse and C. B. Norris, J. Appl. Phys. 45, 5196 (1974).
50. J. F. Verwey, J. Appl. Phys. 43, 2273 (1972).
51. M. H. Woods and R. Williams, J. Appl. Phys. 47, 1082 (1976).
52. T. H. Ning, J. Appl. Phys. 47, 1079 (1976).
53. J. M. Aitken and D. R. Young, IEEE Trans. Nucl. Sci. NS-24, 2128 (1977).
54. Z. A. Weinberg, Appl. Phys. Lett. 27, 437 (1975).
55. R. C. Hughes, Phys. Rev. B 15, 2012 (1977).
56. R. C. Hughes, Appl. Phys. Lett. 26, 436 (1975).
57. F. B. McLean, G. A. Ausman, H. E. Boesch, and J. M. McGarrity, J. Appl. Phys. 47, 1529 (1976).
58. F. B. McLean, H. E. Boesch, and J. M. McGarrity, IEEE Nucl. Sci. NS-23, 1506 (1976).
59. J. R. Srour, S. Othmer, O. L. Curtis, and K. Y. Chiu, IEEE Nucl. Sci. NS-23, 1513 (1976).
60. O L. Curtis and J. R. Srour, J. Appl. Phys. 48, 3819 (1977).
61. J. M. Aitken, D. J. DiMaria, and D. R. Young, IEEE Trans Nucl. Sci. NS-23, 1526 (1976)
62. G. W. Hughes, R. J. Powell, and M. H. Woods, Appl. Phys. Lett. 29, 377 (1976).
63. G. W. Hughes and R. J. Powell, IEEE Trans. Nucl. Sci. NS-23, 1569 (1976).
64. P. M. Solomon and J. M. Aitken, Appl. Phys. Lett. 31, 215 (1977).
65. M. Lax, J. Phys. Chem. Solids 8, 66 (1959).
66. M. Lax, Phys. Rev. 119, 1502 (1960).
67. G. Ascarelli and S. Rodriguez, Phys. Rev. 124, 1321 (1961).
68. V. L. Bonch-Bruevich and E. G. Landsberg, Phys. Stat. Sol. 29, 9 (1968).
69. J. S. Blakemore, Phys. Rev. 163, 809 (1967).
70. G. A. Dussel and K. W. Böer, Phys. Stat. Sol. 39, 375 (1970).
71. G. A. Dussel and R. H. Bube, J. Appl. Phys. 37, 2797 (1966).
72. R. H. Bube, (1974) Electronic Properties of Crystalline Solids, Academic, New York, Chap. 9.
73. A. F. Tasch and C. T. Sah, Phys. Rev. B 1, 800 (1970).
74. M. R. Belmont, Thin Solid Films 28, 149 (1975).
75. C. H. Henry, J. Electron. Mat. 4, 1037 (1975).
76. C. H. Henry and D. V. Lang, Phys. Rev. B15, 989 (1977).
77. T. H. Ning, C. M. Osburn, and H. N. Yu, Appl. Phys. Lett. 26, 248 (1975).
78. T. H. Ning, J. Appl. Phys. 47, 3203 (1976).
79. D. J. DiMaria, F. J. Feigl, and S. R. Butler, Phys. Rev. B 11, 5023 (1975).
80. J. M. Fanet and R. Poirier, Appl. Phys. Lett. 25, 183 (1974).
81. P. Solomon, J. Appl. Phys. 48, 3848 (1977).
82. L. L. Rosier and C. T. Sah, Solid-St. Electron. 14, 41 (1971).

178

83.   F. J. Feigl, S. R. Butler, D. J. DiMaria, and V. J. Kapoor, (1976) in Thermal and Photosti-
        mulated Currents in Insulators, edited by D. J. Smyth, Electrochemical Society, Princeton,
        N.J., pp. 118-134.
84.   D. J. DiMaria, Ph.D. thesis (Lehigh University, 1973), unpublished.
85.   R. F. DeKeersmaecker, D. J. DiMaria, and S. T. Pantelides, paper BA5, this conference.
86.   L. L. Rosier and C. T. Sah, J. Appl. Phys. 42, 4000 (1971).
87.   D. J. DiMaria, F. J. Feigl, and S. R. Butler, Appl. Phys. Lett. 24, 459 (1974).
88.   P. C. Arnett and D. J. DiMaria, J. Appl. Phys. 47, 2092 (1976).
89.   R. J. Powell, J. Appl. Phys. 40, 5093 (1969).
90.   J. H. Thomas and F. J. Feigl, J. Phys. Chem. Solids 33, 2197 (1972).
91.   V. J. Kapoor, F. J. Feigl, and S. R. Butler, J. Appl. Phys. 48, 739 (1977).
92.   D. J. DiMaria and P. C. Arnett, IBM J. Res. Develop. 21, 227 (1977).
93.   S. M. Sze, (1969) Physics of Semiconductor Devices, Wiley-Interscience, New York, Chap.
        9.
94.   B. H. Yun, Appl. Phys. Lett. 25, 340 (1974).
95.   P. C. Arnett and B. H. Yun, Appl. Phys. Lett. 26, 94 (1975).
96.   R. J. Powell and C. N. Berglund, J. Appl. Phys. 42, 4390 (1971).
97.   D. J. DiMaria, Z. A. Weinberg, J. M. Aitken, and D. R. Young, J. Electron. Mat. 6, 207
        (1977).
98.   D. J. DiMaria, J. Appl. Phys. 48, 5149 (1977).
99.   M. Shatzkes and M. Av-Ron, J. Appl. Phys. 47, 3192 (1976).
100.   D. J. DiMaria, Appl. Phys. Lett. 31, 680 (1977).
101.   D. J. DiMaria and D. R. Kerr, Appl. Phys. Lett. 27, 505 (1975).
102.   D. J. DiMaria and F. J. Feigl, Phys. Rev. B 9, 1874 (1974).
103.   R. J. Powell, Appl. Phys. Lett. 31, 290 (1977).
104.   V. J. Kapoor, F. J. Feigl, and S. R. Butler, (1976) in Thermal and Photostimulated
        Currents in Insulators, edited by D. J. Smyth, Electrochemical Society, Princeton, N.J., pp.
        135-148.
105.   J. H. Thomas and F. J. Feigl, Solid-St. Commun. 8, 1669 (1970).
106.   S. R. Bulter, F. J. Feigl, Y. Ota, and D. J. DiMaria, (1976) in Thermal and Photostimulated
        Currents in Insulators, edited by D. J. Smyth, Electrochemical Society, Princeton, N.J., pp.
        149-161.
107.   V. J. Kapoor, F. J. Feigl, and S. R. Butler, Phys. Rev. Lett. 39, 1219 (1977).
108.   J. H. Thomas, Ph.D. thesis (Lehigh University, 1970), unpublished.
109.   D. J. DiMaria, unpublished.
110.   D. R. Young, E. A. Irene, D. J. DiMaria, H. Z. Massoud, and R. F. DeKeersmaecker, The
        Electrochemical Society Spring Meeting, Seattle, Washington, 1978, (unpublished).
111.   E. Harari, Appl. Phys. Lett. 30, 601 (1977).
112.   P. Solomon and N. Klein, J. Appl. Phys. 47, 1023 (1976).
113.   E. P. EerNisse and G. F. Derbenwick, IEEE Trans. Nucl. Sci. NS-23, 1534 (1976).
114.   G. F. Derbenwick and B. L. Gregory, IEEE Trans. Nucl. Sci. NS-22, 2151 (1975).
115.   C. W. Gwyn, J. Appl. Phys. 40, 4886 (1969).
116.   E. Harari, S. Wang, and B. S. H. Royce, J. Appl. Phys. 46, 1310 (1975).
117.   R. F. DeKeersmaecker and D. J. DiMaria, unpublished.
118.   D. J. DiMaria and J. M. Aitken, unpublished.
119.   L. I. Chen, K. A. Pickar, and S. M. Sze, Solid-St. Electron. 15, 979 (1972).
120.   D. Frohman-Bentchkowsky, ISSCC Dig. of Tech. Papers 15, 80 (1971).

DYNAMIC BEHAVIOUR OF MOBILE IONS IN SIO$_2$ LAYERS

M.W. Hillen
Technical Physics Lab, State University Groningen
The Netherlands.

ABSTRACT

Two measuring methods are very suited to study the behaviour of mobile ions
in silicon-dioxide: the thermally stimulated ionic current (TSIC) technique
and the triangular voltage sweep (TVS) technique. As these methods are based
on completely different theoretical assumptions it is useful to know the
limits of their applicability. Experimental work will be shown in which the
theories in their simplest form are used to analyse the results. Contamination experiments indicate clearly the influence of Na$^+$ ions on the first and
K$^+$ ions on the second peak in both TVS and TSIC curves. HCl oxidation seems
to decrease the activation energy for emission of Na$^+$ ions at the silicon-
silicon dioxide interface.

INTRODUCTION

TSIC and TVS measurements are done using an MOS capacitor of which the temperature and the gate voltage can be changed as a function of time. In both
cases the displacement current is measured.
The thermally stimulated ionic current is measured when the temperature is
raised and the gate voltage is held constant. Commonly two peaks are found
in the temperature region 30 - 400 ^{o}C. As the position of these peaks is
dependent on sign and value of the gate voltage, and the background-curves
have to be subtracted, several measurements are necessary to characterize
the trapping sites.
Triangular voltage sweep curves result from quasi-static C-V measurements at
elevated temperatures. Again two peaks are found dependent in place and shape
on temperature and sign and value of the sweeprate. When the ionic current
is not zero at the end of the sweep a waiting time has to be included to
have a well defined starting situation.

TSIC MEASUREMENTS

The emission theory which is commonly used to analyse TSIC curves considers
only the trapping sites at both the silicon and the metal interface. From
these traps mobile ions are emitted when they acquire a sufficient energy
and then driven immediately to the other side of the oxide where they are
trapped again.
Hickmott (1) explained several TSIC curves using a single level theory. The
current is expressed as:

$$I = -q \frac{dn(t)}{dt} = q \frac{n(t)}{\tau_o} \exp(-\frac{E_o}{kT(t)}),$$

where n is the number of trapped ions, E_o the activation energy of the trap
and τ_o the emission time constant. From two current values E_o and τ_o are
determined from which a theoretical curve is found which fits the experimental one at the rising side (Fig. 1).
A better fit is obtained by assuming a spread in activation energies,
corresponding to the amorphous nature of the oxide. When a gaussian distribution is assumed (2) a theoretical curve is found which fits the

180

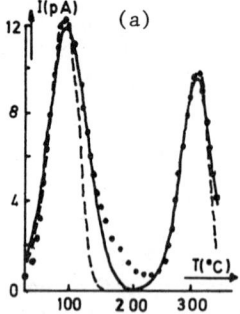
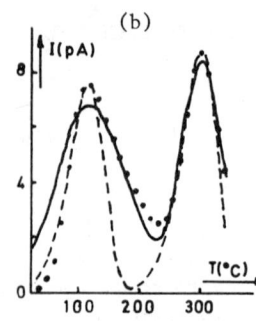

Fig. 1. Comparison of experimental and theoretical TSIC curves to the surface trapping model for negative (a) and positive (b) gate voltage. (---): single level; (——):Gaussian spread; ••: experimental values.

experimental one well but the value of maximum activation energy E_0 is too small to be consistent with an emission dominated process and the emission time constant is physically unrealistic. Boudry suggested to use a physically more plausible emission time constant of 10^{-12} s, related to the atomic frequency factor. The results of this calculation are also shown in Fig. 2. The fit remains reasonably good, because the model is not very sensitive to the value of τ_0, but the activation energies are now consistent with an emission limited process.

Boudry (4) showed also that the fit can be optimized by not constraining the energy distribution of the traps. This results in asymmetric distributions. From the simple emission model no field depend-
ence of the TSIC curves is expected. However it is found experimentally that the peak-maximums are displaced to lower temperature when the gate voltage is increased, resulting in lower values of activation energy. Sometimes this barrier reduction seems to be consistent with a Poole-Frenkel reduction as is shown in Fig. 2 for some results of Hickmott (1), but several other measurements can not be explained by this effect. The energy reduct-ion due to the field is difficult to describe because insufficient knowledge of the nature of the traps, the value of the dielectric constant and the influence of the field caused by the ions.

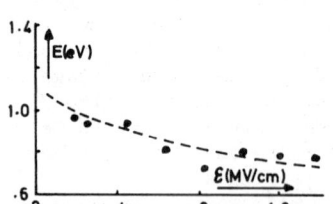

Fig. 2. Field dependence of activation energy. (---): Poole-Frenkel barrier reduction; •:experiments from Hickmott (1).

This last effect can be investigated by stressing the sample before the measurement not with the opposite gate voltage but with V_G = 0. From the resulting TSIC curves it appears that a lot of ions have already drifted to the other interface due to space charge effects alone. The very shallow traps at the silicon interface are completeley depleted, even while the value of ψ_{ms} opposes the drift towards the aluminium interface.

In Fig. 3 TSIC curves of oxides grown in different HCl atmospheres are presented. It appears that the contamination level in HCl-grown oxides is not less than in conventional grown oxides. The concentration of mobile ions is even higher for a 2 % than for a 0.5 % HCl oxidation. However this is not the case for the shallow traps at the silicon interface. Here a higher Cl concentration results in less trapped ions at an energy that seems to be lower than in comparable not HCl-grown oxides. These results reinforce the impression that Cl⁻ is active at the Si-SiO$_2$ interface.

181

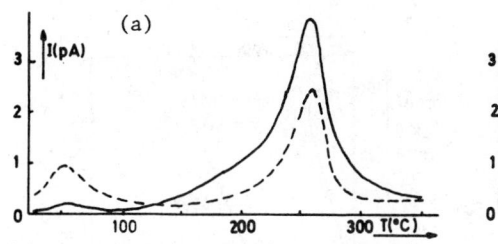 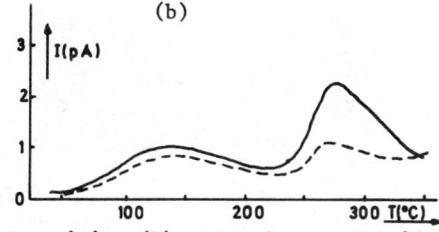

Fig. 3. TSIC curves at negative (a) and positive (b) gatevoltage of oxides grown in different HCl atmospheres. (——): 2 %; (---): 0.5 %.

TVS MEASUREMENTS

In the TVS theory the quasi static distribution of ions in the oxide, which is assumed not to contain any trapping sites is calculated from the continuity and the Poisson relation. This results in a U-shaped distribution of mobile ions and a current which is determined by their charge centroïd. This theory is compared to the experiments in Fig. 4 and it is clear that the model in which only positive mobile ionic charge is assumed (5) seems to be the correct one.

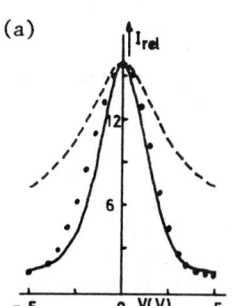 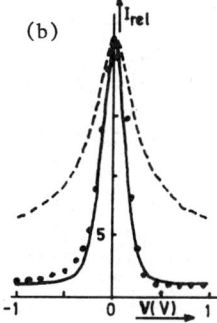

Fig. 4. Comparison of experimental and theoretical TVS curves at 220 ^{0}C, for a negative (a) and positive (b) sweeprate of 30 mV/s. (——): only positive mobile charge (5); (---): also negative fixed compensating charge (6); •:experimental values.

The resulting TVS curve is of course a symmetrical one centered around $V_G = 0$, when the workfunction difference is neglected. In the experiments this is only the case for the first peak which is symmetric and remains of the same shape for temperatures above 200 ^{0}C and not too fast sweeprates. The second peak in TVS curves becomes nearly symmetric at a temperature of 400 ^{0}C and at a sweeprate of 3 mV/s (Fig. 5).
A comparison of the characteristic times for drift and emission, which have the form $t = t_0 \exp (E/kT)$, to the time that is needed to record a TVS peak at a certain sweeprate is made in Fig. 6. Anticipating on the results of the contamination experiments it is assumed that Na^+ and K^+ cause the first and second peak respectively.
This plot suggests that the characteristic time for emission exceeds the transit time for both type of ions at temperatures up to 400 ^{0}C. Furthermore influence of drift on the emission current is only to be expected for K-ions. The characteristic emission time is negligible to the measuring time at 200 ^{0}C for Na (which is also found experimentally) and at 400 ^{0}C for K (which, as this is not found experimentally, would suggest a slightly higher value of τ_o).

182

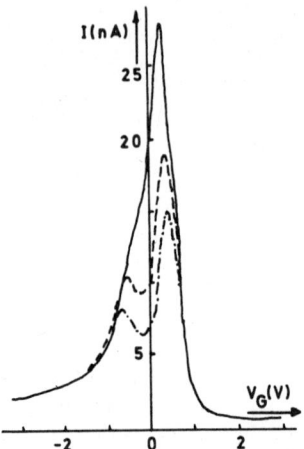

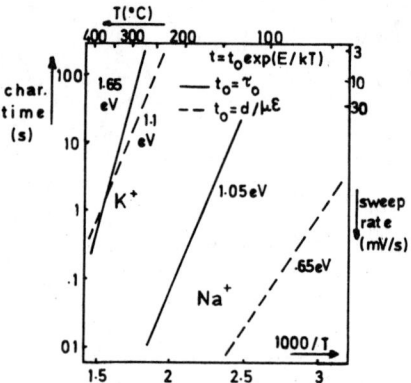

Fig. 6. Characteristic times for emission and drift as function of temperature. (——): emission, activation energies from the model with gaussian spread (2); (---) drift, activation energies according to Stagg (7).

Fig. 8. The first and second peak in TVS curves at a sweeprate of -30 mV/s and different temperatures. (-.-): 348 ^{0}C; (---): 367 ^{0}C; (——): 398 ^{0}C.

4.CONTAMINATION EXPERIMENTS

The occurence of two peaks in both TVS and TSIC curves suggests the presence of two ionic species with different activation energies for emission. Na^+ and K^+ are mostly considered as causing the peaks and therefore some contamination experiments using these ionic species were carried out. TSIC curves of wafers contaminated with KCl or NaCl solution are presented elsewhere (2) and indicate clearly the influence of Na on the first and K on the second peak. To introduce a well known number of mobile ions Na and K were introduced by implantation into the oxide before metallization. The number of introduced ions was $7 \cdot 10^{11}$ cm^{-2} , which is sligthly more than the number of mobile ions which is caused by the aluminium evaporation from a tungsten filament. The implantation energy was 20 keV which gives a projected range of 350 $\overset{o}{A}$ for Na in the oxide. After annealing (450^0, 30 min, N_2) TSIC curves were recorded and again a clear influence of Na^+ on the first and K^+ on the second peak is established (Fig. 7).
A very strange effect is observed when Na and K ions are implanted into the aluminium. The same number of ions was introduced but at an energy of 50 keV, which is not high enough for the Na^+ ions to pass through the metallization. After the same anneal TSIC curves were recorded which show a contamination level which is less than the level from the not implanted sample (Fig. 8). This effect was earlier reported by Fritzsche (8) for implantation of Na^+ and K^+ through a thin semi transparant metal layer, from which it was concluded that radiation damage must be responsible for the gettering effect. Indeed both ions seem to get trapped in very deep traps which are not emptied at temperatures below 400 ^{0}C.

CONCLUSIONS

The TSIC and TVS techniques are very suitable to study ionic contamination in oxides and the influence of oxidation in HCl atmospheres. Contamination experiments show that the first and second peak in both TSIC and TVS curves

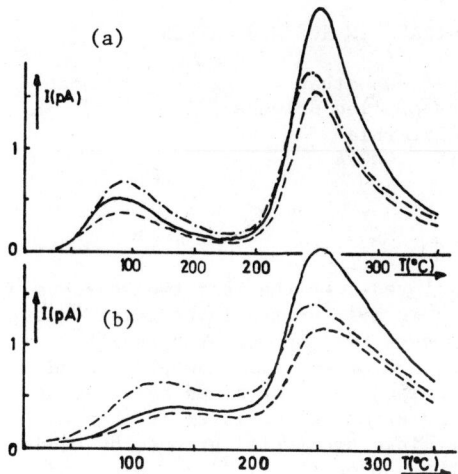

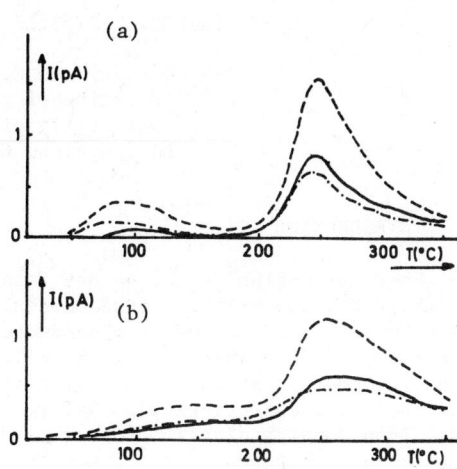

Fig. 7. TSIC curves of implanted samples for negative (a) and positive (b) gate voltage. Implantation into the oxide of 20 kV Na^+/K^+ ions. $(-.-)$: Na implant $(8 \cdot 10^{11} \, cm^{-2})$; (———): K implant $(6 \cdot 10^{11} \, cm^{-2})$; $(--)$: no implant. Oxide thickness 1300 Å.

Fig. 8. TSIC curves of implanted samples for negative (a) and positive (b) gate voltage. Implantation into the aluminium of 50 kV Na^+/K^+ ions. $(-.-)$: Na implant $(4 \cdot 10^{11} \, cm^{-2})$; (———): K implant $(1 \cdot 10^{11} \, cm^{-2})$; $(---)$: no implant. Aluminium thickness 1 μm.

must be attributed for the most part to mobile sodium and potassium ions respectively.

The emission model with a spread in activation energy describes both peaks in TSIC curves well. The drift model which is used for the explanation of TVS measurements is only applicable at temperatures above 200 ^{0}C for Na^+ and above 400 ^{0}C for K^+ ions. A HCl atmosphere during oxide growth influences mainly the mobile Na^+ ions at the $Si-SiO_2$ interface. They are trapped at a slightly lower energy and their number is smaller when the HCl concentration was higher. Implantation of Na^+ and K^+ ions into the aluminium has a strong gettering effect on the mobile ions in the oxide.

Acknowledgement: I am grateful to H. Kalter of Philips Research Laboratories, Eindhoven for preparing the HCl grown oxides and to J.F. Verwey and G. Greeuw for critically reviewing the paper.

REFERENCES

(1) T.W. Hickmott, J.Appl.Phys. 46, 2583 (1975).
(2) P.K. Nauta and M.W. Hillen, to be published in J. Appl.Phys.(March 1978).
(3) P.K. Nauta, 2nd Symp. on Solid St. Techn., Münster (March 1977).
(4) M.R. Boudry and J.P. Stagg, ESSDERC abstract A3.7, Brighton (Sept. 1977).
(5) H.M. Przewlocki and W. Marciniak, Phys. Stat. Sol. (a) 29, 265 (1975).
(6) N.J. Chou, J. Electrochem.Soc. 118, 601 (1971).
(7) J.P. Stagg, Appl.Phys.Lett. 31, 532 (1977).
(8) C. Fritzsche e.a., Rad. Eff, 7, 87 (1971).

PHOTO-INJECTION STUDIES OF TRAPS IN HcℓH$_2$O OXIDES

J. Dorosti and C. R. Viswanathan
School of Engineering and Applied Science
University of California
Los Angeles, Calif. 90024

INTRODUCTION

Photo-injection technique has been employed over the last twelve years to study the properties of the oxide layer as well as the interfacial layer between the oxide and the semiconductor layers by a number of research workers after the initial work by Williams and others.[1,2] Powell[3] showed that information about the charge distribution can be obtained by studying the photo i-v characteristics before and after the oxide layer is charged up. Trap density distributions were determined by several groups, by studying the photo-response as a function of the photon energy.[2,4,5] Ning and Yu[6] used hot-electron techniques to charge up the oxide layer and studied the charge-trapping characteristics by observing the shift in the flat-band (or threshold voltage) continuously. In this work we report experiments on photo-injection of charges in the oxide layer of a MOS capacitor with a HcℓH$_2$O oxide. We use the model of Ning and Yu to determine the trapping characteristics by monitoring the flat-band voltage of the MOS capacitor as a function of time.

EXPERIMENT

MOS capacitors were fabricated by thermally oxidizing p-type 1–3Ω-cm silicon wafers having 100 orientation. Two different types of oxides were grown for the purpose of this investigation. In the first case, the oxide layer was grown by a dry-oxide process at 1175°C. In the second case, oxide layers were grown at 900°C and 1175°C with a vapor mixture of hydrochloric acid and water flowing through the oxidation furnace. This oxide will be referred to as HcℓH$_2$O oxide and the former will be referred to as dry oxide throughout this paper. Some samples were annealed in dry nitrogen gas for two hours at 850°C or for 50 minutes at 1000°C. The rest of the samples was not annealed. After stripping the oxide off the back-side of the wafer, the back surface was metallized. The metallization on the front side was performed so as to form MOS capacitors in which the gate electrode comprised of a thin optically transparent area with a partially overlapping thick electrode for contact purposes.

U-V light from a deuterium lamp was focussed on to the transparent electrode of the MOS capacitor and the MOS capacitor was biassed positive or negative during irradiation. A Keithley electrometer was connected in series with the bias supply and the MOS capacitor to measure the injection current as shown in Fig. 1.

In a typical photo-injection experiment, the photo-injection was interrupted at certain time-intervals and high frequency C-V measurements or in many cases just the flat-band voltage measurements were made, and the photo-injection was resumed. In virgin samples as well as in samples that were charged up by photo-injection experiments, photo i-v measurements were made by irradiating the sample with a low intensity uv light source and measuring the photo-current as a function of the bias voltage. This measurement will be referred to as photo i-v measurements in the rest of the paper.

KINETICS OF TRAPPING

Ning and Yu[6] studied the effect of injection of optically induced hot electrons in the oxide layer of a MOS capacitor. Following their model for the kinetics of carrier capture we can write

$$n_i(x,t) = N_i(x)[1-\exp(-\frac{\sigma_i}{q}\int_o^t j(t')dt']$$ (1)

where n_i is the density of charge carriers trapped in the ith trap center, N_i is the density of the ith trap center characterized by a capture cross-section σ_i, j is the current density in the oxide layer. Equation (1) can be rewritten as

$$n_i(x,t) = N_i(x)[1-\exp(-\sigma_i N_{inj}(t))]$$ (2)

where N_{inj} is the number of injected carriers per unit area and is given by

$$N_{inj} = \int_o^t [j(t')/q]dt'$$ (3)

The shift in the flat-band voltage due to charge injection can be written as

$$\Delta V_{FB} = \frac{q}{C_{ox}} \int_o^t \frac{x}{d} (\sum_i n_i(x))dx$$ (4)

where d is the oxide thickness. If it is assumed that j is independent of time, we can simplify equation (4) using equation (2) as

$$\Delta V_{FB} = \frac{q}{C_{ox}} \sum_i \frac{\overline{x_i}}{d} N_{Ti} \left(1-e^{-t/\tau_i}\right)$$

where $N_{Ti} = \int_o^d N_i(x)dx$, the total number of ith trap center per unit area,

$\overline{x_i}$ = average distance of the ith trap center from the silicon interface and $\tau_i = \frac{q}{\sigma_i j}$

Equation (5) relates the shift in the flat-band voltage to the amount of charge injected. If only one trapping center with a time constant τ_i were to be dominant, then the plot of log $(d(\Delta V_{FB})/dt)$ versus time will be a straight line with a slope equal to $1/\tau_i$. If different trapping species with various time constants were to be present, then the plot of log $(d(\Delta V_{FB})/dt)$

versus time will be characterized by many linear portions.

RESULTS

Fig. 2a gives the C–V curves of a p-type MOS sample with an $HC\ell/H_2O$ oxide layer of thickness 2.7 microns taken at various stages of photo-injection with a positive bias on the gate electrode. As time progresses the C–V curve shifts in the direction of positive voltage corresponding to negative charging due to electron injection from the silicon substrate. Fig. 2b gives the corresponding C–V curves for a sample at various stages of photo-injection under a negative bias voltage. It can be seen that initially the C–V curve shifts in the direction of positive voltage and at still longer invervals of time, the C–V curve shifts in the direction of negative voltage. This behavior can be understood with the following argument. Under negative bias on the gate electrode during photo-injection electrons are injected from the gate electrode while holes are injected from the silicon electrode. However, a net negative charge is trapped initially due to a larger value of the trapping cross section for electrons in comparison with the trapping cross-section for holes. The C–V curve shifts therefore in the positive direction. However as time progresses, more and more holes get trapped near the silicon interface and the C–V curves shift in the negative direction.

A typical plot of the shift in flat-band voltage as a function of time for the thick oxide MOS capacitor photo-injected under a negative bias is shown in Figure 3. From plots similar to this, values of $d(\Delta V_{FB})/dt$ can be obtained and plotted on a semilog paper to obtain $\log (d(\Delta V_{FB})/dt)$ versus time. Figure 4 gives the plot of $\log (d(\Delta V_{FB})/dt)$ versus time for a $HC\ell/H_2O$ sample with a (thin) oxide layer (thickness = 1000°A) obtained at values of times corresponding to hole trapping under negative bias photo-injection. The plot shows two linear portions corresponding to two species of hole trapping centers.

Table 1 gives a summary of the results obtained on various samples.

DISCUSSION OF RESULTS

In both thin and thick $HC\ell/H_2O$ oxides, two species of electron traps appear to exist. Annealing does not appear to make any difference in the values of σ_i within experimental errors. The thick $HC\ell/H_2O$ sample also appears to exhibit two species of electron traps but the smaller cross-section seems to be much smaller than the one obtained with thin oxide samples. The thick dry oxide samples also exhibit similar electron trap characteristics. The value of the larger capture cross-section agrees with the observation of $\sigma = 3.3 \times 10^{-13} cm^2$ by Ning and Yu.[6] We do not have an explanation for why the thick $HC\ell/H_2O$ oxide samples have two orders of magnitude lower electron trap density than the other samples.

By plotting $\log (d(\Delta V_{FB})/dt)$ in the region corresponding to the negative shift in the C–V curves, we obtain hole trapping characteristics. Two types of hole traps are obtained in the thin $HC\ell/H_2O$ oxide samples. In the thick $HC\ell/H_2O$ sample, only the type of hole trap with a cross-section corresponding to the smaller of the two obtained in the thin $HC\ell/H_2O$ oxide, is observed since the range of observation was limited only to this due to experimental difficulty. It is possible the species with larger capture cross-section also exist. The important observation that must be made is no

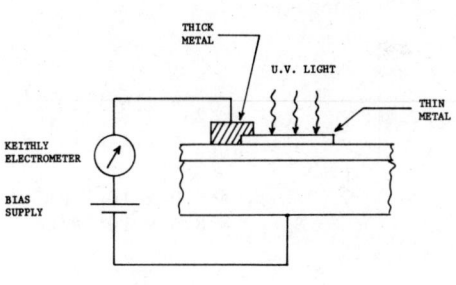

FIG. 1: EXPERIMENTAL ARRANGEMENT

TABLE 1

SAMPLE	Type of Carrier	σ_i cm²	N_i cm⁻²
Thin Hcℓ/H₂O Oxide Unannealed	Electrons	3.7×10^{-13}	4.4×10^{11}
	"	7.6×10^{-14}	2×10^{13}
	Holes	1×10^{-16}	2×10^{12}
	"	2.1×10^{-17}	2.2×10^{12}
Thin Hcℓ/H₂O Oxide Annealed	Electrons	1.8×10^{-13}	8.7×10^{12}
	"	9.5×10^{-14}	2×10^{13}
	Holes	5×10^{-17}	8.6×10^{11}
	"	2.6×10^{-17}	1.1×10^{12}
Thick Hcℓ/H₂O Oxide Unannealed	Electrons	1.8×10^{-13}	3×10^{10}
	"	1.9×10^{-14}	4×10^{10}
	Holes	2.7×10^{-17}	5×10^{12}
Thick Dry O₂ Oxide	Electrons	1.2×10^{-13}	1.2×10^{12}
	"	2×10^{-14}	3.5×10^{12}
Thin Dry O₂ Oxide Avalanche inj.	Electrons	1.5×10^{-16}	8×10^{11}
	"	4.2×10^{-17}	1×10^{12}

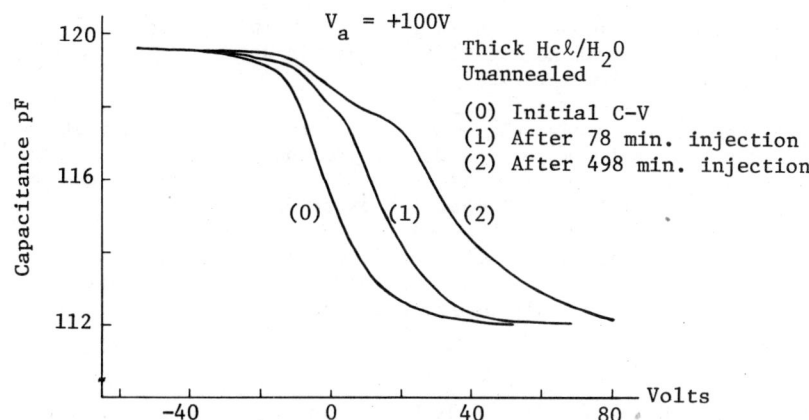

$V_a = +100V$

Thick Hcℓ/H₂O Unannealed

(0) Initial C-V
(1) After 78 min. injection
(2) After 498 min. injection

Fig. 2A: C-V Curves Taken at Various Stages of Injection Under Positive Bias

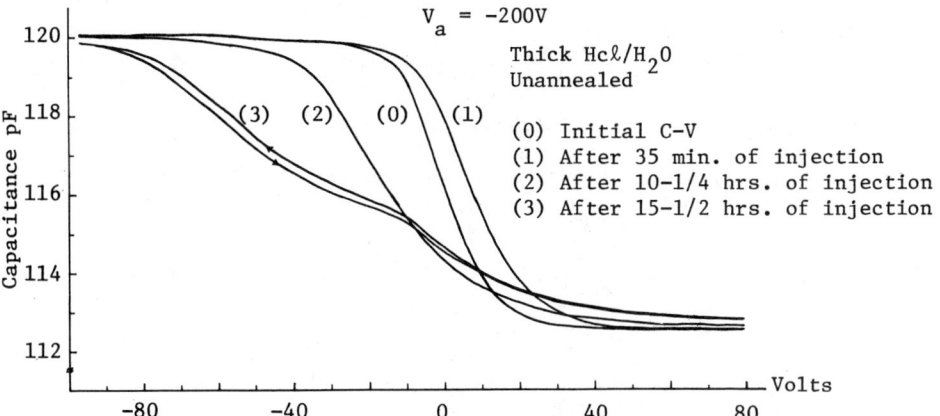

$V_a = -200V$

Thick Hcℓ/H₂O Unannealed

(0) Initial C-V
(1) After 35 min. of injection
(2) After 10-1/4 hrs. of injection
(3) After 15-1/2 hrs. of injection

Fig. 2B: C-V Curves Taken at Various Stages of Injection Under Negative Bias

188

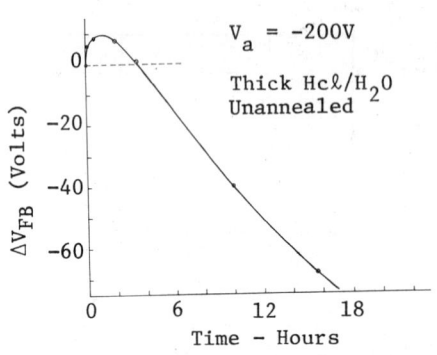

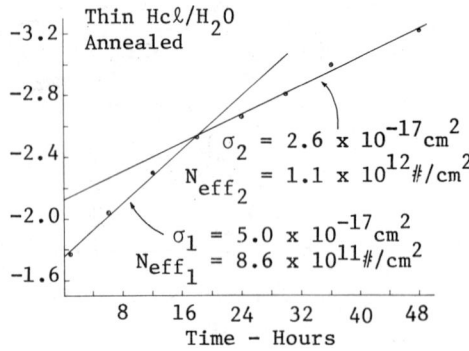

Fig. 3: Shift in the Flat-Band Voltage as a Function of Time Under Continuous Photoinjection

Fig. 4: Plot of $\ell n \left(\frac{d}{dt} \Delta V_{FB}(t) \right)$ Versus Time

negative C-V shift was observed in dry oxide samples corresponding to hole trapping. Hence it was not possible to obtain data on hole trapping characteristics in dry oxides.

For comparison, we also give electron trapping characteristics of thin dry oxide p-type MOS samples which were avalanche injected in Table 1. The capture cross-section that is observed for electrons is much smaller than those obtained in photo-injection experiments.

Charge distribution in the oxide layer was obtained using the photo i-v technique developed by Powell,[3] both before and after photo-injection. They corroborate the observations made in the photo-injection experiments.

ACKNOWLEDGEMENTS

Grateful thanks are due to J. Maserjian and the Jet Propulsion Laboratory staff for making available the experimental facility for this work.

REFERENCES

1. R. Williams, Phys. Rev. 140, A569 (1965).

2. A. M. Goodman, Phys. Rev. 144, 588 (1966).

3. R. J. Powell and C. N. Berglund, J. Of Appl. Phys., Vol. 42, No. 11, 4390 (1971), also R. J. Powell, IEEE Trans. Nucl. Sci., NS-17, 41 (1970).

4. C. R. Viswanathan and S. Ogura, Proc. IEEE, Vol. 57, No. 9, 1552 (1969).

5. D. J. DiMaria, et. al., Appl. Phys. Lett. 24, 459 (1974).

6. T. H. Ning and H. N. Yu, J. of Appl. Phys., Vol. 45, No. 12, 5373 (1974).

PHOTODEPOPULATION OF ELECTRONS TRAPPED IN SiO$_2$ ON SITES RELATED TO AS AND P IMPLANTATION*

R. F. DeKeersmaecker, D. J. DiMaria, and S. T. Pantelides
IBM T. J. Watson Research Center, Yorktown Heights, NY   10598

ABSTRACT

Ion implantation has been used to incorporate As and P into the thermally grown SiO$_2$ layer of metal-silicon dioxide-silicon structures. These impurities increase the electron trap density in the oxide layer proportionally to the ion fluence (1-3 x 10^{13} cm^{-2}). Avalanche injection from the silicon substrate was used to populate the electron trapping sites. It is shown that the negative charge is removable from the trapping centers under illumination with photons between 3 and 5.6 eV. From the detrapping experiments, we determined a photoionization threshold of $\approx$ 4 eV both for As and P related centers, and a spectrally resolved effective photoionization cross section; the latter quantity is defined as the convolution of the photoionization cross section with the optically accessible trap distribution in the energy gap of the SiO$_2$. A simple model is presented which suggests that the observed detrapping originates from levels corresponding to substitutional P and As at O sites in the SiO$_2$.

INTRODUCTION

Photodepopulation spectroscopy in metal-insulator-semiconductor (MIS) structures has recently been used to study both the spatial and energy distribution of optically accessible charges in the insulator film [1]. Several variations of the technique exist, all relying upon the optical stimulation of electrons, trapped in deep levels in the insulator, into conducting states from which they drift towards an electrode.

For MIS structures with SiO$_2$ as the insulator, literature is available on a 2.4-2.5 eV deep electron trap, possibly related to sodium incorporated at elevated temperatures in an immobile configuration in the oxide film [2]. Few studies, however, have been carried out to characterize other charge trapping centers. Several ions such as phosphorus and arsenic were recently demonstrated to exhibit electron trapping characteristics when implanted into the SiO$_2$ layer [3]. It was determined that P implantation results in a dominant electron trap with a capture cross section of 3x10^{-17} cm^2 [3], whereas for As related centers the dominant trap has a capture cross section of $\approx$ 10^{-15} cm^2 [4]. The trapping sites are in a net neutral state before electron capture. In both cases, the integrated trap density increases with the ion fluence. Good agreement was found between the trapped electron distribution centroid and both theoretical predictions and experimental observations of the ion distribution centroid [3].

In this work, we report on spectrally resolved detrapping measurements for charged MOS structures, from which a photoionization threshold and an effective photoionization cross

*This work was supported in part by the Defense Advanced Research Projects Agency, and monitored by the Deputy for Electronic Technology (RADC) under contract No. F19628-76-C-0249 and by the Office of Naval Research under contract No. N00014-76-C-0934.

section spectrum were determined. We also present a simple theoretical model which suggests that the observed levels are those of substitutional P and As at O sites.

EXPERIMENTAL

Sample Fabrication

The starting material was p-type <100> silicon with a resistivity of 0.1-0.2 Ωcm. The wafers were oxidized at 1000°C in a "dry" oxygen ambient to oxide thicknesses ranging from 560 to 1430 Å as determined by ellipsometry. Then P^+ or As^+ implantation was performed at room temperature with energies of 20 to 80 keV and fluences of 1 to 3 x 10^{13} cm^{-2}. The ion current at target during implantation was of the order of 1 x 10^{-6} A. All wafers were then cleaned and annealed in nitrogen at 1000°C for 30 min. Using a shadow mask, semitransparent aluminum electrodes (100-150 Å thick) with an area of 0.0052 cm^2 were deposited to form MOS capacitors. Finally, all devices were given a post-metallization annealing treatment at 400°C in forming gas for 20 min.

Electron Trapping and Photodetrapping

The oxide traps were charged by avalanche injection of hot electrons from the silicon substrate driven into deep depletion [5]. The amplitude of the 50 kHz ramp wave used for this purpose was constantly adjusted in order to keep the average dc injection current constant. The currents ranged from 5 x 10^{-10} to 3 x 10^{-9} A. Some of the injected electrons were trapped in the SiO_2, causing a shift in flat-band voltage, determined from 1 MHz differential capacitance-voltage characteristics.

After reaching a given charge level, the avalanche injection was stopped and the sample mounted in a set-up for photoelectric measurements, consisting of a 900 W xenon arc lamp in combination with a 500 mm grating monochromator (Bausch and Lomb). The incident photon flux at the sample position was measured over the spectrum using a thermopile in combination with an electronic chopper. An electric shutter was used to control the illumination time which was usually 5 min. The change in flat-band voltage induced by this illumination at room temperature was monitored to within 1 mV with an automatic tracking system. The spectrum was scanned step-by-step in the direction of increasing photon energies.

The detrapping experiments are performed with zero gate bias in order to avoid injection from the contacts at photon energies greater than the Si-SiO_2 or Al-SiO_2 energy barrier, by using the internal fields in the SiO_2 layer due to the trapped negative charges as a potential barrier against this electron injection. This situation is preserved throughout the entire detrapping experiment, since only small portions of the total charge are removed at each step.

PHOTODETRAPPING ANALYSIS

If first order kinetics (neglecting charge retrapping) is assumed, the local depopulation of occupied traps under illumination is governed by the following equation:

$$\delta n_t (x,E,t) / \delta t = -F_p(x,\hbar\omega)\, \sigma_p(x,E,\hbar\omega)\, n_t(x,E,t), \qquad (1)$$

where $n_t(x,E,t)$ is the trapped electron concentration per unit energy, $F_p(x,\hbar\omega)$ is the local photon flux in the SiO_2 layer and $\sigma_p(x,E,\hbar\omega)$ is the trap photoionization cross section. The photon flux F_p is a function of both the photon energy $\hbar\omega$ and the position x in the SiO_2 layer due to the optical interference phenomenon [6]. This interference gives rise to a standing wave pattern in the SiO_2 layer, which depends upon photon energy.

The photoionization cross section σ_p is a function of the trap energy level E in the SiO_2 band gap (E=0 and E_g at the top of the SiO_2 valence band and the bottom of the conduction band, respectively) and of the photon energy $\hbar\omega$ since it includes the transition probability to a final state $E+\hbar\omega$, and may be position dependent through variation of the electric field due to the presence of charge in the SiO_2 layer. The field dependence of σ_p was experimentally found to be weak and is, therefore, neglected here.

It was concluded from discharging experiments [4], that the charge centroid $\bar{x}$ (measured from the Al-SiO_2 interface) is constant in time and that, if an energy spectrum of trapping centers is present, they all have the same spatial charge distribution centroid. These two observations allow us to separate the variables determining n_t, i.e.

$$n_t(x,E,t) = n_t^o(t) \, n_t^1(E) \, n_t^2(x),\tag{2}$$

where n_t^1 and n_t^2 are normalized distributions. For the spatial distribution $n_t^2(x)$ a Gaussian is used with the charge centroid as median value and the same standard deviation as for the ion distribution.

Equation 1 is integrated over the oxide thickness L and over the SiO_2 energy gap, using the expression for the flat-band voltage shift:

$$\Delta V_{FB}(t) = q(\bar{x}/\varepsilon)\int_0^L \int_0^{E_g} n_t(x,E,t)dEdx,\tag{3}$$

where ε is the static permittivity of SiO_2 and q is the electron charge. At a particular photon energy ($\hbar\omega$), only centers between $E=E_g-\hbar\omega$ and $E=E_g$ can be depopulated, if thermal broadening of the trapping levels and two-photon processes are disregarded. We thus get:

$$\frac{d[\Delta V_{FB}(t)]/dt =}{-\Delta V_{FB}(t)}\int_0^L F_p(x,\hbar\omega)n_t^2(x)dx\int_{E_g-\hbar\omega}^{E_g}\sigma_p(E,\hbar\omega)n_t^1(E)dE.\tag{4}$$

Let

$$\chi(\hbar\omega) = \int_0^L F_p(x,\hbar\omega)\,n_t^2(x)\,dx\tag{5a}$$

and

$$\Sigma(\hbar\omega) = \int_{E_g-\hbar\omega}^{E_g}\sigma_p(E,\hbar\omega)\,n_t^1(E)\,dE,\tag{5b}$$

where $\chi(\hbar\omega)$ is the convolution of the photon flux with the spatial distribution of the trapping centers and $\Sigma(\hbar\omega)$ is the convolution of the photoionization cross section with the optically accessible trap distribution over energy. The latter quantity will be viewed as an effective photoionization cross section. If a single monoenergetic trap is involved, then $\Sigma(\hbar\omega)$ reduces to $\sigma_p(\hbar\omega)$.

The solution to eq. 4 can be approximated as:

$$[\Delta V_{FB}(0)-\Delta V_{FB}(t)]/\Delta V_{FB}(0) = \chi(\hbar\omega)\Sigma(\hbar\omega)t\tag{6}$$

if

$$t \ll [\chi(\hbar\omega)\Sigma(\hbar\omega)]^{-1}.\tag{7}$$

If long discharging times were used, the charge centroid would ultimately be determined by the minimum in the standing wave pattern of the light. However, since the illumination interval t was kept small compared to the discharging time constant, we may disregard the effect of light interference upon the charge centroid.

Since the variation in flat-band voltage shift due to illumination for a period t can be measured and the quantity $\chi(\hbar\omega)$ can be calculated knowing the sample geometry, the optical constants of the various materials of the multilayer structure, the incident photon flux and the light energy used, the effective photoionization cross section $\Sigma(\hbar\omega)$ can be determined from Eq. 6.

RESULTS

The photo I-V technique which is a sensitive method of determining both the density and the centroid of oxide charges [7], was initially used to ascertain that the negative oxide charge is removed by exposure to light, and not compensated by positive charge [4]. It was also experimentally verified that the discharging phenomenon obeys first order kinetics, and that the discharging time constant is long compared to the illumination time [4].

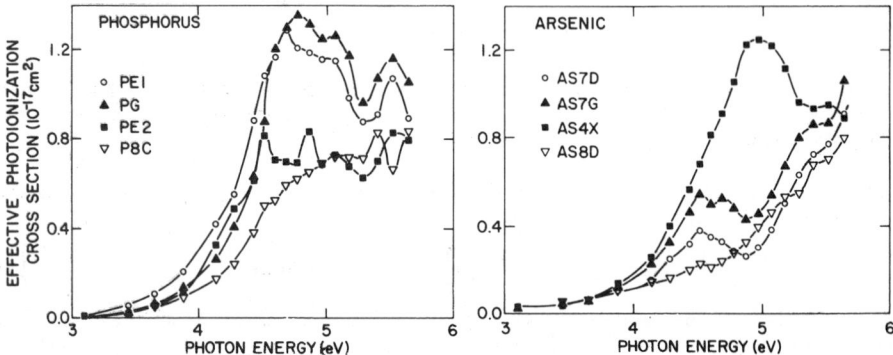

Fig. 1. Effective photoionization cross section spectra for P and As implanted SiO$_2$.

Figure 1 displays the effective photoionization cross section spectra, for P and As implanted samples, for which the experimental parameters are summarized in Table 1. All the samples were implanted with a fluence of 1 x 10^{13} cm^{-2}, except for wafer P8C, which was implanted with 3 x 10^{13} P+/cm^2. In the latter case, partial penetration of the P ions into the Si substrate was taken into account. In view of the low doping level of the SiO$_2$ introduced by the ion implantation, we used the optical constants for unimplanted SiO$_2$ in the calculation of the standing wave pattern. Incorporating the finite band-pass of the monochromator (50 Å full width at half maximum) into the analysis reduces the magnitude of $\Sigma(\hbar\omega)$ in Fig. 1 by 25-40 %, the larger correction being for higher photon energies, without changing the observed structure in the spectra.

Table 1. Experimental Parameters for the Samples.

Sample	Ion	Oxide thickness (Å)	Aluminum thickness (Å)	Ion energy (keV)	Charge centroid (Å)	Ion standard deviation (Å)
PE	P	1283	135	30	420	142
PG	P	1277	140	40	545	181
P8C	P	583	106	25	290	121
AS7D	As	1433	120	60	470	118
AS7G	As	1415	120	80	570	149
AS4X	As	1272	150	60	465	118
AS8D	As	559	134	20	185	51

DISCUSSION

The nature of the traps that exist in the ion-implanted samples cannot be determined in a direct way. Because of the high annealing temperature used after implantation, it is very likely that all major structural damage in the SiO$_2$ is healed. Therefore, we consider trapping levels related to the implanted ions themselves rather than due to the damage.

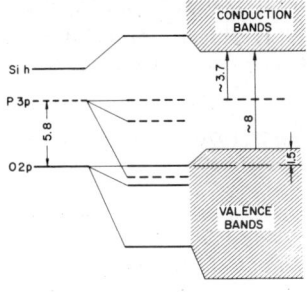

Fig. 2. Bond-orbital energy-level structure for SiO$_2$:P. The gap marked ~8 eV corresponds to the onset of absorption (not the band gap), so that 3.7 eV is the predicted onset of absorption of the impurity.

At first, one might expect that P would occupy Si sites and form phosphate-type clusters, as is the case when P is diffused-in instead of implanted. That is possible, but P and As at Si sites would act as donors and not as electron traps and would, therefore, go undetected in our experiments. On the other hand, a simple tight-binding model for substitutional impurities in SiO$_2$, based on the bond-orbital description of the bulk material [8], reveals that P and As are very likely to be substitutional at O sites. An energy level diagram describing P at an O site is shown in Fig. 2. The energy separation between the O 2p and P 3p levels is about 5.8 eV [9]. The P 3p orbitals bond with the neighboring Si hybrid orbitals and, as a result, two levels are expected at midgap. The lower level contains two electrons, whereas the upper level contains only one, and would, therefore, act as an electron trap. Photoionization of this electron would have a threshold at approximately 3.7 eV, which is close to that observed.

Above threshold, a resonance corresponding to the atomic P 3p → 4s transition is expected, in analogy with the optical spectrum of the pure material [10]. Such a resonance may be present in the data of Fig. 1. According to the same model, As at an O site would behave just like P with the only exception that the level in the gap will be approximately 0.4 eV lower in energy. The data of Fig. 1 show that the As threshold is 0.2-0.3 eV higher than that of P, in agreement with the theoretical prediction. We feel that these results suggest that the observed traps are P and As at O sites, but we cannot at present exclude other possibilities.

If our model is correct, we expect the photoionization cross section above threshold to be insensitive to implantation energy and oxide thickness. As observed in Fig. 1, the analysis indeed compensates for implantation energy, but not completely for oxide thickness. Our experimental oxide thickness range allowed us to perform accurate flat-band voltage measurements on MOS devices, but on the other hand was shown to introduce a strong sensitivity due to minute thickness variations in the interference calculation. In view of this, the structure in the spectra has to be considered preliminary. Experiments are in progress with thinner oxide layers (200-300 Å), in which case the influence of oxide thickness variations upon the photoionization cross section spectrum should be minimized.

ACKNOWLEDGEMENTS

The authors wish to express their gratitude to D. R. Young for stimulating discussions; B. Yun for the design of the flat-band voltage tracking system; the Silicon Process Studies Group for the sample preparation; E. D. Alley for the metallizations; and J. A. Calise and F. L. Pesavento for the experimental assistance.

REFERENCES

1. F. J. Feigl, S. R. Butler, D. J. DiMaria, and V. J. Kapoor, in (1976) Thermal and Photostimulated Currents in Insulators, ed. D. J. Smyth, The Electrochemical Society, Princeton, p. 118.
2. S. R. Butler, F. J. Feigl, Y. Ota, and D. J. DiMaria, in (1976) Thermal and Photostimulated Currents in Insulators, ed. D. J. Smyth, The Electrochemical Society, Princeton, p. 149.
3. D. J. DiMaria, D. R. Young, R. F. DeKeersmaecker, W. R. Hunter, and C. M. Serrano, The Electrochemical Society Fall Meeting, Atlanta, 1977, Abstract No. 212 (unpublished).
4. R. F. DeKeersmaecker and D. J. DiMaria (unpublished).
5. E. H. Nicollian and C. N. Berglund, J. Appl. Phys. 41, 3052 (1970).
6. D. J. DiMaria and P. C. Arnett, IBM J. Res. Develop. 21, 227 (1977).
7. D. J. DiMaria, J. Appl. Phys. 47, 4073 (1976).
8. S. T. Pantelides and W. A. Harrison, Phys. Rev. B 13, 2667 (1976); S. T. Pantelides, J. Vac. Sci. Technol. 14, 965 (1977).
9. F. Herman and S. Skillman, (1963) Atomic Structure Calculations, Prentice-Hall, Englewood Cliffs.
10. S. T. Pantelides, "The Optical Absorption Spectrum of SiO_2," these Proceedings.

CHEMICAL STATE OF PHOSPHORUS IN
DEPOSITED SiO$_2$ (P) FILMS

A. N. Saxena
Data General Corporation, 433 N. Mathilda Ave.
Sunnyvale, CA 94086

and

R. A. Powell
Stanford Electronics Laboratories, Stanford
University, Stanford, CA 94305

Deposited silicon dioxide films (vapox) doped with phosphorus are of utmost importance in LSI technology. The chemical state of P and its concentration in the vapox films affect their thermal flow and etching properties. These properties of the vapox films play a major role in determining the final yield of the LSI circuits. In spite of the technological importance of vapox, relatively little fundamental work on the chemical state and properties of P in vapox has been reported.[1] Recent unexpected observations of P pile-up near the Si/SiO2 interface of thermally grown SiO$_2$ on phosphorus-doped Si point up a corresponding lack of basic understanding of the P/SiO2 system in general.[2,3] The present preliminary study was undertaken to investigate the chemical state of P and its concentration in vapox by Auger electron spectroscopy (AES), electron microprobe analysis (EMA), ESCA and colorimetric techniques.

Three kinds of samples were used in these preliminary studies: (1) CVD vapox doped in-situ with P (oxide thickness 1000-5000 Å, bulk phosphorus concentration of 3.5-7.7% by weight); (2) thermal SiO$_2$ and (3) thermal SiO$_2$ which was ion implanted with P[31]. The energy and the dose of P[31] implantation were 80 KeV and 3 x 10[16] cm^{-2}, respectively. The ion implanted sample was not annealed so that a larger number of interstitial phosphorus sites will be obtained as compared with the substitutional sites. Thus, both the chemical states of phosphorus could be observed in these samples, as described below.

Table 1 shows the P concentration in the two CVD samples as measured by AES, EMA and colorimetric techniques. P concentrations were obtained with AES by first measuring the peak-to-peak height of the Si LVV (76 eV), P LVV (116 eV) and O KLL (510 eV) signals in the Auger dN/dE derivative spectrum and then using elemental sensitivity factors[4]. The P concentrations measured by the different techniques were mutually consistent to within about 10%.

196

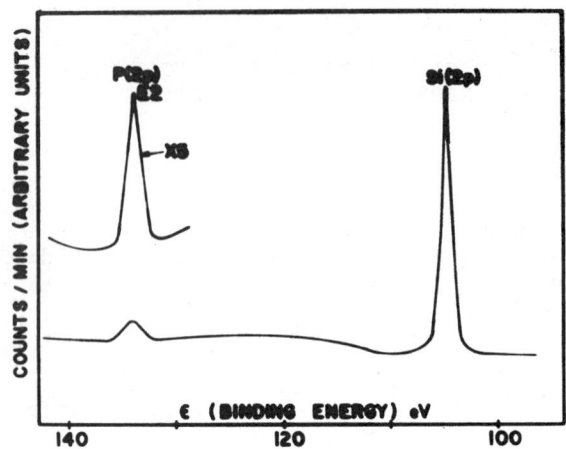

Fig. 1.   ESCA data of the CVD vapox sample: counts/min.
(arbibrary units) <u>versus</u> ε, binding energy, in
eV.   It shows only one phosphorus peak, E2, at
133.5 eV.

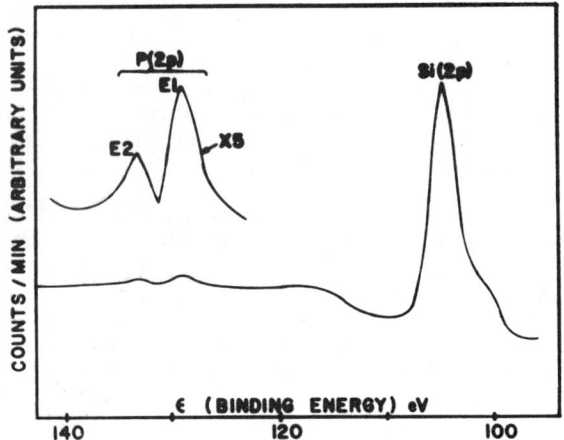

Fig. 2.   ESCA data of the ion implanted sample: counts/min.
(arbitrary units) <u>versus</u> ε, binding energy, in eV.
It shows two phosphorus peaks, E1 at 129.1 eV, and
E2 at 133.3 eV.

TABLE 1   Concentration of Phosphorus in Vapox
Determined by Various Techniques

Characterization of Mole % of PH3 in PH3 + SiH4 Mixture (Pre-Mixed Gases)	Weight % of Phosphorus		
	Auger	Electron Microprobe	Colorimetric
6.5	---	3.5	4.4
15.0	7.6	6.6	8.3

ESCA measurements made on both CVD vapox and phosphorus-implanted SiO_2 samples (made using an HP5950A ESCA spectrometer at Surface Science Laboratories, Palo Alto, CA) revealed that P was present in different chemical states in these samples.   Fig. 1 shows the Si(2p) and P(2p) ESCA peaks obtained from the CVD vapox sample (doped with about 8% P by weight) and Fig. 2 shows the same from the ion implanted sample.   The latter was sputtered with Ar^+ (1 KeV, 25 $\mu A/cm^2$) until the peak in the implanted P distribution (calculated to lie about 800 Å below the SiO_2 surface) was reached.   As can be seen from Fig. 1, the CVD sample displays only one peak associated with phosphorus, E2 (133.5 eV).   In Fig. 2, the implanted sample displays E2 (133.3 eV) and an additional P peak, E1, located at about 4 eV less binding energy (129.1 eV).   The peak height of E1 is larger than that of E2 in Fig. 2 for the unannealed implanted sample.   This is consistent with the fact that the number of interstitial P (corresponding to E1) sites are greater in number than the number of substitutional P (corresponding to E2) sites in this sample. These two peaks are tentatively interpreted as due to phosphorus incorporated substitutionally and interstitially.   The substitutional sites are postulated to be those occupied by P in place of Si in the SiO_4 tetrahedra.

In contrast to the ESCA data, only one phosphorus peak, corresponding to the LVV Auger transition, was observed in the AES spectrum from the implanted SiO_2.   Presumably the energy difference between the Auger peaks associated with the two chemical states of P was too small to be detected with our instrumentation.   These measurements were made using a Varian scanning Auger spectrometer (Model 981-2730) equipped with a single pass cylindrical mirror analyzer.   With this instrumentation, a difference of about 0.3 eV should have been measurable.   As mentioned earlier a large chemical shift (4 eV) was observed in the ESCA measurements.   We note that Chang et al[1] have reported observing two P LVV peaks during sputter profiling studies of P-doped vapox.

It is known that under electron bombardment, lattice oxygen can be desorbed from SiO_2 and the surface becomes enriched in "free" Si.[5,6] It has also been reported that bulk dopants such as P and Cl can outdiffuse to the surface of SiO_2 under electron bombardment.[1,7]   By using P-doped vapox, one has the opportunity to study both these effects in the same sample.   In Fig. 3, the Auger data on such a P-doped vapox sample are given which show the Si in SiO_2 peak at 76 eV, "free" Si peak at 92 eV, and the P

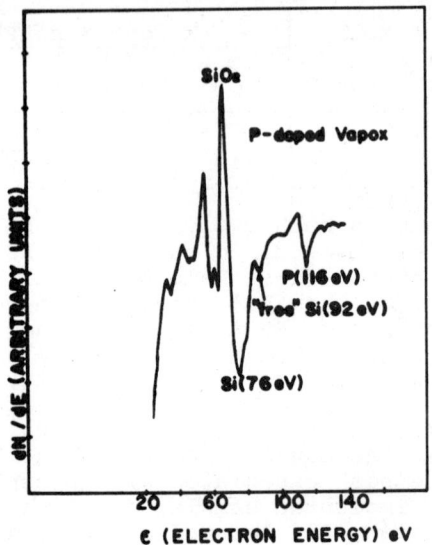

Fig. 3.   Auger data of the P-doped vapox showing the Si in
SiO2 peak at 76 eV, "free" Si peak at 92 eV, and
the P peak at 116 eV.

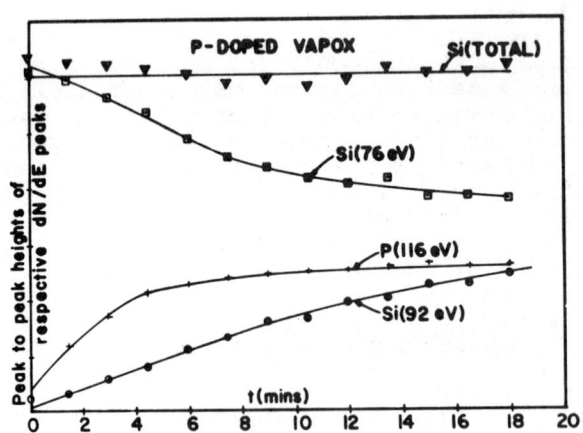

Fig. 4.   Growth of "free" Si at 92 eV and P at 116 eV peaks,
and decrease of Si in SiO2 at 76 eV peak, under
electron bombardment.

peak at 116 eV. Their variations under electron bombardment were studied.

Fig. 4 shows the effect of electron bombardment (2 KeV, 3 μA, 0.3 A/cm^2) on the vapox sample doped with about 8% P by weight. Initially, both the P and "free" Si peaks increase linearly with the time of electron bombardment (i.e., with total exposure). However the P signal quickly saturates while the "free" Si peak continues to grow, although at a slower rate after about 9 minutes. When the current density was reduced, the growth rates of both the P and Si peaks were reduced proportionately. The sum of the Si 92 eV and Si 72 eV (associated with the oxide) peak heights remain constant with exposure reflecting the conservation of surface Si atoms. The growth kinetics of the Si peak in thermal oxide was identical to that observed in the vapox doped with P under the same electron bombardment conditions (energy, current density, etc.). The presence of P then does not appear to affect the growth kinetics of free Si in vapox for the concentrations of P used here. After about 18 minutes, the electron beam was turned off and the vapox allowed to remain in UHV for 12 hours. When the electron beam was then turned on, the strength of the P and Si peaks were unchanged. P did not diffuse back toward the bulk at room temperature in the absence of the electron beam.

More work is needed to understand this interesting system and better characterize the behavior of P under electron bombardment. We are presently investigating the dependence of the outdiffusion rate of P on the electric field strength at the surface and whether this outdiffusion is enhanced by the electron induced chemical reduction in the SiO_2 surface region. Also, the effect of implanted P dose and various annealing conditions on the two chemical states of P in vapox is under evaluation.

The authors wish to thank J. Verma and R. Corso for preparing the vapox samples, J. Lee for doing phosphorus implantations, and W. E. Spicer and A. Schifrin for helpful discussions.

1. C. C. Chang, A. C. Adams, G. Quintana and T. T. Sheng, J. Appl. Phys. 45, 252 (1974).

2. S. A. Schwarz, C. R. Helms and W. E. Spicer (to be published J. Vac. Sci, Technol.).

3. J. S. Johannessen, W. E. Spicer, J. F. Gibbons, J. D. Plummer and N. J. Taylor (to be published J. Appl. Phys.).

4. Handbook of Auger Electron Sprectroscopy, 2nd ed., L. E. Davis, N. C. MacDonald, P. W. Palmberg, G. E. Riach and R. E. Webster (Physical Electronics Industries, Inc., Eden Prairie, MN, 1976).

5. B. Carriere and B. Lang, Surface Sci. 64, 209 (1977).

6. J. S. Johannessen, W. E. Spicer and Y. E. Strausser, Appl. Phys. Lett. 27, 452 (1975).

7. N. J. Chou, C. M. Osburn, Y. J. van der Meulen and R. Hammer, Appl. Phys. Lett. 22, 380 (1973).

SPECTROSCOPIC AND STRUCTURAL PROPERTIES OF NITROGEN
DOPED LOW-TEMPERATURE SiO_2 FILMS

Gordon Wood Anderson, William A. Schmidt, and
James Comas[*]
Naval Research Laboratory, Washington, D. C. 20375,
U.S.A.

ABSTRACT

The stoichiometry, contamination, chemical bonding, and electrical properties
of nitrogen doped, low-temperature SiO_2 films prepared by the pyrolytic decomposition of silane in an ammonia rich atmosphere on InSb and Si substrates have been studied by attenuated total reflection (ATR), Auger, electrical, ellipsometry, nuclear reaction analysis, optical transmission,
Rutherford backscattering (RBS), and secondary ion mass spectroscopy (SIMS)
techniques. The films were quite similar to thermal SiO_2 films in their
physical properties as demonstrated by each of the above techniques and had
uniform Si and O concentrations. The N concentration in the films was
between 1 to 3% of the O concentration atomically, depending on growth parameters. The optical results demonstrated conclusive evidence of Si-H bonding but inconclusive evidence of Si-N bonding in the films.

INTRODUCTION

Nitrogen doped SiO_2 films grown by chemical vapor deposition at low-temperatures are of importance in the study of the fundamental physical and chemical properties of doped SiO_2 films and also have important practical uses in
the development of advanced infrared sensor arrays. Making reliable metal-
insulator-semiconductor (MIS) devices on narrow bandgap semiconductors such
as InSb is essential to developing these arrays. The major problem in
making the MIS devices is growing a thin film insulator with good physical
properties on the semiconductor substrate. This must be done at a low
enough temperature to be compatible with the semiconductor material, about
$200^{\circ}C$ in the case of InSb. Empirically, the films used successfully on InSb
contained Si, O, and N in an SiO_xN_y composition where $x \simeq 2$ and $y \ll x$.
The films were grown by the pyrolytic decomposition of silane in the presence
of oxygen and ammonia.

The electrical properties of MIS devices made using these nitrogen doped,
low-temperature SiO_2 insulating films have been studied extensively
(Ref. 1-4), but the stoichiometric, chemical, and other physical properties
of these films have only recently been studied (Ref. 5). The results of a

---

[*]On leave at the School of Electrical Engineering, Cornell University,
Ithaca, New York 14853, U.S.A.

series of studies of the spectroscopic and physical properties of these films on InSb and Si substrates carried out by ATR, Auger, electrical, ellipsometry, nuclear reaction analysis, optical transmission, RBS, and SIMS techniques are presented in this and related papers (Ref. 5). This work is the first detailed effort to characterize the chemical bonding, stoichiometric, and structural properties of these films.

INSULATOR GROWTH

The films were grown by the pyrolytic decomposition of silane in the presence of oxygen and ammonia at temperatures in the range 200-250°C. At elevated temperatures in the presence of oxygen, a film consisting of both indium oxide and antimony oxide grows on clean InSb surfaces, but the indium oxide grows faster than the antimony oxide. In addition, most of the free Sb and antimony oxide evaporate leaving a film consisting primarily of In_2O_3 (Ref. 6). Indium from the substrate diffuses more readily than Sb through the In_2O_3 film and forms more In_2O_3. Consequently, an Sb rich layer remains at the oxide/substrate interface (Ref. 6) and very likely has a degrading effect on the electrical properties of MIS devices using such an oxide and an InSb substrate. This potential sequence of events is thought to bring about the necessity of growing passivating layers on InSb at low-temperatures.

Three general reactor designs were used (Fig. 1), and the best results were obtained using the designs A and C shown in Fig. 1. In each case in Fig. 1, a silane and nitrogen gas mixture was introduced at one port and an ammonia, oxygen, and nitrogen gas mixture was introduced at the second port. The film growth rate decreased with increasing ammonia flow rate (Ref. 5). The ammonia thus acted as a modifying agent which retarded the pyrolytic decomposition of the silane possibly by inhibiting an intermediate reaction of the total silane decomposition reaction (Ref. 7). Detailed information on the growth process and growth process parameter variations is given in Ref. 5.

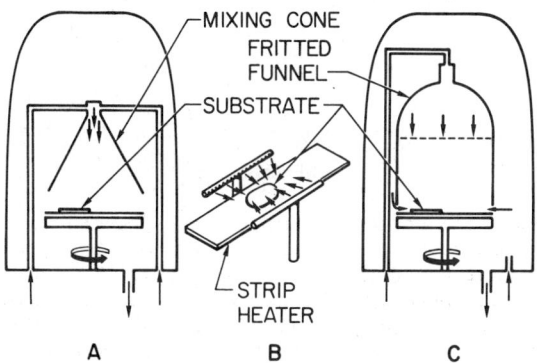

Fig. 1.   Schematic illustrations of the three reactor designs used in this study. Stationary strip heater and gas ports in design B also enclosed in bell jar similar to those shown in designs A and C.

RESULTS AND DISCUSSION

The experimental results indicated that the low-temperature SiO_xN_y films were very similar to thermal SiO_2 films in their physical properties but had a small N concentration, 1 to 3% of the 0 concentration on an atomic basis (Ref. 5). The quantitative value of the N concentration was obtained by nuclear reaction analysis. Thus these films are more appropriately thought of as doped SiO_2 films rather than silicon-oxy-nitride films. This point was indicated directly or indirectly by several experimental methods including ATR, Auger spectroscopy, ellipsometry, nuclear reaction analysis, optical transmission, RBS, and SIMS. In addition, the SIMS results indicated that the films were uniform in composition and very similar to SiO_2 in stoichiometry (Ref. 5).

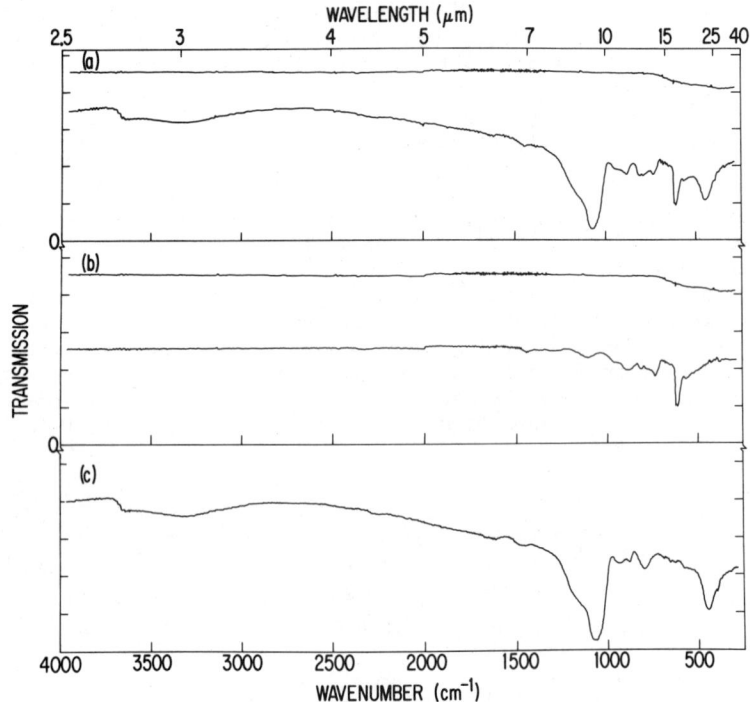

Fig. 2. Room-temperature optical transmission spectra: (a) nitrogen doped SiO_2 films on both sides of Si substrate (lower curve) and 100% transmission line (upper curve); (b) clean Si substrate (lower curve) having approximately same thickness as coated substrate [spectrum shown in (a)] and 100% transmission line (upper curve); (c) relative transmission spectrum of SiO_2 coated Si substrate [spectrum shown in (a)] with clean Si substrate [spectrum shown in (b)] in reference beam showing sharp band at about 880cm⁻¹.

The optical properties of the films were very similar to those of thermal SiO_2 films (Ref. 5). An unresolved question is whether or not Si-N bonding occurs in these low-temperature films, and new data (Fig. 2) demonstrate this point. A weak, narrow band was observed in the spectra of these films [Fig. 2(a) and 2(c)] at about $880cm^{-1}$ at the energy at which a broad Si-N vibrational band is observed in amorphous Si_3N_4 films (Ref. 8 and 9). This band was quite sharp [Fig. 2(a) and 2(c)], however, and its sharpness casts some doubt on this identification. The absorption at this energy also could be due to a weak, sharp Si-H vibrational band (Ref. 8). No other absorption bands were observed which could be attributed to Si-N type vibrations. The two bands at about $3255cm^{-1}$ and $3600cm^{-1}$ observed in the ATR spectra (Ref. 5) were observed in transmission as well [Fig. 2(a) and 2(c)] and are attributed to either N-H or O-H type vibrations or to both. The ATR band at about $3255cm^{-1}$ was observed at about $3320cm^{-1}$ in direct transmission [Fig. 2(a) and 2(c)]. The data shown in Fig. 2 were taken at room-temperature using a Perkin-Elmer 512 spectrophotometer in the double beam mode.

Thus, these nitrogen doped low-temperature SiO_2 films are uniform in composition and are very similar to thermal SiO_2 films in their physical characteristics. Moreover, these films are significantly promising for use in MIS electronic devices, and successful integrated circuits using InSb as the substrate and these films as the insulator have been prepared (Ref. 4).

ACKNOWLEDGEMENTS

The authors thank H. M. Day, J. K. Hirvonen, and L. Plew for carrying out several experiments and W. D. Baker, R. E. Dehl, E. D. Palik, M. L. Rebbert, N. S. Saks, and L. E. Smith for numerous, valuable discussions.

REFERENCES

1. C. E. Hurwitz and J. P. Donnelly, Planar InSb Photodiodes Fabricated by Be and Mg Ion Implantation, Solid-State Electronics 18, 753 (1975).
2. J. Shappir, S. Margalit, and I. Kidron, p-Channel MOS Transistor in Indium Antimonide, IEEE Trans. Electron Devices ED-22, 960 (1975).
3. J. C. Kim, Interface Properties of InSb MIS Structures, IEEE Trans. Parts, Hybrids, and Packaging PHP-10, 200 (1974).
4. J. C. Kim, "InSb MOS Detector," Final Technical Report, U.S. Army Electronics Command, Night Vision Laboratory, Contract No. DAAK02-73-C-0006, February, 1975, Department of Defense Documentation Center, AD No. B004428L; J. C. Kim, "Fabrication and Evaluation of InSb CID Arrays," Final Report, Naval Electronics Systems Command, Defense Advanced Research Projects Agency, and Naval Research Laboratory, Contract No. N00014-75-C-0124, August, 1976, National Technical Information Service, AD No. A030022/8GI; and J. C. Kim, W. E. Davern, and D. Colangelo, "Continued Development of Indium Antimonide CID Arrays," Final Report, Naval Electronics Systems Command and Naval Research Laboratory, Contract No. N00173-76-C-0128, May, 1977, National Technical Information Service, AD No. A039990.
5. G. W. Anderson, W. A. Schmidt, and J. Comas, Composition, Chemical Bonding, and Contamination of Low Temperature SiO_xN_y Insulating Films, J. Electrochem. Soc. 125, 424 (1978); and G. W. Anderson, W. A. Schmidt, and J. Comas, Spectroscopic, Structural, and Electrical Properties of Low-Temperature SiO_xN_y Insulating Thin Films, in (1978) Proceedings of the Symposium on Thin Films: Interfaces and Interactions, Ed. by J.E.E. Baglin and J. M. Poate, Electrochemical Society, Princeton, N. J., U.S.A.

204

6. C. W. Wilmsen, Oxide Layers on III-V Compound Semiconductors, <u>Thin Solid Films</u> 39, 105 (1976); and A. J. Rosenberg and M. C. Lavine, The Oxidation of Intermetallic Compounds. I. High Temperature Oxidation of InSb, <u>J. Phys. Chem</u>. 64, 1135 (1960).
7. H. J. Emeléus and K. Stewart, The Oxidation of the Silicon Hydrides. Part I, <u>J. Chem. Soc. (Lond.)</u>, 1182 (1935).
8. E. A. Taft, Characterization of Silicon Nitride Films, <u>J. Electrochem. Soc</u>. 118, 1341 (1971).
9. Yu. N. Volgin and Yu. I. Ukhanov, Vibrational Spectra of Silicon Nitride, <u>Opt. Spektrosk</u>. 38, 727 (1975).

SOME OBSERVATIONS OF DEFECTS IN AMORPHOUS SiO_2 FILMS

E. A. Irene

IBM Thomas J. Watson Research Center
Yorktown Heights, New York 10598

INTRODUCTION

The atomic structure of amorphous SiO_2 is well known (see for example Refs. (1) and (2)) and found to be essentially invariant despite such diverse preparation techniques as glow discharge, thermal oxidation and SiO_2 glass blown from the melt (2). However, it is also known that for many amorphous materials, the use of different methods of preparation oftentimes results in different physical properties (3) for a given material. This phenomenon is quite commonplace within the field of crystalline materials and the defect structure of the material is frequently and justifiably used to explain the variability of such properties as electrical conductivity, density, stress, strength, diffusivities, dielectric breakdown strength etc.

The present study is a presentation of a variety of defects observed in amorphous SiO_2 films by transmission electron microscopy (TEM). In order to demonstrate that a large number of different defects can be observed, a variety of techniques common to the MOS field were employed to prepare the SiO_2 films: thermal oxidation of single crystal Si , polycrystalline Si, heavily B and P doped Si; thermal oxidation in one and five hundred atmospheres of O_2, in H_2O and HCl plus O_2; chemically vapor deposited (CVD) SiO_2; SiO_2 doped with Au and NaCl; and SiO_2 with high temperature annealing. For convenience the defects to be shown are categorized as processing, substrate and impurity related. In most cases the appearance of defects in SiO_2 can be related a change in an important electrical property. Correlations are made where data is available.

What emerges from this study is a picture of the morphology of SiO_2 films in which the variability of many properties of the SiO_2 films can be explained by the defects grown into the films by particular treatments. This picture is by no means complete yet it is compelling enough to warrant further research. The concept of describing an amorphous material such as SiO_2 in terms of its defect structure appears to be valid.

RESULTS AND DISCUSSION

Processing Related Defects. Fig. 1 a-c shows a comparision of perfect SiO_2 (a) with SiO_2 which contains Si inclusions (b and c). The size and number of the inclusions are larger for films grown at higher oxidation temperatures and for films exposed to high temperature inert gas anneals. The presence of HCl in the oxidation ambient also seemed to increase the number of Si inclusions (Fig. 4c).

The presence of this defect correlates with an increase in SiO_2 dielectric breakdown defect density (4). Osburn and Ormond (4) found that the defect density increased with the temperature of oxidation and to a greater extent for post oxidation inert gas anneals.

The occurrence of the Si inclusions can be explained by considering a chemical transport (5) mechanism in which Si is converted to gaseous SiO at the Si - SiO_2 interface by the reversible reaction:

$$Si(s) + SiO_2(g) = 2 SiO(g)$$

and then diffuses away from the interface where the SiO disproportinates producing Si deposits in the SiO_2 film. Studies of the sublimation of SiO_2 (6,7) have demonstrated that gaseous SiO is a predominat species above Si - SiO_2 mixtures under both inert and oxidizing conditions.

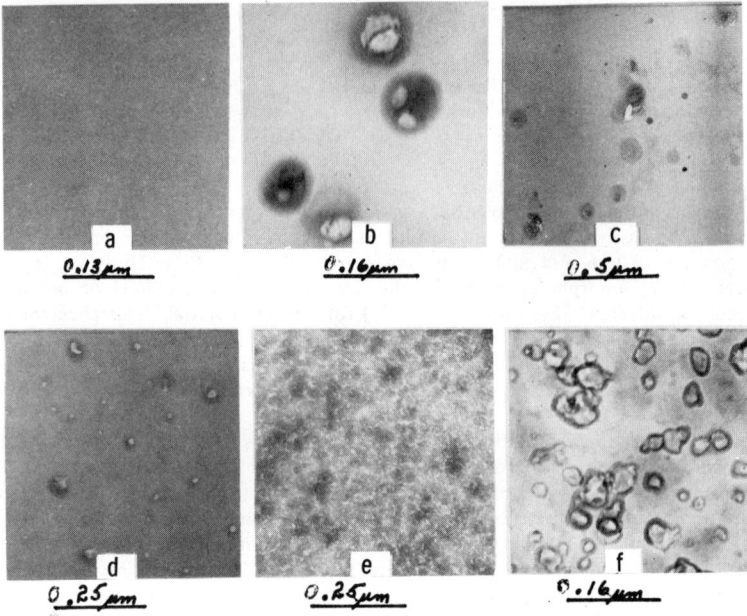

Fig. 1 TEM results on SiO_2 prepared by a variety of techniques: a) 1100°C SiO_2 perfect area, b) with Si inclusions, c) annealed at 1150°C for 24 hrs., d) CVD SiO_2, e) Si rich SiO_2, f) a defective area of SiO_2 prepared at 800°C and 500 atm O_2.

Preliminary thermochemical calculations using equilibrium values for the appropriate constants indicate that the amount of Si produced by the above reaction at 1100°C is marginally sufficient to exceed the saturation pressure for Si. However, under real conditions where the system is away from equilibrium the situation may be even better and further studies are in progress.

For the case where HCl is present a parallel chemical transport mechanism is envisioned in which the well known silicon chlorides are produced which diffuse and disproportionate to produce Si as for the oxygen case. The chemical transport reactions of Si via oxygen and chlorine are well known in the literature (5).

CVD SiO_2 films generally exhibit inferior electrical properties (in terms of increased breakdown defect densities, electron trapping and conductivity). Two types of defects have been identified which may explain the diminished quality of CVD SiO_2. The first type is an occlusion of particles of SiO_2 in the SiO_2 film (Fig. 1 d). Particulate of this type is observed to some degree in all CVD films and is believed (8) to originate from gas phase reaction, nucleation and particle growth in the CVD reactor in an upstream position relative to the substrates. The occluded particles provide surfaces which disrupt the continuity of the chemical bonding in the film thereby creating trapping sites and distortions in the applied fields. The second type of defect is due to Si enrichment of the SiO_2 film (Fig. 1e). Si rich SiO_2 films have been found to consist of two phases (9), a Si phase and an SiO_2 phase.

Preferential etching of the Si from the SiO_2 (9) has shown that the Si phase is a connected phase. This connectivity accounts for the increased conductivity for Si rich films (9).

Recently, the thermal oxidation of Si using high pressure oxygen has received attention (10). The technique provides a considerable oxidation rate enhancement and is therefore technologically important. Although these films possess acceptably low charge and surface state levels, recent prelimiary measurements have shown that some samples have a high breakdown defect density (11). Fig. 1f shows that defective samples have numerous disk like amorphous inclusions. The origin and nature of these defects is at present unknown. However, this particulate is undoubtedly related to the breakdown defect density.

Substrate Related Defects. The oxidation of polycrystalline Si (Poly) provides an oxide (Poly-ox) which has more breakdown defects, larger currents at any given field and more trapping than SiO_2 prepared by oxidation of single crystal Si. This has been attributed to asperities on the Poly surface (12,13). Fig. 2 shows that Poly-ox has variations in thickness which replicates the previous Si grain structure. It is believed that the undulations in the SiO_2 are caused by oxidation occuring in the grain boundary thereby causing Si to recede from the boundary region. The absence of Si causes oxidation to cease in the boundary regions. Since the surface area of the thinner regions is far smaller than the normal area, large thickness errors can be expected from conventional measurement techniques. Consequently, applied fields based on voltage and film thickness will be low compared with the actual field in the boundary regions hence the larger currents and number of defects are explainable.

Impurity Related Defects. Previously (14) it was found that thin SiO_2 (<200Å) grown with H_2O in the oxidizing ambient had fewer breakdown defects than SiO_2 grown in dry O_2 (Fig. 3). It was also observed (14) that the growth kinetics of these thin films in H_2O was decidedly more parabolic. Since the parabolic oxidation mode is attributed to a diffusion of oxidant limitation, the H_2O grown films are considered to be better diffusion barriers hence more protective. Fig. 4a shows that the thin films dissolve heterogenously. This means that the films contain small (<50Å) inhomogeneities. The existence of micropores in SiO_2 was hypothesized to explain the dielectric breakdown, TEM and oxidation kinetics results (14).

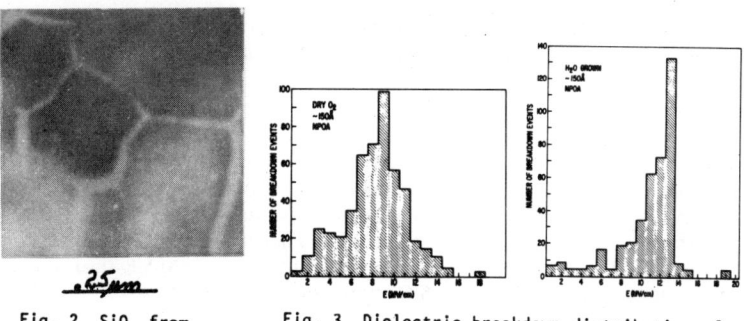

Fig. 2 SiO_2 from thermal oxidation of Polycrystalline Si.

Fig. 3 Dielectric breakdown distributions for 150Å SiO_2 prepared using a) ultra dry O_2 b) H_2O in N_2.

Figure 4b shows the result of B_2O_3 phase separation from an SiO_2 phase during oxidation of heavily B doped Si (15). This phase separation is expected based on studies of the metastable liquidus boundary in this system (16). Since the amount of phase separation is small, the overall SiO_2 reliability implications are unclear. However, due to the phase boundaries, enhanced trapping is anticipated.

Fig. 4c shows the previously mentioned Si inclusions due to HCl. Although benefical effects in terms of charge levels are attributed to HCl oxidations, it is predicted that large amounts of HCl plus high temperatures will increase dielectric breakdown defects.

Fig. 4d and e show the result of Au and NaCl reactions with SiO_2 films. Au reacts to from reduced areas (17). NaCl reacts leaving regions which are preferentially attached by the etchant used to prepare these TEM samples. For both cases the damage to SiO_2 is extensive and will undoubtedly increase the breakdown defect density and this has been reported for the NaCl case (18).

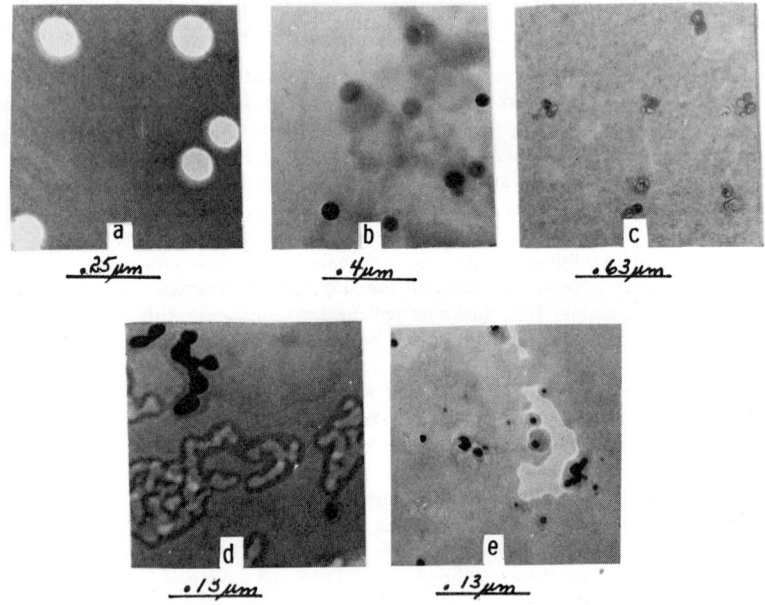

Fig 4 TEM results on SiO_2: a) 150Å SiO_2, b) B doped, c) prepared with HCl in O_2, d) Au and e) NaCl doped and annealed.

CONCLUSIONS

A large number of defects have been observed in SiO_2 films. The defects are related to processing, the substrates used and to impurities. In many cases the defect structure has caused a degradation of the dielectric reliability. The defect structure of amorphous SiO_2 appears to be a valid means to describe the behavior of some of the physical properties of the material. Further studies relating other SiO_2 properties of the defect structures are in progress.

ACKNOWLEDGEMENT

The author acknowledges D. Dong and J. Kucza for the CVD films and HCl oxides respectively, Drs. R. J. Zeto for the SiO_2 prepared using 500 atm O_2, D. J. DiMaria and C. M. Osburn for helpful discussions of asperity and breakdown effects and to Dr. D. R. Young for critically reviewing this manuscript. This research was partially supported by the Defense Advanced Research Projects Agency and monitored by the Deputy for Electronic Technology, RADC, under Contract F19628-76-C-0249.

REFERENCES

(1)   B. E. Warren, X-Ray Determination of the Structure of Liquids and Glass, J. Appl. Phys., 8, 645 (1937).

(2)   M. V. Coleman and D. J. D. Thomas, The Structure of Silicion Oxide Films, Phys. Stat. Sol., 22, 593 (1967).

(3)   R. Roy. Classification of Non - Crystalline Solids, J. of Non-Crystalline Solids, 3, 33 (1970).

(4)   C. M. Osburn and D. W. Ormond, Dielectric Breakdown in Silicon Dioxide Films on Silicon, J. Electrochem. Soc., 119, 597 (1972).

(5)   Schäfer, H. (1964) Chemical Transport Reactions, Academic Press, New York.

(6)   R. F. Porter, W. A. Chupka, and M. G. Inghram, Mass Spectrometric Study of Gaseous Species in the Si-SiO$_2$ System, J. Chem. Phys., 23, 216 (1955).

(7)   K. F. Zmbov, L. L. Ames, and J. L. Margrave, A Mass Spectrometric Study of The Vapor Species over Silicon and Silicon Oxides, High Temperature Science, 5, 235 (1973).

(8)   S. Zirinsky and E. A. Irene, Selective Studies of Chemical Vapor-Deposited Aluminum Nitride - Silicon Nitride Mixture Films, J. Eletrochem. Soc., 125, 305 (1978).

(9)   D. Dong, E. A. Irene and D. R. Young, Preparation and Some Properties of Chemically Vapor Deposited Si Rich SiO$_2$ and Si$_3$N$_4$ Films, J. Electrochem. Soc., in press.

(10)   R. J. Zeto, C. G. Thornton, E. Hryckowian and C. D. Bosco, Low Temperature Thermal Oxidation of Silicon by Dry Oxygen Pressure above 1 Atm, J. Electrochem Soc., 122, 1409 (1975).

(11)   C. M. Osburn and R. J. Zeto, unpublished results.

(12)   D. J. DiMaria and D. R. Kerr, Interface Effects and High Conductivity in Oxides Grown From Polycrystalline Silicon, Appl. Phys. Lett., 27, 505 (1975).

(13)   R. M. Anderson and D. R. Kerr, Evidence for Surface Asperity Mechanism of Conductivity in Oxide Grown On Polycrystalline Silicon, J. Appl. Phys., 48, 4834 (1977).

(14)   E. A. Irene, Silicon Oxidation Studies: Some Aspects of The Initial Oxidation Regime, J. Electrochem. Soc., submitted for publication.

(15)   E. A. Irene and D. Dong, Silicon Oxidation Studies: The Oxidation of Heavily B and P Doped Single Crystal Silicon, J. Electrochem. Soc., in press.

(16)   R. J. Charles and F. E. Wagstaff, Metastable Immiscibility in the B$_2$O$_3$-SiO$_2$ System, J. Amer. Ceram. Soc., 51 16 (1968).

(17)   E. I. Alessandrini, D. R. Campbell, and K. N. Tu, Interfacial Reaction in MOS Structures, J. Vac. Sci. Technol., 13, 55 (1976).

(18)   C. M. Osburn and D. W. Ormond, Sodium Induced Barrier - Height Lowering and Dielectric Breakdown in SiO$_2$ Films on Silicon, J. Elctrochem Soc., 121, 1195 (1974).

210

MEASUREMENT OF HYDROGEN PROFILES IN SiO_2 BY A NUCLEAR
REACTION TECHNIQUE.

D.D. Allred,* C.W. White, G.J. Clark,** B.R. Appleton,
Solid State Div., Oak Ridge National Lab., Oak Ridge, TN.
I.S.T. Tsong,
Mat. Res. Lab., Penn. State Univ., University Park, PA.

ABSTRACT

The nuclear reaction $^1H(^{19}F,\alpha\gamma)^{16}O$ was used to measure the hydrogen concentration and depth profile in natural and synthetic quartzes and in an SiO_2 film
grown by oxidation of Si in an HCl ambient. A program was developed for extracting true hydrogen profiles from the γ-ray yield measured as a function of
projectile energy. Many samples possess a region (.1 to .3 μm thick) at the
sample's surface where the hydrogen concentration is much higher than that at
1 to 4 μm. This is believed to be due to hydration of the surface. These results provide a plausible explanation for the disagreement observed in other
methods of hydrogen analysis. Evidence was found for hydrogen mobility in
crystalline SiO_2 under ion beam bombardment.

INTRODUCTION

The concentration and depth distribution of hydrogen in SiO_2 is a matter of
some importance. Work beginning with Griggs and Blacic (1) has demonstrated
that the introduction of small amounts of water into the quartz structure will,
under certain conditions, reduce the mechanical strength of quartz by an order
of magnitude. This observation clearly has important implications in the study
of tectonic processes.

Tsong et al. (2) used IBSCA (Ion Beam Spectrochemical Analyzer) to determine
the hydrogen content of a number of silicates. The results show values which
are one to two orders of magnitude above those determined by infrared absorption for similar samples. There is, of course, the possibility that hydrogen
may be present in another form (3,4). It is also difficult to allow for changes
in the molar extinction coefficient with band frequency in complex OH spectra
in infrared determinations (5).

In view of the important implications associated with high hydrogen content in
quartz we have investigated these discrepancies by the use of a technique which
is not sensitive to the bonding or chemical state of the hydrogen to provide an
estimate of the average hydrogen content near the surface and at a depth of a
few microns for a wide range of natural and synthetic SiO_2 samples and to measure the depth distribution of hydrogen in several of these samples. The samples used, with one exception, were bulk silica samples several mm thick.

*NSF Energy-related Postdoctoral Fellow now at Opt. Sci. Cen., Univ. of Arizona.
**Present address: Mineral Physics Laboratory, C.S.I.R.O. North Ryde, Australia.

TECHNIQUE

The use of isolated resonances in nuclear reaction cross sections for quantitative hydrogen analysis and depth profiling has been described elsewhere (6).
Consequently, only a brief description of the technique will be given here. The
reaction $^1H(^{19}F,\alpha\gamma)^{16}O$ exhibits a strong resonance at 16.44 MeV in the laboratory system with a peak cross section of 0.5 barn and a width of about 90 keV
producing 6.1, 6.9 and 7.1 MeV γ-rays from the de-excitation of the residual
excited ^{16}O nucleus. It is often convenient to regard the nuclear resonance
used for depth profiling as a delta function probe. Then, if a material containing hydrogen is irradiated with ^{19}F ions at an energy slightly greater than
the resonance energy, the ^{19}F ions will be slowed down until at a certain depth,
X_R, the resonance energy will be reached and the resonance reaction will occur,
producing γ-rays of the appropriate energy. In this ideal case, the yield of
γ-rays will be proportional to the hydrogen concentration at the depth X_R. The
depth X_R at which the resonance occurs is determined by the stopping power dE/dX
for ^{19}F ions in the sample under analysis. A depth profile of the hydrogen concentration is built up when the incident ion energy is raised in steps, so that
at each step the resonance occurs deeper within the solid.

The experimental apparatus is described in ref. 6. A 30 nA $^{19}F^{+3}$ beam from the
ORNL Tandem Van de Graaff collimated to .4 cm diameter was employed. NaI(Tl)
detectors were used for γ-ray detection; and materials of known hydrogen concentration were used for calibration. All of the SiO_2 samples except the one thin
film had a thin carbon overlayer (~500 Å thick) deposited to avoid charge buildup during bombardment with the ^{19}F analyzing beam.

RESULTS AND ANALYSIS

Figure 1 shows the γ-ray as a function of beam energy (and depth) in a natural
smoky quartz single crystal. The profile exhibits three distinct features.
They are (a) a peak centered at 16.45 MeV due to hydrogen surface contamination,
(b) a broad region below the surface of approximate thickness 2000 Å where the
γ-ray yield is high and (c) a region deep in the crystal where the γ-ray yield
is low. These three regions were seen in most of the bulk quartz samples examined.

If the reaction's cross section had a delta function shape, the curve in Fig. 1
would give the hydrogen depth profile. The $^1H(^{19}F,\alpha\gamma)^{16}O$ reaction does not
have this structure (other nuclear probes also share several of these characteristics) in that:
    (1) Straggling and the finite resonance width limits the depth reso-
lution ΔX to about 200 Å at the surface of quartz and to about 300 Å at depths
of 4000 Å For example, the surface of all non-amorphous samples we examined
possessed a layer of 1-2 X10^{16} H atom cm^{-2} contamination. Since the measured
width of the surface peak was approximately 95 keV, which is in agreement with
the known resonance width, all that can be said about the thickness of this
layer is that it is >200 Å.
    (2) The off-resonance cross section is non-zero. As a result, the
γ-ray yield in the region (labeled C) of Fig. 1 is partially (2/3 of the yield
at 17.4 MeV) due to off-resonance nuclear reactions occurring with hydrogen on
the surface (A) and in the near-surface region (B).
    (3) There are other nuclear resonances between ^{19}F and ^1H. According
to our measurements the total cross section below 16.1 MeV is 1/3 the total
cross section between 16.1 and 17.25 MeV. An unfolding procedure which originated with this group (6) uses this fact to estimate the average hydrogen con-

212

centration in the 1.5 to 4.0 μm depth range.

The corrected depth profile for smoky quartz is shown in Fig. 2. Detailed hydrogen profiles were not obtained for all samples. On some samples, yield data was only taken at one energy below the resonance (usually 16.1 MeV) and one (usually 17.25 MeV) or more above the resonance. To aid in the interpretation of the yield data in terms of hydrogen concentrations for these samples, we assumed a near surface region of constant hydrogen concentration h_1 and a deep or bulk region of constant hydrogen concentration h_2. If the surface hydrogen contamination was not measured, it was assigned the value 1.5×10^{16} H atom cm^{-2},[6]. The boundary between the near surface and bulk regions was measured to be about .13 μm for suprasil and vitreosil, .2 μm for X_0, and .24 μm for smoky; .3 μm was used for other samples, but results are largely independent of the value chosen. Our observations above together with those of Lanford (ref. 7) suggest that this model is satisfactory to the first order.

The data analysis procedure is further discussed in ref. 6. The results of these analyses for a set of natural and synthetic quartz samples are given in Table 1. A computer program, called UNFOLD, was written to aid in unfolding samples such as smoky and X_0 where three or more yield measurements had been obtained.

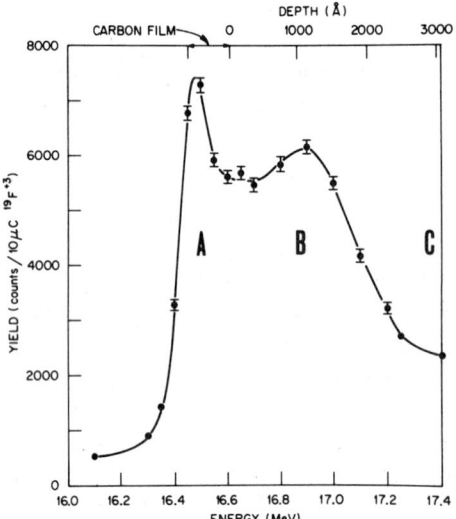 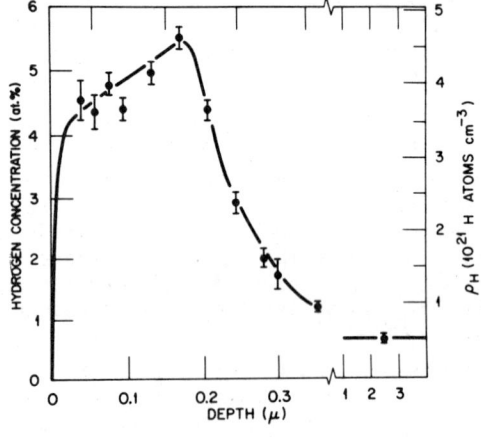

Fig. 1: γ-ray yield for a smoky natural quartz single crystal.

Fig. 2: Unfolded hydrogen profile for the smoky natural quartz crystal of Fig. 1.

DISCUSSION

A source of uncertainty not included in the results arises from hydrogen mobility under ion beam bombardment in crystalline quartz samples. Of the twelve samples mentioned in Table 1, Bell, X_0, W_2, W_4, Smoky and Toyo were single crystals and quartzite was polycrystalline. The mobility of light ions under irradiation in both amorphous and crystalline materials is well established (8). The yield of γ-rays as a function of integrated beam charge is shown in Fig. 3 for a range of amorphous and crystalline. For the fused silica samples (vitreosil and suprasil) there was no evidence for hydrogen mobility. The quartz and opal samples show evidence of hydrogen mobility in that the γ-ray yield dropped as a function of integrated charge. The decrease in yield indicated mobility out of the beam path; there was no evidence for mobility into the beam path. This effect implies that the hydrogen values measured are lower limits for crystalline samples. Data, such as Fig. 3, indicate that the results may in fact be as much as 40% low. No such uncertainty is to be associated with the results quoted for the fused silica samples. The diffusion process is probably driven by an electric field in the crystal resulting from the electric charge deposited in the crystal by the ^{19}F beam. This could be checked by damaging samples with a neutron flux prior to analysis. We anticipate that the hydrogen preferentially diffuses along the C axis (9). It was surprising to see no evidence of hydrogen mobility in the fused silica samples, vitreosil and suprasil, as sodium mobility in amorphous SiO_2 films under irradiation has been observed (8).

Table 1: Hydrogen in SiO_2 (10^{20} H atom/cc)

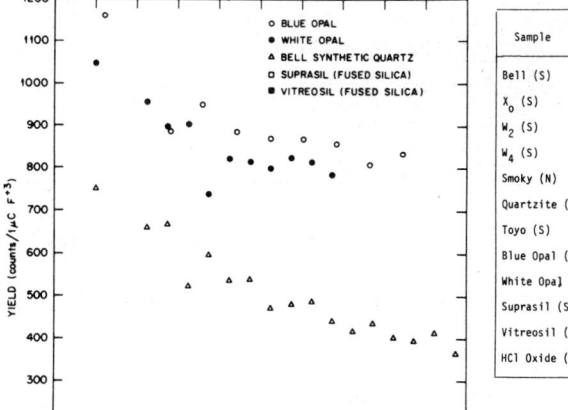

Sample	Near Surface Region (0 to 0.2 μm)	Bulk Region (0.2 μm to 4 μm)	IBSCA[a]	IR and Thermal
Bell (S)	85.3 ± 8.5	0 ± 0.5	---	---
X_0 (S)	16.5 ± 2.3	1.4 ± 0.8	36.7	2.34[b], 1.1[c]
W_2 (S)	8.2 ± 1.6	7.7 ± 4.8	20.7	.43[d]
W_4 (S)	10.1 ± 1.6	4.2 ± 2.4	9.6 - 28.7	0.19[d]
Smoky (N)	39.3 ± 5.3	5.3 ± 1.6	9.6	~.003[e]
Quartzite (N)	109. ± 12.	69. ± 15.	6.4 - 22.3	---
Toyo (S)	4.0 ± 1.6	3.1 ± 2.4	4.8	---
Blue Opal (N)	72. ± 11.	89. ± 16.	---	---
White Opal (N)	84. ± 12.	69. ± 16.	---	---
Suprasil (S)	25. ± 4.	1.3 ± 0.8	27.	---
Vitreosil (S)	27. ± 4.8	0.8 ± 0.5	16.	---
HCl Oxide (S)	2.9 ± 0.8	0.0 ± 0.7	---	---

S denotes synthetic sample, N denotes natural sample.
[a]Tsong, McLaren and Hobbs (1976)(Ref. 2)   [d]Hobbs, et al. (1972)(Ref. 11)
[b]Griggs (1967)(Ref. 8)   [e]Brunner, et al. (1961)(Ref. 12)
[c]Godbeer and Wilkins (1977)(Ref. 10)

Fig. 3: Apparent hydrogen concentration as function of integrated charge.

Despite these uncertainties, repeated measurements indicated that the general form of the profiles shown in Figs. 1 and 2 is correct, i.e., the profiles consist of a region just below the surface of high hydrogen concentration and a region in the bulk of the material with a much lower concentration. As discussed by Lanford (7), this is probably produced by water diffusing in from the surface of the crystal and interacting with the SiO_2 lattice to produce SiOH. The thickness of the high hydrogen concentration region is probably equal to the depth to which the water has diffused into the silica sample, and thus is a function of the exposure time of the sample's surface.

The hydrogen concentrations in the bulk 1-4 μm region of the samples listed in Table 1 are generally lower than those quoted by Tsong et al. (2) using the ISBCA technique, but are still somewhat higher than the IR results. The ^{19}F technique does not discern, however, whether there is hydrogen in the sample in excess to that bonded as SiOH and seen in the OH spectra. The hydrogen concentrations quoted for the surface region in Table 1 are much higher than those found using IR techniques and in better agreement with those quoted by Tsong et al., although there is still considerable scatter between the results. These results imply that Tsong et al. observed such high concentrations because only the surface of the samples was examined while the IR measurements were averaged through the whole sample. There are also possible systematic errors in the IBSCA results due to hydrogen contaminants in the vacuum system and in the ion beam.

HCl oxide in Table 1 is an approximately .15 μm SiO_2 film grown on silicon in an HCl ambient. Unlike other samples there was no detectible hydrogen contaminant on its surface, which might be due to the manner in which the film was prepared or to the fact that no graphite layer was deposited on it.

(1)   D.T. Griggs and J.D. Blacic, Quartz: Anomalous Weakness of Synthetic Crystals, Science 147, 292 (1965).

(2)   I.S.T. Tsong, A.C. McLaren and B.E. Hobbs, Determination of Hydrogen in Silicates Using the Ion Beam Spectrochemical Analyzer: Application to Hydrolytic Weakening, Am. Mineral 61, 921 (1976).

(3)   R.A. Weeks and M.M. Abraham, Electron Spin Resonance of Irradiated Quartz: Atomic Hydrogen, J. Chem. Phys. 42, 68 (1965).

(4)   D. Chakroborty and G. Lehmann, On the Structure and Orientations of Hydrogen Defects in Natural and Synthetic Quartz Crystals, Phys. Stat. Sol. (a) 34, 467 (1976).

(5)   A. Kats, Y. Haven and J.M. Stevels, Hydroxyl Groups in α-quartz, Phys. Chem. Glasses 3, 69 (1962).

(6)   G.J. Clark, C.W. White, D.D. Allred, B.R. Appleton, F.B. Koch and C.W. Magee, The Application of Nuclear Reactions for Quantitative Hydrogen Analysis, Nucl. Instr. and Meth., 143 (1977).

(7)   W.A. Lanford, Glass Hydration: A Method of Dating Glass Objects, Science 196, 975 (1977).

(8)   D.T. Griggs, Hydrolytic Weakening of Quartz and Other Silicates, Geophys. J.R. Astron. Soc. 14, 19 (1967).

(9)   D.V. McCaughan and R.A. Kushner, (1974) Impurity-Movement Problems in Analysis Methods Using Particle Bombardment, (Eds.) P.F. Kane and G.B. Larrabee, Characterization of Solid Surfaces, Plenum Press, New York.

(10) W.C. Godbeer and R.W.T. Wilkins, The Water Content of Synthetic Quartz X_0, Amer. Mineral. (To be published).

(11) B.E. Hobbs, A.C. McLaren and M.S. Patterson (1972) Plasticity of Single Crystals of Synthetic Quartz. (Ed.) H.C. Heard et al., Flow and Fracture of Rocks, Geophysical Monograph, 16, 29 (1972).

(12) G.O. Brunner, H. Wondratschek and F. Laves, Ultrarotuntersuchungen uber den Einbau von H in Naturalichen Quartz, Z. Elektrochem. 65, 753 (1961).

INTERACTION OF DISSOLVED MOLECULAR HYDROGEN WITH A VITREOUS SILICA HOST[†]

J. Vitko, Jr., Charles M. Hartwig, and P. L. Mattern
Sandia Laboratories, Livermore, California 94550

ABSTRACT

Measurements of the solubility of molecular hydrogen in vitreous silica and the observation of Raman and induced IR spectra of dissolved H_2 and D_2 molecules have been used to evolve a simple picture of the gas-glass interaction. This model successfully predicts the isotope and temperature dependence of the shift in the stretching frequency of the dissolved molecule from that of the isolated molecule. A more fundamental approach describes the H_2, D_2-SiO_2 interaction in terms of measured Lennard-Jones parameters. Calculated values of the solubility parameters and the hydrogen matrix shift depend markedly on the assumed host structure, suggesting that experimental determination of these parameters constitutes a sensitive probe of the local atomic structure in amorphous materials.

INTRODUCTION

For all practical purposes hydrogen[*], helium, and neon are insoluble in crystalline quartz. However, these same gases are readily dissolved in the more open structures characteristic of the high temperature crystalline polymorphs of silica[1] and in vitreous silica.[2-5] Interactions between the dissolved molecules and the surrounding host network are manifested in the heat of solution,[2-4] and for hydrogen, in the appearance of an induced infrared (IR) spectrum with phonon sidebands[6,7] and in the broadening and shifting of the hydrogen Q band.[8,9] These manifestations of the gas-glass interaction constitute a sensitive probe of the distribution and character of free volume in silica[5,8,9] and, as such, have received considerable attention. In this paper we review some of our recent experimental and theoretical efforts in understanding the interaction of hydrogen with silica.

SOLUBILITY[2,5]

D_2, He, and Ne solubility in vitreous silica has been measured at elevated pressure and ambient temperatures using PVT and TGA techniques.[2-4] In all cases, the concentration of dissolved molecules, C, was given by the Langmuir isotherm, a model predicated on a system of V_0 equivalent and non-interacting solubility sites, each capable of accommodating at most one of the dissolved molecules. Under these conditions

$$C = KFV_o/(1+KF), \tag{1}$$

[†]Work supported by the U. S. Department of Energy
*Hydrogen as used in this context refers to both protium, H_2, and deuterium D_2.

where K is an equilibrium constant and F is the fugacity of the surrounding gas. Typically[3,4] $V_o \simeq 10^{21}$ sites/cm^3. K may be expressed as[10]

$$K(T) = \left(\frac{h^2}{2\pi m k_B T}\right)^{3/2} \frac{Q_{vib}}{k_B T} \; e^{-E/k_B T} , \qquad (2)$$

where h is Planck's constant, k_B is Boltzman's constant, T is the temperature of the sample, and Q_{vib} the center of mass vibrational partition function of a particle of mass m in a well of minimum potential E. One simple model[1,2,10] describes the solute molecule in terms of a 3-D isotropic harmonic oscillator of frequency ν and well depth E. For this model

$$K(T) = \left(\frac{h^2}{2\pi m k_B T}\right)^{3/2} \frac{1}{kT} \left(\frac{e^{-h\nu/2k_B T}}{1-e^{-h\nu/k_B T}}\right)^3 e^{-E/k_B T}. \qquad (3)$$

The use of K(T) of eq. (3) in the Langmuir isotherm provides a satisfactory description of the measured temperature dependent solubilities. Typically $-E \simeq 1-3$ kcal/mole and $\nu \simeq 1-3 \times 10^{12}$ Hz.[3-5] These low interaction energies are suggestive of physical rather than chemical interactions and are in accord with Raman studies[5,8] which show that dissolved hydrogen undergoes no major alterations in structure or symmetry and retains considerable rotational freedom.

SPECTROSCOPY OF MOLECULAR HYDROGEN IN VITREOUS SILICA

A. Raman[5,8]

The room temperature Raman Q band (molecular stretching mode) and S_1 band (combined stretching and rotation) spectra of H_2 dissolved in vitreous silica are shown in Fig. 1. Similar spectra are observed for D_2 in vitreous silica. These spectra demonstrate the molecular nature of the solubility process and suggest that the dissolved molecule retains considerable rotational freedom. This information and the observation that the depolarization ratio of the dissolved molecule is the same as that of the isolated molecule indicate that the dissolved molecule has undergone no major alterations in structure or symmetry.

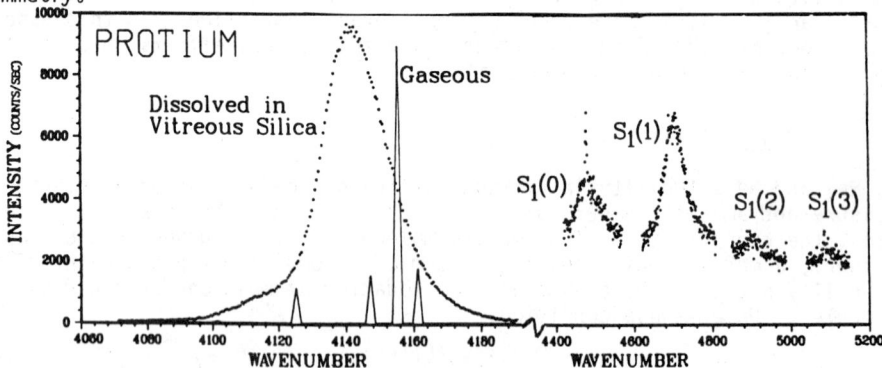

Fig. 1. Raman spectra of H_2 dissolved in vitreous silica. The dotted lines are the Q band and the vibrational-rotational spectrum of H_2 dissolved in silica. The solid line is a Q band spectrum of gaseous H_2.

Weak interactions between the dissolved molecule and the host matrix do, however, significantly broaden the Q and S band transitions and cause them to occur at a somewhat lower frequency than for the isolated molecule. This frequency shift is often referred to as a matrix shift and is defined as

$$\Delta\nu_s = \nu_{dissolved} - \nu_{isolated} \cdot \qquad (4)$$

B. Induced Infrared (IR)[6,7]

Interactions between the dissolved hydrogen and the host matrix lower the symmetry of the molecule and result in an induced IR activity. To within experimental uncertainty, the induced IR Q band is identical to that observed by Raman scattering (Fig. 2), suggesting that the induced absorption is due to the lack of inversion symmetry at the solution site.

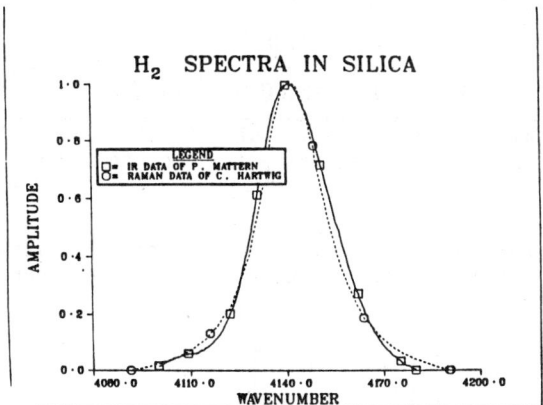

Fig. 2. A comparison of the induced IR Q band to that obtained via Raman. Observed lineshapes and frequencies agree to within experimental uncertainty. This suggest that the induced absorption is due to lack of inversion symmetry at the solution site.

MODELING[9]

Assuming that H_2, He, and Ne interact primarily with the host oxygen ions[1,9] (and that these ions are similar to the oxygen ions in H_2O), the gas-glass interaction was modeled using a pairwise sum of experimentally-determined H_2O-gas Lennard-Jones potentials[11]. Host atom positions were assumed to be those of a static, unrelaxed beta cristobalite structure. Uncertainties regarding the exact structure of this crystalline phase prompted calculations[9] for both the 'ideal' (straight Si-O-Si bond) structure and one of the six equivalent bent Si-O-Si bond structures.[12] For both structures, the calculated eigenstates and eigenenergies of the dissolved molecule were nearly those of the isotropic 3-D harmonic oscillator of frequency ν and contained in a well of minimum potential E. In comparing calculated to experimental values, greater import was assigned to values of E. This judgment acknowledges limitations in the calculated value of ν (due to the inadequacy of the static lattice approximation), and in the experimental value of ν (due to the relative insensitivity of the fitting procedure to this parameter. Table 1 compares calculated parameters to those determined from solubility measurements. The slight difference in the size of the solubility sites between the two structures

(that of the ideal structure being slightly larger) is seen to have significant impact on the results. Reasonable agreement is obtained only for the
'ideal' structure.

TABLE 1 Comparison of Gas Solubility Parameters calculated for
a Beta Cristobalite Structure with those measured in Vitreous
Silica (from ref. 9)

System	E (Kcal/mole)			$\nu(10^{12}Hz)$		
	$Calc^a$	$Calc^b$	Exp^c	$Calc^a$	$Calc^b$	Exp^c
D_2	-2.42	7.16	-2.42	11.25	18.6	3.13
He	-0.23	2.44	-1.05	6.30	9.63	2.82
Ne^{22}	-1.59	1.60	-2.00	2.70	4.11	1.35

(a) 'ideal' beta cristobalite structure; (b) one of six
Si-O-Si bond structures (ref . 12); (c) from ref. 3,4.

To calculate the hydrogen matrix shift one notes that the dispersion constant
A and the equilibrium intermolecular separation R_o of the Lennard-Jones pair
potential V_{LJ},

$$V_{LJ} (R,X) = A(X) \left[\frac{R_o^6(X)}{2R^{12}} - \frac{1}{R^6} \right] ,$$
(5)

both depend on the molecular vibration coordinate X defined by

$$X = r - r_o ,$$
(6)

where r_o and r are the internuclear separation of the isolated molecule in its
ground vibrational state and of the dissolved molecule respectively. Expanding in a Taylor series about the internuclear coordinate X=0, yields

$$V_{LJ} (R,X) = A(0) \left[\frac{R_o^6(0)}{2R^{12}} (1 + S_1 \frac{X}{r_o}) - \frac{1}{R^6} (1 + S_2 \frac{X}{r_o}) \right] ,$$
(7)

$$S_1 \equiv 6 \frac{r_o}{R_o} \frac{dR_o(X)}{dX} + S_2, \text{ and } S_2 = \frac{r_o}{A} \frac{dA(X)}{dX} .$$

by thinking of hydrogen as a helium atom in which the protons (and neutrons)
have been moved a distance r_o apart, $\frac{dR_o}{dX}$ may be approximated by

$$\left.\frac{dR_o(X)}{dX}\right|_{x=0} \approx \frac{R_o(0, \text{for } H_2\text{-}H_2O) - R_o(0, \text{for } He\text{-}H_2O)}{r_o} \qquad (8)$$

Use of experimental values[11] of R_o in eq. (9) indicate $\frac{dR_o}{dx}$ is approximately zero. Thus

$$S_1 = S_2 \equiv S . \qquad (9)$$

A value of $S = 0.9$ for H_2-H_2O is obtained from perturbation theory. This Lennard-Jones formulation of the gas-glass interaction then yields $\Delta\nu_s$ -28 cm^{-1}, 189 cm^{-1} for the ideal and bent bond structures respectively. The measured low temperature value (20°K) for the H_2 shift is -24 cm^{-1}. Again reasonable agreement is obtained only for the 'ideal' structure.

The import of these calculations lies not in their compatibility with a beta cristobalite model of glass structure but in the implication that solubility parameters and matrix shifts may be used to discriminate between various models of the vitreous state.

FREQUENCY SHIFT CALCULATIONS[8]

The preceding calculations of frequency shifts (performed for a solute molecule in the CM ground state) become exceedingly complicated when finite temperatures necessitate the inclusion of excited CM states. To enable such calculations, the dissolved molecule is described as an anharmonic oscillator (of reduced mass μ, force constant k, and anharmonicity g) interacting with a perturbing potential V obtained by summing the pair potential of equation (7) over all oxygen atoms in the host matrix. The Hamiltonian, H, is then

$$H = \frac{1}{2}\mu\dot{X}^2 + \frac{1}{2}kX^2 - gX^3 + V(R')(1 + SX/r_o) , \qquad (10)$$

where R' is the displacement of the solute CM from the minimum of V, and V in the simple harmonic oscillator approximation is given by

$$V(R') = E + 2\pi^2\nu^2 mR'^2 . \qquad (11)$$

A straight forward application of second order perturbation theory then yields

$$\Delta\nu_s(R') = \frac{C}{\mu^{1/2}} V(R') \text{ where } C = \frac{3gS}{2\pi k^{3/2} r_o} . \qquad (12)$$

Averaging V(R') over all thermally populated CM states results in a temperature dependent shift, $\Delta\nu_s(T)$, given by

$$\Delta\nu_s(T) = \frac{C}{\mu^{1/2}} [E + \frac{3}{2}(\frac{1}{2} + \frac{1}{e^{\beta h\nu}-1})h\nu], \ \beta = (kT)^{-1} . \qquad (13)$$

Thus at high temperatures

$$\Delta\nu_s(T) = \frac{C}{\mu^{1/2}} (E + \frac{3}{2}k_BT), \qquad (14)$$

i.e., $\Delta\nu_s$ (a) increases linearly with T with slope proportional to $\mu^{-1/2}$, and (b) $\Delta\nu_s \equiv 0$ at $T_o = -\frac{2}{3}E/k_B$. A more complicated mass dependence is predicted for the $T \rightarrow 0$ limit, i.e.,

$$\Delta \nu_s(T=0) = \frac{C}{\mu^{1/2}} [E + \frac{3}{4} h\nu].$$ (15)

The measured temperature dependence of the Q band frequency of H_2[6,8] and D_2[8] in vitreous silica is shown in Fig. 3. To relate these data to the predicted frequency shift one needs to establish the reference frequency, $\nu_{isolated}$ of eq. 4. Because of the asymmetry of the dissolved Q bands, the H_2 shift is referenced to the Q(1) transition the dominant line in the gaseous spectrum over the entire temperature range studied. In the case of D_2 the dominant gaseous Q(J) transition varies over the temperature range studied. Therefore, to minimize referencing ambiguities, measured Q band frequencies were related to frequency shifts only over the range 160-300 K, a range in which the Q(2) is dominant.

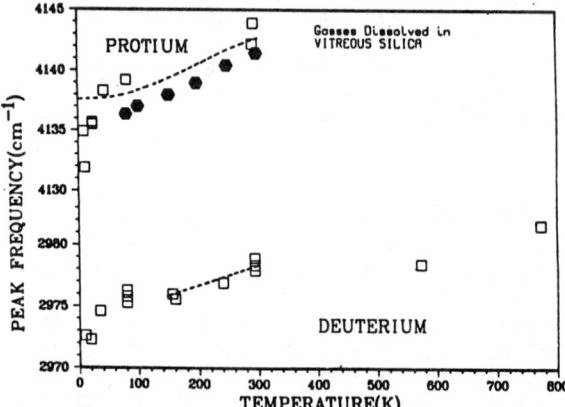

Fig. 3. The temperature dependence of the IR (solid points) and Raman (open points) Q band of H_2 and the Raman Q band of D_2 in vitreous silica. The dotted line is the theoretical prediction.

Figure 3 compares the frequency shifts predicted by eq. 14 to those actually observed. These predictions use values of E and ν derived from solubility data[4] and a value of C derived from the measured high temperature slope, $\frac{d\nu_{dissolved}}{dT}$, for H_2 in vitreous silica. Except for one point at 11 k, the calculated and observed values agree to within experimental error. As predicted, (a) the high temperature shifts increase linearly with T and (b) the measured H_2 slope is 1.44 times that of D_2. Extrapolation of room temperature shifts to $\Delta \nu_s = 0$ yields $T_o = 820$ K for H_2 and 811 K for D_2, in excellent agreement with the value of 820 K obtained from solubility data. The measured low temperature shift ratio, $\Delta \nu_s(H_2)/\Delta \nu_s(D_2)$, is 1.1 ± 0.25 vs 1.35 predicted from solubility data.

ACKNOWLEGEMENTS

The authors are indebted to J. E. Shelby and C. F. Coll for permitting us to incorporate their studies in this review and for numerous invaluable discussions.

REFERENCES

[1] R. M. Barrer and D. E. W. Vaughan, Trans. Faraday Soc. 63, 2275 (1967).

[2] J. F. Shackleford, P. L. Studt, and R. M. Fulrath, J. Appl. Phys. 43, 1619 (1972).

[3] J. E. Shelby, J. Appl. Phys. 47, 135 (1976).

[4] J. E. Shelby, J. Appl. Phys. 48, 3387 (1977).

[5] Charles M. Hartwig, J. Appl. Phys. 47, 956 (1976).

[6] P. L. Mattern, Bull. Am. Phys. Soc. II 21, 226 (1976).

[7] J. Vitko and C. F. Coll, Bull. Am. Phys. Soc. II 21, 226 (1976)

[8] Charles M. Hartwig and J. Vitko, Jr., to be published.

[9] J. Vitko, Jr. and C. F. Coll III, to be published.

[10] e.g. see D. Olander, J. Chem. Phys. 43, 785 (1965).

[11] R. W. Bickes, Jr., G. Duquette, C. J. N .van den Meijdenberg, A. M. Rulis, G. Scoles, and K. M. Smith, J. Phys. B: Atom. Molec. Phys. 8, 3034 (1975).

[12] A. J. Leadbetter, T. W. Smith and A. F. Wright, Nature Phys. Sci. 244, 125 (1973).

HYDROGEN IN SiO_2 FILMS ON SILICON

A. G. Revesz
COMSAT Laboratories, Clarksburg, Maryland 20734 (USA)

INTRODUCTION

It has been known for a long time that hydrogen, often in the form of water vapor, in the oxidizing or annealing ambient greatly affects the properties of Si/SiO_2 interface structures [1]. Infrared spectroscopy supplies direct evidence of OH and SiH groups in SiO_2 films. This paper places the evidence in proper perspective by discussing (a) the effects of preparation conditions and (b) the important differences between grown SiO_2 films and fused silica with respect to SiH groups and hydrogen-bonding, as well as (c) the implications concerning the properties of the Si/SiO_2 interface.

INDIRECT EVIDENCE OF HYDROGEN

Hydrogen, mostly in the form of water vapor, has a significant influence on the oxidation of silicon [1]. Thus, for example, the growth of 100 nm SiO_2 film at 1000°C requires 100 min in dry O_2 but only 6 min in H_2O; the oxidation rates are parabolic and linear with time, respectively. The pre-exponential part of the linear rate constant varies from ~5 x 10^{-6} to 8 x 10^2 cm s^{-1} as the hydrogen (water) concentration increases from a presumably negligible value (when the oxidation tube is cold) to that corresponding to 1 atm steam; concomitantly, the activation energy decreases from ~1.6 to 0.7 eV. Obviously, hydrogen in the oxidation ambient profoundly influences the generation of defects associated with oxide growth, particularly at the Si/SiO_2 interface where the oxidation occurs.

Indirect evidence of hydrogen also emerges from various effects of oxidation conditions and post-oxidation treatments on the electrical properties of the Si/SiO_2 interface. Thus, protons are considered responsible for some very fast ion migration observed under positive bias stress [2]; negative bias instability [1,3] and irradiation effects [4] have also been attributed to hydrogen in the SiO_2 film. Furthermore, the importance of hydrogen has been pointed out, among others, in connection with EPR [5,6] and cathodoluminescence [7] studies.

DIRECT EVIDENCE OF HYDROGEN

Radioactive analysis of tritium, which has been introduced into the SiO_2 film during its growth or by a post-oxidation treatment, provides direct evidence of hydrogen in SiO_2 films. For example, the density of tritium in steam-grown oxide is ~8 x 10^{19} cm^{-3}, except in the region ~60 nm from the Si/SiO_2 interface where it rises sharply toward the interface to reach ~2 x 10^{20} cm^{-3} [8]. The disadvantages of this technique are that unintentional H-contamination cannot be detected and the various possible forms of hydrogen present in the oxide film cannot be distinguished.

Infrared Spectroscopic Evidence of SiH and OH Groups

These disadvantages can be eliminated by using multiple internal reflection infrared spectroscopy as demonstrated by Beckmann and Harrick (9). They have shown that, depending on the preparation conditions, both OH and SiH groups (absorbing at 3600 and 2350 cm^{-1}, respectively) can be present in SiO_2 films. In fact, SiH groups were found in all the films; their concentration was particularly large in the films grown in "dry" oxygen (with 60 ppm H_2O). The SiH concentration increases sharply toward the Si/SiO_2 interface as shown in Fig. 1, which is qualitatively the same behavior found by the tritium analysis. Note that SiH concentration is larger in the oxide film grown in "dry" O_2 than in the steam-grown film. In contrast, SiH has never been detected in fused SiO_2 which is another form of noncrystalline SiO_2. Thus, the presence of SiH in *all* SiO_2 films grown on silicon must be related to the involvement of defects in the growth process and the essential role of hydrogen in the defect reactions previously inferred from the oxidation kinetics.

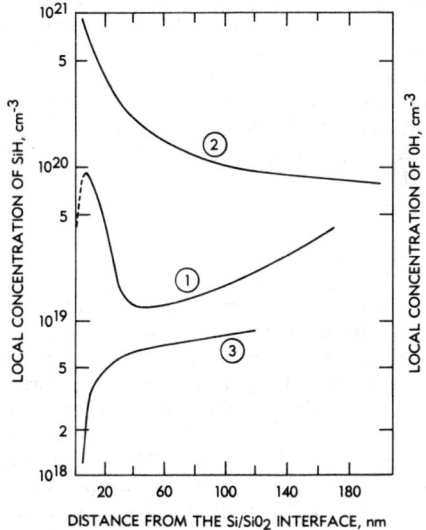

Fig. 1. The distribution of SiH and OH groups in SiO_2 films Curve 1: SiH concentration profile in an SiO_2 film grown in "dry" (60 ppm H_2O) O_2; curve 2: SiH concentration profile in an SiO_2 film grown in steam; curve 3: OH concentration profile in the steam-grown film. Based on reference 9.

The distribution of OH groups in the vicinity of the Si/SiO_2 interface in the steam-grown oxide exhibits a behavior which is opposite to that of SiH groups (see Fig. 1). These concentration profiles can be qualitatively explained by the diffusion of OH and/or H_2O through the oxide and oxidation at the Si/SiO_2 interface:

$$Si + OH(H_2O) \rightarrow SiO_2 + H \tag{1}$$

followed by a reaction between H and an Si atom with an unsaturated electron (dangling bond, represented below by a dot) at the Si surface or in the inter-face region of the oxide:

$$Si\cdot + H \rightarrow SiH \tag{2}$$

The concentration profile of SiH further removed from the interface in the steam-grown oxide is rather puzzling.

Recently, Hughes and Pankey (10) have found by IR spectroscopy significant variation in the concentration of SiH and OH in SiO_2 films prepared under varying but reasonably well defined conditions. An important result is that the concentration depends on the nature of the tube in which the oxidation or post-oxidation heat treatment in N_2 was performed: oxide films grown in a conventional resistance-heated silica tube furnace contained ~3 x 10^{21} and ~2 x 10^{22} SiH/cm^3 (average rather than localized concentration) before and after heat treatment in N_2, respectively, whereas films grown and heat treated in a resistance-heated *silicon* tube furnace contained ~2 x 10^{21} SiH/cm^3. Apparently, serious H-contamination occurred in the silica tube furnace during both the oxidation and heat treatment. This contamination was less, but still appreciable, in the silicon tube. The OH concentration exhibited the same trend but its value was about an order of magnitude lower than that of SiH. In view of the uncertainties of the quantitative evaluation of the IR spectra (9) and possible instrumental errors, the absolute values of these concentrations must be viewed with caution, and the data should be considered as indicators of trends. This caveat does not diminish the importance of these results which clearly demonstrate that *hydrogen is present in all the analyzed SiO_2 films* and that the nature of the H-related defects as well as their concentration depend strongly on the preparation conditions.

Hydrogen-Bonding in SiO_2 Films

Depending on the preparation conditions, the hydrogen in both SiH and OH can form a hydrogen-bond with another oxygen (9,10). In addition to the presence of SiH groups in grown SiO_2 films, this feature also distinguishes this form of noncrystalline SiO_2 from fused silica where H-bonded OH groups have never been found. Hydrogen-bonding shifts the absorption peak of SiH and OH to lower frequencies; the extent of this shift in grown SiO_2 films is 230-500 cm^{-1}. This shift is significantly larger than that characteristic of H-bonding involving bridging oxygen in some siloxane compounds (45-85 cm^{-1}) (11) but is less than the one in silicate glasses which results from H-bonding between OH and nonbridging oxygen: OH...ONa (640-800 cm^{-1}) (12). It is, however, comparable to the shift resulting from H-bonding between SiOH groups adsorbed on silica surfaces (13).

On the basis of this comparison, it is suggested here that H-bonding of OH groups in SiO_2 films involves (a) another OH or (b) a nonbridging oxygen. In the first case, the OH groups (very probably as SiOH) either form pairs or are aligned preferentially along structural channels because their concentration is too low (10^{19}-10^{21} cm^{-3}) for a random distribution. Since the energy of the H-bond (~0.3 eV) is the same as the activation energy of various transport processes occurring in grown SiO_2 films (e.g., water permeation at 50-250°C (14), dielectric relaxation due to injected protons (15), annealing of charge trapped at OH groups (16)), the OH groups could be aligned along structural channels. These channels are unique for grown SiO_2 films (1); hence, there is no transport process in fused SiO_2 characterized by such a low activation energy. In the second case, the nonbridging oxygen involved in H-bonding probably results from the oxide growth process during which these defects are generated by an interaction between molecular oxygen and the Si-O network (1). Apparently, H-bonding does not occur in fused SiO_2 because structural channels and/or nonbridging oxygen defects do not exist and the formation of H-bonded OH pairs is unlikely.

Hydrogen-bonding in SiO_2 films affects the stability of SiH and OH groups. To wit, a heat treatment at 500°C in N_2 eliminated all the SiH groups that were not H-bonded from an anodically grown oxide film, but did not affect the H-bonded SiH groups in thermally grown films (9). It was also observed that irradiation caused a shift from 3700 to 3600 cm^{-1} of the absorption peak of those OH groups which were not H-bonded (in films grown in a silica tube furnace), whereas the absorption peak (at 3200 cm^{-1}) of H-bonded OH groups (in films prepared in a silicon tube furnace) remained unaffected (10).

EFFECTS OF HYDROGEN ON THE Si/SiO$_2$ INTERFACE

Since the Si-H bond is relatively easy to break and the dissociation may be associated with hole trapping (3,4), it is probable that the relatively large concentration of SiH groups at the Si/SiO$_2$ interface is responsible for many phenomena. For example, the Si/SiO$_2$ structures in which the SiO_2 films had a higher density of SiH groups (prepared in a silica tube furnace) exhibited 6-7 V shift in the flat-band voltage upon irradiation, whereas those with a lower SiH density (prepared in a silicon tube furnace) exhibited only ~1 V shift (10). SiH groups at the Si/SiO$_2$ interface can also lead to a gross over-estimation of the so-called "excess silicon" (17) there since Auger analysis cannot distinguish between Si and SiH. The accumulated direct and indirect evidence clearly demonstrates that the concentration and distribution of hydrogen, particularly as SiH, in SiO_2 films are very sensitive to oxidation and annealing conditions. The many contradictory observations concerning the effects of these conditions on various properties of Si/SiO$_2$ interface structures evidence that the problem of controlling the H-contamination has not yet been solved. This lack of control as well as the meager understanding and appreciation of the complex role of hydrogen has important consequences for the Si/SiO$_2$ interface technology. It is also important to recognize that *no definitive statements concerning this interface can be made without considering the presence of hydrogen in SiO$_2$ films.*

REFERENCES

(1) The chemical aspect of SiO_2 films on silicon have been reviewed in A. G. Revesz, J. Non-Cryst. Solids 11, 309 (1973).

(2) S. R. Hofstein, IEEE Trans. Electr. Dev. ED-14, 749 (1967).

(3) K. P. Jeppson and C. M. Svensson, J. Appl. Phys. 48, 2004 (1977).

(4) A. G. Revesz, IEEE Trans. Nucl. Sci. NS-24, 2102, (1977).

(5) A. G. Revesz and B. Goldstein, Surf. Sci. 14, 361 (1969).

(6) Y. Nishi, Jap. J. Appl. Phys. 10, 52 (1971).

(7) J. P. Mitchell and D. G. Denure, Solid State Electronics 16, 825 (1973).

(8) P. J. Burkhardt, J. Electrochem. Soc. 114, 196 (1967).

(9) K. H. Beckmann and N. J. Harrick, J. Electrochem. Soc. 118, 614 (1971).

(10) H. L. Hughes and T. Pankey, private communication.

(11) G. Englehardt and H. Kriegsmann, Z. anorg. allg. Chem. 336, 286 (1965).

(12) H. Scholze, Glastech. Ber. 32, 142 (1959).

(13) T. Fripiat and A. Jelli (1965), in Proceedings of the VII International Congress on Glass, Gordon and Breach, New York.

(14) G. L. Holmberg, A. B. Kuper, and F. D. Miraldi, J. Electrochem. Soc. 117, 677 (1970).

(15) R. Nannoni and M. J. Musselin, Thin Solid Films 6, 397 (1970).

(16) E. H. Nicollian, C. N. Berglund, P. F. Schmidt, and J. M. Andrews, J. Appl. Phys. 42, 5654 (1971).

(17) J. S. Johannessen, W. E. Spicer, and Y. E. Strausser, Appl. Phys. Letters 27, 452 (1975).

ESR CENTERS AND CHARGE DEFECTS NEAR THE Si/SiO$_2$ INTERFACE

Edward H. Poindexter, Edwin R. Ahlstrom, and Philip J. Caplan
US Army Electronics Technology and Devices Laboratory (ERADCOM)
Fort Monmouth, N.J. 07703

ABSTRACT

The esr center P$_b$ has been examined as a function of thermal oxidation and annealing conditions in single-crystal silicon wafers. It was observed that P$_b$ concentration was not correlated with fixed surface-state charge density Q$_{ss}$, despite some similarities in their behavior. In freshly oxidized wafers, P$_b$ typically ranges up to 2×10^{12} cm^{-2}, and Q$_{ss}$ up to 5×10^{11} cm^{-2}. The concentration of P$_b$ is highly dependent on cooling rates; concentration does not depend on oxide thickness and P$_b$ may be observed even in native oxides. The g-value of the orientation-averaged P$_b$ signal (2.0064) favors its assignment to Si^{III} in a silicon environment, as in crushed silicon (2.0055), rather than to the well-studied E' centers (2.0008) found in various forms of damaged SiO$_2$. In our thermal oxides, E' centers were unobservable, and thus cannot be the main source of Q$_{ss}$, despite their nominal positive charge. Preliminary evidence shows a correlation of P$_b$ with initial, unannealed "fast" interface states N$_{st}$, supporting the idea that Si^{III} is the origin of these states.

INTRODUCTION

Extensive studies of the Si/SiO$_2$ interface by electrical methods have not defined the origins of interface defects (1), which are crucial in advanced integrated circuit devices. The wealth of esr studies of silicon, quartz, and glass suggests application of esr to the interface. Nishi (2) observed and tentatively identified three main esr signals from thermally oxidized silicon wafers, P$_a$, P$_b$, P$_c$. The P$_b$ signal, visible at 300 K (3), seems to have the greatest significance for the defect structure of the Si/SiO$_2$ interface. Our study was prompted by Nishi's tentative identification of P$_b$ with the E' centers of damaged silica, and by the implied connection between P$_b$ and surface charge Q$_{ss}$ evidenced by similar depth profiles. We have attempted to confirm these conclusions or offer alternatives.

EXPERIMENTAL DETAILS

The samples were from n- and p-type Monsanto double-polished wafers, 100-300 ohm-cm. They were cleaned and oxidized in a quartz-tube diffusion furnace by standard semiconductor procedures. For C-V measurements, one-micron aluminum was evaporated onto the etched or sandblasted back surface and 1.2 mm dots were evaporated onto the oxide. A PAR system was used at 1.0 MHz. Samples consisted of 4 to 6 pieces 4x20 mm, with [111] or [100] face. A Varian 4501 with dual cavity was used for spin concentration, and an E-line Century for anisotropy and line structure. An H-P 5480 signal averager was used. All runs were at 300 K.

RESULTS AND DISCUSSION

Character of Esr Signals

Samples with [111] face showed P_b signals (P_b[111]) with up to 20:1 signal-noise without averaging. A typical P_b[111] signal is shown in Fig. 1a. The line shows reproducible structure which is symmetrical about the center, and is superimposed on a wider low-amplitude signal which contributes to an anisotropic asymmetry of the line base. This latter signal is mostly residual silicon damage at g=2.0055. In Fig. 1b, the signal has been accumulated 400 times; much detail is seen, nearly all from the silicon. Hoped-for symmetrical hyperfine lines are not evident in the clutter, with the possible exception of the close-in blips noted above.

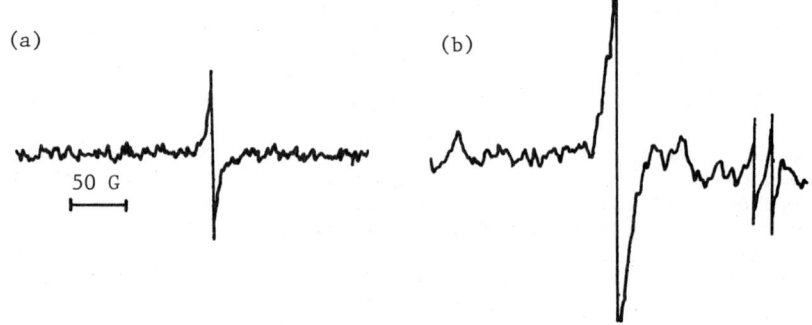

(a)

(b)

50 G

Fig. 1. P_b signal from [111] silicon (a) one trace (b) 400 traces.

Signals from the [100] face are different (Fig. 2). The s-n ratio is much worse than P_b[111] signals; 16 to 64 traces were used. The line has several ill-defined components, with narrowest, least-structured appearance at $\sim 25°$-$30°$ from [100] face axis. Resolution is impossible at this time.

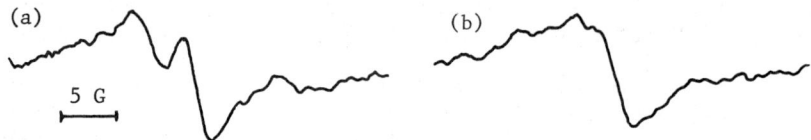

(a)

(b)

5 G

Fig. 2. P_b signal from [100] silicon (a) $H_o\perp$[100] face (b) $H_o\perp$[110].

Anisotropy of P_b[111] (2) is well confirmed in our studies. Within errors, our signal shows [111] axial symmetry, $g_{\parallel}$ =2.0013, $g_{\perp}$ =2.0081. The g-map is shown in Fig. 3. Only one limb is observed, for spin centers with major axis normal to the interface. The line width of P_b[111] varies from 1 G on [111] to 3 G at 90°. For P_b[100] (not shown) one possible but arbitrary deconvolution gives 2 components with g-anisotropy like P_b[111].

Identification of Centers

The anisotropy and g-values of P_b[111] are like other centers observed in

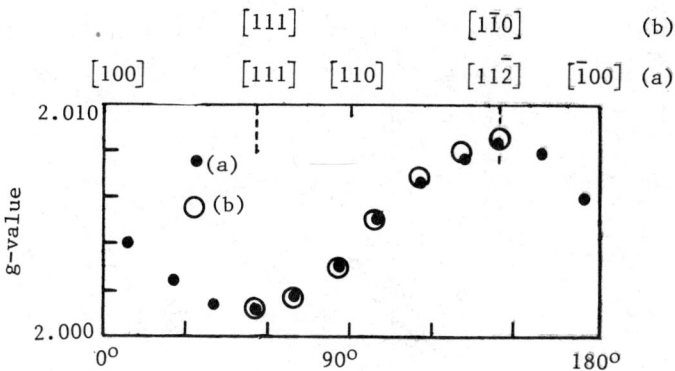

Fig. 3. g-anisotropy for P_b [111], rotation plane (a) $[11\bar{2}]$ (b) $[1\bar{1}0]$.

irradiated silicon (4-6) which have been ascribed to Si^{III} with unbonded orbital facing into a vacancy. In a study of the Si-E center (vacancy plus phosphorus), Watkins and Corbett (5) calculate the anisotropy for a Si^{III} fragment ($\Delta g_{||} \sim 0$; $\Delta g_{\perp} \sim +0.017$). They ascribe the deficiency vis-a-vis observed values ($g_1=2.0005$, $g_2=2.0112$, $g_3=2.0096$) to delocalization, and the hfs is consistent. Other well-characterized centers having similar anisotropy are the negatively-charged divacancy (G-7) (6), and the low-temperature form of the five-vacancy cluster (4). In view of the similarity of P_b[111] to these Si^{III} defects, we assign it to Si^{III} bonded to silicon, not oxygen, with unpaired p-rich orbital normal to the interface. The variation in line width may reflect different g-values in different sites.

P_b[111] is not like the E' center of damaged quartz (7-8), which has a lower g_{av} and much weaker, prolate anisotropy ($g_1=2.0018$, $g_2=2.0005$, $g_3=2.0003$).

The identification of P_b[100] is not clear at present. We have been unable to resolve its map into combinations of single Si^{III} signals or bent-bond signals (like the Si-B1 center, structurally plausible for the [100] face).

Relation to Interface Charge Defects

It is well known that Q_{ss} in thermal oxides is controlled by the final annealing temperatures (Deal triangle) (1). Figure 4 shows a series of p- and n-type samples grown at 1000 C in dry O_2, but annealed for an hour in O_2 at various temperatures. For both p and n, Q_{ss} values were found to follow the trend of the Deal triangle; the magnitude of P_b[111] did not. This is contrary to our first tests, where correlation existed between P_b and Q_{ss} (9). (Earlier samples were erratically cooled.)

The values of Q_{ss} increase with longer cooling time (1). Table I shows the variation of P_b[111] with cooling rate. After oxide growth, one sample was given the standard fast pull, and the other was cooled slowly in situ by turning off the furnace, while the oxygen continued to flow. Also compared are two samples annealed one hour in N_2 after oxidation, and respectively slow- and fast-cooled in N_2. The P_b results are different from Q_{ss}. The sample slow-cooled in O_2 had Q_{ss} about three times higher than the other samples, as expected. But the P_b signal is lower in the slow-cooled samples, whether the oxidation is followed by a nitrogen anneal or not. Nishi (2)

230

likewise observed that faster quenches generally enhanced P_b signals.

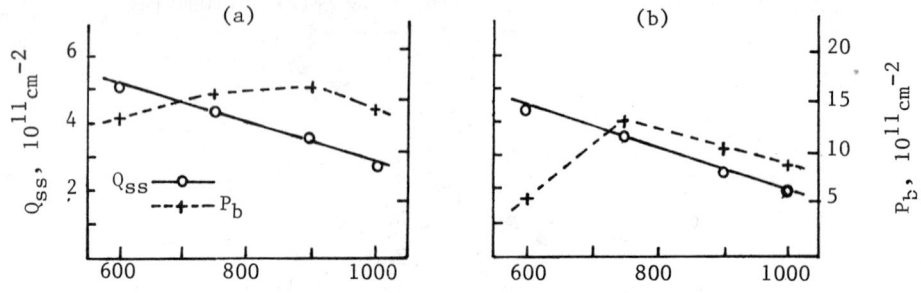

Fig. 4. P_b signal vs Q_{ss} on $[111]$ (a) p-Si (b) n-Si.

The dependence of $P_b[111]$ centers on grown-oxide thickness reflects their site and origin. The spin concentrations shown in Table II are constant over two orders of magnitude in thickness. Moreover, we have recently observed P_b with nitrogen-annealed native oxides. Our results disagree with Nishi's report (2) of large increase in P_b from 2000 Å to 4000 Å, and are inconsistent with the depth profile obtained by etching (2), which peaks at $\geq$ 150 Å. Constant P_b suggests association with the immediate interface region only.

A seemingly oxide-located signal might come from small islands of silicon in the oxide or from centers in the silicon surface whose paramagnetism disappears when the last remnants of the oxide overburden are removed. A stress-related center would behave in the latter fashion (10). Si^{III} with unpaired orbital directed perpendicular to the interface is much more reasonable chemically than preferential Si^{III} orientation far out in the oxide. In this view, neither P_b centers nor unobserved E' centers are the major source of Q_{ss}. (Quasi-E' centers--a doubly-charged pair of Si^{III}, or isolated singly-charged Si, invisible to esr--are not ruled out.)

Table I. P_b vs cooling rate.

Cooling Conditions	Q_{ss} (Rel.)	Spin Conc. (Rel.)
O_2 fast cool	3.5	3.4
O_2 slow cool	11.	1.5
N_2 fast cool	3.5	3.6
N_2 slow cool	3.5	0.6

Table II. P_b vs oxide thickness.

Oxide Thickness (Å)	Spin Conc. (Rel.)
100	2.9
300	3.2
530	3.7
1,560	3.7
3,400	4.3
12,760	3.7

Lack of correlation of P_b and Q_{ss} must be viewed in respect to usual device processing regimens. Determination of Q_{ss} is assumed to be made after aluminum dot deposition and annealing--this reduces N_{st} (1) and allows a tractable C-V plot of the presumably unaffected fixed charge. The concentration of interface states N_{st} in freshly oxidized wafers is of the same order as

P_b, 10^{12} cm^{-2}. Esr spectra, however, are taken on wafers which have not been "alnealed." We prepared a sample with substantial P_b signal, deposited aluminum over the sample surface, and "alnealed" it. After removing the aluminum, there was no detectable esr. Nishi (2) has noted that water vapor during oxidation affects P_b and N_{st} similarly. Finally, weaker P_b signals on [100] faces correlate with lower N_{st} values typical of [100] wafers.

In summary, we believe that the P_b[111] signal arises from Si^{III} at the interface, and that it is not the source of Q_{ss}. Furthermore, silica E' centers, not observed by esr in oxidized wafers, are likewise not a major source of Q_{ss}. Presently, the observed evidence favors a closer relation between P_b and N_{st} than between P_b and Q_{ss}. If substantiated, this would help to confirm the concept that Si^{III} is a source of initial interface states.

REFERENCES

1. B.E. Deal, The current understanding of charges in the thermally oxidized silicon structure, J. Electrochem. Soc. 121, 198C (1974).

2. Y. Nishi, Study of silicon-silicon dioxide structure by electron spin resonance I, Japan J. Appl. Phys. 10, 51 (1971).

3. P.J. Caplan, J.N. Helbert, B.E. Wagner, and E.H. Poindexter, Paramagnetic defects in silicon/silicon dioxide systems, Surface Science 54, 33 (1976).

4. Y.-H. Lee and J.W. Corbett, EPR studies in neutron-irradiated silicon: a negative charge state of a nonplanar five-vacancy cluster (V_5^-), Phys. Rev. B8, 2810 (1973).

5. G.D. Watkins and J.W. Corbett, Defects in irradiated silicon: electron paramagnetic resonance and electron-nuclear double resonance of the Si-E center, Phys. Rev. 134, A1359 (1964).

6. G.D. Watkins and J.W. Corbett, Defects in irradiated silicon: electron paramagnetic resonance of the divacancy, Phys. Rev. 138, A543 (1965).

7. D.L. Griscom, E.J. Freibele, and G.H. Sigel, Jr., Observation and analysis of the primary Si^{29} hyperfine structure of the E' center in noncrystalline SiO$_2$, Solid State Commun. 15, 479 (1974).

8. F.J. Feigl, W.B. Fowler, and K.L. Yip, Oxygen vacancy model for the E' center in SiO$_2$, Solid State Commun. 14, 225 (1974).

9. E.H. Poindexter, J.N. Helbert, B.E. Wagner, and P.J. Caplan, Esr signatures of defects near the Si/SiO$_2$ interface, IEEE Trans. ED-24, 1217 (1977).

10. E.P. EerNisse, Viscous flow of thermal SiO$_2$ Appl. Phys. Letters 30, 290 (1977).

CHAPTER V
DEFECTS AND IMPURITIES IN α-QUARTZ AND FUSED SILICA

DEFECTS AND IMPURITIES IN α-QUARTZ AND FUSED SILICA

D.L. Griscom
Naval Research Laboratory, Washington, DC 20375

INTRODUCTION

Numerous properties of α-quartz and fused silica are strongly influenced by the presence of impurities or defects. The purpose of this paper is to survey the physics of <u>point</u> defects either intrinsic to these materials or associated with some of the most commonly encountered impurities or dopants. Extended defects such as dislocation loops in quartz (1), will not be considered. Of all the experimental techniques used to investigate point defects in SiO_2, electron spin resonance (ESR) is the most specific in terms of providing detailed information concerning the nature of the defect and its immediate atomic surroundings. Before the advent of performing optical spectroscopy in 100 m lengths of fused silica fiber waveguides (2), ESR was also the most sensitive method available. But in view of the immense uncertainties in interpreting optical data, ESR studies will receive the primary emphasis here. The application of ESR of course implies the presence of unpaired electrons. Since ionizing or particle irradiation is usually necessary to produce such paramagnetic states, a focus on radiation-induced defects is inevitable. Nevertheless, such studies provide considerable insight into the atomic scale structure which exists at the precursor sites prior to electron or hole trapping.

Even within the limited emphases defined above, it should be recognized that the literature is enormous. For example, a review article dealing with one impurity-related defect alone, the aluminum center in α-quartz, has recently appeared containing 125 references (3)! Length restrictions on the present paper impose a high degree of selectivity and likewise preclude detailed developments of the experimental and analytical techniques. Very brief introductions to the crystallography of α-quartz and the ESR formalism will be provided, however. These will be followed by successive sections dealing first with impurity and intrinsic centers in α-quartz and then those centers occurring in fused silica. An attempt will be made to reduce most results to general terms while at the same time collating a number of details of interest to specialists. The final section will treat the few good correlations which have been established between the ESR and optical spectra, and some possible implications of these results will be explored.

CRYSTALLOGRAPHY OF α-QUARTZ

α-Quartz is the crystalline polymorph of SiO_2 which is stable below 573°C. It is composed of identical, slightly distorted SiO_4 tetrahedra (Fig.1) connected together at the corners so that each oxygen "bridges" two silicons with a bond angle of ~144°. The crystal structure is trigonal, belonging to the proper point group 32 (D_3). Three-fold screw axes (parallel to the crystalographic c-axis) pass down the centers of small open channels which can accomodate interstitial cations at for example positions α and β in Fig.1(4). There are 3 silicons and six oxygens per unit cell; each silicon lies on one of the 3 two-fold (piezoelectric) a-axes. The oxygens, which do not lie on symmetry axes, are pairwise equivalent; in Fig.1, oxygens O(I) are 1.598Å from the central silicon while oxygens O(II) are 1.616Å distant(5).

There exist left-hand and right-hand modifications of α-quartz which are distinguished on the basis of optical rotation but not by ESR. However, interpretation and communication of the ESR results is critically dependent on knowing the handedness of the specimen and relating the data to a well-defined coordinate system (see below and, especially, Ref. 6).

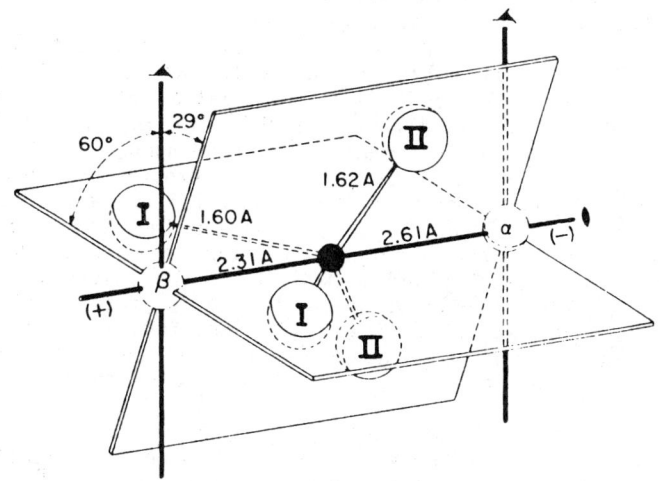

Fig. 1. Tetrahedron of oxygen atoms (open circles) surrounding a silicon atom (filled circle) in right-handed α-quartz. (After Ref. 4).

INTERPRETATION OF THE ESR SPECTRA

In the ESR experiment, the absorption of a fixed quantum $h\nu$ of microwave energy is recorded as a function of the magnitude of an applied d.c. magnetic field $\bar{B}$. For single crystal samples, a series of spectra are generally obtained for different discrete orientations of $\bar{B}$. Such angular dependence studies can then be compared with the known point group symmetries of the crystal leading to firm conclusions regarding the location of the paramagnetic center within the unit cell. In the case of α-quartz, a defect at an arbitrary point will be chemically equivalent to defects at 5 other locations related by 3-fold and 2-fold rotations about the respective symmetry axes in Fig.1. These six sites will yield inequivalent ESR spectra for an arbitrary orientation of $\vec{B}$, but degeneracies occur and some of the spectra overlap when the direction of $\vec{B}$ coincides with a symmetry axis. On the other hand, if the defect itself lies on a symmetry axis the number of inequivalent sites is reduced below six.

In addition to such crystallographic information, the ESR spectrum generally conveys a wealth of data concerning the interactions of the defect with its surroundngs. This "microscopic" information is summarized most compactly by use of the so-called "spin Hamiltonian" $\mathcal{H}$ (7). For present purposes it is possible to specialize to the case of a single unpaired electron (S=1/2) interacting with a single magnetic nucleus of spin I. Then

$$\mathcal{H} = \beta_e \ \vec{B} \cdot \vec{\vec{g}} \cdot \vec{S} + \vec{I} \cdot \vec{\vec{A}} \cdot \vec{S} + \vec{I} \cdot \vec{\vec{P}} \ \vec{I} - g_N \beta_N \vec{B} \cdot \vec{I}, \qquad (1)$$

where β_e and β_N are the electron and nuclear Bohr magnetons and $g_N \beta_N$ is the nuclear magnetic moment. $\vec{S}$ and $\vec{I}$ are respectively the electron and nuclear spin vectors. In eq. 1, the first term expresses the Zeeman interaction of the electron magnetic moment with the applied field, the second term the electron-nuclear hyperfine interaction, the third term the interaction of the nuclear quadrupole moment with the electric field gradient at the nucleus, and the fourth term the nuclear Zeeman interaction. The various "tensors" in eq. 1 need not be coparallel and are not in every case independently diagonalizable(7). Diagonalization of the first two terms leads to the resonance condition $\vec{B}(\nu, m_I, \theta, \emptyset)$ for the principal set of 2I+1 lines, where m_I is the nuclear magnetic quantum number and the angles θ and $\emptyset$ relate the direction of $\vec{B}$ to the crystal axes. The second two terms generally lead to small shifts in the principal lines while admitting a new series of "forbidden" transitions(8). Fitting the resonance condition to the observed ESR angular dependences yields numerical values for the diagonal elements of the g and A matrices. The values of the hyperfine parameters give precise information on the ground state wavefunction of the unpaired spin (see, e.g., Ref.7 or the simplified discussion in Ref.9), while the g tensor can be related in the ideal case to the energies of possible optical transitions which characterize the defect (e.g., Ref.10).

In powders or glasses, the observed ESR spectrum represents an angular average. However, accurate measurements of some spin Hamiltonian parameters can be accomplished by computer averaging of $|\vec{B}(\nu, m_I, \theta, \emptyset)|$ in order to simulate the observed line shape(11,12). The information which comes from crystal symmetry is lost, of course, but in the case of glasses it is possible to determine statistical distributions of parameters resulting from vitreous state disorder(9,11,12).

IMPURITY CENTERS IN α-QUARTZ

The aluminum center. The "aluminum center" is the generic name given to a wide class of defects in α-quartz associated with a substitutional aluminum (3). The "bare" $[Al]^o$ center (unassociated with interstitial cations) was first reported in irradiated synthetic and natural "smoky" quartz by Griffiths, Owen, and Ward(13) and interpreted in some detail by O'Brien (14) in 1955. O'Brien's basic model, illustrated in Fig.2, has stood the test of time.

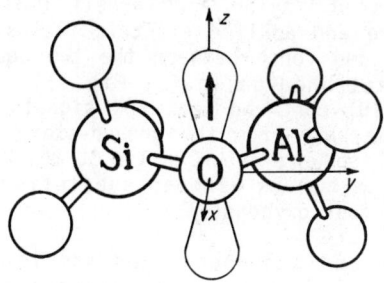

Fig. 2.   The "aluminum center" in α-quartz.

In essence, the bare [Al]^0 defect comprises a hole trapped in a non-bonding 2p orbital of an oxygen bridging between a substitutional aluminum and a neighboring silicon, the orbital of the unpaired spin being perpendicular to the plane defined by the Si-O-Al bond. The quadrupole and ^{27}Al hyperfine interactions have approximate axial symmetry about the O-Al direction. The hyperfine interaction is small, proving that there is little if any hole localization on the aluminum.   ESR yields only the absolute magnitudes of the hyperfine coupling constants, whereas the algebraic signs of these quantities are needed to draw more specific inferences. Barker (15) carried out an electron-nuclear-double-resonance (ENDOR) experiment which demonstrated that the diagonal elements of the A matrix all have the same sign. Although there is presently some disagreement in the literature regarding which sign is correct for the [Al]^0 center in α-quartz, the data and calculations of Stapelbroek, Gilliam, and Bartram (16) pertaining to the analogous defect in tetragonal GeO$_2$ may prove decisive.   The latter authors showed that the ^{27}Al coupling constants for [Al]^0 in tetragonal GeO$_2$ must be negative and arise from a negative isotropic core polarization term (7) and a smaller, positive dipole-dipole contribution.   The magnitude of the dipole-dipole coupling is evidently reduced due to a lattice relaxation which results in an Al-O(defect) separation substantially larger than the normal Al-O bond length.   In order to account for the measured hyperfine anisotropy in α-quartz, Schirmer (17) has calculated a relaxed Al-O(defect) distance of 2.4Å.   O'Brien's ESR study (14) showed that the sign of the quadrupole coupling constant P must be opposite from that of the hyperfine parameters. Assuming with Stapelbroek et al. (16) and Shirmer (17) that the hyperfine coupling constants are negative, a positive P results; this in turn accords with a simple calculation of field gradients at the Al nucleus carried out by O'Brien (14).

Because of the chemical inequivalence of oxygens O(I) and O(II) in Fig. 1, only one of these two sites can serve as the ground state of the [Al]^0 center.   The question of which site acts in this capacity has remained controversial, however, due presumably to the perennial problem of defining crystal coordinates.   Although several recent publications (10,18) have ascribed the bare [Al]^0 center to a hole trapped on O(I), Mackey, Boss, and Wood (19) assigned this defect to O(II). Weil (20) has carried out studies supportive of the latter conclusion. Whichever interpretation is correct, the experiments of Taylor and Farnell (18) employing electron-electron double resonance and applied electric fields have shown that the hole actually hops back and forth between the two equivalent ground-state oxygens on each AlO_4 tetrahedron (c.f., Fig. 1). Equally important, Schnadt and Schneider (10) observed weak ESR signals clearly attributable to the excitation of the hole from the ground-state pair of oxygens onto the inequivalent pair, which were shown to lie 30 meV higher. This completed the proof that the aluminum is indeed substutional, surrounded by a distorted tetrahedron of four oxygens(3).

By irradiating alkali- and hydrogen-doped samples at 77K, Mackey(21) was able to identify a series of cation-compensated aluminum centers $[Al, e^+ / M^+]^{1+}$, where e^+ indicates a trapped hole, $M^+ = Li^+$, Na^+, or H^+, and the superscript 1+ indicates that the overall defect structure has a net positive charge. The precursor of this defect (in the unirradiated material) is thus inferred to be $[Al^{(-)}/M^+]^0$. Mackey $\underline{et\ al}$(19) gave evidence that $[Al, e^+/H^+]^{1+}$ comprises a hole trapped on an $\overline{O(I)}$ oxygen with the proton located near the position α in Fig.1, but displaced inward toward the aluminum. Upon warming, the compensating ion M^+ (when M=Na or Li) was found to leave the aluminum center site, creating a bare $[Al]^0$ defect(21) (see Fig.3).

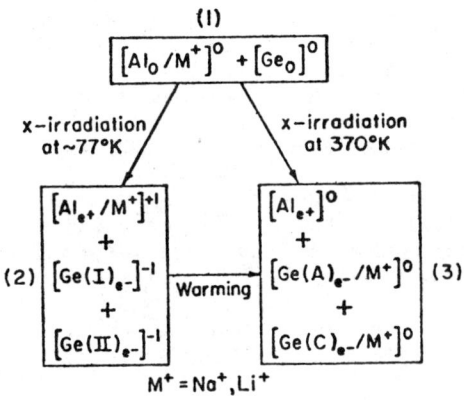

Fig. 3 Kinetic relationships among paramagnetic centers in germanium-doped α-quartz at low x-ray doses. (After Ref. 21.).

An interesting recent development has been the reporting by Nutall, Weil, and Claridge (22) of a double-hole aluminum center in α-quartz x-irradiated at 77K. Although not yet fully characterized, this defect appears to consist of two holes trapped on a pair of equivalent oxygens at a substitutional Al site. The spins of these two holes are aligned forming a triplet state, which requires for its full description an additional term in eq. 1 quadratic in the electron spin (7,22).

Germanium-related centers. Anderson and Weil(23), Haven, Kats, and Van Wieringen(24), and Mackey(21) have reported on a broad class of defects associated with silicon-substituted germanium in α-quartz. Two "bare" centers [Ge(I,II)e$^-$]$^{-1}$ and two alkali-compensated centers [Ge(A,C)e$^-$/M$^+$]^0 were described (M$^+$=Na$^+$,Li$^+$). In all cases the g shifts were found to be negative and this fact was adduced to support the trapped-electron natures of these centers(23). Further evidence came from samples irradiated at 77K which initially exhibited bare Ge(I) and Ge(II) centers plus alkali-compensated aluminum (hole-type) centers but which upon warming displayed bare aluminum centers plus alkali compensated Ge(A) and Ge(C) centers(21). From this behavior emerged the picture of an interstitial M$^+$ ion initially charge-compensating the precursor aluminum site prior to its trapping a hole but subsequently being released and finding its way to the germanium site where it charge-compensates (and thus stablizes) the trapped electron (see Fig.3). That the germanium involved is actually substitutional is attested by the fact that each type of Ge center shows only three magnetically inequivalent sites for an arbitrary orientation of $\hat{B}$(23), and in each case the principal axes of the g-tensor lie in or near the special planes illustrated in Fig.1. An important additional constraint on the natures of these substitutional germanium centers comes from the large, nearly isotropic hyperfine coupling observed with 7.6% abundant ^{73}Ge(21,24); this is indicative of a high degree of localization of the unpaired spin in a germanium sp hybrid orbital. The relationships between Ge(I) and Ge(II) and between Ge(A) and Ge(C) are not apparent, although g-tensor similarities suggest that [Ge(A)e^{-1}/M$^+$]^0 may result from the capture of an interstitial M$^+$ ion at the site of a [Ge(II)e$^-$]$^{-1}$ defect(21). Very recently, Laman and Weil(25) have reported an Ag$^+$-compensated Ge(A) center with large spin density on the silver nucleus. The symmetry of the g and A tensors shows the silver to be located on two-fold axes, probably quite near the position β in Fig.1.

In the cases of the four different germanium centers discussed above, none of the principal axes of the g tensors lie particularly close to an Si-O bond direction in the quartz structure. On the other hand, Weil(26) has reportd a defect, the GeHLi$_2$ center, for which the principal axes of the smallest g shift and of the "unique" ^{73}Ge hyperfine splittings are virtually coincident and lie close to an Si-O(I) direction (refer to Fig.1). The GeHLi$_2$ defect also displays a hyperfine interaction of intermediate strength with hydrogen (verified by deuterium substitution) and very weak hyperfine interactions with two interstitial Li$^+$ ions; this center is produced by x-irradiation at 300 K but not 77K(26). Weil's tentative model for GeHLi$_2$ involves a silicon-substituted Ge^{2+} precursor site compensated by two Li$^+$ ions. This site is hypothesized to capture a radiation-generated hydrogen atom which in turn accepts an electron from the germanium, forming Ge^{3+}. According to calculations(26), the unpaired spin is ~50% localized in a germanium hybrid orbital (4s,4p) directed along the Si-O(I) bond. It is interesting to note that centers with very similar g tensors and virtually identical ^{73}Ge hyperfine tensors (but no ^{1}H or ^{7}Li splittings) have been observed by Feigl and Anderson(6) in samples of germanium doped α-quartz following room-temperature x-irradiation and certain bleaching schedules. Calculations performed by the latter authors using somewhat different values for $|\psi_{4s}(0)|^2_{Ge}$ and $<r^{-3}_{4p}>_{Ge}$ than employed in Ref.(26) led to the conclusion that the unpaired spin is ~90% localized in a germanium hybrid orbital projecting along the Si-O(I) direction. On this basis, Feigl and Anderson defended the thesis that there must be an oxygen vacancy at the O(I) site in question, i.e., it was argued that the latter centers (designated [Ge(III,IV)e$^-$]) are E$_1'$ centers (see following Section).   This contrasts with Weil's model for the GeHLi$_2$ center which assumes no structural faults.   One perceives the possibility that both defect types (GeHLi$_2$ and Ge(III,IV)) are characterized by the <u>same</u> arrangement of oxygens, but this has not been asserted by any of the original investigators and does not necessarily follow from the remarkably similar g and A matrices.

<u>Stable multiple-alkali centers.</u>   Lorenze and Feigl(27,28) have demonstrated yet another class of multiply compensated silicon-substituted germanium defects which they designated ß centers.   Experiments involving electrolytic diffusion of monvalent impurities (Na$^+$, Li$^+$, H$^+$) and and subsequent observation of the associated hyperfine multiplets in ESR led to the specification of these defects as [Ge, e$^-$/Na$^+$, M, N], where M, N = Na or Li. In principle,the ß centers would represent just another variation of Weil's GeHLi$_2$ center,except that in the ß-center case the ^{73}Ge hyperfine splitting is an order of magnitude smaller, the g shifts are an order of magnitude greater, and the half-life at room temperature is more than two orders of magnitude longer.   Detailed kinetic studies(28) showed that Weil's proposed model of defect formation (see above) is inappropriate to the ß centers. Rather, a case was made for the sequential trapping scheme illustrated in Fig.4(28), where the precursor Ge site is uncompensated.   It would appear that the ß centers are the principal stable electron traps in heavily x-irradiated α-quartz with low Ge/Al ratios.

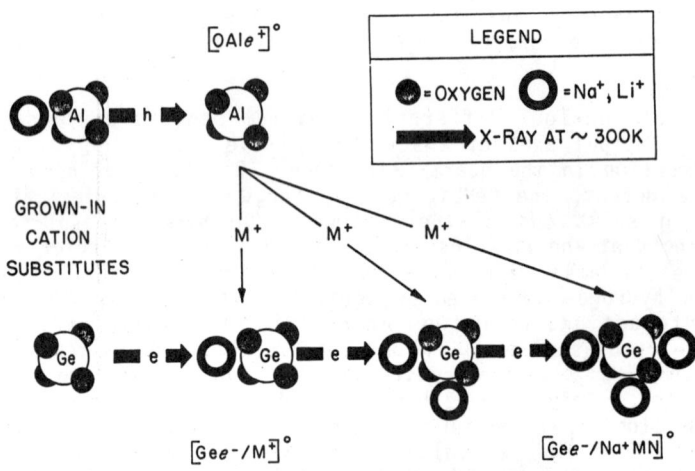

Fig. 4. Kinetic relationships among paramagnetic centers in germanium-doped α-quartz of low Ge:Al ratios at high x-ray doses. (After Ref. 28.)

<u>Atomic hydrogen.</u> Protons are a ubiquitous impurity in both synthetic and natural α-quartz (29). Weeks and Abraham(30) conducted an ESR study of atomic hydrogen in α-quartz produced by γ irradiation at 77K following

electron irradiation at 300K. Spectral symmetry, coupled with the observed weak superhyperfine interactions with two inequivalent ^{29}Si sites, places the H^0 on a two-fold axis somewhat displaced from the α or β positions in Fig.1. Warming above ~100K destroys the center. Perlson and Weil (31) report observing a number of different atomic hydrogen sites in irradiated natural as well as synthetic α-quartz and they give refined values for the principal-axis components of the g and ^{1}H A matrices of the site described by Weeks and Abraham.

Other impurities. Brief mention will be accorded a number of other substitutional impurities which may serve as electron or hole trapping sites. In the hole trap catagory, Schnadt and Schneider(10) showed that gallium evidently performs the same role as aluminum. On the other hand, Rinneberg and Weil(4) demonstrated that an important class of electron traps in natural rose quartz is the result of substitutional Ti^{4+} ions which react with radiation-produced hydrogen atoms (see above) to form proton-compensated Ti^{3+} centers. Two types of Ti^{3+} centers were discerned corresponding to the two possible proton positions in Fig.1; both are created after x irradiation at 77K (forming H^0) and subsequent warming above 120K. Irradiation of rose quartz at room temperature gives rise to analogous Li compensated species. Two well defined Fe^{3+} centers in α-quartz have been extensively characterized by ESR: $\underline{S}_1$(32,33) which dominates in natural amethyst has been assigned to a silicon-substituted position; $\underline{I}$ (34,35) which is more prevalent in synthetic brown quartz is generally regarded as occupying an interstitial site, although the data do not preclude its being a second type of substitutional center (35). The amethyst color--difficult to reproduce in synthetic crystals--is evidently the result of natural radiations, with $\underline{S}_1$ and $\underline{I}$ serving as precursors to the actual color centers (36).

DEFECT CENTERS IN QUARTZ

E' centers. In 1956, Weeks(37) reported an ESR signal in neutron-irradiated α-quartz characterized by small negative values of $\Delta g = g$-g(free electron); the g tensor exhibited approximate axial symmetry with $|\Delta g_\perp| > |\Delta g_\parallel|$(38-40). Because this resonance was quite stable and its line shape and intensity bore no relationship to the type or quantity of impurities known to be in the samples, it was attributed to defects intrinsic to the quartz lattice. Nelson and Weeks(38,39) termed such defects E' centers and distinguished two varieties, E_1' and E_2', which appeared to be correlated with optical bands at 6.2 and 5.4 eV, respectively. Both were tentatively assigned to electrons trapped at oxygen vacancies. Silsbee(40) examined the E_1' center in some detail and recognized two significant facts: First, an observed low-intensity doublet structure with a large (400 Gauss) splitting must be due to the hyperfine interaction of the E'-center electronic spin (S=1/2) with the 4.7%-abundant ^{29}Si nucleus (I=1/2). Second, the unique axes of the g and hyperfine tensors are nearly coincident and lie close to Si-0 bond directions in the quartz lattice. Estimation of the quantities $|\psi_{3s}(0)|^2_{Si}$ and $<r_{3p}^{-3}>_{Si}$ for the free silicon atom and comparison with the "strong" ^{29}Si hyperfine data for the E_1' center suggested that the unpaired spin is in an orbit comprising 24% Si 3s and 63% Si 3p states (40). This picture fits the now-generally-accepted model of an unpaired electron in a dangling $\underline{sp}^3$ orbital of a silicon bonded to only three oxygens.

There were problems with this model, however, which seem to have been resolved only recently. It was noted early on(38) that the E_1' center exhibited additional, weak hyperfine splittings with two nonequivalent ^{29}Si nuclei which were not readily explainable(40). Indeed, when one considered the oxygen vacancy model, it was anticipated that the unpaired electron would spend nearly equal time on each of the two adjacent silicons, leading to relatively strong-hyperfine interactions with ^{29}Si nuclei at both sites. This anticipation contrasts with the experimental evidence for a high degree of localization on just one silicon. A unified theory explaining most aspects of the experimental data was finally proposed by Feigl, Fowler, and Yip(41). Basically, the theory pictures the E_1' center as an (oxygen) vacancy featuring strongly asymmetric relaxation of the two adjacent silicons (see Fig.5). Molecular orbital calculations for the Si_2O_6 cluster of Fig. 5b by Yip and Fowler(42) using Madelung-type potentials for α-quartz verified that the total system energy can be minimized by relaxing Si_I into the plane of oxygens V, VI, and VII while relaxing Si_0 slightly in the direction of the vacancy (the relaxation of Si_0 illustrated in Fig. 5c is ~10 times that calculated). The unpaired spin resides in the Si_0 hybrid orbital projecting into the vacancy. The calculations also gave the correct order of magnitude and approximate directions for the two weak hyperfine interactions, which were attributed to silicons I and V in Fig. 5c.

Feigl and Anderson(6) have pointed out that the unique axes of the g and strong-hyperfine tensors of the E_1' center lie along Si-O(I) directions in the quartz structure (refer to Fig.1) while those of the E_2' center

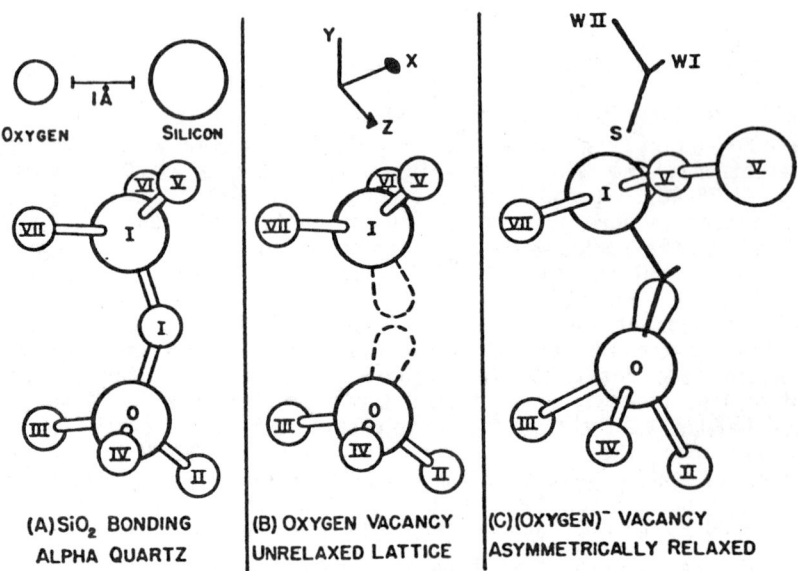

(A) SiO₂ BONDING ALPHA QUARTZ

(B) OXYGEN VACANCY UNRELAXED LATTICE

(C) (OXYGEN)⁻ VACANCY ASYMMETRICALLY RELAXED

Fig. 5. Asymmetric relaxation model for E_1'-center formation at oxygen vacancies in α-quartz, according to Feigl, Fowler, and Yip. (After Ref. 41.)

fall more nearly parallel to Si-O(II) directions. Thus, the latter authors ascribed the two types of E' center to the two possible modes of the structural relaxation at the oxygen vacancy, i.e., if the unpaired electron traps on Si_0 in Fig.5 an E_1' center results, while trapping on Si_I (with Si_0 relaxing to planarity) yields an E_2' center. Yip and Fowlers's calculations(42) indicate that the energy minima corresponding to these two possible states should be separated by only ~ 0.1 eV. Evidently, the E_1' center has a lower energy than the "bare" E_2' defect. Weeks(43) characterized the E_2' center in considerable detail, showing among other things that it is normally associated with a proton which presumably is a factor in its stabilization. Weeks(43) also noted a very low intensity signal which could have been the "bare" E_2' center. According to Yip and Fowler's analysis, this would be a thermally populated excited state of the E_1' center.

As mentioned in the previous Section, a series of defects attributed to silicon-substituted germanium E' centers have been reported by Feigl and Anderson(6). Two of these appear analogous to the silicon E_1' and E_2' centers while a third Ge center seems to be cryptically related to the E_1'.

There is a strong school of thought that the precursor of the E' center in x- or γ-irradiated α-quartz is a pre-existing neutral oxygen vacancy, i.e., a vacancy containing two (paired) electrons(6,42). (In the nomenclature of Weeks and Nelson(38) this would be denoted an E'' center.) In this view, the E' center arises from the trapping of a hole at the site of a "stretched" Si-Si bond spanning the vacancy, followed by the asymmetric relaxation portrayed in Fig.5. It should be remarked, however, that acceptance of this model for quartz does not preclude the possibility that the E' center may be a trapped-electron center in certain forms of fused silica (see below).

Other defects. No other defects intrinsic to the α-quartz structure have been at all well characterized by ESR. A group of resonance lines centered near g = 2.009 were reported by Weeks(37) in neutron bombarded quartz; these conceivably may arise from species associated with displaced oxygens.

IMPURITY CENTERS IN FUSED SILICA

A wide variety of impurities which are normally rejected from the quartz structure are nevertheless readily accomodated into fused silica. Whereas a few percent of impurities can be sufficient to completely arrest the growth of α-quartz(44), fully tens of percent of the same impurities can be incorporated into silica-based glasses. The following compilation will emphasize impurity centers which have been studied at low concentrations in fused silica or whose local surroundings are adjudged to be independent of the concentrations of both the dopant and any codopant. This emphasis will thus tend to exclude the so-called "complex" glasses, for which the reader is referred to reviews by Lell, Kreidl, and Hensler(45) and Bishay(46).

The aluminum center. A six-line ESR spectrum in irradiated fused natural quartz, apparently characteristic of a hyperfine interaction with ^{27}Al, was first reported by Fröman, Pettersson and Vänngård(47). Subsequently, similar resonances were obtained by Lee and Bray(48) in irradiated Al_2O_3-SiO_2 glasses containing up to 10 wt% alumina. In both investigations, a parallel was noted between the mean g values and hyperfine splittings of the glass spectra and the corresponding parameters for the $[Al]^0$ center in α-quartz.

More recently, Schnadt and Räuber(49) made a direct comparison of the "aluminum center" ESR spectra of powdered aluminum-doped quartz and a glass prepared of the same material. An easily notable discrepancy in the region of the largest g value (g_3 in Fig. 6) was attributed to random distortions in the vitreous environment causing a distribution in g_3 values and hence a smearing out of this part of the glass spectrum(49). A preliminary computer line shape simulation (Fig. 6e) by Stapelbroek and Griscom(50) supports this notion and shows the mean g values for Al-doped silica to be much closer to those for the $[Al]^0$ center in tetragonal GeO_2(16) than to those for the $[Al]^0$ center in α-quartz. The significance of the latter finding is not yet clear.

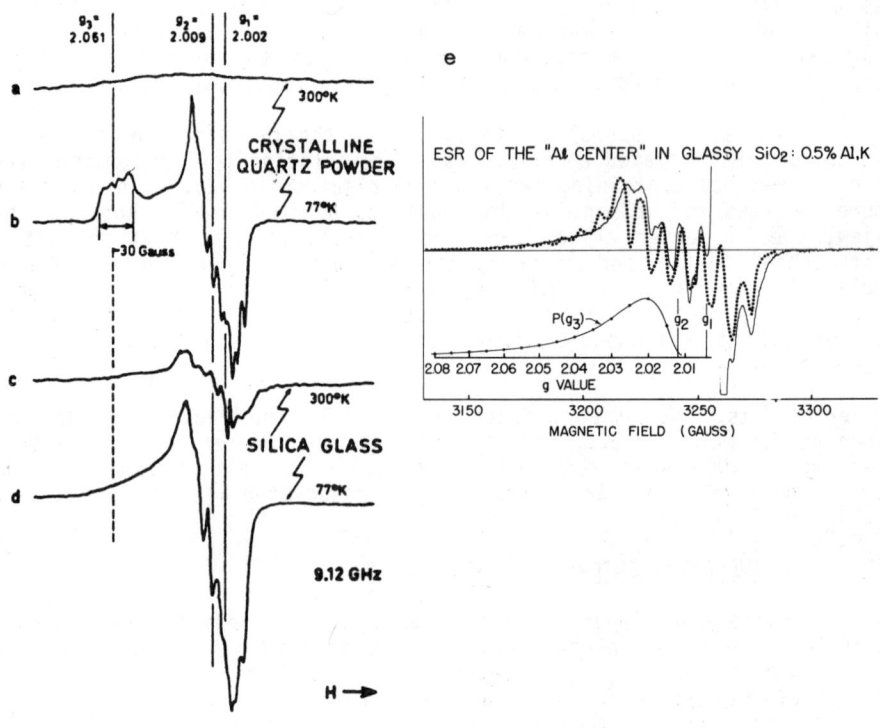

Fig. 6. ESR spectra of aluminum-associated trapped hole centers in x-irradiated α-quartz and fused silica. (A,b,c, and d after Ref. 49; e after Ref. 50.)

Schnadt and Räuber raised an interesting point with regard to Fig. 6a thru d. Namely, they explained the (reversible) disappearance of the quartz spectrum at room temperature in terms of thermally activated relaxation line broadening. Their linewidth-vs.-temperature data established an activation energy of 70 meV, which agrees well with the difference in energy determined(10) for hole trapping at the O(I) and O(II) positions of Fig. 1 (see above). It was then reasoned (49) that random distortions in the vitreous environment would cause all four oxygens surrounding the aluminum to become inequivalent, with the energy barrier for hopping between sites being >>70 meV. This would suppress hole hopping and thus explain the observability of the silica-glass spectrum at 300K (Fig. 6c).

Boron centers. Boron is one of those elements which evidently defies incorporation into α-quartz but which readily enters silica glass. Because of its application to low-loss fiber optic waveguides, borosilicate glass has been the object of recent studies. Griscom, Sigel, and Ginther (51) carried out an ESR investigation of γ-irradiated high-purity glasses of the composition $B_2O_3 \cdot 3SiO_2$. Only three defect species were detected; these were the well-known silicon E' center, a previously unreported boron E' center and the so-called boron-oxygen hole center (BOHC), familiar from studies of alkali borate materials(52). The best interpretation of the BOHC seems to be that it is the analogue of the $[Al]^0$ center in α-quartz, i.e., a hole trapped on an oxygen bridging between a boron and a silicon both in tetrahedral coordination (51). This follows in part from the same types of g and   hyperfine tensor analyses as have been applied to $[Al]^0$ defect (e.g., in Ref. 10). Hyperfine analysis was also decisive in showing that the boron E' center results from electron trapping at the sites of normal (three-coordinated) borons in accordance with certain predictions of the Feigl-Fowler-Yip theory(41). On the other hand, it was noted that the concentrations of silicon E' centers in γ-irradiated $B_2O_3 \cdot 3SiO_2$ glasses were 10-40 times those induced in equal masses of high purity silicas by the same γ-ray dose(51). From the foregoing facts it was inferred that a few borons (>200ppm) are tetrahedrally coordinated at the expense of the formation of an equal number of three-coordinated silicons. The former are negatively charged and thus trap holes (forming BOHCs); the latter are positively charged and hence trap electrons (forming E' centers). The trapped-electron nature of the E' center inferred for this glassy material contrasts with the trapped-hole picture of the E' center in α-quartz evinced above. This presents no contradiction; rather, it illustrates how the natures of radiation-induced defects are strongly dependent on the precise atomic arrangements and chemical bonding in the materials in which they occur.

Germanium centers. Also in the fiber optic context, Friebele, Griscom, and Sigel(53) investigated the defect centers which appear in γ-irradiated silica glasses doped with $\sim$ 10 mol % GeO_2. In as much as the observed ESR spectra were enhanced in samples heated in hydrogen and diminished in samples heated in oxygen prior to irradiation, the involvement of an oxygen vacancy was inferred. This conclusion was further reinforced by the fact that the g values measured for one particular defect type were virtually identical to those for the E' center in glassy or hexagonal GeO_2(54) or the $[Ge(III)e^-]$ center in Ge-doped α-quartz (see above and Ref.6). In all, the spectrum was analyzed into four distinct parts on the basis of growth and bleaching studies; each part was ascribed to a separate germanium-related defect denoted Ge(0), Ge(1), Ge(2), and Ge(3). The g tensors (determined by computer simulation), relative concentrations, and stabilities suggested that these centers are germanium E' centers having respectively 0, 1, 2, or 3 nearest neighbor silicon-substituted germaniums. A small concentration of Ge(3) was present in the as-quenched glasses prior to irradiation, underscoring the fact that clusters of Si-substituted germaniums can provide especially deep electron traps in fused silica.

Alkali-associated hole centers. Upon irradiation, alkali silicate glasses display two prominent ESR lines centered near g=2.01; the defects responsible for these spectra were designated HC_1 and HC_2 by Schreurs (48). It is generally agreed that both centers comprise holes trapped on nonbridging oxygens (NBOs). NBOs occur pairwise in the glass network with each added molecule of M_2O (M=alkali). HC_1 predominates over HC_2 at low alkali contents and its intensity is proportional to alkali concentration for fused silica doped with 0.01 to 0.5 mol % M_2O(49). Griscom(12) has analyzed the HC_1 spectrum in terms of a hole trapped on a single nonbridging oxygen; refinements of this work, including a model for HC_2, will appear elsewhere(50). Radiation-induced HC_1 and HC_2 defects are common in complex silicate glasses but often undetectable in fiber-optic grade fused silicas where the alkali levels are substantially lower than 1 ppm.

Hydrogen centers. Hydrogen impurities in commercially available fused silicas range from ~5000 to <20 protons per million silicon atoms, depending on the method of manufacture. As in the case of α-quartz (see above), many of these protons can be converted to hydrogen atoms when x- or γ-irradiations are carried out below 100K. The ESR signature of H^0 is a pair of sharp lines centered on g=2.00 with an approximate 510-Gauss splitting. Whereas the H^0 spectrum anneals out completely above ~120K, several room-temperature-stable doublets of substantially smaller splittings are frequently mentioned in the literature. By means of various deuterium substitution experiments, Vitko (59) has recently demonstrated the hydrogenic origins of a pair of doublets with slittings of 119 and 74 Gauss; these were attributed respectively to a hydrogen-compensated germanium impurity center and an E' center intimately associated with a proton. A variation on one of these models is also likely to explain a 133-Gauss doublet observed in high-purity, high-OH fused silicas following x-irradiation at 77K and subsequent annealing (60). Also observed has been a weak quartet structure of splitting 22.6 Gauss and intensity ratio 1:3:3:1, almost surely arising from trapped methyl radicals (60).

DEFECT CENTERS IN FUSED SILICA

The E' center. On the basis of nearly identical optical bands(38) and ESR powder-pattern line shapes(39), Weeks and Nelson argued that the E' centers observed in α-quartz are also present in a structurally similar form in irradiated fused silicas. This conclusion was further substantiated by Griscom, Friebele, and Sigel(61) who utilized ^{29}Si-enriched samples to confirm their own observation of the primary ^{29}Si hyperfine structure of the E' center in silica. The latter work showed not only that the unpaired spin has the same degree of localization in the Si 3s and 3p orbitals in both the crystalline and glassy forms but that the mean variation in O--Si--(defect) bond angles in the glass is only ~ 0.7°. Thus, even vitreous disorder is insufficient to blur the basic structural identity of the E' center as a pyramidal SiO_3 unit with nearly tetrahedral bond angles and an unpaired electron in the dangling sp³ orbital of the apex silicon.

Beyond the characteristics just mentioned, there are several aspects of the E' center which probably do change on going from the crystal to the glass. Primarily, one must consider the fact that there are no obvious constraints on the separations of the two SiO_3 "half units" of Fig. 5

when one is dealing with a glass. Indeed, three-coordinated silicons present in fused silica as Frenkel defects (complementary to an equal number of NBOs, say) could be quite far apart, in which case they must contain no electrons in the dangling silicon orbitals prior to irradiation. Only after irradiation would they trap electrons to form E' centers. This was argued to be the case in borosilicate glasses (see above) and is probably true also in many stoichiometric pure silicas. On the other hand, in reduced silicas one presumes neutral oxygens have been removed from the network and, since such materials are not paramagnetic as made, the existence of Si-Si bonds is logically deduced. Trapping of holes at the latter sites would form E' centers according to the Feigl-Fowler-Yip theory (Fig.5). Griscom, Sigel, and Friebele(62) have recently given evidence that the E' center may on the average be a hole center in silicas of low-OH content but an electron center in normal "wet" silicas. There still remains the question of whether or not the electron trapping site in wet silicas is due to pre-existing SiO_3 "half units" with empty silicon orbitals. Nelson and Crawford(63) have argued that the γ-ray colorability of fused silicas in the ultraviolet region is perhaps best explained in terms of the rupture of strained Si-O bonds in the network. This picture of E' center formation, which was also supported by the optical studies of Arnold and Compton(64), is illustrated in Fig.7.

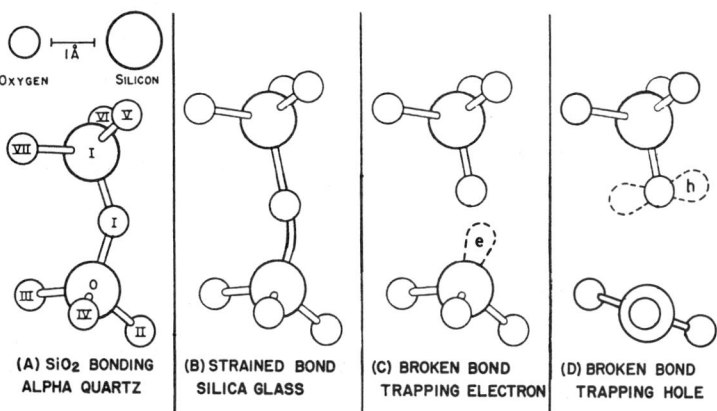

(A) SiO₂ BONDING ALPHA QUARTZ (B) STRAINED BOND SILICA GLASS (C) BROKEN BOND TRAPPING ELECTRON (D) BROKEN BOND TRAPPING HOLE

Fig. 7. Strained bond concept of defect formation in fused silica.

Finally, it should be mentioned that the distinction between the E_1' and E_2' centers in fused silica has not yet been achieved by ESR means; although the optical evidence for their existence as distinct entities seems strong(38). The ESR signature of the E_2' center in silica would consist of a very weak hyperfine splitting with a proton based on the data for quartz(43), but the doublet separation is comparable to the intrinsic linewidth for the E' center in glass and hence difficult to resolve. Thus, the subscripts 1 and 2 are customarily dropped when discussing ESR spectra of E' centers in fused silica.

Oxygen-associated hole centers. Whereas nonbridging oxygens may be difficult to accomodate in the α-quartz structure, they would seem to be a ubiquitous flaw in fused silicas that are not oxygen deficient. NBOs represent negatively charged point defects in the silica network and hence are expected to serve as hole traps. The resulting trapped-hole centers, generically termed OHCs, can generally be recognized on the basis of their g tensors(9). The model for the OHC based on an analysis of the g tensor is that of a hole trapped in a pure nonbonding oxygen 2p orbital orthogonal to the Si-O(defect) bond direction. Recent confirmation of this picture has come from studies of ^{17}O-enriched silica by Stapelbroek and Griscom(50). The latter authors also demonstrated the existence of two types of OHCs, described as "wet" or "dry" in accordance with the type of fused silica wherein they attain the greatest relative abundance. A model for the formation of the "wet" OHC is presented in Fig. 8. It is hypothesized that the "dry" OHC is due to hole trapping on NBOs which are not paired with other NBOs or hydroxol groups in the glass network.

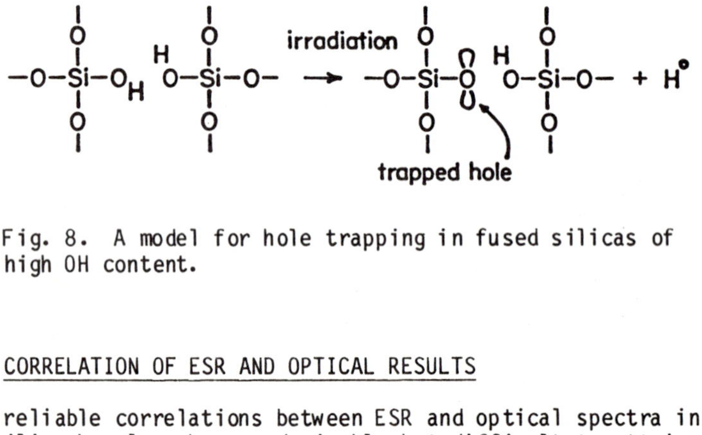

Fig. 8. A model for hole trapping in fused silicas of high OH content.

CORRELATION OF ESR AND OPTICAL RESULTS

Obtaining reliable correlations between ESR and optical spectra in α-quartz or fused silica has long been a desirable but difficult to attain objective. Summarized here are the few apparently successful attempts, together with a brief discussion of one of the more baffling cases, the "aluminum center".

The germanium A and C centers. By means of careful optical and ESR annealing studies of a wide range of samples, Anderson, Feigl, and Schlesinger(65) firmly established a correlation between the Ge(A,C) centers and an optical band at 4.43 eV, thus confirming the earlier conclusion of Anderson and Weil(23).

The aluminum center. The [Al]^0 ESR response has long been believed to be associated with optical absorptions near 2.0 and 2.6 eV(13). However, Weeks (66) carried out a series of irradiation, annealing, and re-irradiation experiments on samples of aluminum-containing synthetic α-quartz and found no correlation between the aluminum center resonance and any optical band between 1.0 and 8.5 eV. Assuming that there is sufficient evidence in the literature to firmly link the aluminum center with the aforementioned optical bands (see, e.g., Ref. 3), Weeks' observations must be explained in terms of some effect of heat treatment on the mechanism whereby the $[Al]^0$ center absorbs light. As noted by Schnadt and Schneider(10), the observed visible absorption spectrum of "smoky" quartz does not correspond to predictions based either on the experimental $[Al]^0$ g tensor or ab initio crystal field calculations. The latter facts, coupled with other results of recent origin (see literature cited in Ref.17), led Schirmer(17) to propose a bound small polaron model for optical absorption by the aluminum center. In this picture, optical absorption is due to charge transfer between the oxygen trapping the hole and one of the other three oxygens surrounding the aluminum. The energy of the absorption is thus ascribed to a self-trapping potential associated with the oxygen which trapped the hole relaxing away from the aluminum. Schirmer(17) argued further that the oscillator strength would be greatly weakened if the hole-trapping oxygen were rendered inequivalent to the other three by some external perturbation. It is conceivable that Weeks(66) generated such perturbed aluminum centers by a process of irradiation, annealing, and re-irradiation, but the whole issue warrents further study.

The E' center. Studying Corning 7943, a dry (and probably oxygen deficient) silica, Weeks and Sonder(67) achieved an excellent correlation between the E'-center ESR signal and a well-defined optical band at 5.85 eV. Based on Nelson and Weeks' earlier work in α-quartz(38), these bands were assigned specifically to the E_1' center. Susceptibility measurements indicated that this one defect was the only paramagnetic state present in significant concentration and made possible the determination of an oscillator strength, f=0.14.

The OHC. Griscom, Sigel, and Friebele(62) examined the annealing behavior of both ESR and optical bands in pairs of high-purity wet (dry) silicas: Suprasil 1 (W1), Spectrosil A (WF), and Corning 7940 (7943). In addition to the previously reported E_1'-center/5.85 eV correlation, Griscom et al also observed a correlation between the total integrated OHC concentration and a well-known (e.g. Ref.68) optical band at 7.6 eV. The latter correlation was particularly striking in the two dry silicas to exhibit OHC spectra (Suprasil W1 and Spectrosil WF); here, the OHC and 7.6 eV signals grew by as much as 100% on annealing to 300C and then decreased together when heated to higher temperatures. Stapelbroek and Griscom (50) have recently shown that it is only the "dry" OHC which initially grows on heating, whence there appears to be a firm correlation between this specific defect and the 7.6 eV optical band. Preliminary estimates of the oscillator strength in Suprasil W1 based on the total OHC population fell in the range, f=0.5--0.75(69). A correction for an optically inactive "wet" OHC would raise these numbers by ~ 30% (using the "wet":"dry" ratio from Ref. 50). Since the ground state of the OHC is a pure O 2p orbital (see above), these results suggest that the 7.6 eV transition is either O 2s → O 2p or O 2p → O 3s. Assuming the OHC ground state to lie within the band gap, the former possibility is ruled out by what is known about the electronic structure of SiO_2 (for a review, see Ref.70). The more likely interpretation of the 7.6 eV band (O 2p→ O 3s) is probably best described as an excitation of a bound Frenkel exciton state of the OHC (see pertinent discussion by Mott, Ref. 71).

248

ACKNOWLEDGMENT

The author is indebted to Dr. M. Stapelbroek for pointing out a number of references and for several helpful discussions. Constructive remarks by Dr. J. A. Weil led to improvements in the final draft.

REFERENCES

(1)   K.H. Ashbee, X-rays and electron microscopy, in "Defects and Their Structure in Nonmetallic Solids", B. Henderson and A.E. Hughs, eds., Plenum Press, New York (1976) p 407.

(2)   E.J. Friebele, R.E. Jaeger, G.H. Sigel, Jr., and M.E. Gingerich, Effect of ionizing radiation on the optical attenuation in polymer-clad silica fiber-optic waveguides, Appl.Phys.Lett. 32,95 (1978).

(3)   J.A. Weil, The aluminum centers in α-quartz, Radiation Effects 26, 261 (1975).

(4)   H. Rinneberg and J.A. Weil, EPR studies of $Ti^{3+}-H^{+}$ centers in X-irradiated α-quartz, J. Chem.Phys. 56, 2019 (1972).

(5)   G.S. Smith and L.E. Alexander, Refinement of the atomic parameters of α-quartz. Acta cryst. 16, 462 (1963).

(6)   F.J. Feigl and J.H. Anderson, Defects in crystalline quartz: associated with germanium impurities, J.Phys.Chem.Solids 31, 575 (1970).

(7)   A. Abragam and B. Bleaney, Electron Paramagnetic Resonances of Transi-tion Ions, Oxford University Press, London (1970).

(8)   J.A. Weil and J.H. Anderson, Direct field effects in electron para-magnetic resonance hyperfine spectra, J.Chem.Phys. 35, 1410 (1961).

(9)   D.L. Griscom, E.J. Friebele, G.H. Sigel, Jr., and R.J. Ginther, Radiation-induced paramagnetic defects as structural probes of pure silica and borosilicate glasses, in The Structure of Non-Cry-stalline Materials, P.H. Gaskell, ed., Taylor & Francis, London, (1977),p.113.

(10) R. Schnadt and J. Schneider, The electronic structure of the trapped hole center in smoky quartz, Phys.Kondens. Materie 11, 19 (1970).

(11) P.C. Taylor and P.J. Bray, Computer simulations of magnetic resonance spectra observed in polycrystalline and glassy samples, J.Magn.Res. 2, 305 (1970).

(12) D.L. Griscom, Defects in non-crystalline oxides, Ref. 1, p. 323.

(13) J.H.E. Griffiths, J. Owen, and I.M. Ward, in Report of the Bristol Conference - Defects in Crystalline Solids, The Physical Society, London (1958), p 81.

(14) M.C.M. O'Brien, The structure of the colour centers in smoky quartz, Proc. Roy. Soc. A231, 404 (1955).

(15) P.R. Barker, Hyperfine parameters of the Al centre in smoky quartz, J. Phys. C 8, L142 (1975).

(16) M. Stapelbroek, O.R. Gilliam, and R.H. Bartram, Transferred hyperfine interactions for trapped-hole centers in tetragonal GeO_2, Phys.Rev.B 16, 37 (1977).

(17) O.T. Schirmer, Smoky coloration of quartz caused by bound small hole polaron optical absorption, Sol. State Comm. 18, 1349 (1976).

(18) A.L. Taylor and G.W. Farnell, Spin-lattice interaction experiments on color centers in quartz, Canad. J. Phys. 42, 595 (1964).

(19) J.H. Mackey, J.W. Boss, and D.E. Wood, EPR study of substitutional-aluminum-related hole centers in synthetic α-quartz, J.Magn.Res. 3, 44 (1970).

(20) J. A. Weil, personal communication.

(21) J.H. Mackey, Jr., EPR study of impurity-related color centers in germanium-doped quartz, J.Chem.Phys. 39, 74 (1963).

(22) R. H. D. Nuttall, J. A. Weil, and R. F. C. Claridge, Double-hole aluminum center in α-quartz, Sol. State Comm. 19, 141 (1976).

(23) J.H. Anderson and J.A. Weil, Paramagnetic resonance of color centers in germanium-doped quartz, J. Chem. Phys. 31, 427 (1959).

(24) Y. Haven, A. Kats, and J.S. VanWieringen, Optical absorption and paramagnetic resonance of colour centers in x-rayed α-quartz containing germanium, Philips Res. Repts. 21, 446 (1966).

(25) F.C. Laman and J.A. Weil, Silver-compensated germanium centers in α-quartz, J. Phys. Chem.Soids. 38, 949 (1977).

(26) J.A. Weil, Germanium-hydrogen-lithium center in α-quartz, J.Chem.Phys. 55, 4685 (1971).

(27) R.V. Lorenze and F.J. Feigl, Defects in crystalline quartz: Electron paramagnetic resonance of multiple-alkali-compensated centers associated with germanium impurities, Phys.Rev.B 8, 4833 (1973).

(28) R.V. Lorenze and F.J. Feigl, Defects in crystalline quartz: Generation of magnetic impurity centers by x-irradiation, (preprint supplied by the authors).

(29) A. Kats, Hydrogen in alpha-quartz, Philips Res. Repts. 17, 133 (1962).

(30) R.A. Weeks and M. Abraham, Electronm spin resonance of irradiated quartz· Atomic hydrogen, J.Chem.Phys. 42, 68 (1965).

250

(31) B. D. Perlson and J. A. Weil, Atomic hydrogen in α-quartz J. Magn. Res. 15, 594 (1974).

(32) D. R. Hutton, Paramagnetic resonance of Fe^{+++} in amethyst and citrine quartz, Phys. Lett. 12, 310 (1964).

(33) T.I. Barry, P.McNamara, and W.J. Moore, Paramagnetic resonance and optical properties of amethyst, J.Chem.Phys. 42, 2599 (1965).

(34) G. Lehmann and W. J. Moore, Optical and paramagnetic properties of iron centers in quartz, J. Chem. Phys. 44, 1741 (1966).

(35) L.M. Matarrese, J.S. Wells, and R.L. Peterson, EPR spectrum of Fe^{3+} in synthetic brown quartz, J.Chem.Phys. 50, 2350 (1969).

(36) G. Lehmann and H. U. Bambauer, Quartz crystals and their colors, Angew. Chem. Internat. Edit. 12, 283 (1973).

(37) R.A. Weeks, Paramagnetic resonance of lattice defects in irradiated quartz, J.Appl.Phys. 27, 1376 (1956).

(38) C.M. Nelson and R.A. Weeks, Trapped electrons in irradiated quartz and silica I, Optical absorption, J.Am.Ceram.Soc. 43, 396 (1960); R.A. Weeks and C.M. Nelson,...II, Electron spin resonance, ibid, 399 (1960).

(39) R.A. Weeks and C.M. Nelson, Irradiation effects and short-range order in fused silica and quartz, J.Appl.Phys. 31, 1555 (1960).

(40) R.H. Silsbee, Electron spin resonance in neutron-irradiated quartz, J.Appl.Phys. 32, 1459 (1961).

(41) F.J. Feigl, W.B. Fowler, and K.L. Yip, Oxygen vacancy model for the E_1' center in SiO_2, Sol.State Comm. 14, 225 (1974).

(42) K.L. Yip and W.B. Fowler, Electronic structure of the E_1' centers in SiO_2, Phys.Rev. B 11, 2327 (1975).

(43) R.A. Weeks, Paramagnetic spectra of E_2' centers in crystalline quartz, Phys. Rev. 130,570 (1963).

(44) C.S. Brown and L.A. Thomas, The effect of impurities on the growth of synthetic quartz, J.Phys.Chem. Solids 13, 337 (1960).

(45) E. Lell, N.J. Kreidl, and J.R. Hensler, Radiation effects in quartz, silica, and glasses, in Progress in Ceramic Science, Vol. 4, ed. J. Burke, Pergamon Press, Oxford (1966).

(46) A. Bishay, Radiation induced color centers in multicomponent glasses, J. Non-Cryst. Solids 3, 54 (1970).

(47) P.O. Fröman, R. Petterson, and R. Vänngård, Electron spin resonance and thermoluminescence in irradiated fused quartz, Ark. Fys. 15, 559 (1959).

(48) S. Lee and P.J. Bray, Electron-spin paramagnetic resonance studies of irradiated alumino-silicate glasses, Phys.Chem. Glasses 3,37 (1962).

(49) R. Schnadt and A. Räuber, Motional effects in the trapped-hole centers in smoky quartz Sol.State Comm. 9, 159 (1971).

(50) M. Stapelbroek and D.L. Griscom, to be published; see also, Oxygen-associated hole centers in high purity fused silicas, this volume.

(51) D.L. Griscom, G.H. Sigel, Jr., and R.J. Ginther, Defect centers in a pure-silica-core borosilicate-clad optical fiber: ESR studies, J. Appl. Phys. 47, 960 (1976).

(52) D.L. Griscom, P.C. Taylor, D.A. Ware, and P.J. Bray, ESR studies of lithium borate glasses and compounds irradiated at 77K: Evidence for a new interpretation of the trapped-hole centers associated with boron, J.Chem.Phys. 48, 5158 (1968).

(53) E.J. Friebele, D.L. Griscom, and G.H. Sigel, Jr., Defect centers in a germanium-doped silica-core optical fiber, J.Appl.Phys. 45, 3424 (1974).

(54) R.A. Weeks and T. Purcell, Electron spin resonance and optical absorption in GeO_2, J.Chem.Phys. 43, 483 (1965).

(55) J.W.H. Schreurs, Study of some trapped hole centers in x-irradiated alkali silicate glasses, J.Chem.Phys. 47, 818 (1967).

(56) S. Urnes, Structure of molten alkali silicates, Trans. Brit. Ceram. Soc. 60, 85 (1961).

(57) G.H. Sigel, Jr., Ultraviolet spectra of silicate glasses: A review of some experimental evidence, J.Non-Cryst. Solids 13, 372 (1974).

(58) D.L. Griscom, Defects in amorphous insulators, in Atomic Scale Structure of Amorphous Solids, G. S. Cargill III and P. Chaudhari, Eds.,to be published.

(59) J. Vitko, ESR studies of hydrogen hyperfine spectra in irradiated vitreous silica, submitted for publication.

(60) M. Stapelbroek and D.L. Griscom, to be published.

(61) D.L. Griscom, E.J. Friebele, and G.H. Sigel, Jr., Observation and analysis of the primary ^{29}Si hyperfine structure of the E' center in non-crystalline SiO_2, Sol. State Comm. 15, 479 (1974).

(62) D.L. Griscom, G.H. Sigel, Jr., and E.J. Friebele, Defect centers in pure fused silica, in Proc. XIth International Congress on Glass, Vol. 1, p.3 (1977).

252

(63) C.M. Nelson and J.H. Crawford, Jr., Optical absorption in irradiated quartz and fused silica, J.Phys.Chem. Solids 13, 296 (1960).

(64) G.W. Arnold and W.D. Compton, Radiation effects in silica at low temperatures, Phys.Rev. 116, 802 (1959).

(65) J.H. Anderson, F.J. Feigl, and M. Schlesinger, The effects of heating on color centers in germanium-doped quartz, J.Phys.Chem. Solids 35, 1425 (1974).

(66) R.A. Weeks, Paramagnetic resonance and optical absorption in gamma-ray irradiated alpha quartz: The "Al" center, J.Am.Ceram.Soc. 53, 176 (1970).

(67) R.A. Weeks and E. Sonder, The relation between the magnetic suscepti-bility, electron spin resonance, and the optical absorption of the E_1' center in fused silica, in Paramagnetic Resonance II ed., W. Low, Academic Press, New York (1963) p. 869.

(68) E.W.J. Mitchell and E.G.S. Paige, The optical effects of radiation induced atomic damage in quartz, Phil.Mag. [8] 1, 1085 (1956).

(69) D.L. Griscom and G.H. Sigel, Jr., unpublished; (refined measurements are in progress).

(70) D.L. Griscom, The electronic structure of SiO_2: A review of recent spectroscopic and theoretical advances, J.Non-Cryst. Solids 24, 155 (1977).

(71) N.F. Mott, Silicon dioxide and the chalcogenide semiconductors; similarities and differences, Adv.Phys. 26, 363 (1977).

A GERMANIUM TRI-HYDROGEN CENTER IN α-QUARTZ

F. C. Laman and J. A. Weil
Department of Chemistry and Chemical Engineering
University of Saskatchewan, Saskatoon, Canada S7N 0W0

ABSTRACT

Single crystal α-quartz (natural, as well as synthetic germanium-doped) was subjected to hydrogen-diffusion, in which H^+ ions were introduced using H-containing graphite electrodes and an electric field of ca. 2000 V/cm applied along the crystal c-axis at 775°K. After x-irradiation of these crystals at room temperature (RT), a paramagnetic center (denoted as GeH_3) stable at RT was observed at RT by electron paramagnetic resonance. An EPR spectrum taken at frequency 9.8 GHz with the magnetic field $\vec{B} \parallel \vec{c}$, clearly shows the involvement of two H ion types: one with hyperfine splitting of 8.706 mT and two with much smaller hyperfine splittings of 0.055 and 0.048 mT respectively. Analysis of the spin-Hamiltonian $\mathcal{H}_s$ for GeH_3, combined with comparison with $\mathcal{H}_s$ measured for an analogous center $GeHLi_2$ observed in undiffused quartz, leads to the following model: $Ge^{3+}H^-H^+H^+$, involving Ge^{3+} substitutional for Si^{4+} in the SiO_2 lattice, with a hydride ion bonded to the Ge^{3+} impurity ion and the two H^+ ions bonded to oxygen ions directly neighboring the Ge^{3+} ion.

INTRODUCTION

The work reported herein originates from our interest in the nature of hydrogen sites in α-quartz, and in the possible occurrence of +2 ions substitutional for +4 silicon ions in the lattice. We shall discuss the effect of proton exchange on a compound paramagnetic center $GeHLi_2$ [1], which gave an analogous center to be denoted by GeH_3.

EXPERIMENTAL

Natural (rose) quartz and synthetic germanium-doped crystals, cut into rectangular plates [ca. 3 ($\parallel \vec{c}$) x 10 x 15 mm], were used. Electrodiffusion [2], using hydrogen-impregnated graphite electrodes with the electric field $\vec{E}$ (ca. 2000 V/cm) applied parallel to the crystal c-axis (8-77 hours in air at temperatures near 775°K), replaced Li^+ and Na^+ with H^+. Subsequent x-irradiations, to create the paramagnetic centers, were done at room temperature (RT) as described previously [1]. Irradiation at 77°K and subsequent warm-up did not produce GeH_3 (or $GeHLi_2$ in lithium-containing crystals). One synthetic Ge-doped (∿ 2000 ppm) crystal showed GeH_3 (and $GeHLi_2$) signals even before diffusion. Electron paramagnetic resonance (EPR) spectra were taken at RT with

a Bruker B-ER 418S spectrometer operating at ca. 9800 MHz. The crystals were
rotated about one ($\equiv \vec{a}_1$) of their three two-fold crystal symmetry axes, with
magnetic field $\vec{B} \perp \vec{a}_1$. Rotation angles could be determined to within 0.1°.

RESULTS AND ANALYSIS

In general, the EPR characteristics of the new center GeH_3 [3] proved to be
remarkably similar to those of $GeHLi_2$ [1]. With $\vec{B} \parallel \vec{c}$, the primary doublet of
GeH_3 (at g = 1.9958) shows a ^1H splitting of 8.706 mT at 296°K, each member
being split further into four almost equally intense lines (see Fig.). These
four lines are grouped into two ^1H hyperfine doublets with splittings of 0.055
and 0.048 mT, respectively. The max-min linewidth is 0.0085 mT. The primary
^1H hyperfine splitting is temperature dependent, although less so than the
corresponding splitting for $GeHLi_2$, increasing by 0.11 mT in the range 211°K
to 311°K. Similarly, the smaller ^1H splittings vary, with values at 211°K of
0.061 and 0.045 mT, respectively.

At some orientations of $\vec{B}$, weak "spin-flip" transitions [1] were observable,
allowing definite identification of ^1H nuclei as giving rise to the smaller
splittings. Crystal rotation (32 settings) with $\vec{B} \perp \vec{a}_1$ gave line positions
fitted to the spin-Hamiltonian

$$\mathcal{H}_s = \beta_e \vec{S} . \bar{g} . \vec{B} + \sum_i \{ \vec{S} . \bar{A}_i . \vec{I}_i - \beta_n \vec{I}_i . \bar{g}_i . \vec{B} \} \tag{1}$$

in which the summation i is over the three ^1H nuclei involved and $S = I_i = 1/2$.

The fitting was done by exact matrix diagonalization, as in [1]. Using 154
line positions (and frequencies), a root-mean-square deviation of 0.0015 mT
was achieved. Results for the matrices $\bar{g}$ and $\bar{A}_i$, assumed to be symmetric,
are presented in Table 1. Spectral simulations at several angles, using
these matrices, together with first-derivatives of Lorentzian lineshapes,
gave excellent agreement with measured spectra. The matrix set given here is
one of a set of six equivalent ones, since the proper point group symmetry D_3
of α-quartz causes appearance (for general $\vec{B}$ orientation) of six symmetry-
related spectra, and is comparable to the set given for $GeHLi_2$ [1].

Unfortunately, so far (because of the decreased EPR sensitivity available
with the diffused plates), the hyperfine matrix for the low-abundance
^{73}Ge isotope has not become available for GeH_3.

DISCUSSION

The center $GeHLi_2$ was interpreted [1] as consisting of a germanium 3+ ion,
substitutional at a silicon 4+ ion site within an oxygen tetrahedron, bonded
to a hydride ion and containing in addition two peripheral almost equivalent
lithium 1+ ions. This center is thought to be formed from a diamagnetic parent
center containing Ge(2+), plus two charge-compensating Li(1+) ions, by reaction
with atomic hydrogen (created by x-irradiation):

$$
\begin{array}{ccc}
O \quad Li^+ \quad O & & O \quad Li^+ \quad O \\
\diagdown \quad \diagup & & \diagdown \quad \diagup \\
Ge^{2+} \quad + \quad H^0 \quad \rightarrow & & Ge^{3+} \quad H^- \\
\diagup \quad \diagdown & & \diagup \quad \diagdown \\
O \quad Li^+ \quad O & & O \quad Li^+ \quad O
\end{array}
\tag{2}
$$

Our findings for GeH_3 are consistent with this model. Several points are noteworthy:

(1)   The matrices $\bar{A}_i$ for the hydride ions existing respectively in GeH_3 and $GeHLi_2$ are very similar in principal values and axis directions.

(2)   However, matrix $\bar{g}$ in Table 1 discloses that the symmetry of GeH_3 is lower (at least in its excited states) than that of $GeHLi_2$. Thus for the latter, but not the former, the principal axes of $\bar{g}$ have directions interpretable in terms of the undistorted GeO_4 tetrahedron, i.e., bond directions of α-quartz.

(3)   The matrices $\bar{A}_i$ for the two protons of GeH_3 are very similar and disclose that almost zero spin density exists on these nuclei (Table 1). On slight adjustment of their principal values, to make the matrices equivalent, a symmetry operator relating them can be obtained, and represents a reflection plane (containing the crystal two-fold axis through the Ge ion) with its normal at $\theta = 40°$, $\phi = 90°$.

Our interpretation of the decreased symmetry of GeH_3 is that the two protons (unlike the Li^+ ions in $GeHLi_2$) are attached to oxygen anions of the tetrahedron. Similar hydrogen bonding is evident when comparing EPR parameters of TiO_4 centers compensated respectively with H^+ and Li^+ ions [4]. The observed greater stability of GeH_3 compared to $GeHLi_2$ seems consistent with this idea, as is the somewhat startling simultaneous presence of H^+ and H^- within a single center.

It is worth noting that we have also observed the complex overlapped spectra from the "mixed" centers $Ge^{3+}H^-H^+Li^+$ and $Ge^{3+}H^-Li^+H^+$, after partial electrolytic replacement of lithium. No centers corresponding to $GeLi_3$ have been seen, consistent with our model.

We hope to publish further details concerning GeH_3 later.

ACKNOWLEDGEMENTS

This research was carried out under the auspices of the National Research Council of Canada. We thank Argonne National Laboratory (U.S.A.) for the loan of certain equipment.

TABLE 1 Matrices $\bar{g}$ and $\bar{A}_1$ (mT) of GeH_3 and their principal
values, isotropic components and principal directions,
for right α-quartz ($P3_221$) at room temperature (296°K)

	matrix			principal values	principal directions x_1	x_2	x_3
$\bar{g}$	1.99973	0.00223	0.00070	2.00115	0.85201	0.47062	0.22934
		1.99646	0.00134	1.99628	0.47363	-0.50625	-0.72068
			1.99580	1.99456	0.22306	-0.72265	0.65423
				1.99733			
$\bar{A}_H{}^-$	8.7200	-0.0409	-0.0258	8.9294	0.2249	-0.8516	-0.4736
		8.8442	0.1337	8.7089	-0.9741	-0.2091	-0.0866
			8.6767	8.6026	0.0253	-0.4808	0.8765
				8.7470			
$\bar{A}_H{}^+$ (1)	0.0984	0.0420	0.0272	-0.0738	-0.2689	0.6322	0.7267
		-0.0445	-0.0100	-0.0405	-0.0904	0.7346	-0.6725
			-0.0551	0.1131	-0.9589	-0.2465	-0.1404
				-0.0004			
$\bar{A}_H{}^+$ (2)	0.0953	-0.0221	-0.0419	-0.0709	-0.2692	-0.6042	-0.7500
		-0.0489	-0.0098	-0.0396	-0.0612	0.7879	-0.6128
			-0.0480	0.1089	-0.9611	0.1191	0.2490
				-0.0005			

x_1, x_2 and x_3 are components along unit vectors $\vec{i}$, $\vec{j}$ and $\vec{k}$ forming a right-
handed coordinate system; $\vec{i}$ is along the crystal two-fold axis $\vec{a}_1$ and $\vec{k}$ is
along $\vec{c}$. Positive values for x are associated with the end of the two-fold
axis developing a negative charge on compression. To compare with Table III
of ref. 1, change signs of Y_{13}, Y_{23}, x_1 and x_2.

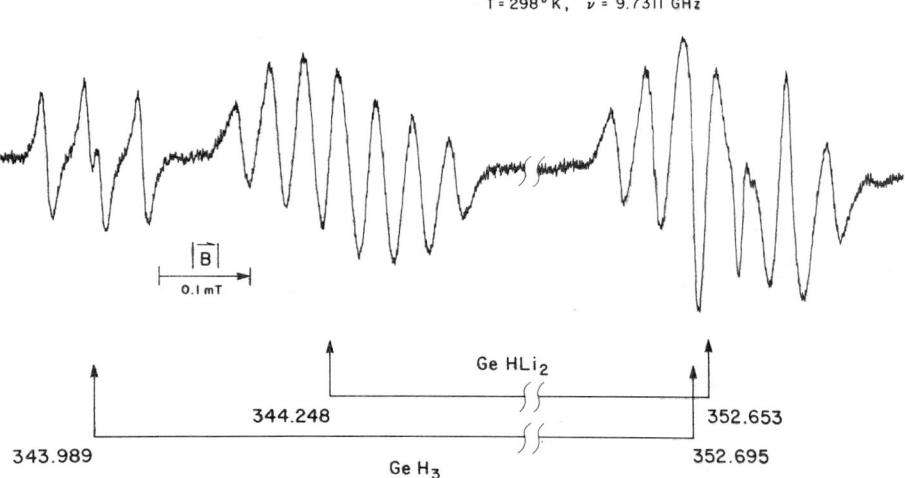

Fig. The EPR spectrum of GeH$_3$ and GeHLi$_2$ simultaneously present in an α-quartz crystal, measured with $\vec{B} \parallel \vec{c}$.

REFERENCES

[1] J. A. Weil, Germanium-Hydrogen-Lithium Center in α-Quartz, J. Chem. Phys. 55, 4685 (1971).

[2] A. Kats, Hydrogen in α-Quartz, Philips Research Rpts. 17, 133-195, 201-279 (1962).

[3] R. V. Lorenze and F. J. Feigl, Defects in Crystalline Quartz: EPR of Multiple-Alkali-Compensated Centers Associated with Ge Impurities, Phys. Rev. B8, 4833 (1973); see footnote 24.

[4] H. Rinneberg and J. A. Weil, EPR Studies of Ti^{3+}-H$^+$ Centers in X-Irradiated α-Quartz, J. Chem. Phys. 56, 2019 (1972).

ELECTRON PARAMAGNETIC RESONANCE STUDIES ON Aℓ
CENTERS IN VITREOUS SILICA*

Keith L. Brower
Sandia Laboratories, Albuquerque, NM 87185

ABSTRACT

Two new aluminum spectra have been observed with EPR in irradiated vitreous silica. Under electron irradiation at 4 K, the low temperature (LT) Aℓ spectrum is produced, and it changes into the high temperature (HT) spectrum upon annealing between 230 and 330 K. The HT Aℓ spectrum is also produced directly upon irradiation at room temperature. Each spectrum is characterized by six well resolved ^{27}Aℓ (I=5/2) hyperfine lines ($A_{isotropic}$(LT) = 0.00330 cm^{-1} and $A_{isotropic}$(HT) = 0.0410 cm^{-1}). These spectra are the most intense in type I and II vitreous silicas which contain Aℓ as a natural impurity. The difference in these spectra may reflect the presence of a neighboring charge compensator, such as an alkali, which becomes mobile at ≈270 K. It is possible that these spectra correspond to Aℓ E_1'-type centers with the unpaired electron strongly localized on a substitutional Aℓ which is bonded to only three oxygen atoms.

PAPER

This paper presents our preliminary electron paramagnetic resonance (EPR) studies on two new aluminum spectra observed in irradiated vitreous silica (1). Previously, only one other aluminum center has been studied in vitreous silica (2,3), and it is believed to be the same defect as that first observed by O'Brien in α-quartz (4) -- namely the aluminum-oxygen-hole center. Although silicon (5,6), germanium (7), phosphorus (8), and boron (9) E_1'-type centers have been studied in glasses, an aluminum E_1' center has not previously been reported. It is well known that Aℓ is a common impurity in vitreous silicas made from natural quartz. In this paper, results are presented which suggest that these new Aℓ spectra arise from Aℓ E_1'-type defects.

After a ^{60}Co γ-irradiation at room temperature of INFRASIL 1, a type I vitreous silica, we observe the EPR spectrum shown in Fig. 1. In addition to the familiar E_1' spectrum and its associated ^{29}Si hyperfine lines (10) and the aluminum-oxygen-hole spectrum, we see new resonances which are part of a 6-line hyperfine spectrum. We believe these new resonances arise from a paramagnetic electron which is highly localized on an Aℓ atom. The reasons for this interpretation are as follows:

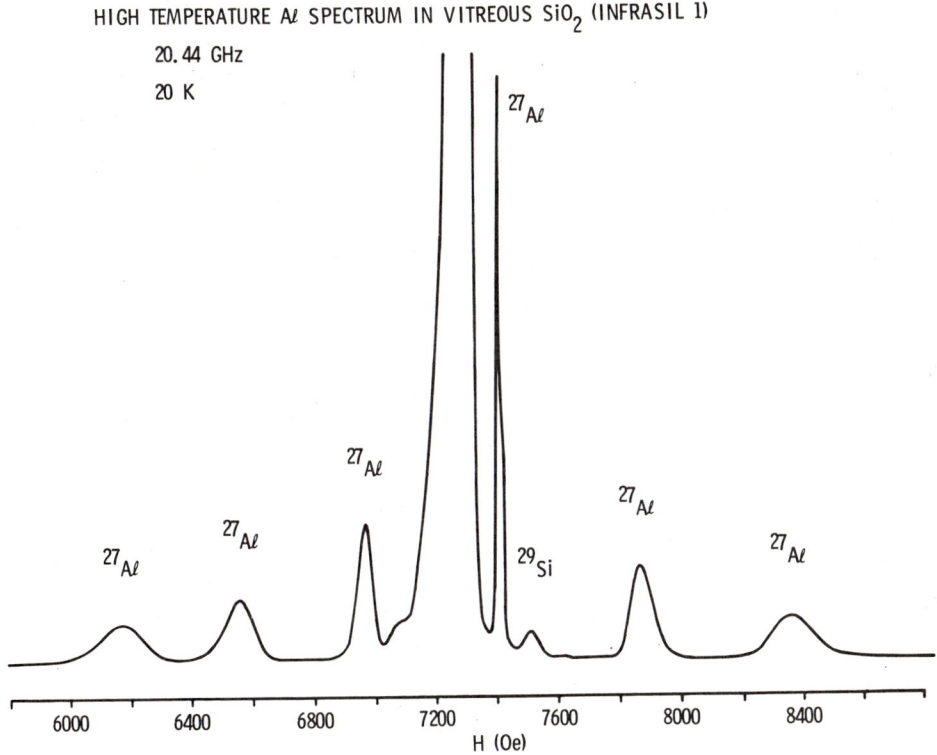

HIGH TEMPERATURE Aℓ SPECTRUM IN VITREOUS SiO$_2$ (INFRASIL 1)

20.44 GHz

20 K

Fig. 1. This spectrum was observed in the dispersion mode under conditions of fast passage in vitreous silica (INFRASIL 1) irradiated at room temperature with ^{60}Co γ-rays to 3 x 10^6 rad.

1. ^{27}Aℓ is 100% naturally abundant with a nuclear spin of 5/2 --- this accounts for the six hyperfine lines.

2. This spectrum is observed in type I and II vitreous silicas which are high in Aℓ content, but we have not observed it in type III or IV vitreous silicas which have low Aℓ content (these centers may still exist, but are below our limit of detectability).

We observe that the HT Aℓ spectrum can be produced at room temperature with either neutrons, electrons, γ-rays, x-rays, or UV light from a He plasma. In Fig. 2, the relative intensity of the HT Aℓ and E$_1'$ spectra are plotted as a function of ^{60}Co γ-ray fluence. Although some differences exist above ≈10^7 rad, the rates of production and the numbers of both kinds of defect are nearly the same for fluences ≤10^7 rad. We have also noticed that for a given irradiation, the intensity of the E$_1'$ spectrum in INFRASIL 1 (type I), OPTOSIL 1, COMMERCIAL TO8 (both type II), SUPRASIL (type III), and SUPRASIL-W 1 (type IV) is approximately proportional to the Aℓ concentration. However, the lack of any resolved splittings in either the E$_1'$ or HT Aℓ spectrum due to electron spin-spin interfactions indicate that these centers are separated from each

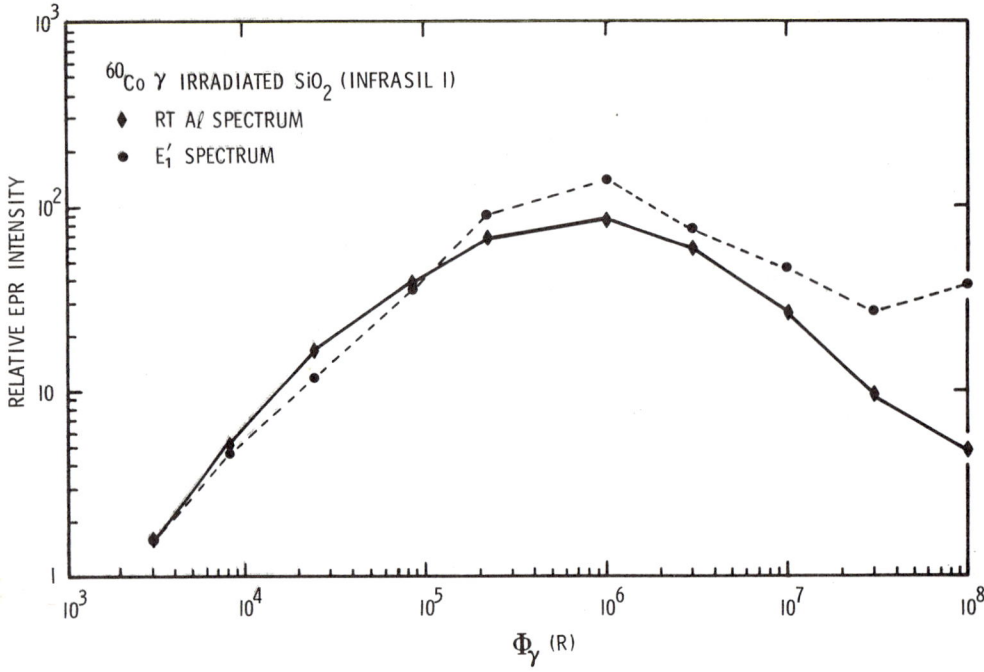

Fig. 2. Relative intensities of the HT Aℓ and
E_1' spectra as a function of ^{60}Co γ-irradiation
at room temperature.

other (and from any other unpaired spins) by distances in excess
of 10 Å.

We have looked for a similar ^{27}Al hyperfine spectrum in neutron
irradiated α-quartz. Although we saw an intense aluminum-oxygen-
hole spectrum (4), we were unable to detect this new Aℓ spectrum
in α-quartz. In unirradiated α-quartz aluminum substitutes for
silicon and is isoelectronic with it. If this new Aℓ spectrum
corresponds to an Aℓ-E_1' defect, then such a center would probably
only be produced if an energetic particle created an oxygen
vacancy next to the Aℓ. Since this is a rather improbable event,
this may be the reason why this spectrum is not detected in
α-quartz.

Since bandgap UV radiation produces the HT Aℓ spectrum in
INFRASIL 1 --- an observation which is in sharp contrast to our
EPR results in α-quartz, it appears that this Aℓ defect exists
in vitreous silica prior to irradiation. In this case, irradiation which causes ionization results in a change in the charge
state of this Aℓ defect making it subsequently paramagnetic.

Electron irradiation at 4 K of INFRASIL 1 followed directly by
EPR measurements at 6 K (no intermediate warmup) yield a low
temperature (LT) ^{27}Al hyperfine spectrum. This LT Aℓ spectrum

has an $A_{isotropic}(LT) = 0.0330$ cm^{-1} whereas the $A_{isotropic}(HT) = 0.0410$ cm^{-1}. Since the atoms are more likely to remain frozen in the network at 4 K, it is very likely that this LT Aℓ defect is the one which corresponds to the Aℓ center in vitreous silica prior to irradiation. The isochronal annealing of this spectrum along with the E$_1'$ spectrum is shown in Fig. 3. Even at 4 K, we still

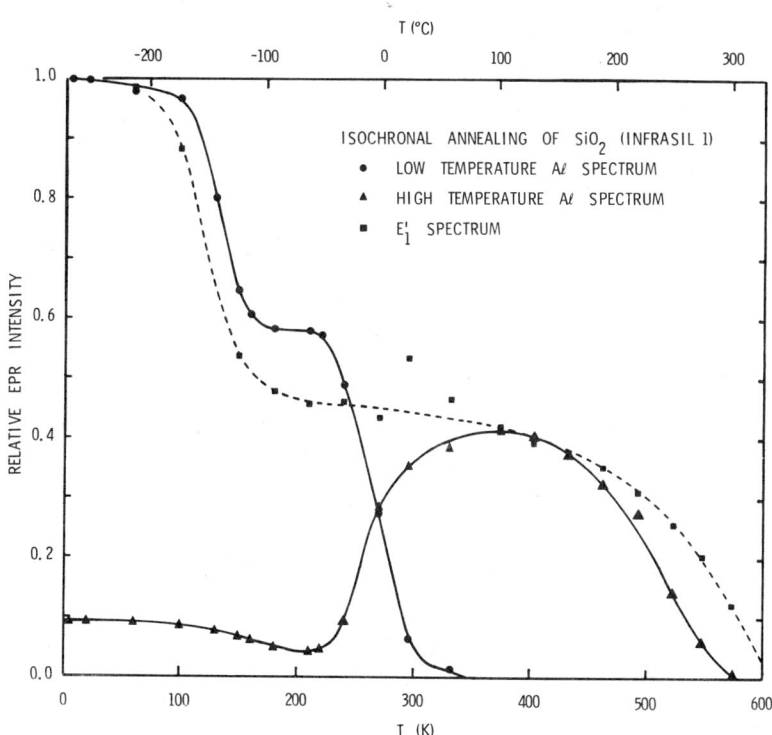

Fig. 3. Isochronal (20 min) annealing of the Aℓ
and E$_1'$ spectra after irradiation with 1.4 MeV
electrons at 4 K to a fluence of 3 x 10^{14} e/cm^2.

get some production of the HT Aℓ center. Between 230 and 330 K the LT Aℓ center appears to transform into the HT Aℓ center. Our suspicion at this time is that an alkali compensator such as Na becomes mobile in this temperature range and readjusts its position in view of the change in the charge of Aℓ. This effect is reflected by a change in the ^{27}Aℓ hyperfine interaction. The reduction in the intensity of all three spectra between 100 and 150 K by a factor of two is not understood; this effect was observed in two different samples. These data indicate that the composite annealing of these two centers correlates rather closely with the annealing of the E$_1'$ spectrum.

ACKNOWLEDGEMENT

The expert assistance of Roger Shrouf in all phases of the experimental work is much appreciated. Discussions with George Arnold, Gerda Krefft, Paul Peercy and John Vitko on various aspects of this work have been appreciated and helpful.

REFERENCES

*   This work was supported by the United States Department of Energy (USDOE) under Contract AT(29-1)789.

1. K. L. Brower, Electron paramagnetic resonance studies on Aℓ centers in fused SiO_2, Bull. Am. Phys. Soc., 23, 460 (1978).

2. P. O. Fröman, R. Pettersson and T. Vänngård, Electron spin resonance and thermoluminescence in irradiated fused quartz, Arkiv for Fysik, Bd 15, nr 40, 559 (1959).

3. S. Lee and P. J. Bray, Electron-spin paramagnetic resonance studies of irradiated alumino-silicate glasses, Phys. Chem. Glasses, 3, 37 (1962).

4. M. C. M. O'Brien, The structure of the colour centres in smoky quartz, Proc. Roy. Soc. A231, 404 (1955).

5. R. A. Weeks, Paramagnetic resonance of lattice defects in irradiated quartz, J. Appl. Phys. 27, 1376 (1956).

6. K. L. Yip and W. B. Fowler, Electronic structure of E_1' centers in SiO_2, Phys. Rev. B11, 2327 (1975).

7. R. A. Weeks and T. Purcell, Electron spin resonance and optical absorption in GeO_2, J. Chem. Phys. 43, 483 (1965).

8. R. A. Weeks and P. J. Bray, Electron spin resonance spectra of gamma-ray-irradiated phosphate glasses and compounds: oxygen vacancies, J. Chem. Phys. 48, 5 (1968).

9. D. L. Griscom, G. H. Sigel, Jr. and R. J. Ginther, Defect centers in a pure-silica-core borosilicate-clad optical fiber: ESR studies, J. Appl. Phys. 47, 960 (1976).

10. D. L. Griscom, E. J. Friebele and G. H. Sigel, Jr., Observation and analysis of the primary ^{29}Si hyperfine structure of the E' center in non-crystalline SiO_2, Solid State Commun. 15, 479 (1974).

OXYGEN-ASSOCIATED TRAPPED-HOLE CENTERS IN
HIGH-PURITY FUSED SILICA

M. Stapelbroek* and D.L. Griscom
Naval Research Laboratory, Washington, D.C. 20375

ABSTRACT

Two distinct oxygen-associated trapped-hole centers are identified in
samples of room-temperature γ-irradiated, high-purity fused silica. One,
which we label the "wet" hole center, predominates in high OH-content
silicas while the other, "dry", hole center is more prevalent in low-OH
silicas. Excellent computer simulations of the low-temperature electron-
spin-resonance spectra are obtained for both high- and low-OH silicas using
only the relative abundance of the "wet" and "dry" hole centers as an
adjustable parameter. In addition, confirmation that the "wet" hole center
is a hole trapped in an oxygen nonbonding 2p orbital is obtained from the
^{17}O-hyperfine structure in ^{17}O-enriched samples of fused silica.

INTRODUCTION

The electron-spin-resonance (ESR) spectra of intrinsic radiation-induced
paramagnetic defect centers in pure fused silicas have been known for over
two decades. As early as 1956, Weeks (1) examined the room temperature ESR
of neutron-irradiated fused silica. On the basis of g shifts, he tentatively assigned a sharp easily saturable resonance as due to trapped-
electron centers (the well-known E' centers) and a broad resonance was
tentatively assigned to intrinsic trapped-hole centers (hereafter referred
to as oxygen-associated-hole centers or OHCs (2)). Since then numerous ESR
(3-7), optical (8, 9), and theoretical (10, 11) studies have appeared
concerning the E' centers in both crystalline α-quartz and fused silica.
These studies have firmly established the model for the E' center as an
unpaired electron residing in a dangling sp^3 orbital of a silicon which is
bonded to only three oxygens.

In contrast, the broad OHC resonance has been nearly neglected in the
literature and a definitive model has not been established. Nelson and
Weeks (12) concluded that the 7.6 eV optical absorption band induced in
γ-irradiated fused silica is associated "with an electron-spin-resonance
system with a g value characteristic of a trapped hole" (presumably the
OHC resonance). More recently, however, Friebele, Griscom, and Sigel (13, 14)
reported that the room-temperature OHC ESR resonance is a superposition of
several different centers and that only part of the hole center ESR spectrum
could be correlated with the 7.6 eV absorption band. They also observed
significant differences in the annealing behavior of the OHCs between high-
OH (wet) and low-OH (dry) silicas. In particular, a large increase in both
the hole center ESR resonance and the 7.6 eV optical band was observed in
room-temperature γ-irradiated dry silicas after a 200° C anneal (14).

---

*NRC-NRL Resident Research Associate

We report here additional results on the OHC-type centers in both wet and
dry high-purity fused silicas. ESR measurements at low temperatures have
enabled the identification of two distinct paramagnetic hole-center resonances in γ-irradiated silica. One center, which we label the "wet" OHC,
predominates in high-OH silicas and the other, the "dry" OHC, is prevalent
in low-OH silicas. Correlation of the "dry" OHC with the 7.6 eV optical
band is also reported herein. In addition, direct confirmation that the
"wet" OHC corresponds to an unpaired spin in a nonbonding oxygen 2p orbital
is presented from the observation of ^{17}O-hyperfine structure in isotopically
enriched samples.

EXPERIMENTAL

All ESR spectra were recorded using a Varian E-9 spectrometer operating at
$\sim$ 9 GHz. The samples studied were either commercially available, high-
purity, wet (Suprasil 1, Spectrosil A) and dry (Suprasil W-1, Spectrosil
WF) fused silicas or wet silicas prepared in-house with and without ^{17}O
enrichment. These "homemade" precipitated silicas were prepared by the
reaction of water with a large excess of $SiCl_4$; evaporating away the excess
$SiCl_4$; sintering the powder at $\sim$ 1200 °C; and fusing in an O_2-H_2 flame.
^{17}O-enriched water obtained from Miles Laboratories was used to obtain
isotopically enriched samples. ^{60}Co γ-irradiations were performed at room
temperature and all samples were irradiated to doses between 1-2 x 10^8 rad
(Si). The ratio of "wet" to "dry" OHCs changed significantly with time
after irradiation and the spectra shown in Figs. 1 and 2 are from samples
aged for several months.

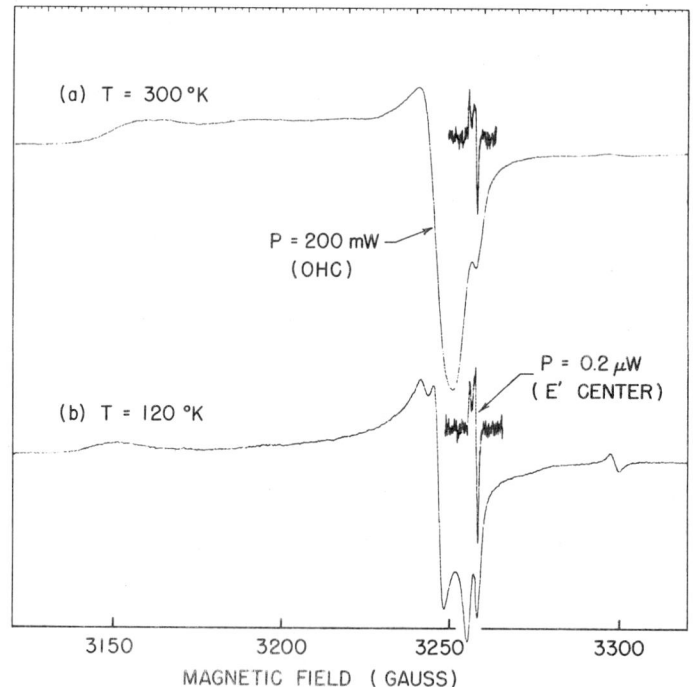

Fig. 1. Room-temperature and low-temperature ESR
spectra of γ-irradiated Suprasil W-1.

RESULTS

The hole-center ESR spectra in room-temperature, γ-irradiated, high-purity
fused silicas have a significant temperature dependence, illustrated in
Fig. 1 for the case of Suprasil W-1. The marked "sharpening" of the spectra
at low temperatures has enabled us to identify two distinct OHCs with the
help of computer simulations. By varying only the relative concentrations
of the two OHCs, excellent computer simulations have been generated to fit
the OHC spectra in all silicas studied. The quality of the computer
simulations is illustrated in Fig. 2 for Suprasil 1 and Suprasil W-1 (the
peaks in the low-field tails of the computer simulation are an artifact of
the simulation).

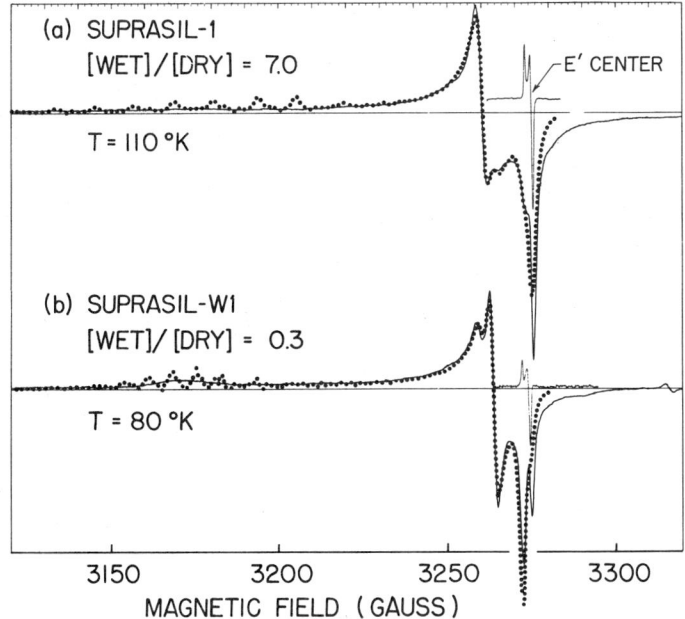

Fig. 2. Computer simulations of the OHC spectra in
Suprasil 1 and Suprasil W-1. The solid lines
are the experimental spectra and the dots are
the computer simulation.

Best-fit g values are g_1= 2.0016, g_2= 2.0070, and < g_3 > = 2.07 for the
"dry" OHC and g_1 = 2.0001, g_2= 2.0095, and < g_3 > = 2.06 for the "wet" OHC.
Gaussian distributions in energy level splittings were assumed to give
rise to the distributions in g_3 for the centers (15).

An isochronal pulse anneal of γ-irradiated dry silicas revealed that only
the "dry" OHC could be correlated with the previously observed (13, 14)
increase in the 7.6 eV optical band at $\sim$ 250 $^{\circ}$C. This is strong evidence
that the 7.6 eV optical band is due to the "dry" OHC.

Observation of ^{17}O-hyperfine structure on the "wet" OHC has provided direct
confirmation that the unpaired spin is localized in an oxygen 2p nonbonding

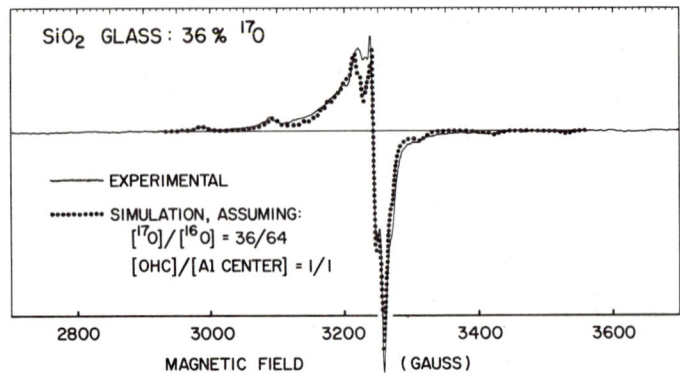

Fig. 3. Low-temperature ESR spectrum and computer simulation of 36% ^{17}O-enriched fused silica.

orbital. The ESR spectrum and associated computer simulation of a γ-irradiated sample, enriched to 36% ^{17}O, is shown in Fig. 3. Unfortunately, Al impurities in the enriched sample led to a significant concentration of Al centers; these were therefore included in the simulation. For this simulation, previously determined g values were not allowed to vary and $(A_\parallel - A_\perp)$ was fixed at the value calculated for ^{17}O$^-$. Therefore, the only adjustable parameters used in the simulation were $A_\parallel$ and the ratio of "wet" OHCs to Al centers. The best-fit ^{17}O-hyperfine parameters are $A_\parallel = (-) 110$ G and $A_\perp = (+) 16$ G. In view of the rigid constraints, the agreement between simulated and experimental spectra is excellent. The magnitude and anisotropy of the hyperfine interaction is direct experimental evidence that the hole is trapped in a nonbonding 2p orbital of an oxygen (16).

CONCLUSIONS

ESR spectra from two distinct trapped-hole centers have been identified in γ-irradiated samples of pure fused silica. The preponderance of "wet" OHCs in wet silicas suggests the model

for the "wet" OHC. The lack of observable proton hyperfine splitting, however, requires that the proton resides at a considerable distance (more than about ≈ 2.5Å) from the unpaired spin localized in the oxygen 2p nonbonding orbital.

For the "dry" OHC, a model of a hole trapped on a nonbridging oxygen without the adjacent OH group is plausible. The excellent correlation between the ESR and optical isochronal pulse anneals indicate that the 7.6 eV absorption band is due to the "dry" OHC.

REFERENCES

(1) R.A. Weeks, Paramagnetic resonance of lattice defects in irradiated quartz, J. Appl. Phys. 27, 1376 (1956).

(2)   D.L. Griscom, E.J. Friebele, G.H. Sigel, Jr., and R.J. Ginther, Radiation-induced paramagnetic defects as structural probes of pure silica and borosilicate glasses, in The Structure of Non-Crystalline Materials, P.H. Gaskell, ed., Taylor & Francis, London (1977).

(3)   R.A. Weeks and C.M. Nelson, Trapped electrons in irradiated quartz and silica: II, Electron spin resonance, J. Am. Ceram. Soc. 43, 399 (1960).

(4)   R.A. Weeks and C.M. Nelson, Irradiation effects and short-range order in fused silica and quartz, J. Appl. Phys. 31, 1555 (1960).

(5)   R.H. Silsbee, Electron spin resonance in neutron-irradiated quartz, J. Appl. Phys. 32, 1459 (1961).

(6)   R.A. Weeks and E. Lell, Relation between E' centers and hydroxyl bonds in silica, J. Appl. Phys. 35, 1932 (1964).

(7)   D.L. Griscom, E.J. Friebele, and G.H. Sigel, Jr., Observation and analysis of the primary ^{29}Si hyperfine structure of the E' center in non-crystalline SiO_2, Sol. State Comm. 15, 479 (1974).

(8)   C.M. Nelson and J.H. Crawford, Jr., Optical absorption in irradiated quartz and fused silica, J. Phys. Chem. Solids 13, 296 (1960).

(9)   C.M. Nelson and R.A. Weeks, Trapped electrons in irradiated quartz and silica: I. Optical absorption, J. Am. Ceram. Soc. 43, 396 (1960).

(10) A.J. Bennett and L.M. Roth, Electronic structure of defect centers in SiO_2, J. Phys. Chem. Solids 32, 1251 (1971).

(11) K.L. Yip and W.B. Fowler, Electronic structure of E_1' centers in SiO_2, Phys. Rev. B 11, 2327 (1975).

(12) C.M. Nelson and R.A. Weeks, Vacuum-ultraviolet absorption studies of irradiated silica and quartz, J. Appl. Phys. 32, 883 (1961).

(13) E.J. Friebele, D.L. Griscom, and G.H. Sigel, Jr., Radiation-induced defect centers in non-crystalline SiO_2, in The Physics of Non-Crystalline Solids, G.H. Frischat, ed., Trans. Tech Publications, Aedermansdorf, Switz. (1977).

(14) D.L. Griscom, G.H. Sigel, Jr., and E.J. Friebele, Defect centers in pure fused silica, Proc. XI International Congress on Glass, Prague (1977), Vol 1, pp. 3-11.

(15) D.L. Griscom, Defects in non-crystalline oxides, in Defects and Their Structure in Nonmetallic Solids, B. Henderson and A.E. Hughes, eds., Plenum, New York (1976), p. 323.

(16) M. Stapelbroek and D.L. Griscom, to be published.

A MODEL FOR POINT DEFECTS IN SILICA

G.N. Greaves*
Pilkington Brothers R. & D. Laboratories,
Lathom, Ormskirk, England

ABSTRACT

We present a new model for dangling bond defects in vitreous silica. In equilibrium the intrinsic states are presumed to be charged and to fall on silicon as well as oxygen sites - forming in pairs and rendering the glass diamagnetic. The trapping of electrons and holes at these sites following irradiation involves distortion of the lattice, so the energy levels of the silicon and oxygen states in the energy gap will depend on occupancy. We analyse the colour centre data for pure silica to deduce energy levels for the different dangling bond states. Finally we describe how the charge on the equilibrium states may be removed by the dissociation of water - the states converting to neutral hydroxyl centres and leaving the glass relatively insensitive to paramagnetic centre formation.

Whilst colour centres in vitreous silica have been extensively studied (1) models have generally been restricted to individual centres and their characteristic optical absorption. The best known is the radiation induced paramagnetic E' silicon defect. This has been modelled as a trivalent silicon with a dangling sp^3 orbital facing onto an oxygen vacancy (2). More recently Mott has employed the charged dangling bond model proposed earlier for chalcogenide glasses (3) to describe dangling bond oxygen centres in silica (4). In this paper we shall attempt to rationalize silica colour centre data in terms of a <u>single</u> defect involving a dangling bond on a silicon atom and a dangling bond on an oxygen atom - the atoms being sufficiently separated to prevent bond reformation. The occupancy of these states will depend on the thermal history and radiation history of the glass. Changes in occupancy will involve atomic relaxation and a pairing of silicon and oxygen energy levels in the energy gap. We will go on to describe the likely modifying effect of water in silica and the way this can quench subsequent E' centre formation.

We begin by discussing the intrinsic states in the pure stoichiometric glass. The model is based on the idea of a defect consisting of a pair of dangling bond states - one on a trivalent silicon site and the other on a separated monovalent oxygen site. Because the cation-anion electronegativity difference is large it is unlikely neutral singly occupied states will be stable. Instead we expect they will decompose into empty positively charged silicon sp^3 states (which we are calling Si^+ states) and doubly occupied negatively charged oxygen 2p states (which we are calling O^- states) - that is, provided the energy released is sufficient to overcome the electron-electron correlation energy of double occupancy. Dielectric screening is small in silica so the energy levels for these charged states will be separated from the band edges and we can expect features in the tail of the optical absorption edge. Following excitation with band gap or ionizing radiation, electrons and holes will be trapped at the cation and anion sites - converting them into neutral metastable states Si^o and O^o. We anticipate considerable distortion of the lattice around these sites (as appears to be the case for impurity centres in silica) so singly occupied energy levels will lie deeper in the energy gap compared to their charged precursors. We propose optical transitions associated with the charged and neutral silicon and oxygen states are responsible for the colour centre absorption and luminescence bands characteristic of vitreous silica.

The notion of the overcoordinated anion site which is fundamental to current models for defects in chalcogenide glasses (3) has been adopted here to help place energy levels in the energy gap of silica for the various states of the intrinsic defect. In particular an empty or singly occupied dangling bond state is expected to draw an adjacent oxygen into 3-fold coordination utilising the lone-pair electrons and leaving the dangling bond surplus to bonding requirements. At silicon sites the appropriate non-bonding state should have sp^3 character - this being the first available state above the valence band in the Si(3)-O(3) complex. We can judge the position of the empty Si^+ state from internal photoemission measurements of the Si/SiO_2 interface (5). It will lie in the vicinity of the silicon conduction and valence bands which are located at 3 and 4 eV respectively below the silica conduction band. Once occupied the change in lattice distortion will push the level closer to the valence band. We propose that this singly occupied Si^o state is the E' ESR centre of irradiated silica (6). Unlike the model of Fiegl, Fowler and Yip (2) the present model does not involve an oxygen vacancy. Turning now to the oxygen states, the singly occupied O^o state should have anti-bonding character as the remaining electrons on the oxygen complex will be in lone-pair or bonding states. Atomic relaxation related to the lone-pair bonding will also be involved but the energy level is still expected to lie near the conduction bond. We propose this is the centre responsible for the ESR oxygen hole signal observed by Friebele and coworkers in unannealed silica fibres at helium temperatures (7). It is also presumed to be present in δ-irradiated silica (8). Following the ideas of Mott (4), the charged oxygen dangling bond state O^- (D^- in his notation) is interpreted as the non-bridging oxygen of glass science. It will have p - like character and will lie close to the uppermost lone-pair valence band - separated by its ionization energy. This will probably amount to about 1eV on account of the low dielectric constant of silica (ϵ = 2.15).

The present model implies equal numbers of states on silicon and oxygen sites - every non-bridging oxygen being compensated by a trivalent silicon. Because the metastable Si^o and O^o states both involve partial bonding with an adjacent otherwise fully coordinated oxygen, it is assumed they will both be associated with similar distortions of the lattice. We can infer the lattice contribution to the energy level of these states from the location of the colour centre absorption bands. In Fig. 1 therefore we plot the complete absorption edge of silica (from various sources) running from the fibre optics minimum at 1 eV to the first major absorption peak at 10.2eV. The solid line refers to the virgin glass (Corning 7940) and the dotted line to the same variety of silica following δ irradiation. For absorption coefficients greater than 10^2 cm^{-1} the edge is characteristic of the pure host material but below this absorption is more strongly influenced by structural imperfections and impurities. The peaks at 2.0 and 5.8 eV have been associated (with reasonable certainty) with the hole and electron ESR signals referred to earlier (9). We can therefore assign these to the metastable O^o and Si^o states respectively. The sharpness of the features

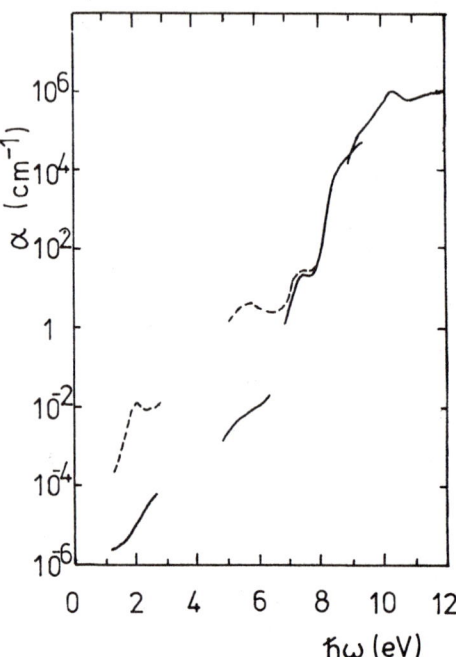

α (cm^{-1})

$\hbar\omega$ (eV)

Fig.1 Optical absorption edge of vitreous silica. Solid curves are for the unirradiated glass and dashed curves are the colour centre bands following γ irradiation of $\sim 10^8$ rad. For further details see Ref.15

implies they involve excitonic transitions rather than transitions from valence band states. In fact UPS spectra indicate the top-most feature in the valence band density of states is approximately 3 eV across which is clearly wider than either of the two colour centre bands. Accordingly we can say 2.0 and 5.8 eV locate the O^o and Si^o levels respectively below a suitable exciton line.

Of considerable interest is the recent observation by Appleton and coworkers (11) of a well defined shoulder close to 7.5 eV in annealed silica. This comes at the same energy as the high energy colour centre band in irradiated silica identified by several authors (1, 12). It has not been associated with an ESR signal and is therefore a natural choice for locating the equilibrium charged states Si$^+$ and O$^-$. Moreover it appears to be split with the two main features at 7.6 and 8.2 eV, so we might attribute these to the Si$^+$ and O$^-$ states respectively. Further confirmation of the origin of this shoulder comes from the observation of photoconduction at these energies (11). The implication here must be that excitons generated at silicon and oxygen equilibrium states readily ionize and contribute to the conduction because the states themselves are charged.

Fig 2. shows the proposed arrangement of silicon and oxygen levels in the energy gap. The characteristic red (1.8 eV) and blue (2.7 eV) luminescence bands of irradiated silica are also included for completion. Taking a (Mott-Wannier) exciton binding energy of 1.6 eV ($\epsilon = 2.15$, $m^* = 0.5m_e$) we deduce an energy gap of about 10.5 eV. This agrees with Mott's recent estimate (4) and is further evidence that the 10.2 eV peak is excitonic. From the scheme of intrinsic levels given in Fig.2 the energy separating charged from neutral states is approximately 5 eV at silicon sites and 6 eV at oxygen sites - indicating similar distortions of the lattice. If we assume thermal levels (dotted lines) lie midway between optical levels as shown then donor silicon states lie above acceptor oxygen states - this ensures the stability of charged dangling bond states over neutral states.

The effect of water on the optical properties of unirradiated silica is similar to that of a conventional modifier: the band edge moves in towards the visible (13). Heat treating water-free silica (e.g. Corning 7943) in water has the same effect (14). It would appear non-bridging oxygens are generated in either case and that these are compensated by protons in an analogous way to alkali cations in a silicate glass. The presence of hydroxyl centres is confirmed by the O-H bond stretching band at 0.46 eV. Interestingly the strength of this feature is found to be inversely proportioned to the radiation induced 5.8 eV band (14). Clearly water has an 'annealing effect' on intrinsic defects in silica and behaves in some respects analogously to hydrogen in amorphous Ge - apparently saturating dangling bond states. In the context of the present model water is expected to dissociate amongst the charged intrinsic states - hydroxyl ions going to Si^+ states and protons to O^- states:

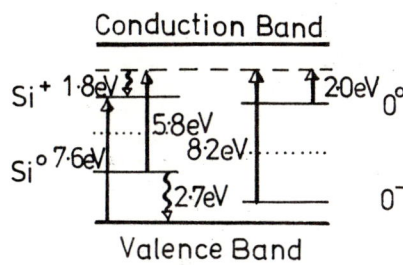

Fig.2 Arrangement of the various energy levels of the primary intrinsic defect proposed for vitreous silica. Absorption data is taken from Fig.1 and luminescence from Ref.1

$$ \left. \begin{array}{c} \overset{|}{-} Si^+ \quad \overset{H_2O}{\underset{\downarrow}{}} \quad O^- \overset{|}{-} \\ \overset{|}{} \quad OH^- + H^+ \end{array} \right\} \longrightarrow \quad \overset{|}{-} Si \overset{|}{-} OH \qquad HO \overset{|}{-} $$

The nett effect will be a reduction in the density of charged states on silicon and oxygen sites and their replacement by non-bridging hydroxyl centres —— OH. Like alkali modified non-bridging oxygen centres we expect they will be close in energy to the O^- level. They will therefore effect the shift in the absorption edge prior to radiation. Moreover, as there will be fewer Si^+ states, fewer paramagnetic centres will form on radiation and the 5.8 eV band will be weakened as observed.

A fuller description of this model is currently in press (15).

*Now at: S.R.C. Daresbury Laboratory, Warrington, WA4 4AD, U.K.

REFERENCES

(1) E.Lell, N.Kreidl and R.Hensler, Radiation effects in quartz, silica and glasses, Progress in Ceramic Science 4, 1 (1966) Pergamon Press. G.A. Sigel, Ultraviolet spectra of silicate glasses, J. Non-Crystalline Solids 13, 372 (1973/74).

(2) F.J. Fiegl, W.B.Fowler and L.K.Yip, Oxygen vacancy model for the E' center in SiO_2, Solid St. Comm. 14, 225 (1974).

(3) R.A.Street and N.F. Mott, States in the gap in glassy semiconductors, Phys. Rev. Lett. 35, 1293 (1975). M.Kastner, D.Adler and H.Fritzsche, Valence-alternation model for localized gap states in lone-pair semiconductors, Phys. Rev. Lett. 37, 1504 (1976).

(4) N.F.Mott, SiO_2 and the chalcogenide semiconductors, Adv. Phys. 26, 363 (1977).

(5)   R.J.Powell, Interface barrier energy determination from voltage dependence of photoinjected currents, J. Appl. Phys. 41, 2424 (1970).

(6)   D.L. Griscon, E.J.Friebele and G.H.Sigel, Analysis of the primary ^{29}Si hyperfine structure, Solid St. Comm. 15, 479 (1974).

(7)   E.J.Friebele, G.H.Sigel and D.L.Griscon, Drawing-induced defect centers in fused silica core fiber, Appl. Phys. Lett 28, 516 (1976).

(8)   P.Kaiser, Drawing-induced colouration in vitreous silica fibers, J. Opt. Soc. Am.64, 475 (1974).

(9)   C.M.Nelson and R.A.Weeks, Trapping electrons in irradiated quartz and silica, J. Am. Ceram. Soc. 43, 396 (1960); Vacuum-ultraviolet absorption studies of irradiated silica and quartz, J.Appl. Phys. 32, 883 (1961). E.J.Friebele, D.L.Griscon and G.H.Sigel, Radiation-induced defect centers in non-crystalline SiO_2 Proc. 4th Conf. in Physics of Non-Crystalline Solids(Clausthal-Zellerfeld) 154, (1977). D.L.Griscon, E.J.Friebele, G.H.Sigel and R.J.Ginther, Radiation-induced paramagnetic defects as structural probes in pure silica and borosilicate glasses, Structure of Non-Crystalline Materials 113 (1977). Taylor and Francis.

(10)   B.Fischer, R.A.Pollak, T.H.DiStefano and W.D.Grodman, Electronic structure of SiO_2, $Si_xGe_{1-x}O_2$ and GeO_2 from photoemission spectroscopy, Phys. Rev. B15, 3193 (1977).

(11)   A.Appleton, private communication (1977).

(12)   C.M.Nelson and J.H.Crawford, Optical absorption in irradiated quartz and fused silica, J. Phys. Chem. Solids 13, 296 (1960).

(13)   T.Bell, G.Hetherington and K.H.Jack, Water in vitreous silica, Phys. Chem. of Glasses 3, 141 (1962).

(14)   R.A.Weeks and E.Lell, Relation between E' centres and hydroxyl bonds in silica, J. Appl. Phys. 35, 1932 (1964).

(15)   G.N.Greaves, Colour centres in vitreous silica, Phil. Mag. (in press).

AUGER SPECTRA OF SiO_2 SURFACE DEFECT CENTERS

Klaus Schwidtal
US Army Electronics Technology and Devices Laboratory
(ERADCOM), Fort Monmouth, N.J. 07703

ABSTRACT

It is shown that Auger electron spectroscopy (AES) can reveal the absolute energy level, with respect to the valence band edge, of Si dangling bond electrons in SiO_2. A theoretical model is proposed, and the Auger electron distribution $N_A(E)$ for the $L_{23}VV$ transition band is calculated for a stoichiometric SiO_2 surface, and for a SiO_x surface with unpaired, dangling bond electrons. The latter is characterized by an additional $L_{23}VD$ transition band, where D is the energy level of the unpaired electron. The theoretical $N_A(E)$ curves are compared with experimental $N(E)$ curves for a pristine, and for an electron radiation damaged quartz surface. Good agreement with the theoretical model is obtained, if D is assumed to lie 7.2 eV above the valence band edge.

INTRODUCTION

Radiation-induced intrinsic defect centers in SiO_2, or on its surfaces or interfaces have been extensively studied by optical, and electron spin resonance (ESR) techniques (1). This paper proposes a theoretical model, and presents preliminary experimental evidence on how these defect centers can be observed by AES.

The best-known intrinsic defect is the E_1' center (3) (or E_s' center, when it is a surface defect (4)), which is an O^- vacancy and is characterized by an unpaired electron, strongly localized in a nonbonding sp^3 orbital on a silicon bonded to just three oxygens. The unpaired electron, or defect electron, is less tightly bound to the Si atom than the other valence electrons, and its energy level in the energy-level scheme lies in the upper part of the band gap (3).

AES should be uniquely useful in revealing the absolute energy level of localized, occupied surface defect states. Because of the strong localization of the E_1' or E_s' center unpaired electron, this electron can be expected to participate in Auger transitions of the respective Si atom. Auger transitions involving valence electrons reflect the whole valence band, and do not just yield the energy difference between two levels of the proper separation without information on the absolute energy level of either, as most other spectroscopies do. Therefore, comparing a core-band-band Auger transition region with and without a defect state should reveal the absolute energy position of that defect state with respect to the valence band.

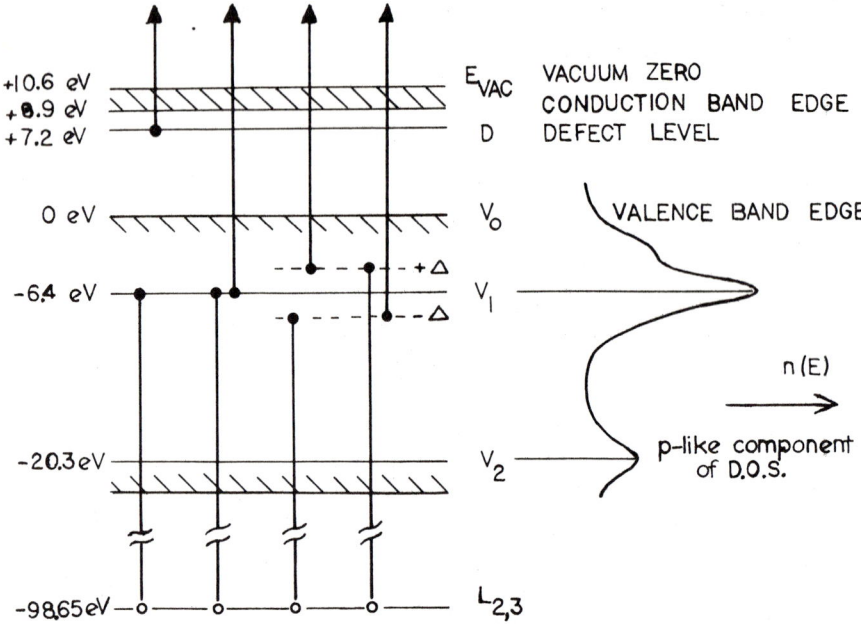

Fig. 1. Energy-level diagram showing a $L_{23}VD$ Auger transition at left, and three $L_{23}V_1V_1$ transitions.

THEORETICAL MODEL

For Si and SiO_2, the $L_{23}VV$ transition is the Auger process of highest probability. Such a solid-state Auger process is schematically shown in Fig. 1. The kinetic energy of the three Auger electrons resulting from the three $L_{23}VV$ Auger transitions indicated in Fig. 1 is always

$$E_A = 2 E_V - E_{L_{23}} - E_{VAC}$$

The Auger electron distribution $N(E_A)$ is obtained by integrating over Δ:

$$N(E_A) = \int_0^\infty W_A(E,\Delta) \cdot n(E+\Delta) \cdot n(E-\Delta) \cdot d\Delta$$

where $W_A(E,\Delta)$ is the matrix element for the Auger transition. The only complete one-body calculation (i.e., including matrix elements) of a core-valence-valence Auger line shape reported so far are those by Feibelman et al (5) for the $L_{23}VV$ transition band of a Si (111) surface. They find that in this case

$$W_A(E,\Delta) \cdot n(E+\Delta) \cdot n(E-\Delta) \simeq n_p(E+\Delta) \cdot n_p(E-\Delta),$$

where $n_p(E)$ is the p-like component of the density of states (DOS). $N(E_A)$ is than a self-fold of $n_p(E)$. Assuming that the same relation also holds, at least approximately, for SiO_2 (and justifying this assumption from the agreement with experimental results), the Si $K\beta$ x-ray emission band from SiO_2 (6)

was chosen for $n_p(E)$. $n_p(E)$ is shown as the (local) density of states in Fig. 1. $N(E_A)$ obtained from self-folding $n_p(E)$ is shown in Fig. 2, together with an experimentally obtained $N(E_A)$ for a pristine SiO_2 surface. (Actually, it is $N(E_A)$ superimposed on a background current of "true" secondary electrons). The reasonably good agreement with the theoretical $N(E_A)$ justifies the assumptions made in calculating $N(E_A)$.

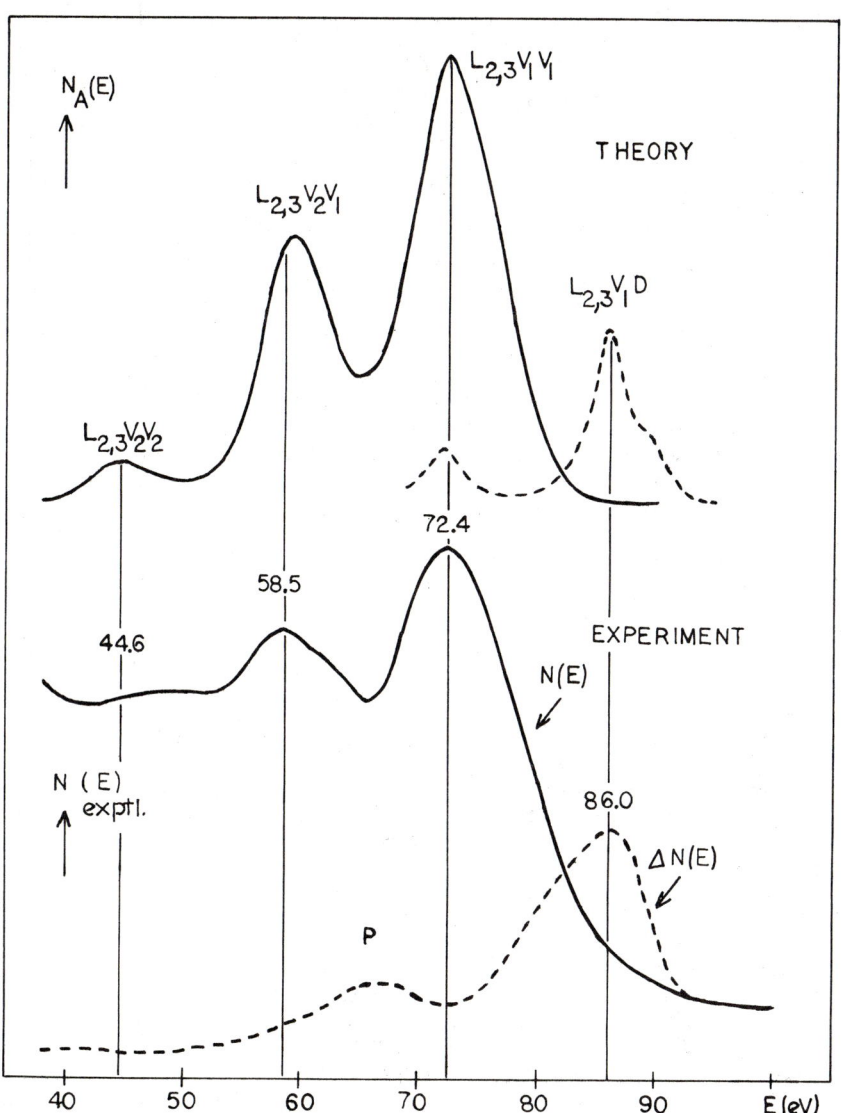

Fig. 2. TOP: The theoretical $L_{23}VV$ (full line) and $L_{23}VD$ (dashed line) Auger transition bands. BOTTOM: Experimental secondary electron distribution $N(E)$ from a pristine SiO_2 surface (full line), and the difference between $N(E)$ from an electron radiation damaged SiO_x surface, and $N(E)$ from the pristine SiO_2 surface.

If a defect level D due to an unpaired electron is included, an additional $L_{23}VD$ transition band is obtained. We assume that, at least in first approximation, n(E) for the defect-containing SiO_x will be the same as for stoichiometric SiO_2. This assumption is essential for our model, because it allows us to treat the $L_{23}VV$ spectrum of SiO_x as a superposition of the $L_{23}VV$ band of stoichiometric SiO_2, and the $L_{23}VD$ band. Conversely, it will allow us to experimentally obtain the $L_{23}VD$ band as the difference between the Auger spectra of SiO_x and SiO_2.

Figure 2 shows the $L_{23}VV$ and $L_{23}VD$ transition bands as separate components. D is assumed to be uniquely valued, and to lie 7.2 eV above the valence band edge, to make it consistent with the experimental results to be described later. We again consider only the p-like component of n(E). With the assumptions made, the shape of the $L_{23}VD$ transition band will be identical to the shape of $n_p(E)$.

The experimentally collected Auger electron current distribution $I_A(E)$ is

$$I_A(E) \propto \int_o^\infty N_A(E,x) e^{-x/d} dx$$

where x is the spatial coordinate perpendicular to the sample surface, x=0 describes the sample surface, and d is called "escape depth." For the SiO_2 $L_{23}VV$ transition considered here, the escape depth is only d $\approx$ 4 Å (7). We should therefore expect to experimentally observe mostly surface defects.

EXPERIMENTAL EXAMPLE

Oxygen deficiency of SiO_2 surfaces can easily be created by electron irradiation. This effect has become known primarily as an annoyance in electron excited AES of SiO_2 surfaces (7). The observed decrease in the oxygen Auger signal is accompanied by changes in the $L_{23}VV$ Auger signal, the most pronounced of which is the appearance of a "Si peak" in the derivative mode (7)(8).

Mitchell and Denure (9) have studied electron irradiation induced defect centers in SiO_2 thin films using cathodoluminescence spectra. They found that electron irradiation greatly enhances an optical emission peak ("C band") at $\approx$ 4500 Å (2.8 eV), which the authors believe to be due to broken Si-O bonds (9). The growth law reported for this peak is the same as that reported for the AES "Si peak" (8), suggesting that both signals arise from the same defect center. Electron radiation damaged SiO_2 surfaces should therefore be a good example for testing our theoretical model.

The experimental Auger electron distribution curve shown in Fig. 2 was obtained from a quartz surface in the "as received" state by taking the Auger spectrum as fast as possible. (The spectrum was actually taken in the derivative mode, and integrated to obtain N(E)). The sample was then electron irradiated until the oxygen signal had decreased by 10%, and the Auger spectrum was taken again as fast as possible. The difference between the two N(E), which should correspond to the $L_{23}VD$ transition band, is shown as $\Delta N(E)$ in Fig. 2. The peak P is a plasmon loss peak. Its energy and intensity are characteristic for a SiO_2 matrix. The maxima of $\Delta N(E)$ and the theoretical $L_{23}VD$ transition band line up if D is assumed to lie 7.2 eV above the valence band edge, as shown in Fig. 1. This D level supplements the "C band" from cathodoluminescence (9) to exactly the SiO_2 band gap. We take this as an indication of the validity of our model, and interpret the C band as arising from transitions from the bottom of the conduction band to the D level.

The principal $\Delta N(E)$ peak is somewhat broader than the theoretical peak. But then, several simplifying assumptions were made in the model. One possible contribution to the broadening would be a finite width of the D level (assumed to be uniquely valued in the model) caused by a difference in the bonding strength for the unpaired electron between the outermost atomic layer and the next following atom layers.

A possible alternative interpretation for $\Delta N(E)$ would be Si islands in SiO_2 (7), because the absorption edges of the $L_{23}VD$ transition band and the Si $L_{23}VV$ transition band about agree (and this information only is reflected in the maximum negative peak of the derivative mode spectrum). However, this should also be reflected in the plasmon loss spectrum, which we found to reflect strictly a SiO_2 matrix.

In summary, we find good agreement between the experimental example and the theoretical model, within the present limits of our general theoretical understanding of Auger transition line shapes.

REFERENCES

1. For a review, see Refs. (2) and (3).

2. E. Lell, N.J. Kreidl and J.R. Hensler, Radiation effects in quartz, silica and glasses, Progr. Ceram. Sci. 4, 1 (1966).

3. K.L. Yip and W.B. Fowler, Electronic structure of E_1' centers in SiO_2, Phys. Rev. B11, 2327 (1975).

4. G. Hochstrasser and J.F. Antonini, Surface states of pristine silica surfaces, Surface Science 32, 644 (1972).

5. P.J. Feibelman, E.J. McGuire and K.C. Pandey, Tight-binding calculation of a core-valence-valence Auger line shape: Si (111), Phys. Rev. Lett. 36, 1154 (1976).

6. G. Wiech, Röntgenspektroskopische Untersuchung der Struktur des Valenzbandes von Si, SiC und SiO_2, Z. Phys. 207, 428 (1967).

7. J.S. Johannessen, W.E. Spicer and Y.E. Strausser, An Auger analysis of the SiO_2-Si interface, J. Appl. Phys. 47, 3028 (1976).

8. B. Carrière and B. Lang, A study of the charging and dissociation of SiO_2 surfaces by AES, Surface Science 64, 209 (1977).

9. J.P. Mitchell and D.G. Denure, A study of SiO_2 layers on Si using cathodoluminescence spectra, Solid-State Electron. 16, 825 (1973).

278

VIBRATIONAL AND ELECTRONIC SPECTROSCOPY OF ION-IMPLANTATION-INDUCED DEFECTS IN FUSED SILICA AND CRYSTALLINE QUARTZ*

G. W. Arnold
Sandia Laboratories, Albuquerque, NM 87185

ABSTRACT

Defects produced by implantation of various atomic species in fused and crystalline SiO_2 were studied using infrared reflection spectroscopy (IRS) and UV-visible spectroscopy. We observe a new vibrational band at 830 cm^{-1} which is tentatively associated with the creation of two nonbridging O atoms in SiO_4 units. Numerous chemical effects were also observed, including evidence for chemical incorporation of Li and anomalously large O-vacancy production for Al^+, B^+ and Si^+ implantation.

RESULTS AND DISCUSSION

IRS Spectra

Crystalline quartz has 24 fundamental optical vibrational modes of which 12 are infrared active (1). For both quartz (Fig. 1) and fused silica (Fig. 3) the principal infrared reflection

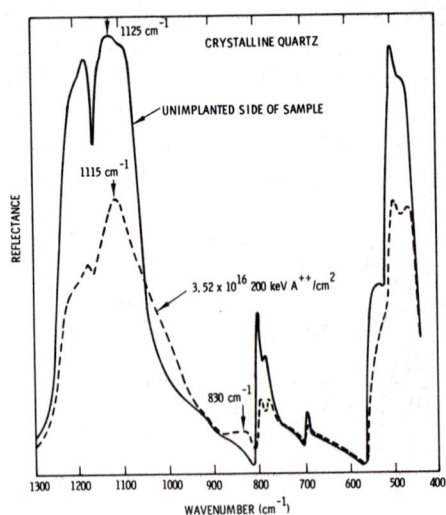

Fig. 1. IRS spectra for 3.52 x 10^{16} 200 keV A^{++}/cm^2 incident on crystalline quartz.

features in the 400-1300 cm^{-1} spectral region are the Si-O-Si stretching vibration near 1100 cm^{-1}, a Si-O-Si bending mode in the Si-O plane near 800 cm^{-1}, and a Si-O-Si rocking mode normal to the basal plane near 500 cm^{-1}. In Fig. 1 and Fig. 3, it is shown that implantation into quartz and fused silica reduces the intensity of these modes, causes a shift to lower wavenumber, and the growth of a band near 1000 cm^{-1}. In addition, a band which has not been previously reported, is produced at 830 cm^{-1} in quartz. The broad band which appears in the spectral region of 950-1000 cm is believed to arise from the formation of nonbridging oxygen (2). The T_d symmetry of the SiO_2 tetrahedron changes to C_{3v} with the formation of one nonbridging oxygen and to C_{2v} for two nonbridging oxygens (3). The band near 1000 cm^{-1} is presumed to arise from the C_{3v} symmetry sites and that at 830 cm^{-1} to the formation of sites of C_{2v} symmetry which can accumulate in appreciable numbers only for very high local damage concentrations. From the spectra there is evidence for disorder for ion fluences as low as 1 x 10^{14} 200 keV A^{++}/cm^2. Isochronal annealing of the quartz sample shows that the 830 cm^{-1} band anneals out in the 600-900°C range and that there is a simultaneous retrograde shift in the Si-O-Si vibrational frequency. Annealing to 900°C, however, does not restore the long-range order of the lattice.

It has been supposed (4) that, in the case of fused silica, the shift in Si-O-Si frequency occurs due to an average decrease in the Si-O-Si bond angle as the Si-Si distance is shortened in the irradiation compacted material. Leadbetter and Wright (5), however, in a recent x-ray diffraction study find that the Si-Si distance does not change upon neutron irradiation. If the Si-Si and Si-O distances do not change with irradiation, then the shift in mode frequencies must be attributed to some other mechanism, possibly to the increased localization of the vibrational energy at the individual SiO_2 tetrahedron as the interstitial concentration increases.

As seen in Fig. 1, the reflectivity near 800 cm^{-1} for crystalline quartz, attributed (7) to Si-O-Si rocking motion in the Si-O plane, has components at 785 cm^{-1} (A$_2$ mode) and at 805 cm^{-1} (E mode). In general, the effect of implantation is to uniformly reduce the intensity of these components and to cause a slight shift to lower energy. For Li$^+$ implantation, however, the effect is quite different in that the 785 cm^{-1} band is degraded in intensity much more than is the 805 cm^{-1} band. This is seen in Fig. 2, where the effect is contrasted with that obtained for the A^{++} implantation. Kleinman and Spitzer (6) associate the 785 cm^{-1} A$_2$ vibration with the tangential motion of Si in the Si-O-Si basal plane accompanying the Si-O-Si rocking motion. The effect of the Li$^+$-ions on the 785 cm^{-1} vibration could be associated with the substitution of Li for Si or with the perturbation of the vibration due to Li residing in a nearby site. The efficiency of production of the 950 cm^{-1} band and of the 830 cm^{-1} band (both of which require nonbridging oxygen) is greater for Li$^+$ implantation as compared to A^{++} implantation in spite of the large mass difference and consequent disparity in the concentration of displaced atoms. This may mean that the Li atom is

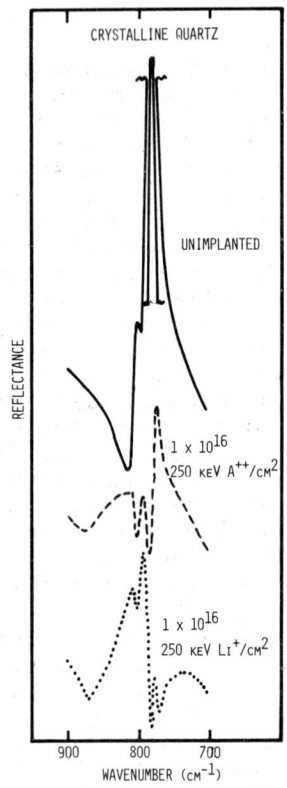

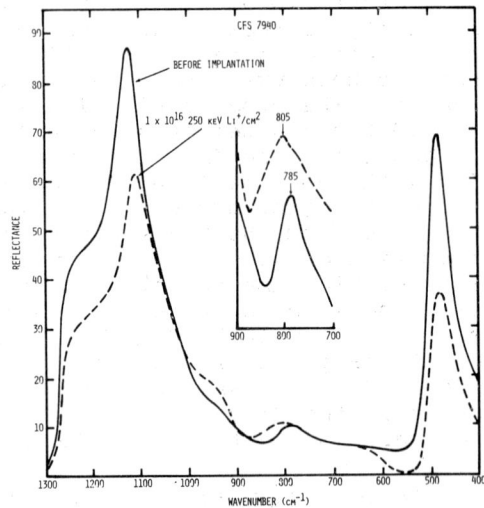

Fig. 3. IRS spectra for
1×10^{16} 250 keV Li^+/cm^2 incident on CFS 7940.

Fig. 2. IRS spectra near
800 cm^{-1} for crystalline quartz
implanted with 250 keV A^{++} and
250 keV Li$^+$. Spectra separated for clarity.

being incorporated chemically, e.g., LiO$_2$ or a lithia silicate
may be formed as the implanted Li$^+$ scavenges the displaced atoms,
thereby decreasing the number of interstitial vacancy recombina-
tions which would ordinarily occur and increasing the intensity
of the 830 cm^{-1} and the 950 cm^{-1} bands.

The effect of Li$^+$ implantation in fused silica is also quite
different as can be seen in Fig. 3. The original peak of the distribution near 785 cm^{-1} can still be discerned after Li$^+$ implantation, but there is a new peak at 805 cm^{-1}, so that the spectral
features are similar to those of crystalline quartz. It is possible that in the region of the collision cascade lithium disilicate nuclei are formed which can nucleate the growth of
crystalline quartz. This is a well-known process in the formation of a glass-ceramic from a complex glass containing lithia
and silica (7).

UV-Visible Spectra

The principal spectral features in the near-ultraviolet for quartz and fused silica are the bands at 2150 A (E_1'-center) and at 2450 A (B_2-center). The E_1'-center is well-established as a center consisting of a localized electron in a (sp^3) orbital on a Si atom adjacent to an oxygen vacancy (8). The non-paramagnetic B_2-center is less firmly established as an oxygen vacancy with two trapped electrons (9). Griscom (8) has shown that the E_1'-center in fused silica formed by ionizing radiation is the same center formed in crystalline quartz by displacing radiation, i.e., it requires an oxygen vacancy. The question that arises is: How is the E_1'-center in fused silica formed by x-irradiation which only results in bond-breaking and not displacement (10)? Jones (11) has suggested that this may result from charge trapping at existing defects. Compton and Arnold (12), however, found that the concentration of E_1'-centers in fused silica x-irradiated at 77°K was about $10^{18}/cm^2$ (Corning 7943) at absorbed doses of about 10^8 rads with no evidence of saturation. Sigel, et al. (13) have also shown that commercial high purity silicas are apparently highly stoichiometric since the generation of E_1'-centers is not appreciably different for glasses heated in oxygen or in argon at 1000°C.

For 25-250 keV H^+-implants, E_1'-center production is linear with energy into displacement processes. Absorption coefficients of $10^3/cm$ can be attained which corresponds, roughly, to defect densities of the order of $10^{18}/cm^3$. Any effect due to ionization is small in comparison to the displacement process. It should be noted that fused silica implantation experiments on refractive index changes (14) and on induced surface stress (15) have shown that defects are generated much more efficiently by the displacing component of the ion energy than by the energy into ionization. The present experiments do not, therefore, help to resolve the question of E_1'-center generation by ionization processes.

For 200 keV A^+-implants, the concentration of E_1'-centers is saturated at displacement energy concentrations produced by ion fluences as low as $1 \times 10^{12}/cm^2$. The saturation exhibited at low fluences for heavy ions is a consequence of the larger diameter ion track produced by the heavier ions which are moving at lower velocities. The damage within the core of the track is presumed to be at saturation levels and the measured absorption coefficient will saturate when adjacent tracks overlap. This occurs at much lower fluence levels for heavy ions than for H^+ ions.

In Fig. 4 is shown a comparison between the optical density produced by $A\ell^+$ as compared to Ne^+ implants. Although the mass difference is not very great, there are apparently larger numbers of oxygen vacancies produced by the $A\ell^+$ implant than for the Ne^+ implant. A similar anomaly exists for B^+ implants. Both B^+ and $A\ell^+$ implants are examples where the chemical nature of the implanted ion itself plays a part in the production of defects. The substitution of either B^{3+} or $A\ell^{3+}$ for Si^{4+} in fused silica requires that one oxygen vacancy be formed for each two such

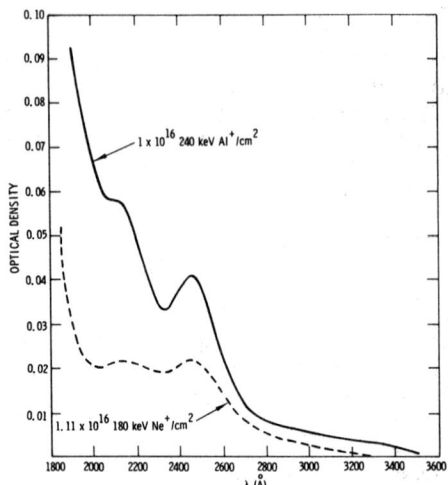

Fig. 4. Optical density vs. wavelength (A) for
240 keV Al^+ and 180 keV Ne^+ incident on CFS 7940.

substitutions. In this way, local enhancement of oxygen vacancy
concentrations can be generated as a result of both the requirements of defect chemistry and by the displacement process. This
interpretation is further substantiated by the similar enhancement where Si^+ is implanted but not for O^+ implantation.

REFERENCES

1. J. F. Scott and S. P. S. Porto, Longitudinal and transverse
   optical lattice vibrations in quartz, Phys. Rev. 161, 903
   (1967).

2. I. Simon, in Modern Aspects of the Vitreous State, ed. by
   J. D. MacKenzie (Butterworths, 1960), p. 120.

3. D. M. Sanders, W. B. Person and L. L. Hench, Quantitative
   analysis of glass structure with the use of infrared
   reflection spectra, Appl. Spect. 28, 247 (1974).

4. I. Simon, Structure of neutron-irradiated quartz and
   vitreous silica, J. Amer. Ceram. Soc. 40, 150 (1957).

5. A. J. Leadbetter and A. C. Wright, The structure of neutron
   irradiated silica, Phys. Chem. Glasses 18, 79 (1977).

6. D. A. Kleinman and W. G. Spitzer, Theory of the optical
   properties of quartz in the infrared, Phys. Rev. 125, 16
   (1962).

7. G. W. Arnold, Near-surface nucleation and crystallization of
   an ion-implanted lithia-alumina-silica glass, J. Appl. Phys.
   46, 4466 (1975).

8. D. L. Griscom, E. J. Friebele and G. H. Sigel, Jr., Observation and analysis of the primary ^{29}Si hyperfine structure of the E´ center in non-crystalline SiO_2, <u>Solid State Commun.</u> 15, 479 (1974).

9. G. W. Arnold, Ion implantation effects in noncrystaline SiO_2, <u>IEEE Trans. Nucl.</u> Sci. NS-20, 220 (1973).

10. G. W. Arnold and W. D. Compton, Radiation effects in silica at low temperatures, <u>Phys. Rev.</u> 116, 802 (1959).

11. C. Jones, Evidence for the ionization of existing defects producing the ionizing radiation density of amorphous SiO_2, <u>Bull. Amer. Phys.</u> 22, 382 (1977).

12. W. D. Compton and G. W. Arnold, Radiation effects in fused silica and $\alpha-A\ell_2O_3$, <u>Disc. Faraday Soc.</u> 31, 130 (1961).

13. G.H. Sigel, Jr., E. J. Friebele, R. J. Ginther and D. L. Griscom, Effects of stoichiometry on the radiation response of SiO_2, <u>IEEE Trans. Nucl. Sci.</u> NS-21, 56 (1974).

14. H. M. Presby and W. L. Brown, Refractive index variations in proton-bombarded fused silica, <u>Appl. Phys. Lett.</u> 24, 511 (1974).

15. E. P. EerNisse, Compaction of ion-implanted fused silica, <u>J. Appl. Phys</u>. 45, 167 (1974).

---

*This work was supported by the United States Department of Energy (USDOE) under Contract AT(29-1)789.

RAMAN STUDIES OF STRUCTURAL DEFECTS IN VITREOUS SiO$_2$

F. L. Galeener, J. C. Mikkelsen, Jr., and N. M. Johnson
Xerox Palo Alto Research Center, Palo Alto, CA 94304

ABSTRACT

This paper discusses the two sharp lines which are seen at 495 cm^{-1} and 606 cm^{-1} in the Raman spectrum of pure fused silica. We conclude that the 495 cm^{-1} line is probably a longitudinal optical mode of the network, although defect origin is not entirely ruled out. More importantly, we conclude that the 606 cm^{-1} is due to a large concentration of non-bridging oxygen defects. We show how the strength of the 606 cm^{-1} line varies with the water concentration and the thermal history of bulk samples of Suprasil fused silica.

INTRODUCTION

Figure 1 shows the polarized Raman spectra of high purity fused silica as a function of water concentration (OH). These spectra, measured on samples as-received from the manufacturer, are discussed in detail by Galeener and Geils (1). In most respects, they are similar to the spectra reported by Stolen and Walrafen (2). The important features for the present discussion are as follows. At high frequencies, there is the 3700 cm^{-1} O-H stretch mode [Fig. 1(c)] which we use to monitor OH concentrations down to the level of 2 ppm by weight. At intermediate frequencies [Fig. 1(b)] the main observation is the absence of any structure other than that due to second order vibrations of the SiO$_2$ network, discussed elsewhere by Galeener and Lucovsky (3). In particular, there is the absence of Si-H stretch modes, which Hartwig (4) has shown to occur at 2285 cm^{-1} in deliberately hydrogenated silica. At lower frequencies [Fig. 1(a)] the features of interest are the 975 cm^{-1} Si-OH stretch mode induced in high water content material, the 606 cm^{-1} line whose intensity is reduced in the high water content as-received material, and the 495 cm^{-1} line whose intensity is unchanged. These latter two lines are the focus of attention in this paper; all other features in Fig. 1 which are not mentioned here have been identified elsewhere (5) as transverse optical (TO) or longitudinal optical (LO) modes of the ideal (perfect) network.

THERMAL STUDIES

Figure 2 shows the variation in strength of the 495 cm^{-1} and 606 cm^{-1} lines with thermal treatment of the sample. The line strength is measured by the area between the defect line and a baseline drawn with a French curve (assuming that the 495 cm^{-1} line has a total width at its baseline of 40 cm^{-1} while the 606 cm^{-1} line has a total width of 100 cm^{-1}). The samples were of two types, those of low water content (solid lines) and those of high water content (dotted lines). All samples were about 3 mm x 6 mm x 12 mm and Raman spectra were obtained from the interior of these materials. Water content was not affected by the heat treatments, as determined by measuring the area of the 3700 cm^{-1} line before and after each treatment. There were no changes in the frequency range 1500 cm^{-1} to 3000 cm^{-1} indicating that neither Si-H bonds nor molecular water were created. Beginning with a new set of samples, treatments (A) through (G) were carried out in the order [1] through [7].

Study of the left half of Fig. 2 reveals that the area under the 495 cm^{-1} line varies erratically about an average value of ~18. For example, in the sequence [3] through [7] the area for the low water content sample is alternately above then below the area for the high water content material. This simply reflects the difficulty in determining the base line for a peak that rides on the steep slope of the main Raman line (whose position varies by a few wavenumbers with thermal treatment). By placing spectra over one another we conclude that the area under the 495 cm^{-1} line is constant within 5%.

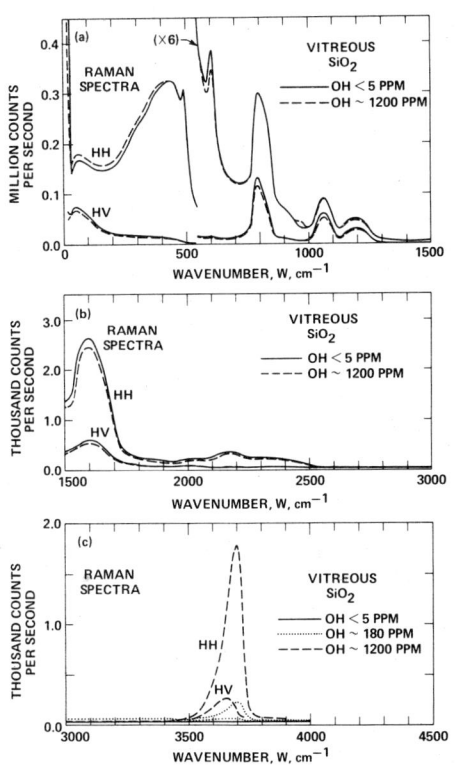

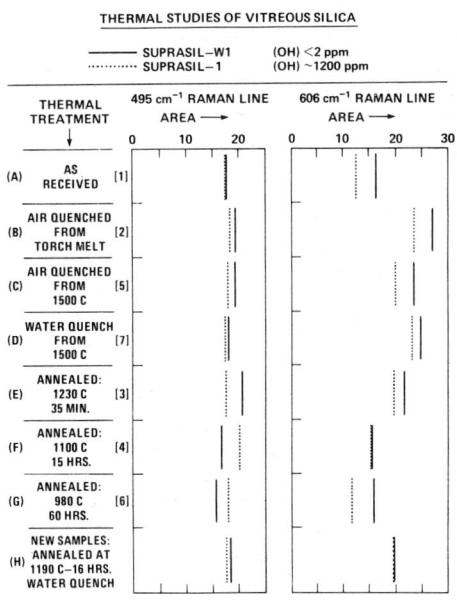

Fig. 1. The OH dependence of the polarized Raman spectra of as-received Suprasil. The sharp lines at 495 cm^{-1} and 606 cm^{-1} in (a) are the features of special interest in this paper.

Fig. 2. The strengths of the 495 cm^{-1} and 606 cm^{-1} Raman lines after different thermal treatments (A) through (H). Note the time sequence of steps, [1] through [7]. The variations in the 495 cm^{-1} line are erratic and small, while the 606 cm^{-1} line strength varies strongly and systematically.

The right half of Fig. 2 reveals clear changes in the strength of the 606 cm^{-1} line. The line strength in high water content material is always less than (or equal to) that for low water content material. Material quenched from high temperatures has a larger 606 cm^{-1} line than material quenched from low temperatures. Figures 2(f) and 2(h) suggest that after sufficiently long periods at a given temperature $\geq$ 1100°C, the strength of the 606 cm^{-1} line is independent of the water concentration. This contradicts the most obvious interpretation of Fig. 2(a), namely that the 606 cm^{-1} defect concentration is lower in high water content samples. Our hypothesis for resolving this is to assume that equilibrium is reached far more quickly in high water content samples, but that the equilibrium concentration is not very dependent on water content. This is consistent with the fact that high water content material has significantly lower viscosity (6). Thus, we suppose that water-containing and water-free as-received material have different 606 cm⁻1 line strength either because they have been annealed at different temperatures, or, if annealed at the same temperature, because the low water content material was not annealed long enough for the defect concentration to achieve equilibrium value. Another possibility is that there is, in fact, a real difference in the equilibrium 606 cm^{-1} defect concentrations in water-containing and water-free samples that have been annealed at temperatures below 1100°C. More data is being obtained, with careful attention to the role of annealing time and quenching speed, and the results will be discussed more extensively elsewhere.

DISCUSSION

Stolen, Krause, and Kurkjian (7) first suggested that the 606 cm^{-1} line might be due to a defect, when they observed that this line increased at least five-fold after the sample had been irradiated with ~10^{20} fast neutrons/cm^2. Their data showed essentially no change in the area of the 495 cm^{-1} line with irradiation, although this line did shift to slightly higher frequency, as did the 606 cm^{-1} line and the main Raman peak. These results were corroborated in an extensive study of neutron effects by Bates, Hendricks, and Schaffer (8). The latter authors reported a seven-fold increase in the 606 cm^{-1} line for an exposure of 2×10^{20} neutrons/cm^2, as well as a possible slight increase in the strength of the 495 cm^{-1} line. They conclude that both of these lines are due to defects that are intrinsic in vitreous (v-)SiO_2, and that both may be associated with the defect formed when a Si-O bond is left broken, i.e., associated with a "non-bridging oxygen defect" (NBOD) of the form $\equiv Si^+ O^- -Si\equiv$. As Bates, et al. (8) point out, their results do not rule out the possibility that the lines are due to other kinds of structural defects. Indeed, from theoretical studies, Laughlin and co-workers (9) have suggested that the 606 cm^{-1} line is due to a "wrong bond" defect of the form $\equiv Si-Si\equiv$, which we refer to as the Si-Si bond defect.

The neutron bombardment studies thus clearly establish the 606 cm^{-1} line as having defect origin. On the other hand, we believe that the very small observed changes in the 495 cm^{-1} line demonstrate that it cannot be associated with precisely the same defect that gives rise to the rapidly increasing 606 cm^{-1} line. Were this so, both lines would rise by the same percentage. Moreover, the small reported changes in strength of the 495 cm^{-1} line may be due to the difficulty of measuring its strength, or they may be due to changes in matrix elements brought about by the compaction of the network under neutron irradiation. If the 495 cm^{-1} is a defect, the defect concentration is very little affected by neutron bombardment; nor is it affected by water content or thermal treatment, as we have shown. At the present, we think it more likely that the 495 cm^{-1} line is an LO mode of the basic net-work, as argued in Ref. 5. Accordingly, we focus the rest of our attention on the 606 cm^{-1} line.

A set of as-received samples of Suprasil 1 and Suprasil W-1 were exposed to 30 Mrad of X-rays from a 60 kV DC machine. This induced the expected UV absorption and strong spin resonance signals, but caused no detectable change in the Raman spectra. Thus, X-irradiation does not induce the 606 cm^{-1} defect, but it may activate associated electronic states.

It can be demonstrated that the 606 cm^{-1} line is not due to Si-Si bonds or O-O bonds, by appeal to measurements made on v-GeO$_2$. Galeener and Geils (1) reported a line at 520 cm^{-1} in the Raman spectrum of v-GeO$_2$ which they believed was analogous to the 606 cm^{-1} line in v-SiO$_2$. This has been confirmed by making measurements on v-GeO$_2$ irradiated by 10^{21} neutrons/cm^2 which show a large increase in the strength of the 520 cm^{-1} line. Raman measurements were also made on non-phase-separated v-GeO$_2$ that was 10% Ge-rich. The defect line at 520 cm^{-1} was unaffected. This proves that the 520 cm^{-1} line in normally stoichiometric material is not due to a small concentration of Ge-Ge bonds, which would increase greatly in the Ge-rich material, nor is it due to a small concentration of O-O bonds which would most probably go to zero in the Ge-rich material. It then follows that the 606 cm^{-1} defect in v-SiO$_2$ is not a wrong bond defect of the Si-Si or O-O type.

Having eliminated wrong bond defects, we next consider the most likely broken bond defects, which are the isolated dangling silicon bond $\equiv Si^+$, the isolated dangling oxygen bond $^-O-Si\equiv$, or their spatial juxtaposition, the previously defined NBOD. Also, in this category are the missing oxygen ($\equiv Si\ Si\equiv$) and the missing silicon defects, both of which might be formed under neutron bombardment. Since the NBOD is locally neutral, it will have lower energy associated with its existence than will the other four (charged) defects. Thus, we expect the NBOD to be the most numerous defect. Also, since the 606 cm^{-1} defect line is the only one observed, it is reasonable to ascribe this line to the existence of NBOD's. We believe this identification is consistent with all experimental observations to date, including thermal history and neutron bombardment.

The question now arises whether the 606 cm^{-1} line is due to the NBOD as an inseparable whole, or whether it can be assigned to only one part of the defect, either the dangling oxygen, or the silicon with the dangling bond.

We eliminate the isolated dangling oxygen by appeal to the OH-dependence data in Fig. 1(a). First, we make the reasonable assumption that the addition of a hydrogen to a dangling oxygen, forming the OH unit, will result in very little change in the frequency or Raman activity of defect vibrational modes previously involving the dangling oxygen, but now involving the dangling OH unit. Since the 975 cm^{-1} line was non-existent before introduction of 1200 ppm OH, the concentration of isolated dangling oxygen units in the water-free material must be very small. Furthermore, if the 606 cm^{-1} line were due to the dangling oxygen half of the NBOD, then this line should have increased dramatically with the introduction of dangling OH units. Instead, it decreased slightly, and after annealing at 1100°C or 1190°C showed the same strength as in the water-free material. Moreover, its concentration can be increased by quenching the glass from a higher temperature, without any concurrent change in the 975 cm^{-1} line.

Although we presently feel that the 606 cm^{-1} line is most likely the signature of silicon atoms with one dangling bond, we are not yet able to eliminate the possibility that it represents a mode of the NBOD as a whole. We are working further in an effort to answer this question.

REFERENCES

(1) F. L. Galeener and R. H. Geils (1977), The Polarized Raman Spectra of OH in Vitreous SiO$_2$ and GeO$_2$, The Structure of Non-Crystalline Materials, ed. P. H. Gaskell, Taylor and Francis, London, 223.

(2) R. H. Stolen and G. E. Walrafen, Water and Its Relation to Broken Bond Defects in Fused Silica, J. Chem. Phys. 64, 2623 (1976).

(3) F. L. Galeener and G. Lucovsky (1976), Second Order Vibrational Spectra of Vitreous Silica, Light Scattering in Solids, ed. M. Balkanski, R. C. C. Leite and S. P. S. Porto, Flammarion, Paris, 641.

(4) C. M. Hartwig, The Radiation-Induced Formation of Hydrogen and Deuterium Compounds in Silica and Observed by Raman Scattering, J. Chem. Phys. 66, 227 (1977).

(5) F. L. Galeener and G. Lucovsky, Longitudinal Optical Vibrations in Glasses: GeO$_2$ and SiO$_2$, Phys. Rev. Lett. 37, 1474 (1976).

(6) See R. H. Doremus (1973) Glass Science, Wiley, New York, 105.

(7) R. H. Stolen, J. T. Krause, and D. R. Kurkjian, Raman Scattering and Far Infra-Red Absorption in Neutron Compacted Silica, Disc. Faraday Soc. 50, 103 (1970).

(8) J. B. Bates, R. W. Hendricks, and L. B. Shaffer, Neutron Irradiation Effects and Structure of Noncrystalline SiO$_2$, J. Chem. Phys. 61, 4163 (1974).

(9) R. B. Laughlin, J. D. Joannopoulous, C. A. Murray, K. J. Hartnett, and T. J. Greytak, Intrinsic Surface Phonons in Porous Glass, Phys. Rev. Lett. 40, 461 (1978).

CATHODOLUMINESCENCE STUDIES OF SiO_2 - Na,Cl,Ge,Cu,Au AND OXYGEN VACANCY RESULTS [*]

Colin E. Jones[+] and David Embree
Lehigh University, Bethlehem, Pennsylvania 18015

ABSTRACT

Cathodoluminescence provides an analytical tool for studying thin films and bulk silicon dioxide. The major problems with the technique are the difficulty in obtaining quantitative data and the fact that few of the luminescence bands have been firmly identified. This paper discusses some of the problems involved and presents data on doping and processing effects on both bulk and MOS thin film SiO_2. Sodium doping has been found to increase the luminescence band at 625 nm but not to affect the 290 nm band tentatively associated with the oxygen vacancy.[1] The 290 nm band is increased in HCl growth of SiO_2 films. Impurity doping has shown bands at 435 nm and 510 nm associated with Ge in crystalline quartz, Cu in thin films gives bands at 425 nm and 525 nm, and Au in thin films gives bands at 365 nm and 520 nm. Defect profiles can be estimated by varying the excitation beam voltage. Reduction in CO increases the 290 nm bands front surface concentration. Both the 290 nm band and the 450 nm band show peaked concentrations at the Si-SiO_2 interface.

INTRODUCTION

Luminescence has the advantages of being a non destructive analytical technique which can be used on both bulk and oxide-semiconductor thin films. Cathodoluminescence does charge SiO_2, but annealing below 300°C usually restores device characteristics.[2] Optical absorption and spin resonance are difficult with thin films since there is not much material to work with. The disadvantages are that it is hard to get quantitative results and that many of the bands have not been firmly identified.

Earlier luminescence studies on SiO_2 by Mitchell and Denure [3] suggested that the band at 290 nm may be a Na defect, 380 nm may be H related, and 450 nm may be the trivalent silicon or oxygen vacancy. Work just published by Koyama, Matsubara and Mouri suggests that the 280 nm and 530 nm bands may be the trivalent silicon defect.[4] A much more detailed study by Jones and Embree showed a correlation of the 290 nm luminescence band with spin resonance and optical absorption data for the oxygen vacancy in neutron irradiated and CO reduced samples.[1]

More data on the association of the 290 nm band with the oxygen vacancy (or trivalent silicon) is presented here along with data on luminescence effects of sodium, chlorine, copper and gold.

[*] Research partially supported by the U.S. Navy Office of Naval Research.

[+] Present address: Honeywell Corporate Technology Center, 10701 Lyndale Avenue, South, Bloomington, Minnesota 55420.

METHODS

An electron gun operated at 2-12 KV, with 10^{-5} amps/cm^2, chopped at 50 Hz, was used to excite room temperature luminescence.(1) The samples are coated with a semitransparent layer of Al 100 to 200 Å thick to ground the surface. Sample preparation descriptions are given in the following text.

LUMINESCENCE STUDIES

Sodium and Chlorine

Sodium doped thin films were grown on silicon by steam oxidation from a NaOH-water solution at 1230°C. The furnace used was a cold wall furnace using a silicon susceptor developed by Butler et al (5,6) to produce research quality SiO_2 films. The sodium is found to concentrate at both interfaces with a bulk average concentration of $\sim 3\times10^{19}$ Na/cm^3.(5,6)

Chlorine doped samples were made in the same design furnace using oxygen gas with 2,3,4 and 6% by volume HCl gas. The 5% sample was examined by α back-scattering and was found to have 7×10^{19} Cl/cm^3 average concentration with 70% of this at the Si-SiO$_2$ interface.(7) The HCl helps purge the system of sodium so the HCl and control sample should be much less than 10^{17}Na/cm^3.(6)

The luminescence at 290 nm in Fig. 1 is seen in the HCl control and the sodium doped films with about the same intensity contradicting Mitchell and Denure's suggestion that this was a sodium defect.(3) The one peak which was increased in the sodium doped samples was the peak at 635 nm. The 290 nm peak is in-creased in the HCl grown samples. Chlorine is not needed for this defect as seen by its presence in the control data. Jones and Embree's correlation of the 290 nm band with the oxygen vacancy (or trivalent silicon) suggest that this defect is produced in greater concentrations in the rapidly growing HCl oxides than in non HCl films.

Germanium

Germanium luminescence has been reported in fused silica at 390-410 nm corres-ponding to an absorption band at 240 nm by Garino-Canina.(8) The present study was made on crystalline quartz from Sawyer Research, Eastlake, Ohio. The sample had $\sim$ 4000 ppm Ge, 200 ppm Al, tens of ppm Na and Li. Spin reso-nance in this sample has been reported by Robert Lorenze.(9,10) Strong lum-inescence at 435 nm and 510 nm shown in Fig. 2 grows in with the electron beam irradiation in the Ge doped samples. The bands do not always have the same intensity ratios and are distinct. Optical absorption shown in Fig. 3 is also increasing in this sample during irradiation with bands at 190 nm, 275 nm and 500 nm. The spin resonance data after room temperature irradia-tion shows the neutral Al_{sub} center, Ge^-Na^+, Ge^-Li^+, and smaller concentra-tions of oxygen vacancy centers.(9,10)

Copper and Gold

Thin films of SiO_2 grown on Si had Cu or gold evaporated on the surface and annealed in. The luminescence data is shown in Fig. 4. A four hour 600°C anneal for the copper doped sample produced a large increase in the 425 nm band and a smaller increase in 525 nm luminescence band. A 20 min. 800°C anneal produced the very strong 525 nm band.

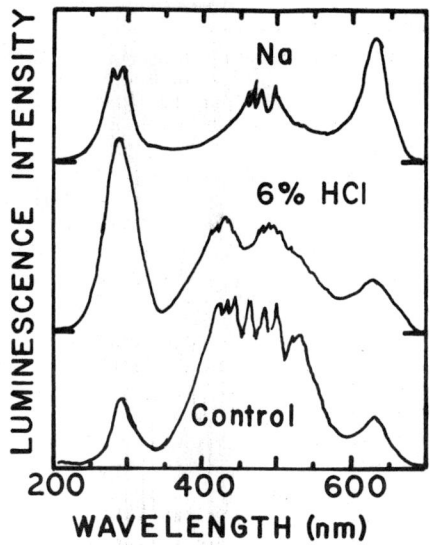

Fig. 1 HCl and Na doped thin films

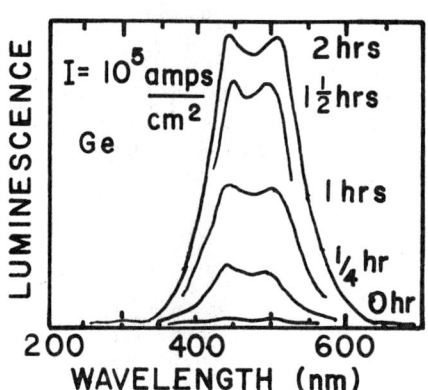

Fig. 2 Ge doped crystalline SiO$_2$

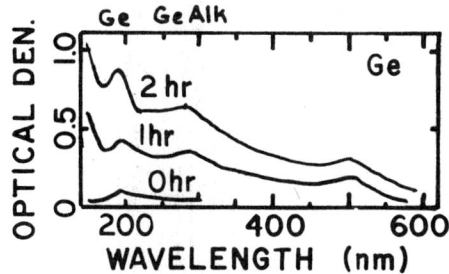

Fig. 3 Ge doped crystalline SiO$_2$

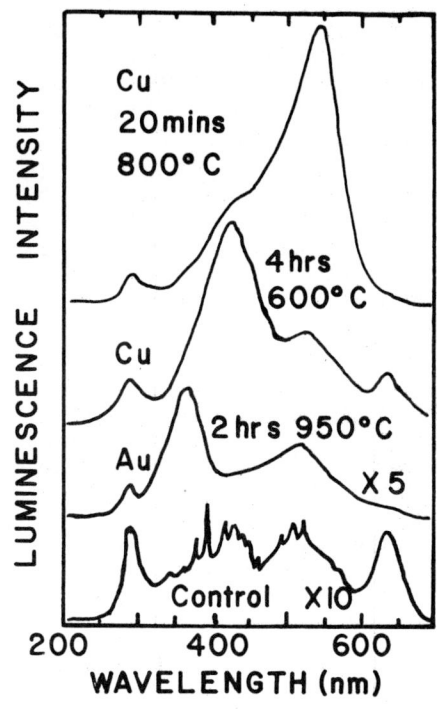

Fig. 4 Cu and Au doped films

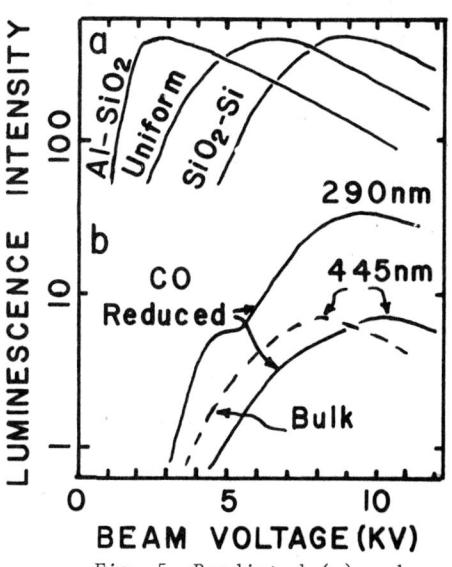

Fig. 5 Predicted (a) and
measured intensities (b)

Gold annealed at 950°C shows luminescence bands at 365 nm and at 520 nm. Trukhin and Silin have reported similar two band luminescence from Ag and Cu doped crystalline SiO_2.(11)

Luminescence Intensity

Implantation with 65 KeV B or 180 KeV P through SiO_2 films only increased the preirradiation luminescence. Device fabrication has a similar effect. Doping with Ge, Cu, or Au all shift the bands slightly but they also increase the luminescence intensity. This implies that the doping or implantation disorder is breaking a local selection rule which has caused the SiO_2 luminescence to be partially forbidden.

Impurity Profiles

Two KeV electrons barely penetrate the 200Å Al sample coating while 10 KeV electrons lose energy fairly uniformly through 1600Å SiO_2 films. It is poss-ible to use this effect to get crude depth profile information from the lum-inescence.(11,12) The relative number of electron hole pairs created as a function of beam energy and depth into the sample was calculated using standard electron energy loss data corrected for 45° incidence.(12,13) The expected luminescence intensity vs beam voltage from a uniform defect distribution, a center at the Al-SiO_2 interface and a defect at the SiO_2-Si interface is shown in Fig. 5 along with actual data for a bulk neutron irradiated sample and a CO reduced sample.

The luminescence from the bulk sample is approximately what is expected for a uniform defect distribution. The 290 nm band in the CO reduced sample shows a low voltage peak fitting the production of oxygen vacancies at the outer surface by the CO reduction. This appears with a uniform and a Si interface concentration effect seen at higher voltages. The 450 nm band was not affect-ed by the CO reduction and it appears only at higher voltages corresponding to a SiO_2-Si interface concentration. These profiles are similar to those reported by Mitchell and Denure obtained by etch back techniques.(3)

CONCLUSIONS

1. The 290 nm band is increased in HCl growth. It does not seem to depend on Na. It increased at the outer surface in CO reductions. These data fit the correlation with the oxygen vacancy suggested by Jones and Embree.(1)

2. A band at 635 nm does seem to be Na dependent.

3. Doping with Ge gives strong luminescence at 435 nm and 510 nm. Cu gives bands at 425 nm and 525 nm, and Au gives bands at 365 nm and 520 nm.

4. The luminescence intensity vs beam voltage can give some information about the depth profiles for defects.

ACKNOWLEDGEMENTS

The authors would like to thank Drs. Sid Butler, Frank Feigl, Dan Leenov, K.T. Tsao, Fred Fowkes, and Frank Hielscher of Lehigh University for their help in supplying samples and in discussing the data.

REFERENCES

1. Colin E. Jones and David Embree, Correlation of the 4.77 - 4.28 eV Luminescence Band in Silicon Dioxide with the Oxygen Vacancy, J. Appl.Phys. 47, 5365 (1976).

2. W. J. Keery, K. O. Leedy and K. F. Galloway, Electron Beam Effects on Microelectronic Devices, Scanning Electron Microscopy/1976, O. Johari (I.T.T. Res. Inst., Chicago) p. 507 (1976).

3. J. Peter Mitchell and Douglas G. Denure, A Study of SiO_2 Layers on Si using Cathodoluminescence Spectra, Solid-State Electronics 16, 825 (1973).

4. H. Koyama, K. Matsubara and M. Mouri, Cathodoluminescence Study of a Silicon Dioxide Layer on Silicon with the Aid of Auger Electron Spectroscopy, J. Applied Phys. 48, 5380 (1977).

5. Y. Ota and S. R. Butler, Reexamination of Some Aspects of Thermal Oxidation of Silicon, J. Electrochem. Soc. 121, 1107 (1974).

6. S. R. Butler, F. J. Feigl, Y. Ota, and D. J. DiMaria, Effects of Grown-in Sodium on Charge Trapping in Silicon Dioxide Thin Films, Thermal and Photostimulated Currents in Insulators, Electrochem. Soc. Inc., p.149(1976).

7. A. Rohatgi, S. R. Butler, F. J. Feigl, H. W. Kramer, and K. W. Jones, Sodium Passivation in HCl Oxide Films on Si, Applied Phys. Letters 30, p. 104 (1977).

8. V. Garino-Canina, Sur un nouveau phénomène au sujet de la bande d'absorption à 2400 A et de la luminescence de la silice vitruise, Comptes Rendus 240, 1331 (1955); and La bande d'absorption à 2420Å de la silice vitreuse: Impureté germanium et perte d'oxygène, Comptes Rendus 242, 1982 (1956).

9. Robert V. Lorenze, An Electron Paramagnetic Resonance Study of the Radiation-Induced Migration and Stabilization of Alkali Ions in Germanium-Doped Quartz, unpublished thesis, Lehigh University (1972).

10. R. V. Lorenze and F. J. Feigl, Defects in Crystalline Quartz: Electron Paramagnetic Resonance of Multiple-Alkali-Compensated Centers Associated with Germanium Impurities, Phys. Rev. B8, 4833 (1973).

11. A. N. Trukhin, S. S. Etsin, and A. V. Shendrik, Luminescence Centers and Electronic Processes in Crystalline and Glassy SiO_2-Ag, Bull.Academy of Sciences, USSR Physical Series, 40, 2329 (1976).

12. David B. Wittry and David F. Kyser, Measurement of Diffusion Lengths in Direct Gap Semiconductors by Electron-beam Excitation, J. Appl.Phys. 38, 375 (1967).

13. C. B. Norris, C. E. Barnes, and W. Beezhold, Depth-Resolved Cathodoluminescence in Undamaged and Ion-implanted GaAs,ZnS, and CdS, J. Appl.Phys. 44, 3209 (1973).

ANOMALOUS DIELECTRIC ABSORPTION IN SiO_2-BASED GLASSES*

Martin A. Bösch
Bell Laboratories, Holmdel, New Jersey 07733

ABSTRACT

At low temperatures silica-based glasses show an anomalous dielectric behavior in the temperature dependent far infrared absorption coefficient: the absorption increases with decreasing temperature. In contrast to the anomalous thermal and elastic properties of vitreous solids below a few degrees K, the dielectric properties are very sensitive to the chemical composition of the material. Frequency and temperature dependent measurements on many different glasses have allowed us to identify the intrinsic and universal behavior in the disordered state. The results can be understood in the framework of the two-level tunneling model.

INTRODUCTION

Vitreous solids exhibit thermal and elastic properties at <u>low</u> temperatures very different from those of pure crystalline solids. The specific heat is anomalously high and dominated by a term with a linear temperature dependence (Ref. 1), indicating the existence of low energy excitations. Such universal features - like the almost quadratic temperature dependence of the thermal conductivity - independent of structural details have led to the assumption of an universal origin inherent in the disordered state. A statistical distribution of localized two-level tunneling systems (Ref. 2) (2LS) can describe phenomenologically these anomalous properties at low temperatures. The 2LS are viewed as structural or electronic (Ref. 3) "centers" with two equilibrium configurations, where transitions between the two sites are possible by quantum mechanical tunneling at low temperature and by thermally activated processes at higher temperature.

In contrast to transport properties the interaction of the electro-magnetic field with the 2LS is expected to be quite sensitive to the chemical composition of the vitreous material. An anomalous dielectric dispersion (Ref. 4) as well as a temperature dependent far infrared absorption-difference (Ref. 5) have been reported but these measurements dealt with systems which have small intrinsic dipole moments so that intrinsic effects associated with the 2LS were masked by the OH^- absorption processes.

---

*Experimental work performed at Cornell University, Laboratory of Atomic and Solid State Physics, Ithaca, N.Y. 14853

EXPERIMENTAL

Transmission measurements have been undertaken in the frquency interval
from 2 to 20 cm^{-1} (vFIR) for temperatures between 0.4 K and 300 K. The
measuring system is based on a lamellar interferometer with large (25 mm)
lamellar constant and a dewar with a liquid He^3 cooled sample section, a
separate He^3 cooled doped Ge detector and proper filtering stages. Some of
the silicate glasses are hygroscopic. The samples have been prepared carefully and cut ultrasonically under exclusion of water. In order to measure
the temperature a calibrated sensor has been mounted inside of the glass
cylinders.

RESULTS AND ANALYSIS

The absorption coefficient α of different silica based glasses at room
temperature is displayed in Fig. 1 on a double-logarithmic scale: it ranges
over two orders of magnitude between the low absorbing fused silica (Ref. 5)
(S:SiO$_2$) and the high absorbing soda modified silica (SS:Na$_2$O·3SiO$_2$).

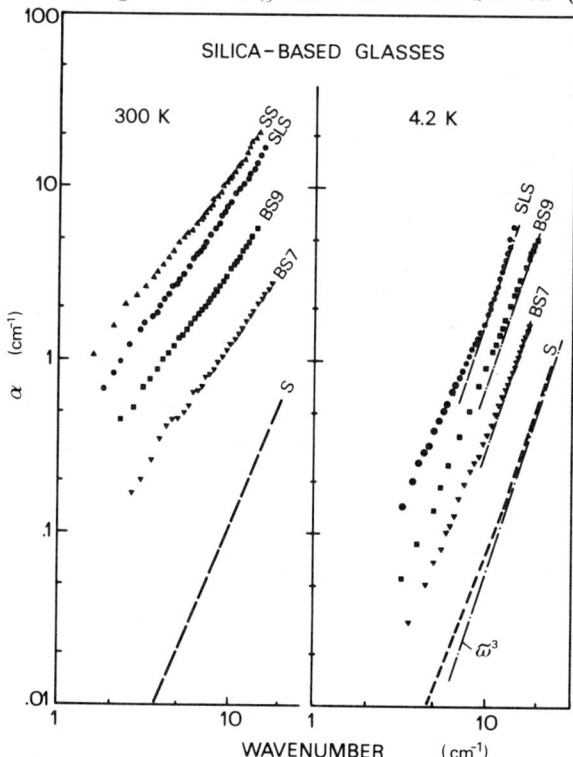

Fig. 1 Absorption coefficient vs. wavenumber ($\tilde{\omega}$) for
silica-based glasses; SS: Na$_2$O·3SiO$_2$, SLS: 17wt%Na$_2$O-
9wt%CaO-72wt%SiO$_2$, BS9: 13wt%B$_2$O$_3$-4wt%Na$_2$O-82wt%SiO$_2$,
BS7: 28wt%B$_2$O$_3$-2wt%Ti$_2$O-70wt%SiO$_2$, S: SiO$_2$ with low
OH$^-$ content (Ref. 5)

No marked change in the character of the frequency dependence occurs
between 300 K and 30 K, although the absolute value of α decreased considerably. A pronounced change in the frequency dependence results with further
reduction of the temperature. At a temperature of about 9 K the absorption
coefficient at any frequency is at a minimum and below this temperature
(T_{min}) the absorption coefficient increases with decreasing temperature
down to the lowest temperature measured (0.4 K). At T_{min} the absorption
coefficient varies as the cube of the frequency for all glasses measured.

The observed behavior indicates a temperature independent absorption
and several temperature dependent absorption mechanisms contributing to
the entire vFIR absorption. The temperature independent contribution can
be explained by the interaction of light with the Debye phonons of the
solid via the elastic strain field of the 2LS. This would produce a FIR
absorption proportional to the phonon density of states. With suitable
correlation (Ref. 6,7) among the centers a power law $\alpha \propto \omega^4$ can be derived
for the background absorption at low frequencies.

The temperature dependence of the absorption coefficient is given in Fig. 2
for a highly absorbing glass.

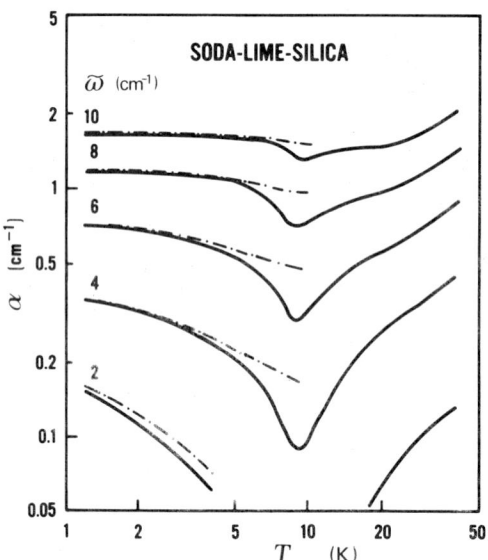

Fig. 2 Temperature dependence of the absorption
coefficient of the soda-lime-silica glass: SLS.
The curves are given for different wavenumbers.

It is obvious that two different processes are responsible for the observed
minimum in the absorption coefficient. The anomalous behavior at the
lowest temperatures (the absorption increases with decreasing temperature)
can be explained by a resonant mechanism. This anomalous absorption is
attributed to the existence of the 2LS, whose density distribution can
thus be probed in the vFIR region. The "tunneling model" (Ref. 2) predicts
an absorption proportional to the density distribution and the population

difference of the 2LS $\alpha_{res} = (4\,\pi^2/\sqrt{\epsilon}\,c)\;n_e\mu'^2\;\omega\;\tanh\,(\hbar\omega/2kT)$

c is the velocity of light and ϵ is the dielectric constant. n_e, the density of states coupling to the electric field is initially assumed to be constant. μ' denotes the induced electric dipole moment of the 2LS. The dashed dotted lines in Fig. 2 represent a fit $(n_e\mu'^2 = 9\times10^{-4}\;(\text{cgs}))$ to the experimental data for the anomalous absorption only. The increasing absorption with increasing temperature is interpreted as a relaxation mechanism among the 2LS and was first used to explain the ultrasonic absorption at low temperatures (Ref. 8).

More accurate measurements of the temperature dependence were possible by varying only the temperature of a given sample, thereby compensating for the temperature-independent background absorption. The temperature dependent part of the absorption coefficient can be expressed as $\alpha(\omega,T) = \alpha_{RES}(\omega,T) + \alpha_{REL}(\omega,T)$. The resonant tunneling contribution is modified in order to account for a decreasing energy dependent quantity $n_e\mu'^2$ in the following way $\alpha_{RES} = (1 - \omega^3/\omega_o^3)\alpha_{res}$. This is a simple approximation, where ω extends only to $\omega_o = 2\pi c \times 15\;\text{cm}^{-1}$. The relaxation process assigned to phonon-assisted tunneling can be written as

$$\alpha_{REL} = A'\int_o^\infty dE\;\frac{\exp(E/kT)}{kT\,(1+\exp(E/kT))^2}\;\frac{\omega^2\tau}{1+\omega^2\tau^2}$$

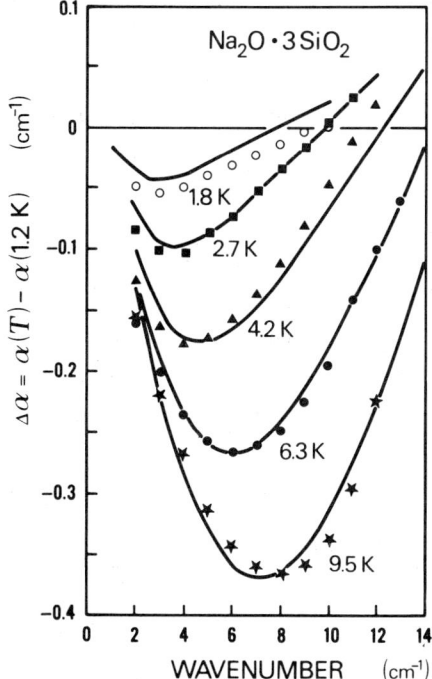

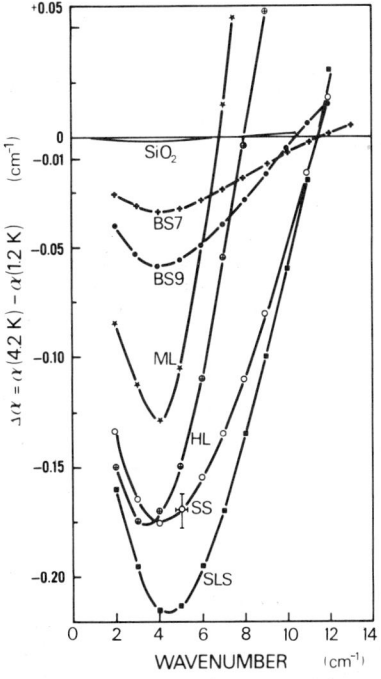

Fig. 3 Absorption difference vs. wavenumber for the soda-silica glass.

Fig. 4 Absorption difference vs. wavenumber for various silicate glasses. ML: 60wt% PbO-10wt% BaO-30wt% SiO_2, HL: 83wt% PbO-10wt% B_2O_3-6wt% SiO_2.

The relaxation time τ is difficult to account for and we will use an approximate behavior only. First we consider a one phonon process at not to low temperatures with $\tau \approx \tau_1/T$, which is independent of the energy E. Second, we simplify again by assuming that the remaining integral is only slightly temperature dependent. In Fig. 3 $\Delta\alpha = \alpha(T) - \alpha(T_R)$ is plotted, T_R is a reference temperature. The experimental data points are compared with a fit of the above model. The parameters are $n_e \mu'^2 \approx 10^{-3}$ (cgs) for the resonant part, the relaxation strength $A = .044/2\pi c$ and $\tau_1 = .135/2\pi c$ for the relaxation process. Although many parameters have been used to fit the data, it should be noted that two different absorption processes have to be considered.

So far we have presented absorption difference data for two highly absorbing silicate glasses only. Fig. 4 displays the diversity in the anomalous properties. Adding network modifiers can change the induced or permanent dipole moment drastically. It should be noted that the lead glasses exhibit a different relaxation behavior.

CONCLUSION

The present results show that an anomalous absorption governs the low temperature dielectric behavior and is characteristic of the disordered state. Although our results support the 2-level tunneling model, a refinement is needed for a full quantitative agreement with the experimental results. The results indicate also that the density of states decreases rapidly for energies corresponding to the upper bond of our vFIR range.

I would like to acknowledge helpful discussions with Prof. A.J. Sievers. This work was supported by the National Science Foundation under contract No. DMR76-81083 through the Cornell University Materials Science Center, Report No. # 2958.

REFERENCES

1. R.O. Pohl, and G.L. Salinger, The anomalous thermal properties of glasses at low temperatures, Ann. N.Y. Acad. Sci. 279, 150 (1976).
2. P.W. Anderson, B.I. Halperin, and C.M. Varma, Anomalous low-temperature thermal properties of glasses and spin glasses, Phil. Mag. 25, 1 (1972); W.A. Phillips, Tunneling states in amorphous solids, J.Low. Temp. Phys. 7, 351 (1972).
3. P.W. Anderson, Model for the electronic structure of amorphous semiconductors, Phys. Rev. Lett. 34, 953 (1975).
4. M. von Schickfus and S. Hunklinger, The dielectric coupling of low-energy excitations in vitreous silica to electromagnetic waves, J. Phys. C 9, L439 (1976).
5. K.K. Mon, Y.C. Chabal, and A.J. Sievers, Temperature dependence of the far-infrared absorption spectrum in amorphous dielectrics, Phys. Rev. Lett. 35, 1352 (1975).
6. E. Schlömann, Dielectric losses in ionic crystals with disordered charge distributions, Phys. Rev. 135, A413 (1964).
7. U. Strom and P.C. Taylor, Temperature and frequency dependence of the microwave absorption in amorphous materials, AIP Conf. Proc. 31, 273 (1976).
8. J. Jäckle, On the ultrasonic attenuation in glasses at low temperatures, Z. Phys. 257, 212 (1972).

XPS STUDY OF SODIUM OXIDE IN AMORPHOUS SiO$_2$*

B. W. Veal and D. J. Lam
Argonne National Laboratory, Argonne, Illinois   60439

ABSTRACT

X-ray photoemission spectroscopy (XPS) studies have been undertaken on amorphous SiO$_2$ and on silicate glasses containing varying amounts of Na$_2$O. The XPS results enable one to distinguish the "bridging" oxygens linking adjacent silicon atoms from the "nonbridging" oxygens that appear in the glass when Na$_2$O is added. Charge readjustment with Na$_2$O addition can also be qualitatively monitored.

INTRODUCTION

Based on a detailed x-ray diffraction study, B. E. Warren (1), in the late 1930's, proposed a model for the structure of sodium silicate glass. The proposed structure is a continuous network of SiO$_4$ tetrahedra with the sodium ions fitting in a random manner in the large holes (2). Because of the extra oxygen atoms introduced with the Na$_2$O, not all the oxygens are joined to two silicon atoms as they are in fused silica. Some of the oxygen atoms are bonded to only one Si atom. These are usually termed "nonbridging" oxygens; and those linking two silicons are termed "bridging" oxygens. As Na$_2$O is added to the SiO$_2$ matrix, the number of nonbridging oxygen atoms should appear in direct proportion to the number of sodium atoms added. For each sodium atom there is one nonbridging oxygen atom. Since the bonding of the bridging and nonbridging oxygen atoms is markedly different, the local charge distributions and hence the XPS binding energies of the oxygen core levels of the bridging and nonbridging oxygen atoms should be different. Furthermore, it would be expected that the core-level binding energies of the silicon atoms located at the center of the tetrahedra might also be sensitive to the number of near-neighbor nonbridging oxygen atoms.

EXPERIMENTAL AND DISCUSSION

The glasses were prepared by melting sodium oxide or sodium carbonate with silica in air in flowing oxygen or in vacuum. Glasses with Na$_2$O concentration higher than 33% were melted in flowing oxygen, quenched between stainless plates and then annealed in the temperature range between 400–550°C to relieve the quenching stress. After preparation, samples were stored in a dessicator placed in an oxygen-restricted glove box environment. Prior to use they were ground on dry emory paper to the appropriate shape for insertion into the spectrometer. Because of the hygroscopic tendency of the Na$_2$O rich glasses, care was taken to keep the glasses away from moisture or extended exposure to atmospheric humidity. After insertion into the spectrometer, the glass surface was scraped with an alumina rod scraper to eliminate the contaminated surface layer that had been exposed to air.

_____

*Work supported by the U. S. Department of Energy.

XPS data were taken on a Hewlett-Packard 5950A spectrometer equipped with an on-line minicomputer supplied by Nicolett Instruments, Inc. With computer control, different spectral regions can be acquired "concurrently". This is an important capability for examining relative compositions of different glass constituents. One can simultaneously monitor several core lines corresponding to different elements without concern for time dependent variations in spectrometer signal intensity. Essentially concurrent data acquisition is obtained since repetitive scans of a single line are always needed for adequate signal-to-noise ratios. Two (or more) spectral regions are concurrently obtained by recording the individual scans of the different regions in alternate sequence.

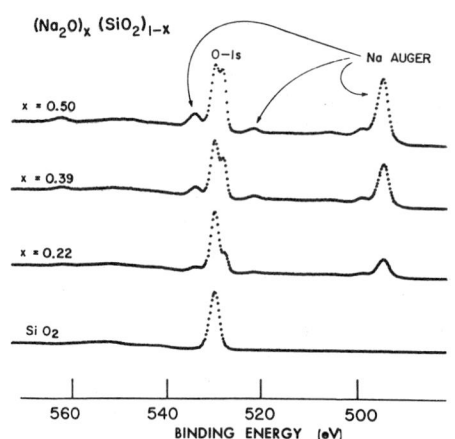

Fig. 1 *The O 1s XPS spectra for SiO₂ and several (Na₂O)ₓ(SiO₂)₁₋ₓ glasses. The low binding component of the O 1s peak scales with sodium concentration.*

Figure 1 shows the O 1s spectra for amorphous SiO_2 and several $(Na_2O)_x(SiO_2)_{1-x}$ glasses. Nearby sodium Auger lines for the sodium silicate glasses are also shown. With the addition of Na_2O to SiO_2, a new peak appears on the low-binding-energy side of the O 1s peak. This peak scales with sodium concentration, as would be expected if the Na_2O were entering the lattice of SiO_2 in the manner suggested by Warren.

In order to obtain quantitative information from the O 1s line, it is helpful to first subtract out the nearby Na Auger structure. This is accomplished by subtracting, on a point-by-point basis, the appropriate portion of an independently measured NaCl spectrum.

The O 1s doublet is then fitted with a sum of two displaced Gaussians to determine the relative intensities and separations of the O 1s component peaks. The half-widths of the two peaks were kept the same for a specific sample. Gaussian fitting parameters thus include the two peak positions, the two peak-intensities and the single linewidth. Figure 2 shows the measured O 1s peak intensities versus x for a series of sodium silicate glasses. For a given glass sample, the O 1s intensities are normalized to the measured intensity of the Si 2p line of the same sample.

For the $(Na_2O)_x \cdot (SiO_2)_{1-x}$ series, the model as described above predicts that, with the addition of Na_2O, the bridging oxygen concentration of SiO_2 is reduced by the number of added units of Na_2O. Thus, the ratio of the bridging oxygen intensity, I_b, to the silicon intensity, I_{si}, is

$$I_b/I_{si} = \frac{2(1-x)-x}{1-x} = \frac{2-3x}{1-x} .$$

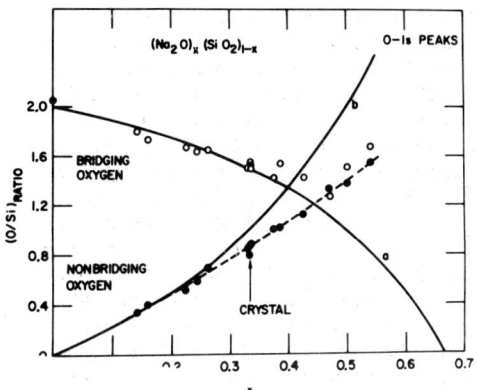

Fig. 2 *The intensities of the bridging and nonbridgin O 1s core lines plotted vs. Na₂O concentration.*

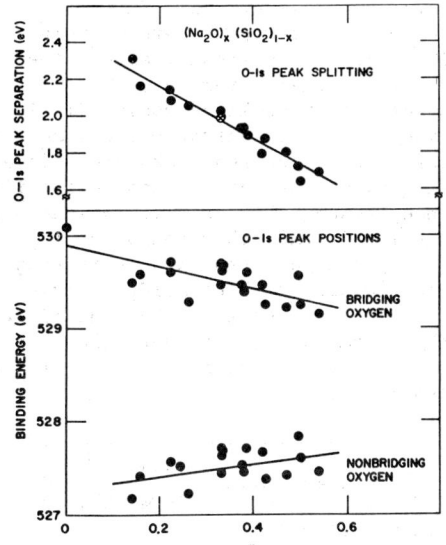

Fig. 3 *Upper graph shows the peak splitting of the O 1s line. The lower curve shows the binding energies of the two components.*

Similarly, the nonbridging oxygen concentration increases at twice the rate that Na_2O is added and is simply proportional to the sodium concentration. The ratio of the nonbridging oxygen intensity, I_n, to the silicon intensity is

$$I_n/I_{Si} = \frac{2x}{1 - x}.$$

These expressions for the bridging and nonbridging oxygen intensities are shown as the solid lines in Fig. 3. Curve (a) corresponds to the bridging oxygen concentration, and curve (b) corresponds to the nonbridging oxygens produced by the presence of Na_2O. The model predicts that, at x = 0.67, the process terminates and all oxygen sites would be nonbridging. For small values of x, the model predictions are nicely confirmed by the XPS data. A check on the validity of the interpretation is provided by the x = 0.33 sample, which can be prepared as an ordered crystal with known crystal structure. For the crystal, one nonbridging oxygen per tetrahedral unit exists. Within experimental error, the crystal result agrees with the model prediction and is indistinguishable from the amorphous x = 0.33 sample. For large values of x, however, the experimental results deviate dramatically from the predicted behavior. The high x samples are more difficult to prepare and are very hygroscopic so that serious experimental errors could be encountered in those data. If, however, the deviations at high x from the model results are confirmed, then it must be that Na_2O in high concentrations enters the lattice in some unknown manner.

Figure 3a shows the separation of the bridging and nonbridging O 1s lines as determined from the Gaussian fits and Fig. 3b shows the binding energies of the component peaks as referenced to the sodium Auger peak taken to occur at 992 eV. The open circle with a cross represents the sodium disilicate crystal results. The Auger peak position is not insensitive to chemical changes in the sample (3) and thus does not provide an ideal internal reference. However, as a function of x, shifts in the Na 2s position relative to the Auger peaks are systematic and small, about 0.4 eV for $0.1 \lesssim x \lesssim 0.6$. For our purposes, either the sodium Auger or Na 2s provides a convenient internal reference. The O 1s of SiO_2 was referenced to the Si 2p line of the crystalline sodium disilicate, used as a secondary standard. As increasing concentrations of Na_2O are added to the $(Na_2O)_x \cdot (SiO_2)_{1-x}$ glass, the two oxygen peaks come closer together with the bridging oxygen peak showing the largest apparent shift. (If one utilized the Na 2s reference, essentially no shift of the nonbridging oxygens is detected while the shift of the other component is enhanced.) The addition of a nonbridging oxygen to a tetrahedral unit thus appears to be reflected in the bonding of the remaining three bridging oxygens. (For small x, unaltered tetrahedra will also be present so that one might expect to observe line broadening with increasing x. However, the electrical conductivity (ionic) of the glasses increases with x and a corresponding narrowing of all the spectral lines is observed.)

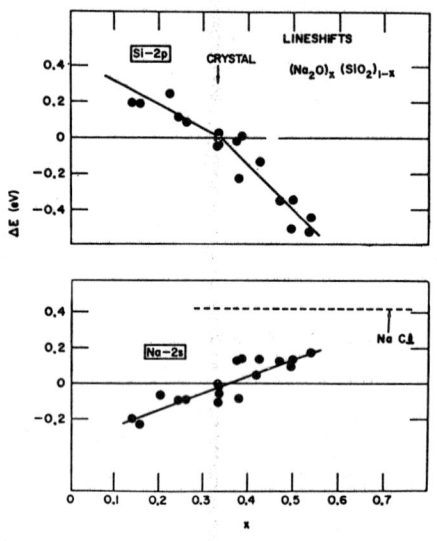

Fig. 4 Upper graph shows the Si 2p binding energy shift relative to $Na_2O \cdot 2SiO_2$ (positive ΔE indicates a shift to higher binding energy). The lower graph shows the Na 2s shift.

A similar effect occurs with the Si atom. Figure 4a shows the Si 2p binding energy shift relative to the crystalline $Na_2O \cdot 2SiO_2$ sample (x = 0.33). The positive sign of ΔE represents a shift to higher binding energy. Again, the data for each sample were referenced to the Na Auger. For increasing x, a substantial shift to lower binding energy is observed. (This shift is even larger if the Na 2s reference were used.) Thus, like the bridging oxygen, the silicon atom, at the tetrahedron center, senses the presence of an adjacent nonbridging oxygen. The shift of both the Si 2p and the bridging O 1s is consistent with the view that electronic charge is transferred toward these atoms with increasing x. The electron contributed (ionically) by the sodium atom to the nonbridging oxygen apparently enables that oxygen to somewhat relax its attactive potential for the electrons associated with the other neighbors of the tetrahedron. Fig. 4b shows the small Na 1s shift relative to the Na Auger.

Significant band structure effects are also occuring in the valence band region as Na_2O is added to the silicate glass. Figure 5 shows valence band data for SiO_2 and several sodium silicates (4). These data are scaled to the silicon concentration. As Na_2O is added, new structure emerges near 5 eV and 10 eV and a substantial increase in intensity is observed for approximately the first 7 eV below E_F. Since the data are normalized to silicon concentration, the regions of significantly increased intensity must be predominately associated with O 2p orbitals. The Si 3s thus appears to contribute most substantially near the bottom of the valence band. Valence band shape changes apparently result from the presence of added nonbridging oxygen atoms.

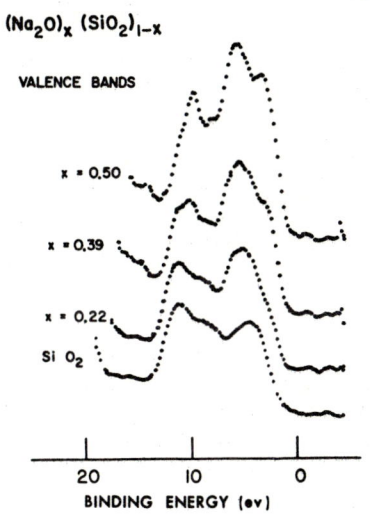

$(Na_2O)_x (SiO_2)_{1-x}$

VALENCE BANDS

x = 0.50

x = 0.39

x = 0.22

Si O_2

20   10   0
BINDING ENERGY (ev)

This study has yielded direct evidence that Na_2O added to the SiO_2 matrix enters the lattice in such a way as to break Si-O-Si linkages, replacing the bridging oxygen site with two nonbridging oxygen atoms. For high x concentrations, however, the preliminary results indicate that Na_2O enters the lattice in some other (as yet undefined) manner. Significant charge readjustment is apparent for both silicon and the bridging oxygen sites as increasing amounts of Na_2O are added.

Fig. 5   Valence band structure for SiO_2 and sodium silicate glasses.

REFERENCES

1.   B. E. Warren and J. Briscoe,J. Am. Ceram. Soc. 21, 259 (1938) and references therein.

2.   See S. Urnes in "Modern Aspects of the Vitreous State, Vol. I", (J. D. Mackenzie, ed.) pp. 10-37, Butterworths, London, for a critical commentary on Warren's work and a discussion of the possibility of phase separation and clustering.

3.   C. D. Wagner, Analytical Chemistry, 47, 1203 (1975).

4.   XPS data for amorphous and crystalline SiO_2 have been reported by B. Fischer, R. A. Pollak, T. H. DiStefano, and W. Grobman, Phys. Rev. B 15, 3193 (1977).

MODIFICATION OF SiO$_x$

Stanford R. Ovshinsky, Krishna Sapru, and Krystyna Dec
Energy Conversion Devices, Inc., Troy, Michigan 48084 U.S.A.

ABSTRACT

We have modified the electrical conductivity and the optical gap of amorphous SiO$_x$ ($x \approx 1.6$) by the addition of various amounts of W, Ni, and Li. The effect of the transition metal elements is quite different from the effect of Li. Comparison of the data for the Li-modified material with that for the unmodified SiO$_x$ strongly suggests that the predominant source of carriers in the latter is due to the nonbridging oxygen atoms.

INTRODUCTION

Although the structure and optical properties have been well studied (1), little is known about the electrical transport in SiO$_x$, including SiO$_2$. The reason for this is the inherent dielectric nature of SiO$_2$. We report here on the enhancement of the bulk conductivity of SiO$_x$ glass brought about by the introduction of controlled amounts of various modifier materials, such as W, Ni, and Li (2). This is part of our ongoing study to elucidate the conduction mechanisms of various amorphous semiconductors (3).

EXPERIMENTAL DETAILS

Films of SiO$_x$ of desired Si/O ratio can be prepared by evaporation or rf sputtering. The electrical properties of these films depend on their composition, which is partly determined by the method of preparation. SiO$_2$ films can be prepared by adding a few per cent of oxygen to the argon sputtering gas. For our purpose of modification, the Si/O ratio is not critical. We have prepared SiO$_x$ films where $x \approx 1.6$ so that optical and electrical measurements can be made with relative ease.

The films for the present work were prepared by rf sputtering. Prior to deposition the chamber was evacuated to 5×10^{-6} torr. A target of SiO$_2$, 3-1/2" diameter was used. Argon pressure of $6\mu m$ and a power density of $1W/cm^2$ resulted in deposition rates of 50-80 Å/min. For modification, small discs of the modifier were glued to the face of the cathode. By varying the area covered by the modifier, the composition was controlled. This technique enabled us to incorporate into the host matrix material a variety of atomic species for modification in a controlled manner. Films of 0.5-2.5 μm thickness were prepared. Either soda-free galss or quartz was used for substrates.

RESULTS

Figure 1 shows the spectral dependence of absorption coefficient of unmodified SiO_x as well as SiO_x modified with different atomic percentages of W.

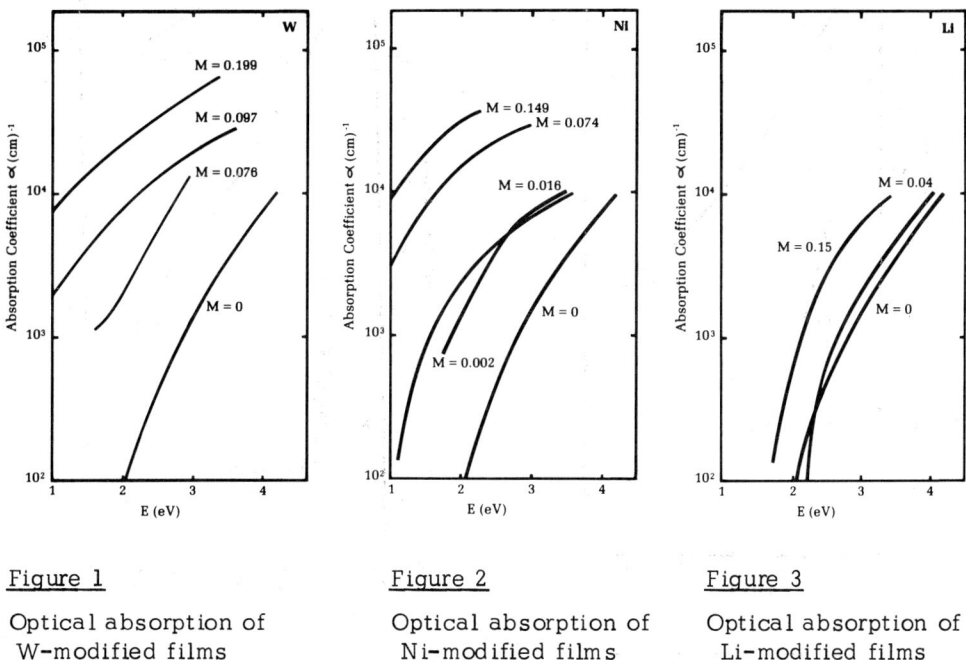

Figure 1

Optical absorption of
W-modified films

Figure 2

Optical absorption of
Ni-modified films

Figure 3

Optical absorption of
Li-modified films

Figures 2 and 3 show similar curves for Ni and Li. The unmodified film has an
energy gap (determined as that photon energy corresponding to an absorption
coefficient of 10^4 cm^{-1}) of 4.2 eV; the gap decreases for increasing amounts
of modifier materials, depending upon their ability to bond structurally. The
corresponding change in electrical conductivity with temperature is shown in
Figures 4, 5, and 6 for W, Ni, and Li, respectively. The data is summarized
in Table I.

Table I

MODIFIER	$\sigma_{RT}(\Omega cm)^{-1}$	ΔE(eV)	E_{opt}(eV)	σ_0(cm)$^{-1}$	$N(E_F)/cm^3/eV$
W(%)					
7.6	3×10^{-11}	0.28	2.8	2×10^{-8}	7×10^{17}
9.7	4×10^{-7}	0.17	2.25	4×10^{-4}	2×10^{18}
19.9	6×10^{-4}	0.09	1.25	2×10^{-2}	3×10^{19}
Ni(%)					
0.2	2×10^{-14}	0.52	3.5	2×10^{-5}	—
1.6	5×10^{-13}	0.36	3.5	9×10^{-7}	—
7.4	7×10^{-7}	0.14	1.69	2×10^{-4}	2×10^{18}
14.9	9.5×10^{-5}	0.11	1.05	7.7×10^{-3}	3×10^{19}
Li(%)					
4.5	2.4×10^{-12}	0.79	4.1	1.4×10^2	
15	5.3×10^{-9}	0.51	3.5	2.1	
Unmodified SiO$_{1.5}$					
	$\sim 10^{-15}$	0.79	4.2	4.8×10^{-3}	

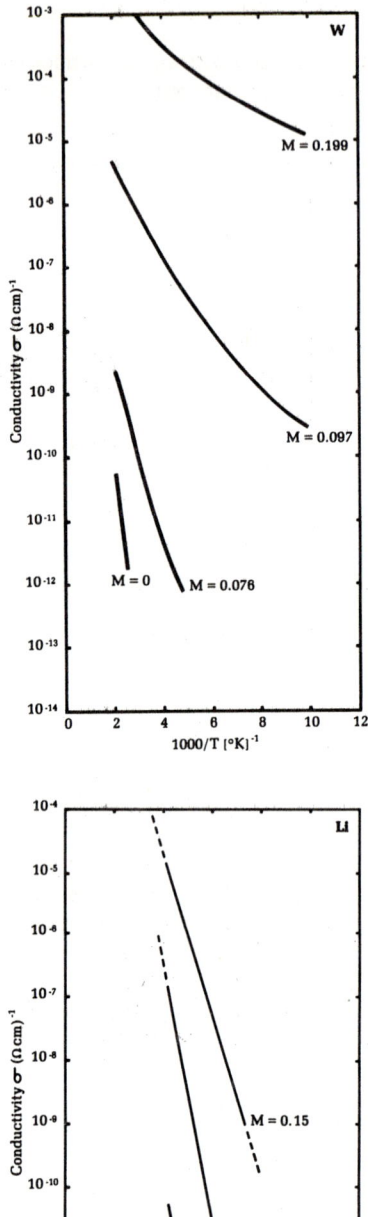

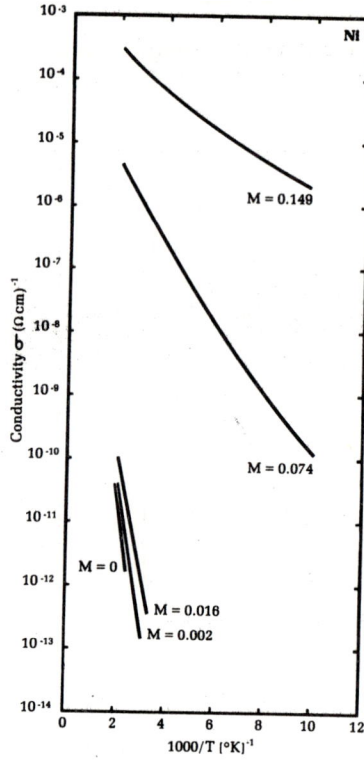

Figure 4 – Electrical conductivity of W–modified films

Figure 5 – Electrical conductivity of Ni–modified films

Figure 6 – Electrical conductivity of Li–modified films

Unmodified SiO_x has a conductivity which obeys the relation, $\sigma = \sigma_0 \exp(-\Delta E/kT)$ over the entire range in which measurements were possible. At high concentrations of W (M> 0.09) and Ni (M> 0.07), deviations from this behavior are evident at low temperatures. Here, the conduction data could be well fit by the Mott relation, $\sigma = \sigma_0 \exp[-(T_0/T)^{1/4}]$. In this case, we can conclude that conduction occurs primarily by means of variable-range hopping near the Fermi energy. The density of localized states at the Fermi energy, given as $N(E_F)$ in Table I, was estimated from the relation $T_0 = \alpha^3/kN(E_F)$, where α^{-1}, the radius of the localized states near E_F, was taken to be about 10Å. In the less modified films, the conductivity was too small to measure at low temperature, so no $T^{1/4}$-law behavior was observed. However, the small values of σ_0 in these cases also suggests the predominance of hopping conduction. We shall return to this point in the next section.

In the sample modified with Li, very different behavior was found. No region in which variable-range hopping predominates was evident. Instead, the activation energy at Li concentrations (~ 4%) remains the same as in the unmodified films, despite a significant increase in σ_0. At higher concentrations, the activation energy decreases somewhat, in a manner roughly the same as the optical gap.

Modification strongly increases the room-temperature conductivity of SiO_x. In the unmodified material, σ_{RT} was too low to measure, although an extrapolated value of $10^{-15}\Omega^{-1}cm^{-1}$ can be obtained from the high-temperature data. Modification with about 20% W increases this value by a factor in excess of 10^{11}.

DISCUSSION

When transition-metal modifiers are added, we modify as well as introduce states in the gap. The electronic states will depend on the equilibrium charge states of the transition-metal atoms. This is determined by the manner in which the ions bind in the amorphous structural matrix as well as by the interaction of the d-bands with the local environment. The Fermi level will almost certainly lie within a narrow d-band (4), leading to a large density of states at E_F, and the predominance of variable-range hopping.

On the other hand, when Li is added, the corresponding electronic state can be expected to be just below the conduction-band edge, the same as it is in Li-doped Si. However, in the case of SiO_x, it is almost certain that the reaction,

$$Li + O \rightarrow Li^+ + O^-$$ (1)

is exothermic. The O^- ions then form nonbridging oxygens in the SiO_x matrix (5). The exothermic nature of Eq. (1) results in a negative effective correlation energy, which tends to pin the Fermi energy. An important conclusion from our data is that the nonbridging oxygens are the primary source of the conducting electrons in unmodified SiO_x. This follows from a comparison of the data on unmodified and Li-modified samples. For M ≤ 4%, the activation energy for conduction is essentially constant, despite large changes in σ_0.

If this were due to a change in the conduction mechanism, for example, from hopping in localized states near an energy E_A to bandlike conduction above the conduction-band mobility edge, E_c, then the activation energy must increase by $E_c - E_A + W$, where W is the hopping energy. Thus, we can conclude that the increase in σ_o upon the introduction of Li is due to an increase in the density of donors, N_d. It is clear from our previous discussion that the source of the conducting carriers must thus be the nonbridging oxygen atoms.

Since the value of σ_o for M≈4% is as high as it is ($\approx 140\Omega^{-1}cm^{-1}$), we can conclude that conduction is bandlike. Hughes (6) has shown that the mobility of electrons in the conduction band of amorphous SiO_2 is about $10cm^2/V$-sec at $550°K$. If we assume that one nonbridging oxygen is created by every Li atom introduced, then we can apply statistical consideration to conclude that the effective mass of electrons in amorphous SiO_x is about 0.7 times the free electron mass. The electronic structure of Li-modified SiO_x is that of a compensated semiconductor with ionized Li donor levels just below the conduction-band edge and nonbridging oxygen levels about 1.6 eV further below. The Fermi energy is 0.8 eV above the latter. We conclude that unmodified SiO_x has bandlike conduction. When modified with transition metals, the conduction mechanism becomes hopping; with Li it remains bandlike.

ACKNOWLEDGEMENT

We wish to thank Professor David Adler for his suggestions.

REFERENCES

(1) M.V. Coleman, and D.J.D. Thomas, The Structure of Silicon Oxide Films, Phys. Stat. Sol. 22, 593 (1967); H.R. Philipp, Optical Properties of Non-Crystalline Si, SiO, SiO_x, SiO_2, J. Phys. Chem. Sol. 32, 1935 (1971).
(2) S.R. Ovshinsky, Amorphous and Liquid Semiconductors, edited by W.E. Spear (Univ. of Edinburgh, Scotland) p. 519, 1977; R. Flasck, et al., Amorphous and Liquid Semiconductors, edited by W.E. Spear (Univ. of Edinburgh, Scotland) p. 524, 1977.
(3) S.R. Ovshinsky, and I.M. Ovshinsky, Analog Models for Information Storage and Transmission in Physiological Systems, Mat. Res. Bull. 5, 681 (1970)[includes earlier references]; S.R. Ovshinsky, Structure and Excitations of Amorphous Solids, edited by G. Lucovsky and F.L. Galeener (AIP, New York) p. 31, 1976.
(4) S.R. Ovshinsky, and D. Adler, Local Structure, Bonding, and Electronic Properties of Covalent Amorphous Semiconductors, to be published in Contemporary Physics, March, 1978.
(5) N.F. Mott, Amorphous and Liquid Semiconductors, edited by W.E. Spear (Univ. of Edinburgh, Scotland) p. 497, 1977.
(6) R.C. Hughes, Charge-Carrier Transport Phenomena in Amorphous SiO: Direct Measurement of the Drift Mobility and Lifetime, Phys. Rev. Lett. 30, 1333 (1973).

INTRINSIC SURFACE PHONONS IN POROUS GLASS

C. A. Murray and T. J. Greytak
Massachusetts Institute of Technology
Cambridge, Massachusetts 02139 U.S.A.

In a recent letter (1), we reported the observation of intrinsic surface phonons in silica. Briefly, we have performed Raman scattering and infrared reflectivity measurements on samples of porous Vycor glass, relying on its large ratio of surface to bulk atoms (approximately 1:10) in order to emphasize features characteristic of the surface.

Porous Vycor glass (2) is manufactured from a phase separated borosilicate glass from which the borate phase, initially comprising about thirty percent of the volume, has been leached out to form a network of 40 Å diameter pores throughout the sample. The remaining glassy matrix is 96% SiO_2 with about 3% B_2O_3 impurities. We have compared the spectra of porous Vycor, preleached Vycor and condensed Vycor with that of pure silica. Although we have identified features in the Vycor spectra which are due to impurities, these will not be discussed here. The important observation is that none of the features which we identify with the intrinsic silica surface phonons are observed in the spectra of preleached or condensed Vycor.

Vitreous silica is a bonded solid in which the local order of the tetrahedral SiO_4 unit is preserved. Thus a good model for the surface is a localized broken bond defect with an unsatisfied Si or O bond. Calculations by Laughlin and Joannopoulos (1), using cluster Bethe-lattice techniques with force constants and matrix elements appropriate for bulk SiO_2, suggest that some of the vibrations associated with these surface broken bonds are not significantly broadened by their interactions with the bulk so that they may appear as well defined peaks in the spectra. In particular, for the oxygen terminated surface they find a sharp mode at approximately 850 cm^{-1} in which the silicon and surface oxygen vibrate out of phase. This stretching mode is strongly Raman active but weak in the infrared. The corresponding wagging mode appears around 300 cm^{-1} in their calculations and is extremely infrared active but not Raman active. In addition near 200 cm^{-1}, they find a shift in Raman active states to lower frequencies due to acoustic-like surface vibrations. These predictions are in excellent qualitative agreement with our experimental results. We observe the infrared active mode at 380 cm^{-1} and the Raman active mode at 980 cm^{-1}, as well as a shift of intensity at low frequencies compared to bulk SiO_2 spectra in the Raman spectrum of porous Vycor. The reader is referred to ref. 1 for more discussion of the details, both experimental and theoretical. In this paper, we wish to provide additional experimental evidence that we are observing intrinsic surface modes. We will concentrate on the Raman spectrum and in particular on the surface Si-O stretch mode at 980 cm^{-1}.

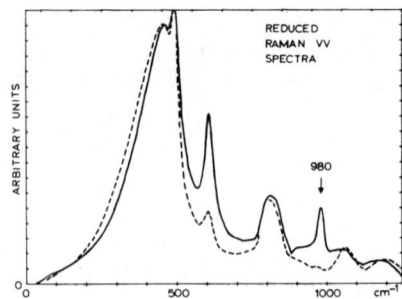

Fig. 1. Reduced Raman spectra of porous Vycor after subtraction of the background (solid line) compared to that of Suprasil (dashed line). The spectra were normalized to be equal at 440 cm^{-1}. Typical noise in the traces is less than the line thickness, although there is an uncertainty of 15% below 100 cm^{-1} due to background subtraction in the porous Vycor trace. The peak at 980 cm^{-1} in the Vycor spectrum is due to an intrinsic surface phonon as discussed in the text.

Figure 1 compares the reduced Stokes Raman spectrum of porous Vycor to that of Suprasil. The reduced spectrum (3) is given by

$$I(\nu) \frac{\nu}{n(\nu,T) + 1} \left(\frac{\nu_0}{\nu_0 - \nu}\right)^4$$

where ν is the Stokes shift from the exiting frequency ν_0, $n(\nu,T)$ is the Bose population factor, and $I(\nu)$ is the observed spectrum with background subtracted. The spectra are normalized to be equal at the 440 cm^{-1} bulk band. Several differences in the two spectra are immediately apparent. Most noticeable are the peaks at 980 cm^{-1} and 607 cm^{-1} in the porous Vycor spectra which are approximately 20 and 5 times larger in integrated intensity than their counterparts in Suprasil.

The peak at 607 cm^{-1} (604 cm^{-1} in Suprasil) has been identified with a broken bond defect (4,5) in silica. It has been shown to grow in intensity with neutron bombardment (4) and decrease in intensity with OH content in the glass (5). We find that about 10^{21} molecules/cc of adsorbed water on the sample of porous Vycor decreases the intensity of this mode by the factor of two. It is not affected to within 5% by adsorbed ammonia. It would be tempting to identify this mode as the vibration of a Si terminated surface where the silicon vibrates in the cage of oxygens as has been suggested in the literature for the bulk mode. This is also a good model for a dangling silicon bond in the bulk. However, Laughlin and Joannopoulos (1) find this dangling Si bond mode strongly infrared active and we do not observe a sharp strong mode in this region in our reflectivity measurements. It also seems unlikely that a silicon dangling bond would remain dangling after the oxygen treatment we have given our samples. Thus we have no definite identification of the peak at this time.

The decrease in intensity around 300 cm^{-1} of the porous Vycor spectrum compared to that of Suprasil we believe to be due to the Raman active surface acoustic modes of Laughlin & Joannopoulos (1). This decrease of intensity should be accompanied by a corresponding increase at lower frequency as it is actually a shift downward in frequency of the phonon density of states caused by surface modes. We observe a 20% increase in the porous Vycor spectra below 150 cm^{-1} compared to both Suprasil and condensed Vycor spectra, however

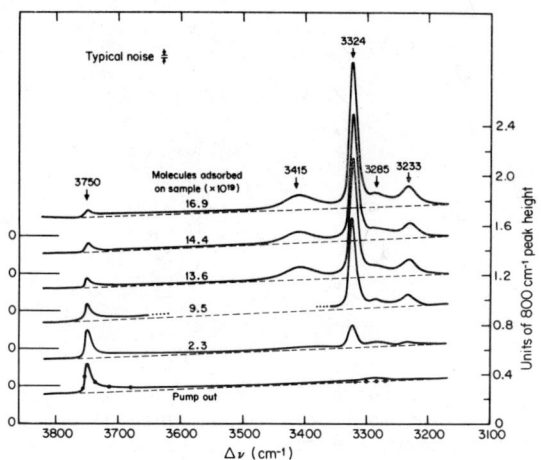

Fig. 2. Raman spectra of porous Vycor in the high frequency region with
increasing amounts of ammonia adsorbed on the sample as marked. The zero for
each trace has been shifted for clarity, and an estimate for the background
under the peaks is shown by a dashed line. The peak at 3750 cm^{-1} is due to
the surface hydroxyl stretching mode of the substrate, while the group of
lines near 3300 cm^{-1} is due to adsorbed ammonia as discussed in the text. The
dots on the lowest curve represent the spectrum of a clean sample, while the
solid line is the spectrum after evacuating ammonia from the sample at room
temperature.

the uncertainty in the intensity below 100 cm^{-1} due to subtraction of the intense elastic scattering in the porous sample is approximately 15%. Note
that the frequency factor used in reducing the spectra de-emphasizes the low
frequency region.

We believe that the sharp mode at 980 cm^{-1} in the porous Vycor spectrum
is due to <u>surface</u> Si-OH stretching vibrations, while the 970 cm^{-1} peak
barely visible in the Suprasil spectrum is due to <u>bulk</u> hydroxyl impurities
bonded to silicon as suggested by Stolen and Walrafen (5). By measuring the
infrared absorption of our Suprasil sample at 2.2μ (6) we determined the bulk
hydroxyl content to be $(3.3\pm.5) \times 10^{19}$/cc. Assuming that the matrix element
for the surface silicon-oxygen stretch mode does not differ appreciably from
that in the bulk and assuming that all the OH is bonded to silicon, we can
determine the number of OH groups in our Vycor sample by scaling the intensities of the Si-OH bands in the two samples. This procedure gives
$(5.2\pm2) \times 10^{20}$/cc hydroxyls in porous Vycor. Our experiments find that both
the silicon-oxygen stretching mode at 980 cm^{-1} and the oxygen-hydrogen stretching mode at 3750 cm^{-1} in the Raman spectrum are affected by the adsorption of
ammonia on the sample, and thus demonstrate that the modes are both surface
features.

The Raman feature due to the O-H stretching mode in the porous Vycor
spectrum at 3750 cm^{-1} is radically different from that in Suprasil. It is
significantly narrower in frequency width, about 10 times more intense in
peak height, and is shifted by 60 cm^{-1} to higher frequency compared with its
bulk counterpart in Suprasil. Furthermore, we find this feature is drastically perturbed by the adsorption of ammonia on the sample in agreement with
previous infrared absorption experiments (1). Figure 2 shows the Raman

312

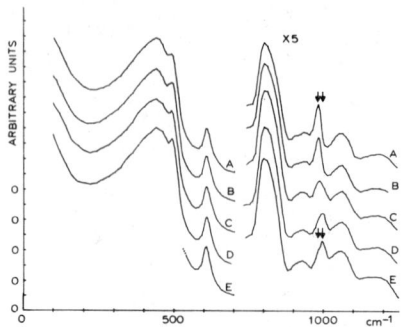

Fig. 3. Stokes Raman spectra of porous Vycor with background subtracted in the low frequency region as increasing amounts of ammonia are adsorbed on the sample as follows: curve A - clean sample, B - 2.0×10^{19} molecules, C - 9.5×10^{19} molecules, D - 13.6×10^{19} molecules, E - 16.7×10^{19} molecules. The sample weighs approximately .5 gram. The arrows point to the modification of the 980 cm^{-1} surface peak due to interaction with the adsorbed molecules as discussed in the text.

spectrum of porous Vycor in the hydroxyl stretching region as ammonia is adsorbed on the sample. The other group of lines in Fig. 2 around 3300 cm^{-1} is due to ammonia vibrations which are perturbed by interaction with the substrate. They will be identified and briefly discussed later in this paper. The substrate OH band at 3750 cm^{-1} decreases in height linearly with the amount of adsorbed ammonia. We can get an upper estimate for the number of OH groups on the surface by assuming each NH$_3$ molecule hydrogen bonds to one OH group and noting that for 1.7×10^{19} molecules of NH$_3$ adsorbed on the sample the OH band is reduced to one-fifth its original intensity. This gives $(7.1\pm1) \times 10^{20}$ OH/cc or $(4.3\pm1) \times 10^{20}$ OH/gm, which is consistent with our less precise Si-O mode intensity ratio estimate. Assuming a surface area of 200 m^2/gm (2), this implies that there are 2 OH/100 Å^2 on the surface, in excellent agreement with determination by chemical means in the literature (7).

In addition to the perturbation of the high frequency OH band with adsorbed ammonia, we see a definite reproducible change in the Si-O stretching peak at 980 cm^{-1}. Figure 3 shows the development of the spectrum in this frequency region. As NH$_3$ is adsorbed on the surface hydroxyl site, the peak at 980 cm^{-1} begins to disappear and a new peak at 997 cm^{-1} grows. This can be interpreted as a shift of $\sim 2\%$ towards higher frequency of the Si-O stretching mode of each surface hydroxyl when it is loaded with NH$_3$. The intensity in this region can be modelled by 2 Gaussian lineshapes, one centered at 980 cm^{-1} with amplitude proportional to the number of unoccupied OH sites (OH 3750 cm^{-1} peak intensity) and the other centered at 997 cm^{-1} with amplitude proportional to the number of unoccupied OH sites.

We will now briefly return to a discussion of the Raman spectrum of adsorbed ammonia in Fig. 2. The peak at 3324 cm^{-1} is the symmetric stretching mode of NH$_3$ shifted to lower frequency by 13 cm^{-1} from the frequency of the gas phase. The magnitude of this perturbation is comparable to that observed in infrared absorption experiments (7). This has been attributed to ammonia hydrogen bonded to an hydroxyl group. The small peak at 3285 cm^{-1} (shift -49 cm^{-1}) may be a more perturbed symmetric stretch vibration due to ammonia chemisorbed on a surface boron site. This line has been observed in

the infrared on porous Vycor (8). This peak remains when the sample is evacuated at room temperature, while the other bands disappear completely and the
OH peak intensity is restored. The two peaks at 3414 cm^{-1} and 3234 cm^{-1} are
shifted by ±90 cm^{-1} from the peak at 3324 cm^{-1} and are definitely associated
with it as sidebands. They have not previously been observed. They may be
associated with a wagging mode of the hydrogen bonded ammonia.

REFERENCES

(1)   R.B. Laughlin, J.R. Joannopoulos, C.A. Murray, K.J. Hartnett, and
      T.J. Greytak, Phys. Rev. Lett. 40, 461 (1978).
(2)   M.E. Nordberg, J. Am. Cer. Soc. 27, 299 (1944).
(3)   R. Shuker and R.W. Gammon, Phys. Rev. Lett. 25, 222 (1970).
(4)   J.B. Bates, R.W. Hendricks, and L.B. Shaffer, J. Chem. Phys. 61,
      4163 (1974).
(5)   R.H. Stolen and G.E. Walrafen, J. Chem. Phys. 64, 2623 (1976).
(6)   D.M. Dodd and D.B. Fraser, J. Appl. Phys. 37, 3911 (1966).
(7)   M.L. Hair, Infrared Spectroscopy in Surface Chemistry (Marcel Dekker,
      New York, 1967) and L.H. Little, Infrared Spectra of Adsorbed
      Species (Academic Press, London, 1966).
(8)   N.W. Cant and L.H. Little, Can. J. Chem. 42, 802 (1964).

The samples of Vycor glass used in our experiments were obtained from
Corning Glass Works, Corning, New York, and the samples of Suprasil from
Amersil Incorporated, Hillside, New Jersey.

POSITRONIUM-SURFACE INTERACTION IN THE PORES OF VYCOR GLASS

S.M. Kim and W.J.L. Buyers
Atomic Energy of Canada Limited, Chalk River, Ontario,
Canada, K0J 1J0

The positronium-surface interaction has been studied experimentally and theoretically. The angular correlation of photons from annihilation of positronium on the internal surfaces of porous Vycor glasses has been measured as a function of pore size in the range 2.2 - 14.0 nm. When the pores were filled with oxygen the angular distribution showed an intense narrow peak, arising from para-positronium annihilation in the pores, superimposed upon a broad distribution. The width of the narrow peak was approximately independent of pore diameter suggesting localization of the positronium in a surface state. The possible states of positronium in a spherical cavity have been studied theoretically as a function of the strength of the surface potential. The positronium-glass interaction that best fits the data was found to be attractive and strong enough to localize the positronium atom near the surface of the pores. The theory gives a good description of the data as shown in Fig. 1 with a strongly attractive surface well, e.g. ~1 eV deep and 0.4 nm wide.

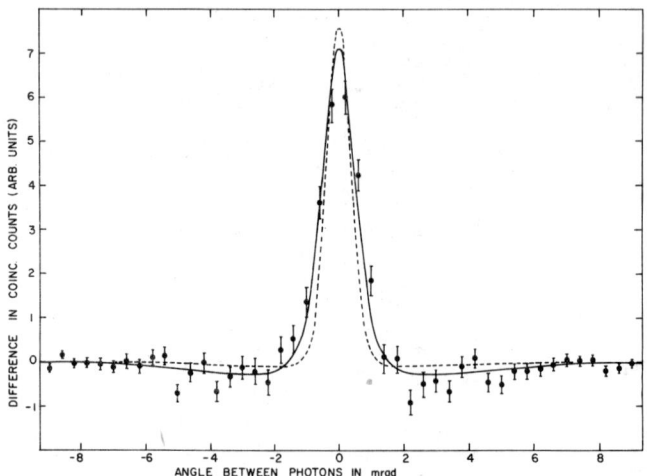

Fig. 1. Theoretical fits to the positronium momentum distribution (difference between the oxygen and vacuum results) for 14.0 nm pores. The full curve was obtained with an attractive surface potential 1 eV deep and 0.4 nm wide and the broken curve without any surface attraction.

The most likely source of the surface attractive potential is the van der Waals force between the positronium atom and the SiO_2 surface molecules. This is small between most atoms but for the highly polarizable positronium atom is much larger. We find that the van der Waals energy between the positronium atom and the SiO_2 molecule of the Vycor glass is about -1.6 eV at a separation of 0.2 nm. In the surface trapping model this configuration is presumably simulated by placing the positronium at the center of a 0.4 nm wide surface well for which our data indicates that an attractive potential of ~1 eV is appropriate. A full report of this work has recently appeared in J. Phys. C 11, 101 (1978).

ION IRRADIATION AND STORED ENERGY IN VITREOUS SiO$_2$

M. Antonini*, A. Manara and P. Lensi
Physics Division, Joint Research Center,
21020 Ispra, Varese, Italy.

ABSTRACT

A large amount of energy release (150 - 250 cal/g) is observed in amorphous silica irradiated with elevated doses of alpha particles and high energy heavy ions. The annealing curve shows two distinct maxima at about 470 and 610°C. The energy stored during irradiation may be attributed to atomic defects, SiO$_2$ defective units, both leading to long range structural distortions, and to local distortions within the Si-O tetrahedra, caused by breakage or modification of the Si-O bond, probably connected with ionizing collisions.

INTRODUCTION

The effects of irradiation on various properties of vitreous silica have been reviewed by various authors (1-3). Recently, it has become of great interest to ascertain the structural and chemical stability of silica based glassy matrices subjected to long term radiation damage, originated by nuclear transmutations and subsequent recoils of heavy radioactive ions, i.e. actinides, incorporated in these materials for waste disposal (Ref.4). Accelerated irradiation tests and post irradiation measurements of stored energy, density changes and leaching rates have been made at various laboratories (Ref.5-7). In order to check the validity of these accelerated tests, a physical model is needed for the radiation damage of silica under heavy irradiation conditions. To this purpose, we have used high energy heavy ions and energetic alpha particles to generate a nominal concentration of displaced atoms of the order of ten per cent or more within a penetration depth of about ten microns.
The determination of stored energy release as a function of the annealing temperature represents a first basic quantity providing some general information on the total amount of damage and

---

*Also: Gruppo Nazionale Struttura della Materia and Istituto di Fisica dell'Università, 41100 Modena, Italy.

on the kinetics of the recovery. Furthermore, stored energy data
obtained after ions irradiation can be compared with analogous
results after neutron irradiation reported by Roux (8) and by Ro
berts et al. (6). To our knowledge, stored energy determinations
after high energy heavy ions irradiation have not previously
been published.

EXPERIMENTAL

The adopted samples were Tetrasil SiO_2 glass disks, containing
a maximum amount of 200 ppm OH groups. The sample dimensions,5mm
dia., 0.1 mm thick, were limited by the size of the beams, by
their small penetration depths and by the thermal gradient developing during irradiation. All samples were first annealed at
600°C for three hours and coated with a 100 Å aluminium layer,to
provide a ground path preventing charge storage and breakdown.
They were then pasted to an aluminium sample holder using a silver conducting adhesive. The heavy ion irradiations were performed at the Variable Energy Cyclotron, A.E.R.E., Harwell, England,
using the irradiation facility described by Worth (9). The irradiation source was a collimated beam of Ni^{+6}ions having an energy of 46.5 MeV. Dose measurements were done by recording the
charge accumulated by the insulated sample holder. Secondary emis
sion of electrons was eliminated by providing a counter potential
of 100 volts. The temperature during irradiations was kept below
15°C by water cooling of the sample. No large temperature gradient was detected at the surface of the sample hit by the beam.
During irradiation, the target was tilted at various angular positions, to provide a uniform distribution of radiation damage
in the total penetration depth, as described by Worth et al.(10).
The 1 MeV alpha particles irradiations were done at the J.R.C.,
Ispra, using a conventional Van De Graaff accelerator, in conditions analogous to those described for heavy ions. Dose and dose
rate parameters adopted during the irradiation are reported in
TABLE 1. The stored energy measurements were performed using DSC
910 and DSC 990 Dupont differential scanning microcalorimeters.
The rate of energy release at a heating rate of 10°C/minute was
recorded between about 100°C and 600°C. Only in one case the upper temperature limit was pushed to 700°C. The sensitivity of the
calorimeter was set at 0.05 millicalories/sec per inch of the recording chart. At this value, the stability of the baseline could
be kept reasonably constant over the explored temperature range.
Eventual changes with temperature of the calibration parameter
were automatically taken into account. Prior to irradiations,the
samples were tested and no structure was found in the DSC spectrum. The behaviour of the specific heat was in agreement with
that reported by the literature.
Two successive runs were performed to detect the energy released
during the annealing of the irradiated samples. A record of the
third run indicated that all the stored energy was released during the first annealing.

TABLE 1   Dose and dose rate irradiation parameters

SAMPLE	RADIATION	TOTAL DOSE $ions/cm^2$	DOSE RATE $ions/cm^2 sec$
$SiO_2(1)$	alpha particles $E = 1$ MeV	1.5×10^{18}	3.0×10^{13}
$SiO_2(2)$ $SiO_2(3)$	Ni^{+6} ions $E = 46.5$ MeV	1.6 5.0 $\times 10^{15}$	4.6 6.9 $\times 10^{11}$

RESULTS AND DISCUSSION

The stored energy release rate is obtained by substracting the spectra recorded during the second annealing from those recorded immediately after irradiation. No change was assumed for the specific heat with irradiation and no change occured of the baseline between the two curves.

Since the specimen could not be uniformly irradiated, the measured values have been normalized to an effective mass, corresponding to the irradiated volume. This procedure requires an accurate knowledge of the ionic penetration depths. The range of heavy ions has been obtained calculating the depth distribution of energy deposition due to electronic collisions and atomic displacements (Ref. 11). A value of 11.2 microns has been found in our conditions. The alpha particle range was determined interpolating the data reported in the tables of Williamson et al. (12).

Typical spectra are represented in Fig. 1 and in Fig. 2. The rate of energy release has a broad maximum at about 470°C. A second hump is also evident in Fig. 2, where a higher temperature was achieved. A maximum in the stored energy release is also observable after high dose neutron irradiation (Ref.6). The most striking feature of the results is however the large amount of total stored energy released per gram, under the assumption that only the fraction of sample subject to irradiation is responsible for the effect. Its value ranges between about 150 and 250 cal/g in the temperature interval from 350 to 520°C. These are much larger values than those obtained in previous measurements, characterized by a larger irradiation temperatures and much lower damage rates. Some tendency towards saturation, giving approximately the same value for the two samples $SiO_2(2)$ and $SiO_2(3)$ (see TABLE 1) irradiated at different doses, is also observed in our results.

Major effects of radiations on amorphous SiO_2, i.e. the compaction of the density (Ref. 3) and the enhancement of the chemical hetch rate (Ref. 13) are interpreted in terms of physical di-

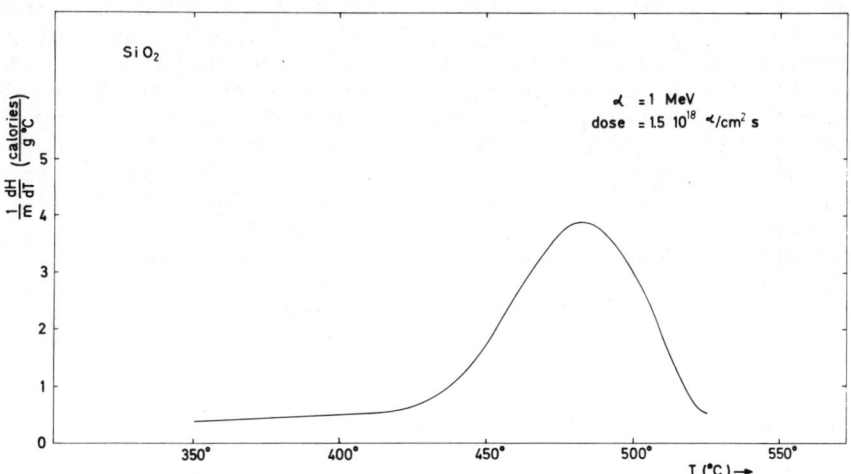

Fig. 1. Rate of energy release after alpha particles
    irradiation

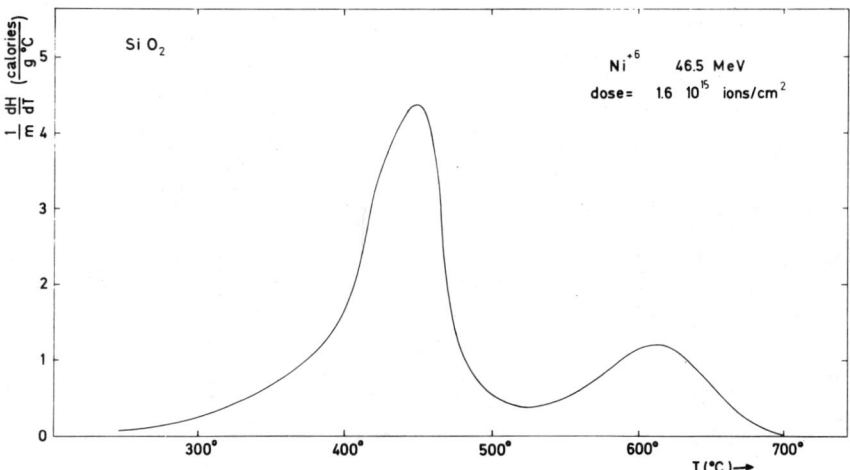

Fig. 2. Rate of energy release after heavy ions
    irradiation

splacements of atoms or of atomic units, leading to a new metastable phase associated with reorientations of the Si-O tetrahedra, and in terms of local distortions caused by cleavage or stra
ining of Si-O bonds associated with ionizing collisions (Ref. 14).
The creation of SiO_2 defective units is in agreement with stored
energy data after neutron irradiation and it seems to account for
the Raman spectra observed by Bates et al. (15). Our results are
also compatible with such a model when assuming that a larger

amount of defects could be present at our smaller temperature. By analogy with the cases of graphite, diamond and silicon carbide, our results might also be compatible with a large number of simple interstitial-vacancy. When assuming a concentration of 5% of atomic defects, a stored energy of 250 cal/g corresponds to about 4 eV per atomic defect, which approximates the Si-O bonding energy. Some structural damage created by ionization collisions, eventually interacting with nuclear collisions, seems compatible with the position of the second hump observed in Fig. 2 (Ref. 14).

In conclusion, the energy stored during irradiation of amorphous silica seems to depend strongly upon the temperature of irradiation and the dose rate. At low temperatures and high dose rates, a large amount of energy anneals below 700°C. The irradiation products can be atomic defects or SiO_2 units and strained or cleaved Si-O bonds. The effects due to atomic displacements and ionization seem to be separable in the energy release spectra. Further measurements are planned at various dose rates.

We wish to thank Dupont De Nemours for providing us with the calorimeters and with technical and scientific assistance.

REFERENCES

(1) E. Lell, N. J. Kreidl, J. R. Hensler, Radiation Effects in Quartz, Silica and Glasses, Prog. Cer. Sci. 4, 1 (1966).

(2) R. Brückner, Properties and Structure of Vitreous Silica. II, J. Non-Cryst. So., 5, 177 (1971).

(3) W. Primak, (1975) The Compacted States of Vitreous Silica, Gordon and Breach, New York.

(4) Management of Radioactive Wastes from the Nuclear Fuel Cycle, Proc. of a Symposium, IAEA, Vienna, 22-26 March 1976, Vol. II.

(5) A. R. Hall, J. T. Dalton, B. Hudson, J. A. C. Marples, Development and Radiation Stability of Glasses for Highly Radioactive Wastes, ibid. pg. 3.

(6) F. P. Roberts, G. H. Jenks, C. D. Bopp, Radiation Effects in Solidified High-Level Waste, Part I, Stored Energy, BNWL-1944 UC 70 (1976).

(7) K. Scheffler, U. Riege, Investigations on the Long-Term Radiation Stability of Borosilicate Glasses against α-Emitters, KFK 2422 (1977).

(8) A. Roux, Energie Emmagasinee dans les Oxydes BeO, MgO, Al_2O_3 et SiO_2 Irradies aux Neutrons, CEA-N-1171 (1969).

(9) J. H. Worth, The Uses of Cyclotrons in Chemistry, Metallurgy and Biology, Butterworths, London (1969).

(10) J. H. Worth, P. A. Clark, J. A. Hudson, An Ion Beam Target that Distributes Irradiation damage Uniformly in Depth with Temperature Control, J. Brit. Nucl. En. Soc., 10, 329 (1971).

(11) M. D. Matthews, Calculations of the Depth Distribution of Energy Deposition by Ion Bombardment, AERE-R7805 (1974).

(12) C. F. Williamson, J. P. Boujot, J. Picard, Tabs of Range and Stopping Power for Charged Particles, CEA-R3042 (1966).

(13) W. Krätschmer, Effects of Heavy Ion Radiation on Quartz Glass, Proc. Int. Conf. on Nucl. Photography and Track Detectors, 10-15 July 1972, Bucharest.

(14) E. P. EerNisse, C. B. Norris, Introduction Rates and Annealing of Defects in Ion-Implanted SiO_2 Layers on Si, J. Appl. Phys. 45, 5196 (1974).

(15) J. B. Bates, Neutron Irradiation Effects and Structure of Noncrystalline SiO_2, J. Chem. Phys. 61, 4163 (1974).

CHAPTER VI

ELECTRONIC STRUCTURE OF THE Si-SiO$_2$ INTERFACE

ELECTRONIC STATES OF Si-SiO$_2$ INTERFACES

R.B. Laughlin and J.D. Joannopoulos
MIT, Cambridge, Massachusetts 02139

D.J. Chadi
Xerox, Parc, Palo Alto, California 94304

ABSTRACT

We have developed a unique method for studying interfaces theoretically which allows us to deal directly with disorder at the interface. Using tight-binding Hamiltonians which give excellent descriptions of the valence and conduction states of α-quartz and silicon, we construct a Bethe lattice for each material and bond them together in a variety of configurations. The method enables us to avoid the problem of lattice mismatching and to deal individually with the effects of different bonding configurations, bond-angle distortions and bonding defects at the interface. The principal result of the theory is that there are no interface states in the gap for an interface which lacks distortions or defects. Bond-angle distortions induce a tail of gap states near the bottom of the silicon dioxide conduction band, an effect which occurs in bulk silicon dioxide as well. Broken-bond defects, on the other hand, lead to distinct interface states in the gap. We specifically identify a state observed experimentally as being due to a silicon dangling-bond defect in the interface.

MOS technology is made possible, to a large extent, by the ability of a layer of silicon dioxide to stabilize a silicon surface, and to eliminate the surface states from it without introducing a large density of interface states in or near the silicon gap. Unfortunately, this elimination is not perfect. There typically remains a residual density of interface states between 10^{11} and 10^{12} per cm^2 (1). These remaining states impede the operation of the device, and it is therefore of great interest to understand their chemical origin. The purpose of this paper is to introduce a new theory of Si-SiO$_2$ interfaces we have developed which can shed some light on this problem.

The great disparity between the lattice constants of Si and SiO$_2$ necessarily causes the interface region between them to be highly disordered; this, in turn, causes the theoretical problem to be extremely difficult. In particular, we expect to find regions in the interface in which the bonding is ideal, regions where the bonds are distorted slightly, and regions where bonds are actually broken, and all three types of region must be included in any realistic theory. Clearly, it would be of great advantage to somehow isolate these regions and study them separately. The method we use to study the interface region, the cluster-Bethe-lattice method, is ideally suited to doing just that.

The cluster-Bethe-lattice method is a way of studying the effects of the local environment of an atom on the electronic states of that atom. It involves removing from the solid a small cluster containing the atom in question and attaching Bethe lattices to all the dangling bonds. A Bethe lattice is an infinite, non-periodic, bonded network of atoms which maintains the atomic coordination found in the actual solid, but which contains no closed rings of bonds. By bonding Bethe lattices to the cluster, we make a surrogate system which is infinite, which maintains the local environment of the atom, and which can be solved exactly.

In Fig. 1 we show how a model for the interface can be constructed using the cluster-Bethe-lattice method. In the upper left and right corners, we show

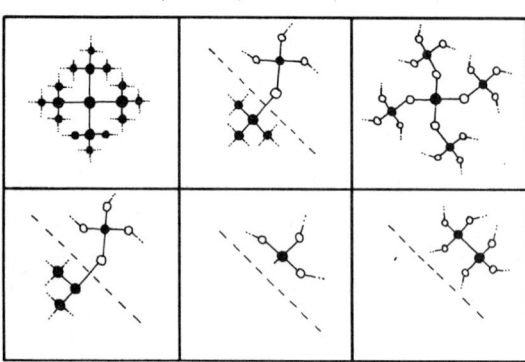

schematic drawings of the silicon and silicon dioxide Bethe lattices, respectively. Given that the electronic structure of the Bethe lattices is very similar to that of bulk silicon and silicon dioxide, we can make an interface simply by breaking a bond in each Bethe lattice and bonding a piece of each together, as is shown in the upper center of Fig. 1. The interface formed in this way is "ideal", in that in contains no distorted bond angles or broken bonds. If we wish to study the effects of bond-angle disorder in the interface, we need only distort a single bond in the Bethe lattice structure, for example the Si-O-Si bond at the center of the interface. If we wish to study the effects of having a dangling Si bond on the silicon side of the interface, we need only break a bond, as shown in the lower left panel of Fig. 1. If we wish to study a dangling Si bond in the

Fig.1 Upper left: Schematic drawing of Si Bethe lattice. Upper right: Schematic drawing of SiO_2 Bethe lattice. These are bonded together to form an ideal interface (upper center). Below are possible bonding defects at the interface: dangling Si bond in the silicon (left), dangling Si bond in the oxide (center) and Si-Si bond in the oxide (right).

oxide, we can sever a bond higher up, as is shown in the lower center panel. Finally, if we wish to study the effects of having Si-Si bonds in the oxide, we can bond two dangling silicon bond defects together, as is shown in the last panel.

We now need a Hamiltonian. We need one that is simple enough so that it is tractable for this complicated problem, yet realistic enough so that direct comparison with experiment can be made. The Hamiltonian we have chosen is an empirical tight binding Hamiltonian (6). It includes parameterized nearest-neighbor silicon-oxygen and oxygen-oxygen interactions which were fit to pseudopotential calculations (2) and to photoemission experiments (3, 4).

We now show the results of calculations performed for three specific types of interface: 1.) ideal (containing no bond-angle distortions or dangling bonds), 2.) disordered (containing bond-angle distortions) and 3.) defective (containing bonding defects). In particular, we look at two types of defect: 1.) dangling bonds and 2.) like-atom bonds.

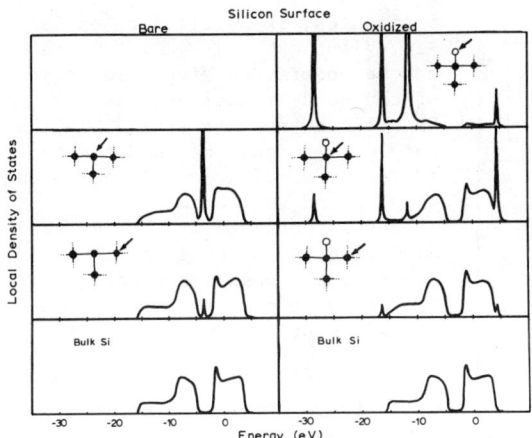

Fig.2 Local densities of states versus energy for bare Si surface (left) and for the same surface with an oxygen atom attached (right).

Ideal Interface (Type 1)

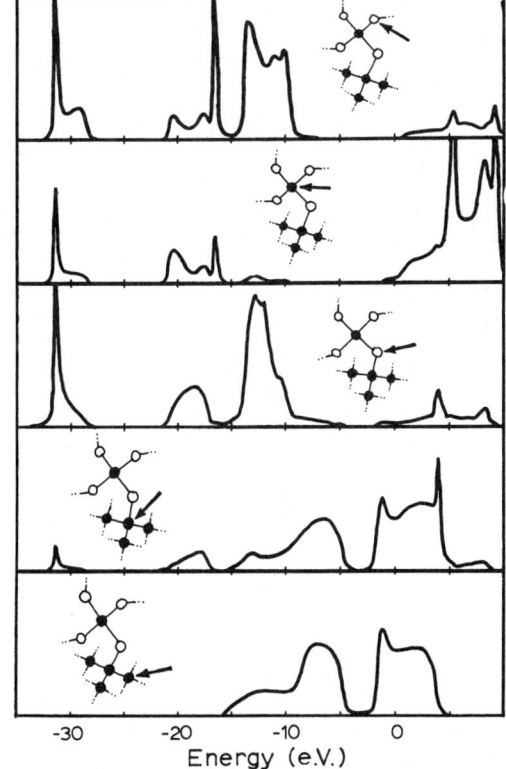

The behavior of the interface when it is ideal can be understood very simply in terms of what happens to a silicon surface when it is oxidized. In Fig. 2 we compare local densities of states near a silicon surface when it is bare (left) and after an oxygen atom has been bonded to the surface (right). As one comes up from the interior towards the bare surface, one sees a sharp surface state growing in the middle of the silicon gap and deriving from the dangling sp^3 hybrid orbital at the surface. This contrasts sharply with the behavior when the surface is oxidized. Now as one approaches the surface, one finds no surface states in or near the gap, but four oxygen-derived surface states away from the gap. These include the oxygen s level (-28 eV), the oxygen non-bonding p level (-13 eV) and a bonding (-16 eV) and anti-bonding (5 eV) pair formed by the interaction of an oxygen p state with the dangling hybrid at the surface. Now, as the oxide is allowed to thicken, these four surface states broaden and shift slightly to become the three major valence bands and the conduction band of the oxide.

In Fig. 3 we show local densities of states calculated for several atoms in and near an ideal interface. On the oxygen atom deepest in the oxide we see the three major valence bands and the conduction band just mentioned. Again, as we approach the interface from the silicon side, we see no interface state growing in or near the silicon

Fig. 3 Local densities of states versus energy for an ideal interface, starting at the top with an oxygen atom, and proceeding layer-by layer through the interface and down into the silicon.

324

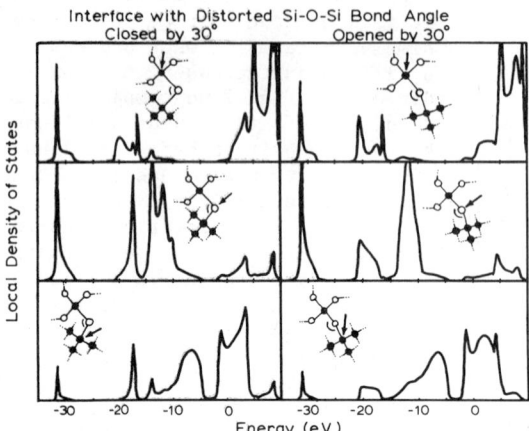

Fig. 4 Local densities of states versus energy for an interface with the same bonding configuration, but with the Si-O-Si angle at the center distorted by -30° (left) and +30° (right).

Interface with Dangling Silicon Bond

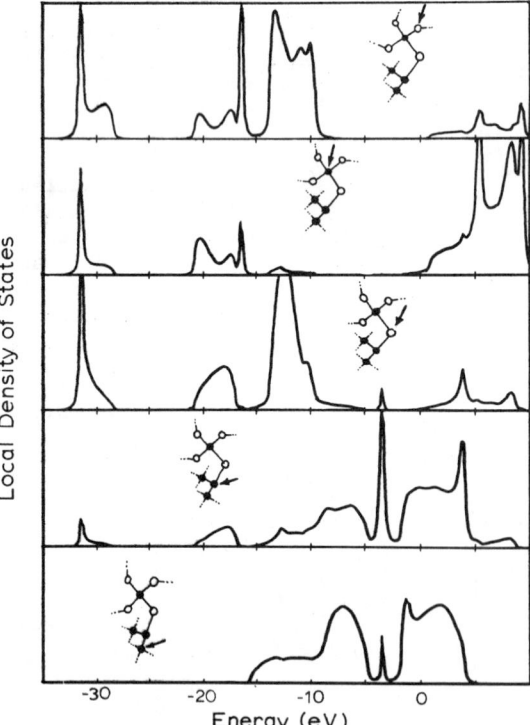

gap. There are, however, interface resonances, in particular, one at -18 eV and one at -14 eV. These can be understood simply in terms of the bond-orbital approach of Pantelides and Harrison (5) in that they derive from the incomplete (3-fold) coordination of the strong-bonding and weak-bonding bond-orbitals contained in the isolated Si-O-Si unit at the interface. However, the most important result in Fig. 3 is that there are no interface states near the silicon gap.

Given that an ideal interface is free of states near the gap, the observed states must be caused by deviations from ideal behavior, specifically, by disorder in the interface or by bonding defects.

In Fig. 4 we investigate explicitly the effects of distorting the Si-O-Si angle at the interface. The left column is the ideal interface with this angle closed down by 30° from its nominal value of 140°; the right column is the ideal interface with the angle opened by 30°. Concentrating on the region near the silicon gap, we see that very little happens at the valence band edge as the angle is opened. Something very important happens at the conduction band edge, however. The small tail near the conduction band edge in the upper panels grows as the angle is increased, indicating that opening the angle pulls states down out of the conduction band and places them on silicon atoms in the oxide. The effect is small because only one angle is distorted. In a real interface, many angles can be

Fig. 5 Local densities of states versus energy for an interface region containing a dangling Si bond on the silicon side of the interface.

opened, causing the state to grow and eventually pop out into the gap. Therefore, we expect that distortions of the Si-O-Si angle near the interface will
cause a tail of localized states near the conduction band edge. This pulling
down of states occurs because the region is becoming more like β-Cristobalite
(5,6), which has a smaller gap than α-quartz, given the same Hamiltonian.
Opening the angle also causes states to shift away from the gap at -16 eV.
This is because the bonding interaction is increasing (moving the strong-
bonding states near -18 eV downward) while the weak-bonding bands (near -14
eV) are moving upward to become degenerate with the lone-pair bands at 180°.
However, the most important result in Fig. 4 is that distortion of this angle
can generate a tail of states near the silicon conduction band edge but not
near the valence band edge.

Another bond-angle distortion one might investigate is that of the O-Si-O
angle in the oxide. This would tend to make a tail of states near the oxide
valence band edge, since the upper part of the valence band is lone-pair-like

Dangling Silicon Bond in Oxide

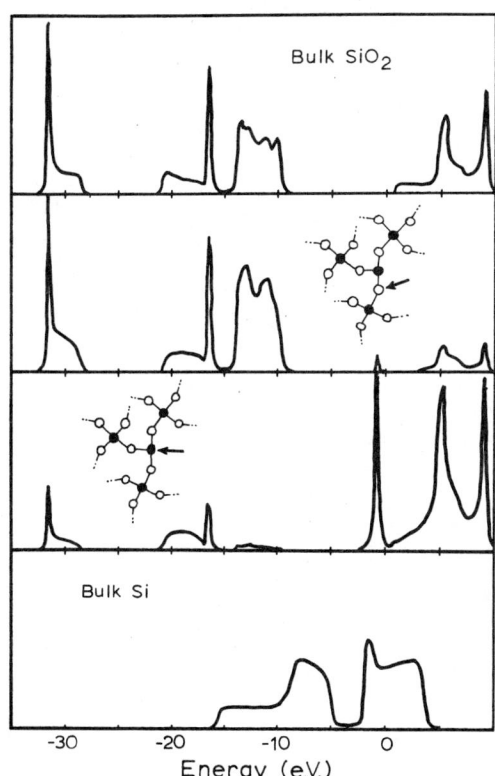

Fig. 6 Local densities of states
versus energy near a dangling Si
bond defect in the oxide.

and sensitive to second-neighbor
oxygen-oxygen interactions (6).
However, this kind of bond-angle
disorder is much less prevalent in
vitreous silica (7) and it is not
of immediate interest since it does
not effect the silicon gap. One
might also have bond-angle disorder
in the silicon. We know, however,
from studies (8) performed on bulk
silicon that this kind of disorder
generates a tail of states near the
valence band edge due to dehybridization, and a much weaker tail of
states near the conduction band edge
due to fluctuations in the second-
neighbor distance. Therefore, we
have established that bond-angle
disorder tends to generate states
near both band edges, those near
the valence band being localized in
the silicon, those near the conduction band being localized in the
oxide.

Localized states can also be generated by bonding defects. There are
two major types of bonding defect
one might find in the interface
region: those deriving from dangling
oxygen bonds and those deriving from
dangling silicon bonds. The former
tend to generate localized states
near the oxide valence band edge
because they are non-bonding p-like.
They are thus not relevant to the
problem of interface states in the
silicon gap. Dangling silicon bonds,
on the other hand, tend to generate
states in and near the silicon gap.

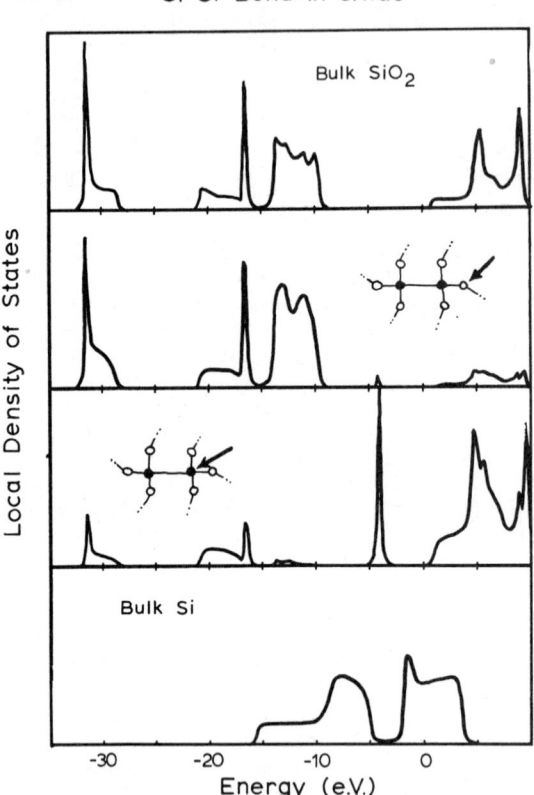

Fig. 7   Local densities of states versus energy for a Si-Si bond in the oxide.

There are two types of dangling silicon bond one might find in the interface: one on the silicon side and one on the oxide side. We have examined both of these as examples of broken-bond defects. In addition, we have investigated the effects of having Si-Si bonds in the oxide, in response to the suggestion that the oxide may be silicon-rich.

In Fig. 5 we show the ideal interface discussed earlier, but with a missing atom on the silicon side. The local densities of states of the various atoms are practically identical with those of the ideal interface, except in and around the dangling-bond defect. There, we see a sharp interface state, highly localized and situated in the middle of the silicon gap at the same energy where the dangling-bond surface state appeared on the bare silicon surface. This state also derives from the dangling sp^3 hybrid at the defect site and is so localized that it is essentially the same state as appears at the silicon surface. From Fig. 5, therefore, we conclude that dangling bonds on the silicon side of the interface can generate localized states deep in the gap. A sharp interface state similar to this one has been observed experimentally (9).

In Fig. 6, we show the effects of having the dangling Si bond on the oxide side of the interface. Again, we see a very localized state associated with the dangling sp^3 hybrid. However, its energy is -1.0 eV, which places it not in the gap, but coincident with the silicon conduction band. We conclude from this calculation that a dangling bond in the oxide will not make a state in the gap, but instead generate a trap state near the conduction band edge.

In Fig. 7, we show a similar calculation performed for a Si-Si like-atom bond in the oxide, formed by connecting two dangling silicon bonds together. As expected, the dangling bond states split into bonding and anti-bonding states, with one member pushed up into the oxide conduction band, and the other bonded down through the gap to the valence band edge. Therefore, we conclude from this calculation that the presence of like-atom bonds in the oxide will not generate states deep in the gap either, but will form trap states near the valence band edge. These will be about 1.5 eV below the silicon conduction band edge.

In conclusion, therefore, we have established the following:

1.) If the interface is ideal (containing no broken or distorted bonds), there are no interface states in or near the silicon gap.

2.) Bond-angle disorder in the oxide tends to generate a tail of states near the silicon conduction band edge, but not near the valence band edge.

3.) The only bonding defect which causes a sharp state deep in the silicon gap is a dangling Si bond on the silicon side of the interface.

4.) Both dangling silicon bonds and silicon-silicon bonds in the oxide can induce trap states near the silicon conduction and valence band edges.

ACKNOWLEDGEMENTS

Two of us (J.D.J. and R.B.L.) should like to thank the Xerox Research Center for their hospitality and support during the period in which this research was performed.

REFERENCES

(1) M. Pepper and N.F. Mott, The Si-SiO$_2$ interface and anderson localization in the inversion layer, Physics of Semiconductors, Proceedings of the 13th International Conference, Rome, 1976, p. 762.

(2) J.R. Chelikowsky and M. Schluter, Electronic states in α-quartz: A self-consistent pseudopotential calculation, Phys. Rev. B15, 4020 (1977).

(3) T.H. DiStefano, Field dependent photoemission probe of the electronic structure of the Si-SiO$_2$ interface, J. Vac. Sci. Tech. 13, 856 (1976).

(4) H. Ibach and J.E. Rowe, Electron orbital energies of oxygen adsorbed on silicon surfaces and of silicon dioxide. Phys. Rev. B10, 710 (1974).

(5) S.T. Pantelides and W.A. Harrison, Electronic structure, spectra and properties of 4:2-coordinated materials. I. Crystalline and amorphous SiO$_2$ and GeO$_2$, Phys. Rev. B13, 2667 (1976).

(6) D.J. Chadi, J.D. Joannopoulos and R.B. Laughlin, this conference.

(7) R.L. Mozzi and B.E. Warren, The structure of vitreous silica, J. Appl. Cryst. 2, 164 (1969).

(8) J.D. Joannopoulos, Theory of fluctuations and localized states in amorphous tetrahedrally bonded solids, Phys. Rev. B16, 2764 (1977).

(9) N.M. Johnson, D.J. Bartelink and M. Schulz, this conference.

THE DEFECT STRUCTURE OF THE Si-SiO$_2$ INTERFACE, A MODEL
BASED ON TRIVALENT SILICON AND ITS HYDROGEN "COMPOUNDS"

Christer M Svensson
Research Laboratory of Electronics, Chalmers University
S-402 20 Gothenburg, Sweden

ABSTRACT

The present paper is an attempt to formulate an unifying model describing the properties of the Si-SiO$_2$ interface defects. We concentrate on the "classical" defects, fixed surface charge, interface traps and radiation induced charge. We postulate the existence of three different forms of trivalent silicon and electrical and chemical properties of these. We show that the model makes it possible to understand most known properties of the interface defects. We can for example understand why fixed charge and interface trap densities follow each other after oxidation but not after hydrogen anneal. We can also understand how holes, trapped in the oxide, can give rise to new interface traps.

INTRODUCTION

In an excellent review from 1974, dr Bruce Deal described the current understanding of charges in the Si-SiO$_2$ structure (1). He raised a number of questions which are still unanswered. Such questions are for example: What is the exact origin of fixed surface charge? What are the relationships among the origins of the various charges? What is the origin and mechanism of the radiation induced fast states?

In the present work I will propose a model which hopefully will shed some light on these questions. The model is based on the idea, first formulated by Revesz (2), that trivalent silicon and its hydrogen "compounds" are the most important defects in the Si-SiO$_2$ system. I am specially interested in understanding the relationships among the various observed effects, and in understanding how oxide holes can create new defects (surface traps)(3).

THE MODEL

The model is based on the assumption of three forms of trivalent silicon in the Si-SiO$_2$ interface, with slightly different chemical and electrical properties. The three forms are silicon surface trivalent silicon, Si$_g^\bullet$, bonded to three other silicons, oxide trivalent silicon, Si$_o^\bullet$, bonded to three oxygens and oxide trivalent silicon close to the silicon surface, Si$_{os}^\bullet$. The three forms are demonstrated in fig. 1.

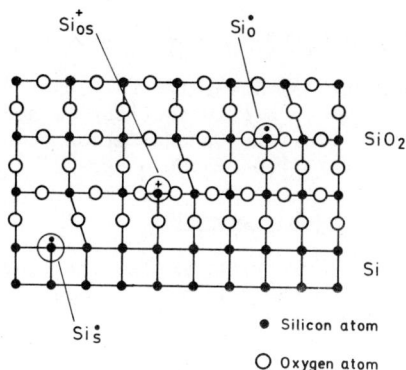

Fig. 1.   The Si-SiO$_2$ interface with the three forms
of Si· defects

We now postulate the following electrical properties of these defects. Si$_s$·
is assumed to be a surface trap. This assumption have been used by many
others (1,2,4,5). Si$_o$· is assumed to be a deep hole trap. This assumption was
also used by Revesz (2) and is compatible with the cluster model calculations
by Bennet and Roth (6). The single trivalent silicon may be equivalent to a
situation like fig. 4a or c in ref. 6. Si$_{os}$· finally, is assumed to be a very
deep hole trap, deeper than Si$_o$· because of its interaction with the silicon
surface. The interaction may be of the image force type, lowering the energy
of the charged defect. The trap level is assumed to be above the silicon conduction band edge, making Si$_{os}$ always ionized to Si$_{os}^+$. This defect is thus
assumed to be the fixed surface charge, as also proposed by Deal (1).

We furthermore postulate the following chemical properties of these defects.
The most important reaction is the reaction with hydrogen (H$_i$ = interstitial
hydrogen)(2):

$$\equiv Si\cdot + H_i \rightleftharpoons Si\,H \tag{1}$$

This reaction may proceed in any direction depending on temperature and
hydrogen concentration. The reaction is assumed to occur for Si$_s$· and Si$_o$·.
Si$_s$ H is assumed to be electrically inactive, i.e. it is not any longer a
surface trap. Si$_o$ H is assumed to be a hole trap, again this is compatible
with Bennet and Roth (6). Si$_o$ H may be equivalent to the hole trap in fig. 5b
and d in ref. 6. We furthermore postulate that the ionized form of Si$_o$ H is
unstable, thus upon hole trapping:

$$\equiv Si_o H + h_o^+ \rightarrow \equiv Si_o^+ + H_i \tag{2}$$

where h$_o^+$ is an oxide hole. This process is similar to the electrochemical
charging of SiOH groups in oxide films by electrons, investigated by
Nicollian et.al.(7). For Si$_{os}^+$ we postulate that reaction (2) always dominate, why Si$_{os}^+$ do not form a hydrogen "compound".

We will furthermore assume that the different forms of trivalent silicon may
exchange with each other, at least at high temperatures:

$$\equiv Si_o\cdot \rightleftharpoons \equiv Si_{os}\cdot \rightleftharpoons \equiv Si_s\cdot \qquad (3)$$

Assuming that the reactions (3) are relatively fast at high temperatures implies that the three defects are in equilibrium among themselves after a high temperature process, i.e. the concentration of one of the defects is proportional to the concentration of the others.

The defect structure after oxide growth

It is obvious from numerous experiments that the conditions during oxide growth is most important for the defect structure. It is generally accepted that the oxidation takes place by diffusion of oxygen related defects to the silicon where they react (1,8). It is generally accepted that this results in an oxide where the outer layer contain excess oxygen and a thin layer, close to the silicon surface, contain excess silicon (1). The excess silicon is then assumed to be related to the surface charge. This general picture gives a qualitative explanation to the so-called "Q_s-oxygen triangle", relating the surface charge, Q_s, to the final oxidation temperature and ambient (1).

It is also well known that hydrogen and water vapor play an important role in the oxide growth (8). There are reason to believe that the oxidizing species are silanol groups, Si-OH, also in most cases of "dry" oxidation (9,10). This means that hydrogen always is availible during oxide growth. We therefore suggest that the excess silicon existing close to the silicon surface, exist both as Si· and SiH. This is in good agreement with infrared spectral analysis, which have shown high concentration of SiH groups piled up near the silicon surface (11).

The fixed surface charge

We have postulated that the surface charge, Q_s, consists of trivalent silicon in the oxide close to the silicon surface, Si_{os}^+. The effect of the silicon surface is thus to lower the energy of a charged defect through, for example, the image force effect. Because of this, Si_{os}^+ will always be charged and can not react with hydrogen. The fact that Si_{os}^+ only can exist very close to the silicon surface explain why Q_s only exist very close to this surface (less than 20 Å from surface) (12).

In our proposed model, Q_s will be directly related to the amount of trivalent silicon near the silicon surface. The model used by Deal (1) to explain the Q_s-oxygen triangle is thus applicable in our case. The fact that our Si_{os}^+ do not react with hydrogen explains why Q_s is unaffected by low temperature hydrogen anneal (1). Montillo and Balk also shown that the water vapor pressure during oxidation or high temperature anneal did not affect Q_s although it did affect the surface trap density (5).

Our model is also compatible with the model for the negative bias stress effect proposed by Jeppson and Svensson (13). In their model, surface charge (Si_{os}^+) and surface traps ($Si_s\cdot$) are created from Si_sH according to:

$$\equiv Si_sH + \equiv Si_{os}-O-Si_o\equiv \rightarrow \equiv Si_s\cdot + \equiv Si_{os}^+ + \equiv Si_oOH \qquad (4)$$

The rate of Q_s growth is in this model proportional to the amount of Si_sH, which according to our present model is proportional to Q_s (Si_{os}^+). This then agree with the experimental fact (1).

The surface trap density

We have postulated that surface traps, N_{st}, consist of trivalent silicon at the silicon surface, $Si_s\cdot$. Thus, both Q_s and N_{st} are trivalent silicon defects. According to our previous assumptions they will then be closely related to each other after oxide growth, which then agree with experimental facts (1).

It is further known that N_{st} can be reduced by low temperature annealing in hydrogen. This is believed to take place through complexing the N_{st} defect with hydrogen (1), or in our model through the reactions

$$\equiv Si_s\cdot + H_2 \rightarrow \equiv Si_s H + H_i \qquad (5)$$

$$\equiv Si_s\cdot + H_i \rightarrow \equiv Si_s H \qquad (1)$$

The active surface trap, $Si_s\cdot$, is thus made inactive by its reaction with hydrogen.

In high temperature annealing in an inert atmosphere or vacuum the fast surface trap density normally increase (1,5). We believe that this occurs through the reverse of the reactions (1) and (5), i.e. through thermal dissociation of the SiH bond.

Surface traps are also created by the negative bias stress effect. This phenomenon was discussed in the previous section. Furthermore surface traps are created as a result of radiation and hole trapping. This will be discussed in the next section.

Radiation effects and hole trapping

It is generally agreed that ionizing radiation forms electron-hole pairs in the oxide and that the trapping of these holes is the dominant mechanism in radiation damage of MOS structures (14,15,16). It has been clearly shown that hole injection only, gives the same effects, including surface trap creation, as different forms of radiation (3,17). We will therefore only discuss hole trapping in the following.

Hole trapping in the $Si-SiO_2$ system occurs only close to the silicon surface, within tens of nanometers from that surface (15,18). The density of these hole traps is very large, $10^{15}-10^{16}\,cm^{-2}$ (15) or, assuming a depth of 10 nm, $10^{21}-10^{22}\,cm^{-3}$. This strongly indicate that trivalent silicon is responsible for these traps, as also suggested by Woods and Williams (15). The concentration is of the same order of magnitude as the observed concentration of SiH groups at the interface, $10^{20}-10^{21}\,cm^{-3}$ (11). Our model therefore explain the strong hole trapping close to the silicon surface through either of the two trapping processes:

$$\equiv Si_o\cdot + h_o^+ \rightarrow \equiv Si_o^+ \qquad (6)$$

$$\equiv Si_o H + h_o^+ \rightarrow \equiv Si_o^+ + H_i \qquad (2)$$

Let us finally discuss the creation of new surface traps as a result of hole trapping (3,17). This is somewhat hard to understand, as the traps are created at the silicon surface, whereas the hole trapping can be effective only about 5 nm away from this surface (otherwise the holes tunnels into the silicon very fast). It can however, be understood in our model. According to reaction (2), interstitial hydrogen is created as a result of hole trapping in the oxide. This H_i is assumed to react with interface Si_sH groups according to the reverse of reaction (5):

$$\equiv Si_s H + H_i \rightarrow \equiv Si_s \cdot + H_2 \qquad\qquad (7)$$

It can be shown that also if all reactions (1), (5) and (7) are active, surface traps may be created in an excess of H_i and annihilated in an excess of H_2 (low temperature hydrogen anneal). In an excess of H_i the Si_s centers acts as a catalyst for the recombination of H_i to H_2, during which process the hydrogen occupation of the Si_s center may be low.

REFERENCES

(1) B. E. Deal, The current understanding of charges in the thermally oxidized silicon structure, J. Electrochem. Soc. 121, 198C (1974)

(2) A.G. Revesz, Defect structure and irradiation behavior of noncrystalline SiO_2, IEEE Trans. Nucl. Sci. NS-18, (No 6) 113 (1971)

(3) R.J. Powell and G.F. Derbenwick, Vacuum ultraviolet radiation effects in SiO_2, IEEE Trans. Nucl. Sci. NS-18 (No 6) 99 (1971)

(4) E. Kooi, The surface charge in oxidized silicon, Philips Res. Rep. 21, 477 (1966)

(5) F. Montillo and P. Balk, High-temperature annealing of oxidized silicon surfaces, J. Electrochem. Soc. 118, 1463 (1971)

(6) A.J. Bennet and L.M. Roth, Electronic structure of defect centers in SiO_2, J. Phys. Chem. Solids, 32, 1251 (1971)

(7) E.H. Nicollian, C.M. Berglund, P.F. Schmidt and J.M. Andrews, Electrochemical charging of thermal SiO_2 films by injected electron currents, J. Appl. Phys. 42, 5643 (1971)

(8) A.G. Revesz, Noncrystalline silicon dioxide films on silicon: a Review, J. Noncryst. Solids, 11, 309 (1973)

(9) A.G. Revesz and R.J. Evans, Kinetics and mechanism of thermal oxidation of silicon with special emphasis on impurity effects, J. Phys. Chem. Solids, 30, 551 (1969)

(10) In most cases dry oxidation involve water vapor, see refs. 5 and 8.

(11) K.H. Beckman and N.J. Harrick, Hydrides and hydroxyls in thin silicon dioxide films, J. Electrochem. Soc. 118, 614 (1971)

(12) R.J. Powell and C.N. Berglund, Photoinjection studies of charge distribution in oxides of MOS structures, J. Appl. Phys., 42, 4390 (1971)

(13) K.O. Jeppson and C.M. Svensson, Negative bias stress of MOS devices at high electric fields and degradation of MNOS devices, J. Appl. Phys., 48, 2004 (1977)

(14) E.H. Snow, A.S. Grove and D.J. Fitzgerald, Effects of ionizing radiation on oxidized surfaces and planar devices, Proc. IEEE, 55, 1168 (1967)

(15) M.H. Woods and R. Williams, Hole traps in silicon dioxide, J. Appl. Phys. 47, 1082 (1976)

(16) G.W. Hughes, R.J. Powell and M.H. Woods, Oxide thickness dependence of high-energy-electron, VUV-, and corona-induced charge in MOS capacitors, Appl. Phys. Lett., 29, 377 (1976)

(17) P.S. Winokur and M.M. Sokoloski, Comparison of interface-state buildup in MOS capacitors subjected to penetrating and nonpenetrating radiation, Appl. Phys. Lett., 28, 627 (1976)

ELECTRONIC STRUCTURE OF A MODEL Si-SiO$_2$ INTERFACE

Frank Herman and Inder P. Batra
IBM Research Laboratory, San Jose, CA 95193

Robert V. Kasowski
Experimental Station, E. I. duPont de Nemours & Co.
Wilmington, DE 19898

ABSTRACT

We are investigating the electronic structure of model Si-SiO$_2$ interfaces with a view to determining the nature of the localized interface states. Using superlattices (repeated Si and SiO$_2$ regions) containing 17 to 23 atoms per unit cell, we have determined the electronic structure of fully-ordered interfaces using both the ETB (extended tight-binding) and the LCMTO (linear combination of muffin-tin orbitals) methods. In this note, preliminary results will be reported for an interface formed by connecting the (100) face of a silicon crystal with the (100) face of an idealized SiO$_2$ crystal with one rotated 45° with respect to the other. One attractive feature of this model is that the interface is formed from two crystals: there is no need to introduce an amorphous transition region. Another is the possibility of introducing chemical and structural imperfections at the interface and taking their presence into account by first-principles methods.

INTRODUCTION

One of the major challenges of contemporary semiconductor physics is understanding the electronic structure and related properties of the Si-SiO$_2$ interface (1-7). A great deal of empirical information is already at hand, but our knowledge of the atomic arrangements and localized electronic states at or near the Si-SiO$_2$ interface is still largely speculative (1,2,7). During the past few years, the electronic structure of various forms of crystalline SiO$_2$ has been investigated by a number of different theoretical approaches (8-11). Simple theoretical treatments of the Si-SiO$_2$ interface itself are just now beginning to appear (12,13).

Perhaps the greatest obstacle to theoretical progress on the Si-SiO$_2$ interface is the complexity of the atomic arrangements in the neighborhood of this interface. There can be little doubt that the lattice mismatch between Si and SiO$_2$ leads to considerable structural disorder near the interface, and that this disorder plays an important role in establishing many of the detailed features of the electronic structure, such as band tails. However, it is not unreasonable to assume -- based on our experience with amorphous semiconductors -- that the gross (or average) features

of the electronic structure are determined by the local atomic
arrangements (nearest neighbor coordinations particularly), and
by whatever chemical impurities may be present nearby. These
gross or average features include the location of the Si gap
within the span of the SiO_2 gap, and the centers of gravity of
bands formed from localized interface states. Accordingly, we
will focus our attention on idealized interface models which
neglect long range disorder but are capable of dealing directly
with various types of local order and chemical imperfections.
In effect, we will neglect the fluctuations in local atomic
arrangements and concentrate on an idealized average connection
between Si and SiO_2 regions.

FULLY-ORDERED INTERFACE MODEL

Because the ratio of the bond lengths of Si-O-Si (in various forms
of SiO_2) and Si-Si (in a silicon crystal) is very nearly equal to
$\sqrt{2}$ (14), a particularly simple way to form an ordered interface
is to connect the (100) face of SiO_2 with the (100) face of Si,
with one of these faces rotated just $45°$ with respect to the other.
As indicated in Fig. 1, this construction leads to a fully-ordered
interface for which half the silicon atoms on the Si crystal
surface also belong to the SiO_2 crystal (which we represent by
idealized β-cristobalite). Each of these shared silicon atoms
is four-fold coordinated; all attached neighbors maintain their
normal bond distances; but because of the $45°$ rotation, the
Si-Si-O bond angles deviate somewhat from ideal tetrahedral values.
We are referring here to the Si(in Si crystal)---Si(interface)
---O(in SiO_2) angle.

In view of the wide range of bond angles occurring in various forms
of SiO_2 (15-17), these angular deviations are not expected to
affect our overall model adversely. Moreover, the average value
of the Si-Si-O bond angles at the interface is very nearly equal
to the ideal tetrahedral value.

If one wishes to "absorb" the $45°$ rotation differently, each of the
shared silicon atoms at the interface can be replaced by a double-
bonded Si_2 molecule, one end of which belongs to the Si crystal
on one side of the interface, and the other end to the SiO_2 crystal
on the other side. The $45°$ rotation is now absorbed by the Si-Si
double bond. This modification was not embodied in the present
study because the deviations from ideal tetrahedral coordination
at the interface were considered to be more acceptable than the
introduction of three-fold coordinated silicon atoms here, within
the spirit of our model.

In any event, the remaining half of the silicon atoms on the Si
crystal surface are only two-fold coordinated, and have two dangling bonds each.

Ordered interfaces of the type just described can be used to model
a variety of assumed interface geometries and chemical constituents.
For example, the dangling bonds on the two-fold coordinated silicon
atoms can be retained, or they can be replaced by single or dual

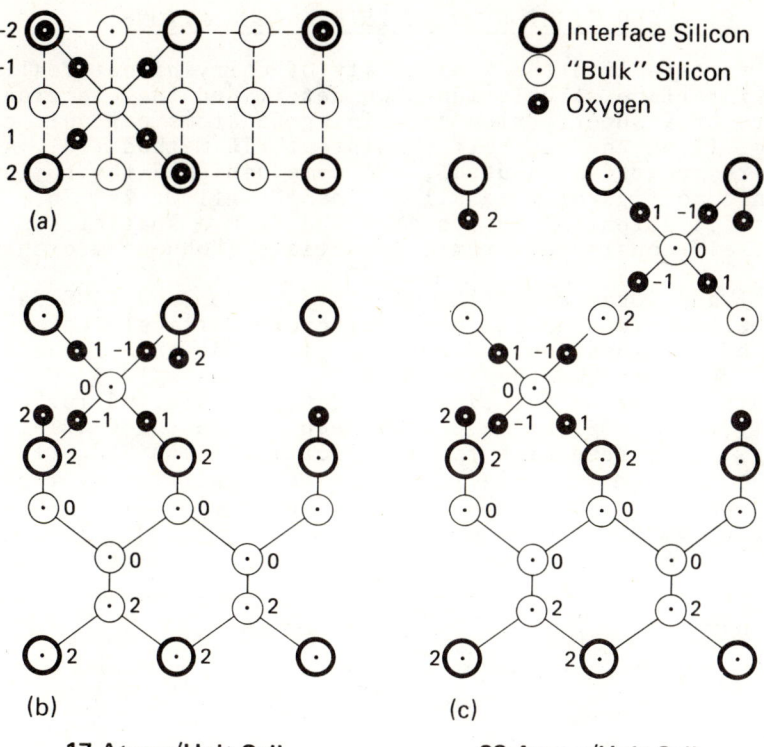

17 Atoms/Unit Cell   **23 Atoms/Unit Cell**

Fig. 1. Si-SiO$_2$ superlattice models. 17 atom
model: top view (a) and front view (b). 23 atom
model: front view (c). Numbers indicate atomic
locations on vertical planes defined in (a). In
(b) and (c) these numbers are placed next to
17 and 23 inequivalent atoms in the unit cell.
Note that oxygen atoms at -1 attach to silicon
atoms at 0 and -2 (rather than 2). In (a) four
of the oxygen atoms shown belong to the SiO$_4$
tetrahedron in SiO$_2$; the other three are bonded
(from below) to three silicon interface atoms.
The remaining three silicon interface atoms are
bonded to the SiO$_2$ and silicon crystal regions.

Si-H bonds or Si-O double bonds. Or the two-fold coordinated Si
atoms could be replaced by O atoms. It is also possible to
incorporate impurity atoms in or near the interface (as we hope
to do in subsequent studies). In the present work, we use Si-O
double bonds to remove the dangling bonds.

Before turning to a discussion of preliminary calculations based
on the ETB (10) and LCMTO (18,19) methods, we note that while we
tried to construct ordered interface models using other crystal
faces, we were never able to obtain as simple a model as the 45°-
rotated (100) face model described above.

EXTENDED TIGHT-BINDING (ETB) CALCULATIONS

Largely to establish the feasibility of carrying out realistic ordered-interface calculations, we determined the electronic structure of a superlattice containing 23 atoms per unit cell (cf. Fig. 1) by the non-self-consistent ETB method (10). All atoms were treated as neutral, and were assigned the following electronic configurations: all Si: $3s3p^3$; all O: $2s^2 2p^4$. The superlattice potential was constructed from a spatial superposition of self-consistent atomic potentials (Kohn-Sham exchange).

The total and local densities of states (TDOS and LDOS) are shown in Fig. 2. Various panels have different vertical scales for clarity of presentation. Fig. 2(b) gives the LDOS for a Si atom located "deep" within the Si crystal. It has a 12.5 eV valence band width and about a 1 eV energy gap, as expected for bulk Si. Similarly, the LDOS for an O atom deep within the β-cristobalite (cf. Fig. 2(f)) resembles the TDOS for idealized SiO_2 (10). The interface Si atom (cf. Fig. 2(c)) has a significantly reduced LDOS compared with the bulk-like Si atom, but <u>no</u> localized interface states are found.

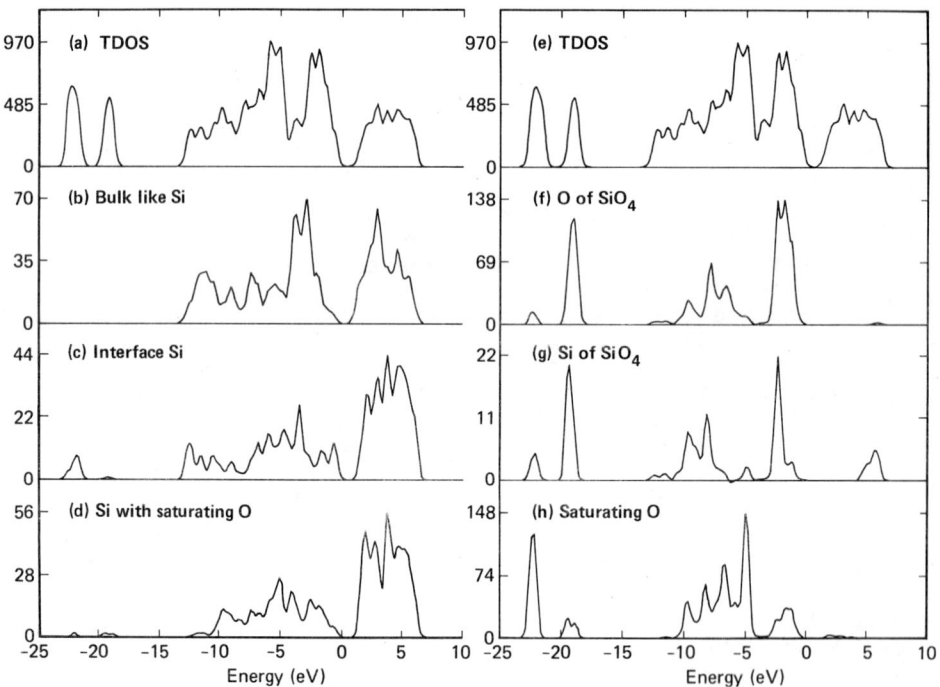

Fig. 2. Total and local densities of states for 23 atom Si-SiO_2 superlattice, as given by ETB calculation. TDOS appears in panels (a) and (e), and selected LDOS in the remaining panels.

In the present model using the ETB method we find no discontinuity
in the valence band maximum (VBM) across the interface. This is
contrary to experiment: interpretations range from discontinuities
of 3.75 eV (3) to 6.2 eV (5). Our finding that the VBM of Si and
SiO$_2$ coincide may be due to the strained Si-Si-O bond at the
interface, to the choice of crystal potential, to the neglect of
charge transfer effects, or to other causes. In any event, this
result deserves further study.

The most important conclusion to be drawn from these preliminary
studies is that there are no localized interface states associated
with our fully-ordered interface model. Another important finding,
established from a detailed examination of the LDOS for various
atoms in the SiO$_2$ region, is that fewer SiO$_4$ tetrahedral can be
used in this region, while still maintaining reasonable isolation
between successive interfaces. This result provides the justification for using the 17 atom superlattice for the LCMTO studies.

LINEAR COMBINATION OF MUFFIN-TIN ORBITALS CALCULATIONS

Using the same approach as in an earlier study of (110) Ge-GaAs
superlattices and interfaces (19), we investigated the electronic
structure of a Si-SiO$_2$ superlattice containing 17 atoms per unit
cell (cf. Fig. 1). In order to partition the entire unit cell into
suitably-shaped space-filling atomic polyhedra, it was necessary
to augment the 11 Si subcells and the 6 O subcells with 8 "empty"
subcells. The present study represents an application of the LCMTO
method (18,19) to one of the most "open" solid state structures
yet treated by this method.

In our initial studies, all atoms were taken as neutral (Si $3s3p^3$,
O $2s^22p^4$). The LCMTO calculations were also non-self-consistent,
but in contrast to the ETB work, the superlattice potential was
determined directly from a superposition of self-consistent atomic
charge densities. Kohn-Sham exchange was used both for the constituent atoms and their superlattice superposition. Preliminary
results suggest that, just as in earlier LCMTO studies (18,19),
the electronic structure of the (Si-SiO$_2$) superlattice is very
nearly equal to a superposition of the electronic structures of
its constituent regions, with the LDOS of the interface Si atoms
located at a compromise position on an energy scale. This arises
when the superlattice exchange potential is determined by the
cube root of the sum of the atomic charge densities, but not when
determined by the sum of the cube roots of the atomic charge
densities (as in the present ETB work). It is important, therefore, to calculate the exchange potential as accurately as
possible, within the framework of the statistical exchange theory.

According to our preliminary results, then, the Si VBM lies about
5.4 eV above the SiO$_2$ VBM, and the SiO$_2$ band gap is 8.3 eV (Kohn-
Sham exchange). For Slater exchange, the band gap is 9.5 eV.
If we assume the following ionic model for SiO$_2$: Si^{+1} ($3s3p^2$);
O$^{-0.5}$ ($2s^22p^{4.5}$), the SiO$_2$ band structure moves upwards by 1.5 eV,
more or less rigidly, with respect to the neutral SiO$_2$ bands and
also with respect to the Si crystal bands. Thus, for Kohn-Sham

exchange, the Si VBM lies about 3.9 eV above the SiO_2 VBM for this ionic model. These estimates for the VBM discontinuity -- 3.9 eV (ionic model) to 5.4 eV (neutral model) -- are consistent with current experimental estimates (3.75 eV (3) to 6.2 eV (5)). To recapitulate, the SiO_2 bands are not changed much relative to one another by the inclusion of ionicity, but their positions relative to the Si bands are shifted upwards by about 1.5 eV for an assumed charge transfer of 0.5 electron to each oxygen atom in SiO_2.

The ETB and LCMTO studies carried out thus far are most encouraging, in that they demonstrate the feasibility of obtaining detailed theoretical information for idealized $Si-SiO_2$ interfaces. Because of its relative simplicity, the present model offers the prospect of incorporating various chemical imperfections at or near the interface, and treating these from first principles.

ACKNOWLEDGMENT

The authors are grateful to Dr. William E. Rudge of IBM San Jose Research Laboratory for helpful advice and many stimulating discussions.

REFERENCES

1. B. E. Deal, J. Electrochem. Soc. 121, 198C (1974).
2. A. Goetzberger, E. Klausmann, and M. J. Schultz, Crit. Revs. Solid State Sci. 6,40 (1976).
3. R. Williams, J. Vac. Sci. Technol. 14, 1106 (1977).
4. S. T. Pantelides, Comments Solid State Phys. 8, 55 (1977).
5. N. F. Mott, Adv. Phys. 26, 363 (1977).
6. C. W. Gwyn, J. Appl. Phys. 40, 4886 (1969).
7. H. F. Wolf, Semiconductors, Wiley-Interscience, N.Y., 1971.
8. S. T. Pantelides and W. A. Harrison, Phys. Rev. B13, 2667 (1976); W. A. Harrison, this conference.
9. J. R. Chelikowsky and M. Schlüter, Phys.Rev.B15, 4020 (1977).
10. S. Ciraci and I. P. Batra, Phys. Rev. B15, 4923 (1977); I. P. Batra, this conference.
11. P. M. Schneider and W. B. Fowler, Phys.Rev.Lett.36,425 (1976); W. B. Fowler,P.M.Schneider, and E. Calabrese, this conference.
12. S. T. Pantelides, J. Vac. Sci. Technol. 14, 965 (1977); M. Long and S. T. Pantelides, this conference.
13. R. B. Laughlin, J. D. Joannopoulos, and D. J. Chadi; K. Hübner; C. M. Svensson; papers at this conference.
14. Using 1.66 A and 2.35 A for Si-O and Si-Si bond distances (15), the Si-O-Si/Si-Si ratio is 2x1.66/2.35=1.41. Also (16).
15. J. E. Huheey, Inorganic Chemistry, Harper and Row, N.Y., 1972.
16. R.W.G. Wyckoff, Crystal Structures, Interscience, N.Y., 1965.
17. A. F. Wells, Structural Inorganic Chemistry, Oxford University Press, 1962, Third Edition.
18. R. V. Kasowski, Phys. Rev. B14, 3398 (1976).
19. F. Herman and R. V. Kasowski, Phys. Rev. B17, 672 (1978).

CHAPTER VII

THE STOICHIOMETRY OF THE Si-SiO$_2$ INTERFACE

CONTINUOUS-RANDOM-NETWORK MODELS FOR THE Si-SiO$_2$ INTERFACE*

Sokrates T. Pantelides
IBM T. J. Watson Research Center, Yorktown Heights, NY 10598

and

Marshall Long[‡]
Dept. of Physics, Yale University, New Haven, CT 06520

ABSTRACT

Continuous-Random-Network (CRN) models have been constructed in order to simulate the atomic arrangement of the Si-SiO$_2$ interface. It was found that models could be constructed with or without an SiO$_x$ layer between the crystalline Si and the amorphous stoichiometric SiO$_2$. The atomic coordinates were relaxed using a computer program and a simple force model. In order to probe the width w of the SiO$_x$ layer, additional O atoms were gradually inserted, thus reducing w, and a normalized elastic energy was monitored. Using a wide range of choices for the force constants, it was found that the elastic energy was always lowered as w → 0.

The stoichiometry of the Si-SiO$_2$ interface is a subject that has attracted considerable attention. The interface has been probed experimentally by a wide variety of techniques, many of which are discussed in a series of papers in these Proceedings. The main objective of such experiments is to determine the width of the SiO$_x$ (x ≠ 2) layer that may exist at the interface. A variety of results has been reported, ranging from less than 4 Å (Refs. 1 and 2) to 8 Å (Ref. 3), 12-15 Å (Ref. 4), 15-20 Å (Ref. 5), and so on. Much of this variation is due to the differences in experimental techniques and interpretation of results. For example, some of the techniques may alter the composition of the sample during measurement. Also, some techniques probe the scattering properties of *atoms*, whereas others probe the behavior of *electrons*, which may see a different effective width. This observation brings up the subject of *how to define* an appropriate width. One possibility is to

---

*Work supported in part by the Office of Naval Research under Contract No. N00014-76-C-0934.

[‡]Participant, IBM Summer Student Program.

define the width strictly on the basis of stoichiometry. Even then a number of alternatives exist. Take, for example, an abrupt (100) interface. One would normally assume that such an interface has no SiO_x layer. However, as Stern[6] has pointed out, the Si atoms in the last crystallographic plane of atoms on the Si side are bonded to two Si atoms and to two O atoms, corresponding to SiO stoichiometry. [For a (111) interface this plane of atoms corresponds to $SiO_{0.5}$ stoichiometry.] Since it is only one plane of atoms, one might define the width to be the distance between atoms on either side of that plane, which is about 3 Å.

Another possibility is to define the width having in mind the behavior of electrons. Clearly, even if one had an abrupt interface, electrons in the vicinity of the first few layers, including core electrons, would not behave as bulk electrons do. One measure of the width could be taken to be the range over which the self-consistent potential is different from bulk values. Calculations by Baraff, Appelbaum, and Hamann have found this width to be very small for the Ge-GaAs interface. No calculations are available for the Si-SiO_2 interface. Other properties, such as the dielectric constant, which depend on excited states, may in fact exhibit differences from bulk values over an even wider range.

In this paper, we attempt to shed some light on the question of the interface stoichiometry by constructing continuous-random-network (CRN) models. This technique has been widely used in modeling the atomic structure of amorphous solids. In fact, pioneering work using CRN models was done for amorphous SiO_2 by Evans and King[8] and by Bell and Dean.[9] Later, the technique was used to study amorphous Si, Ge and other materials.[10-12]

In carrying out the present study we first built several models of amorphous SiO_2. We used the "Framework Molecular Models" manufactured by Prentice-Hall, Inc., of Englewood Cliffs, New Jersey. They consist of sturdy plastic tubing which can easily be cut to size and which fits snugly over the prongs of small metal units, representing atoms. The units used to represent Si atoms have four metal prongs pointing in tetrahedral directions. Oxygen atoms, which ought to have two prongs, were represented by small pieces of bare copper wire bent at an angle of 144°, which corresponds to the observed average angle for amorphous SiO_2.[13] Since, however, the plastic tubing was slightly bendable, the resulting Si-O-Si angles actually have a distribution about 144°. The Si-O bonds were represented by blue tubing cut to 1.6 inches, so that one inch corresponds to 1 Å. Every Si atom was always connected to four O atoms in the tetrahedral directions, and every O atom was always connected to two Si atoms in order to maintain proper stoichiometry. No excessive bending of the plastic tubing connecting atoms was allowed. O atoms were not allowed to approach each other by less than about 2 inches (O nearest-neighbor atoms in crystalline SiO_2 are separated by about 2.5 Å). While in crystalline SiO_2 one finds only six-fold rings (counting an Si-O-Si unit as a "fold"), we allowed from four-fold to eight-fold rings.[9]

For the purpose of modeling the interface, we first had to choose an orientation for the crystalline Si substrate. We chose the (100) orientation because it is the one used most often in devices. We therefore constructed a substrate of several (100)

planes of crystalline Si. Red tubing cut to 2.35 inches was used for the Si-Si bonds
(1 inch = 1 Å). The dimensions of the exposed surface were approximately 2 ft × 3
ft. The objective of the study was to continue this structure and build several layers
of amorphous SiO_2.

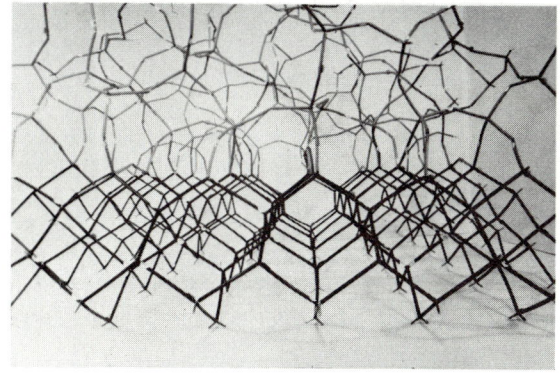

Fig. 1 A photograph of the Si-SiO_2 CRN
model showing a stoichiometrically abrupt
interface.

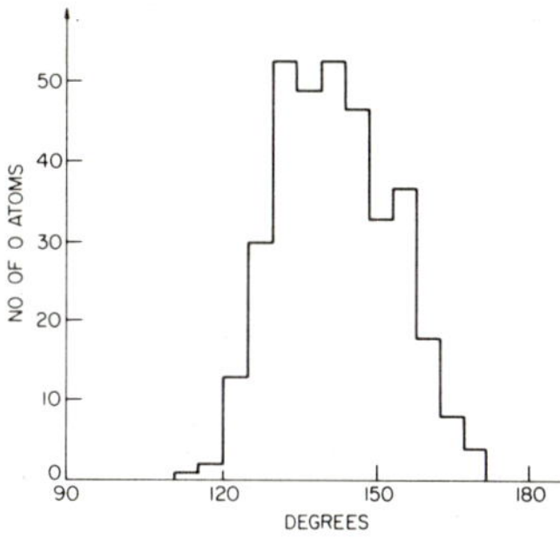

Fig. 2 A histogram of the
Si-O-Si angle distribution.

At the first stage of our interface studies we relied totally on visual tests. Clearly,
we could use both Si-Si bonds (red) and Si-O bonds (blue) and construct an SiO_x
layer of arbitrary thickness.   We could also allow an arbitrary amount of
roughness.[3,14] Our objective was therefore to attempt to construct a model with an
interface region which is as smooth as possible and as thin as possible, without
excessive distortions of the bonds and without dangling bonds.   Such an exercise
would thus not answer the question of how thick the interface region actually is in
real systems (it probably varies from sample to sample), but it would provide an
indication as to how thin it *can* be and perhaps set a lower limit. Our initial work
indicated that an SiO_x region of about 4-6 Å was necessary to avoid excessive
distortions. We finally succeeded, however, in constructing a model with a completely abrupt interface, i.e., without an intermediate layer containing both red and blue
bonds. The distortions of the bonds were similar to those of the other models.   A
photograph of this model is shown in Fig. 1. The conclusion that one reaches is that
an abrupt interface in real samples is not ruled out by geometrical considerations and
should be achievable, unless *energetic* considerations favor an interface with an SiO_x
layer several Ångstroms thick.

In the second stage of our work we attempted to calculate which configuration would
actually be favored by energetic considerations. For this purpose we used one of the
CRN models which had an SiO_x interface region of about 6 Å. The coordinates of
all the atoms were measured with a ruler to an accuracy of better than 0.1 inches.
These coordinates were then used as a starting point in a computer relaxation
program similar to the one used in Ref. 12.   The total force on each atom was
calculated as the sum of three types of forces:

(a) a bond-stretching force with a potential

$$U_s = \frac{1}{2} \alpha \sum_i \left[(R - R_i)^2 - R_0^2 \right]^2$$

where R is the position vector of the atom of interest, R_i is the position vector of a nearest neighbor, R_0 is the measured crystalline bond length, and α is the bond-stretching force constant.

(b) a bond-bending force with a potential

$$U_b = \frac{1}{2} \beta \left[(R - R_j) \cdot (R - R_k) - C \right]^2$$

where R_j and R_k are the position vectors of the two nearest neighbors, and C is a constant such that the potential is minimized for a bond angle of 144°, if the atom at R is an oxygen, and 109.6° if the atom at R is a silicon. β is the bond-bending force constant and is different for Si and O atoms (see below).

(c) a van der Waal's force with a Lennard-Jones potential with a cutoff:

$$U_{LJ} = -\frac{A}{|R-R_i|^6} + \frac{B}{|R-R_i|^{12}} \qquad |R - R_i| \leq R_c$$

$$= 0 \qquad\qquad\qquad\qquad\qquad |R - R_i| > R_c$$

The main role of this force is to keep the O atoms from going too close to one another. The above forces are certainly not the totality of forces on atoms in solids. They are, however, a minimal set needed for a stable solid and have been found to work well within the context of CRN models.[12] The choice of force constants will be discussed later.

In the relaxation procedure, each atom was displaced in turn toward its equilibrium position determined by the total potential $U = U_s + U_b + U_{LJ}$. This was done by computing the local gradient vector and moving the atom in the direction of the force by small steps. The cycle was repeated until the atoms approached a good approximation to equilibrium in both energy and position.

Once the coordinates of the original model were relaxed, a search was made by the computer to locate all the Si-Si bonds lying between 6 and 5 Å from the top layer of the crystalline Si substrate. Oxygen atoms were then inserted at the midpoints of all such bonds and the relaxation program was run again. Fig. 2 shows a typical histogram for the Si-O-Si angle after relaxation. Once equilibrium was reached, the final elastic energy was recorded. Si-Si bonds lying between 5 and 4 Å were then eliminated by inserting oxygen atoms and relaxing. The procedure was repeated for each successive inch until all Si-Si bonds above the top layer of the substrate were eliminated. The result for our initial choice of force constants indicated that the elastic strain energy decreased as the width of the SiO_x region was decreased. The same result was obtained for the elastic distortion energy of the entire system and also for a bond-normalized elastic distortion energy.[15] We therefore repeated the entire procedure for a wide range of force constants within physically acceptable limits. In all cases an abrupt interface was preferred.

The final conclusion of this work is that an abrupt interface is not ruled out by either geometric or energetic considerations at $T = 0°$ K. The model we used is not reliable enough to allow us to conclude that the abrupt interface would in fact be formed under realistic growing conditions at finite temperatures. The effect of impurities, defects, such as dangling bonds, finite temperature, and other factors would have to be included before a more definite conclusion can be reached.

REFERENCES

1. J. Maserjian, **J. Vac. Sci. Technol. 11**, 996 (1974).
2. T. H. DiStefano, **J. Vac. Sci. Technol. 13**, 856 (1976).
3. C. R. Helms and W. E. Spicer, **Solid State Commun.**, in press.
4. S. I. Raider and R. Flitsch, **J. Vac. Sci. Technol. 13**, 58 (1976).
5. W. L. Harrington, R. E. Honig, A. M. Goodman, and R. Williams, **Appl. Phys. Lett. 27**, 644 (1975).
6. F. Stern, **Phys. Rev. B 17** (May 15, 1978).
7. G. A. Baraff, J. A. Appelbaum, and D. R. Hamann, **Phys. Rev. Lett. 38**, 37 (1977).
8. D. L. Evans and S. R. King, **Nature 212**, 1353 (1966).
9. R. J. Bell and P. Dean, **Nature 212**, 1354 (1966); **Phil. Mag. 25**, 138 (1972).
10. D. E. Polk, **J. Non-Cryst. Solids 5**, 365 (1971).
11. P. Steinhardt, R. Alben, M. G. Duffy, and D. E. Polk, **Phys. Rev. B 8**, 6021 (1973).
12. M. Long, P. Galison, R. Alben, and G. A. N. Connell, **Phys. Rev. B 13**, 1821 (1976).
13. R. L. Mozzi and B. E. Warren, **J. Appl. Crystallogr. 2**, 164 (1969).
14. J. S. Johanessen, W. E. Spicer, and Y. E. Strausser, **J. Appl. Phys. 47**, 3028 (1976).
15. The procedure of monitoring only the changes in the strain energy is equivalent to assuming that the extra oxygen atoms at the interface were taken from the outside of the cluster, where they were bonded with no strain at all.

344

STUDIES OF THE Si-SiO$_2$ INTERFACE BY MeV ION SCATTERING

L. C. Feldman, I. Stensgaard, and P. J. Silverman
Bell Laboratories, Murray Hill, N. J. 07974 U.S.A.

T. E. Jackman*
University of Guelph, Guelph, Ontario, Canada

ABSTRACT

Rutherford backscattering and channeling have been used to determine the interface properties of thin layers of SiO$_2$ on Si(110). The results are consistent with a model which assumes either: 1) two non-registered layers of Si and a sharp transition to SiO$_2$ or 2) approximately one non-registered monolayer and ~5 Å of a Si rich oxide. These two models correspond to an interface width of 4 Å and 7 Å respectively.

INTRODUCTION

Rutherford backscattering spectrometry provides a quantitative determination of the stoichiometry of thin films and surface oxides. For interface studies of very thin films on single crystals, the technique is often combined with channeling for improved "depth resolution" and accuracy. Recent theoretical and experimental studies have demonstrated that the single crystal substrate contribution can be accurately predicted, thus allowing increased confidence in the use of the backscattering-channeling technique.[1,2] Since channeling depends on the crystalline nature of the substrate, this technique also tests the single crystal quality of the interface region. We report on such ion scattering experiments using an optimum experimental geometry. Our new understanding of the substrate contribution and the improved sensitivity allow us to draw the following conclusions about the Si-SiO$_2$ interface on Si(110). There is at least one monolayer and no more than two monolayers of disordered Si below the oxide and the interfacial region contains less than 6 Å of non-stoichiometric SiO$_2$.

Ion backscattering in the MeV region has been applied to the study of the SiO$_2$ stoichiometry by other authors.[3-5] The study of the Caltech group[3] has been particularly noteworthy and has guided the work presented here. In the

---

*Resident visitor, Bell Laboratories, Murray Hill, N. J.

regions of overlap our results are in good agreement with that work; the improvements in understanding and experimental technique mentioned above have allowed us to extend the application of the technique to the very thin layer-interface area not previously accessible.

The origin of the backscattering spectrum for the systems of interest is shown schematically in Fig. 1. For thin layers and usual experimental conditions the oxide region cannot be resolved from the substrate. The scattering spectrum consists of a single silicon peak corresponding to scattering from Si in the oxide, non-registered Si in the surface region and a contribution from the first monolayers of crystalline Si and at lower energy, an oxygen peak. Standard techniques can be used to convert these peak areas to atoms/cm^2 with approximately 5% accuracy. Much of this study then involves the determination of the areal density of the Si and oxygen as a function of oxide thickness.

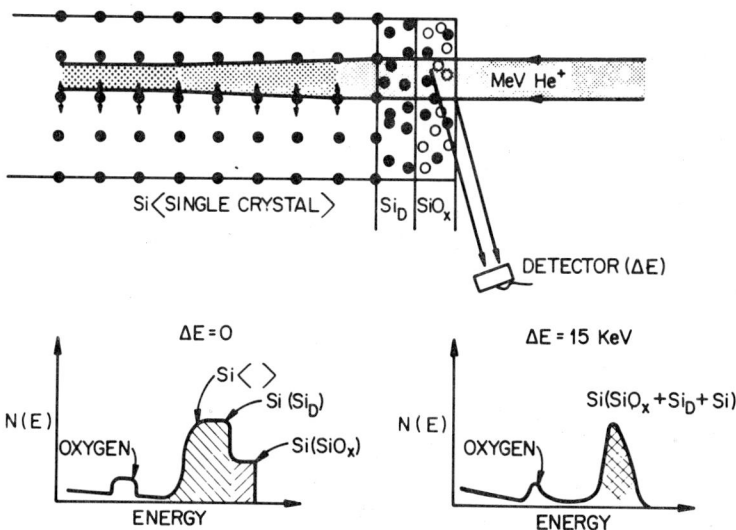

Fig. 1.   Schematic of backscattering spectrum for 1 MeV He incident on an aligned sample consisting of SiO_x, disordered Si and a single crystal substrate. ΔE refers to the detector resolution.

EXPERIMENTAL

MeV ion backscattering systems have been described by a number of authors.[6] In this particular study the major difference has been the use of grazing angle techniques for improved depth resolution as described by Williams.[7] Figure 2 compares the spectrum from a <110> Si sample with ∿15 Å of oxide in the grazing exit angle geometry to the more conventional large angle geometry. Of import is the reduction of background under the Si peak as well as the oxygen region. While the oxygen peak could be extracted reliably in both cases the reduced background increases sensitivity. In the case of the Si peak the improvement is more fundamental since the shape of the background is simply not determined in the 180° case. Using a method of background

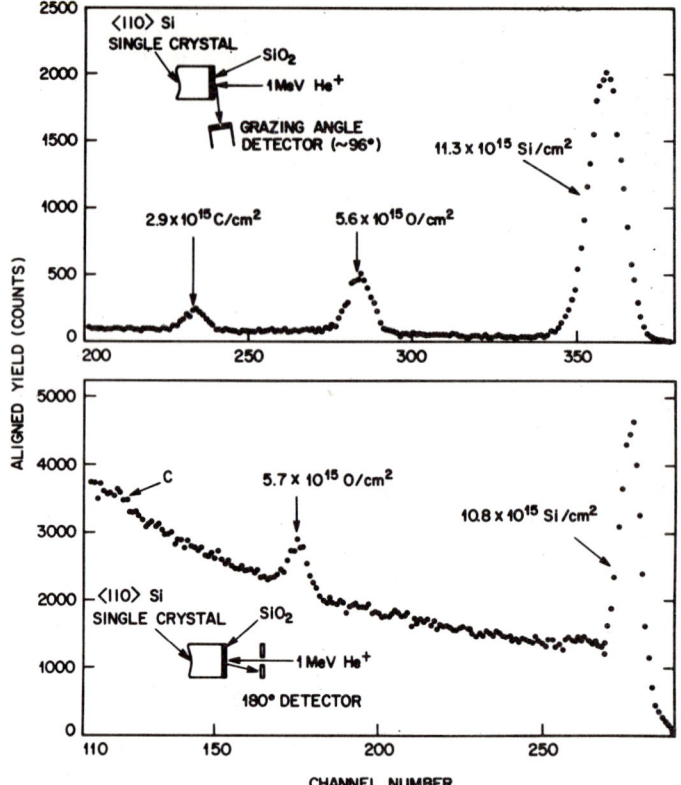

Fig. 2. Comparison of backscattering spectra for a film
of ~15 Å of oxide on Si <110> in the case of:
a) grazing exit angle geometry and
b) conventional large angle geometry.

subtraction described in Ref. 2, we find the peak areas for the two different geometries in reasonable agreement.

Determination of the intensity of the peaks in terms of atoms/cm^2 makes use
of a "standard" which was calibrated by ion scattering techniques and relied
on the stopping power of He$^+$ in Silicon.[8] The 5% uncertainty we ascribe
to this quantity appears to be the largest known error in the measurement
and enters linearly in all areal density determinations. The Si samples
were 10-13 ohm-cm p-type and oxidized by the CO_2+SiH_4 process. In some
cases determinations were made on the native oxide grown in air after
chemical etch.

RESULTS

Typical spectra for two thin oxide coverages are shown in Fig. 3. Note
that in this comparison SiO_2 was removed from the oxide by the chemical

etch. These spectra indicate that for thicknesses greater than 1.5×10^{15} O/cm^2 (corresponding to ~5 Å of SiO or ~3.5 Å of SiO$_2$), the oxide is stoichiometric SiO$_2$.

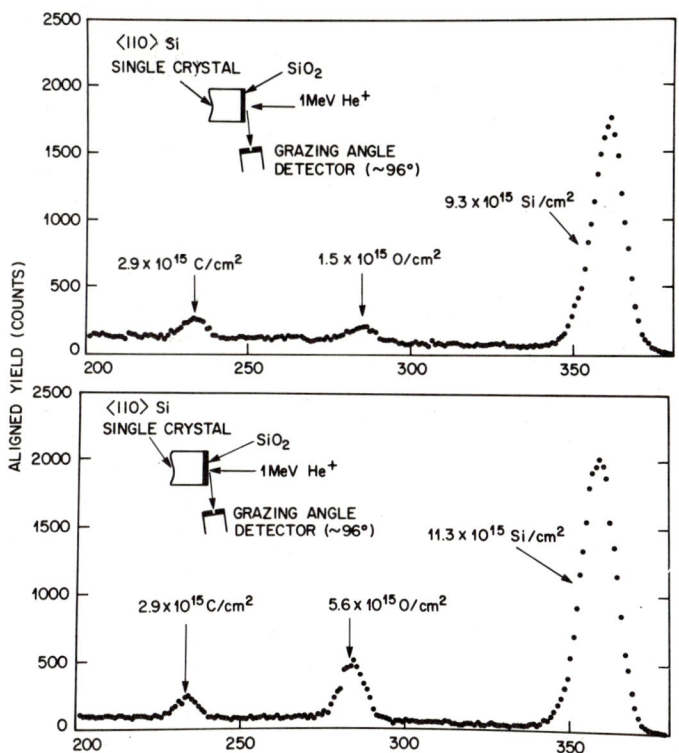

Fig. 3. Two spectra corresponding to: a) 5.6×10^{15} O/cm^2 and b) 1.5×10^{15} O/cm^2. The differences of the areal densities for the Si and Oxygen indicate that SiO$_2$ was removed in this case.

The Si to oxygen intensities are shown in Fig. 4 for a range of oxides up to 790 Å. There is a very good fit on this scale to a straight line given by SiO$_2$+8.6×10^{15} Si/cm^2. This result is very similar to that found in Ref. 3 in which a constant offset of Si was required to fit the results. This offset will be discussed in detail in the next section; however, it is evident that this data cannot be fit by any other reasonable stoichiometric ratio.

A similar plot over a more limited range of oxides is shown in Fig. 5. Note again that the data can be fit well by a straight line of the type described above although a significant number of points fall below the line at very small coverages.

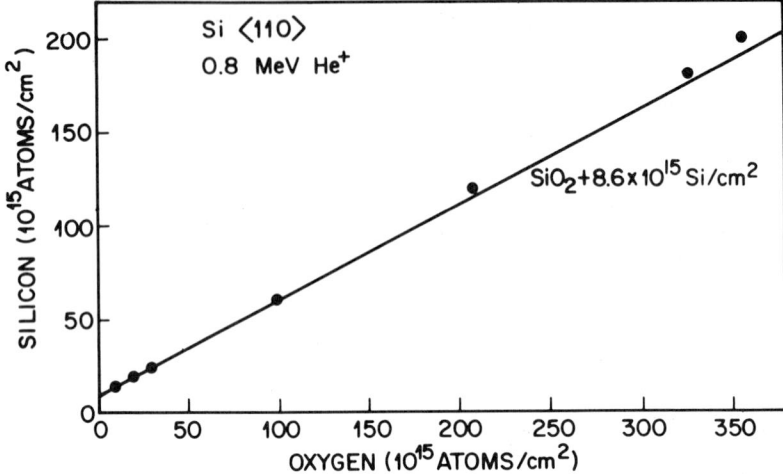

Fig. 4. Si vs. oxygen areal densities for a range of
oxides up to ~790 Å.

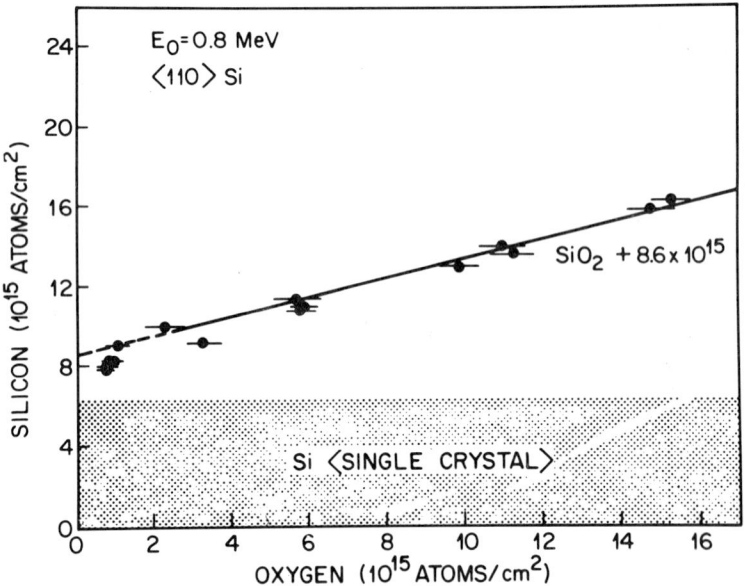

Fig. 5. Si vs. oxygen areal densities for a range of
oxides up to ~40 Å.

DISCUSSION AND CONCLUSIONS

As mentioned before the Si peak arises from contributions from the oxide,
non-registered Si and the first monolayers of the single crystal. We consider this latter contribution. Ion scattering phenomena can be simulated

in a computer study and have been shown to be in good agreement with experiment. The recent interest in the use of this analysis technique for surface studies has spurred further work in the prediction of surface peak intensities (S.P.) using Monte Carlo computer models which include screened potentials, thermal vibrations and many known quantities.[1],[2] For the system studied here the simulation studies indicate that the surface peak for 800 keV He[+] incident on a perfect Si crystal along the <110> direction corresponds to 6.4×10^{15} Si/cm^2, which makes up the bulk of the excess 8.6×10^{15}/cm^2. We acquire confidence in these predictions by noting that there is reasonable agreement between S.P. measurements and theory on Si under U.H.V. (no oxide) conditions.[1] One cannot expect better agreement in light of the fact that the geometry of the surface atoms is not known. Recent U.H.V. measurements on Si(100) show an excess of two monolayers indicating that two monolayers are not registered with respect to the bulk. This is in agreement with the most recent interpretations of the (100) LEED pattern.[9]

The magnitude of the substrate contribution was not appreciated in the work of Ref. 3 and thus erroneously led to the conclusion of an interface with a Si excess of 6×10^{15}/cm^2.

In Fig. 6 we show the results of our measurements compared to the predictions of some simple models. We treat the interface as consisting of

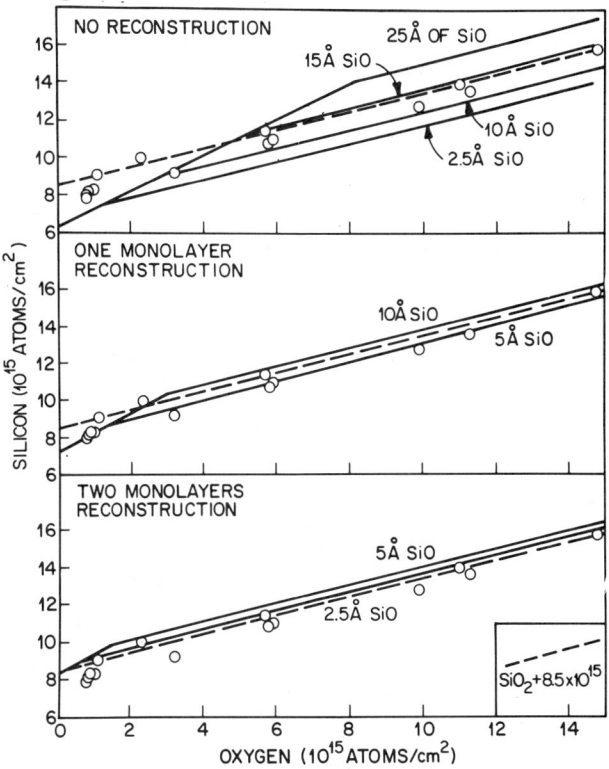

Fig. 6. Comparison of thin oxide data to the predictions of some simple models involving reconstructed Si, SiO and SiO$_2$.

disordered Si, a region of SiO and the remaining oxide being SiO_2. Since one layer of Si(110) corresponds to $.96 \times 10^{15}$ atoms/cm^2 our data may be fit by simply two non-registered layers of Si and then SiO_2. A better fit, which includes the low oxide coverage, corresponds to one non-registered layer and ~5 Å of SiO. The discussion of Fig. 3 indicated that there can be no greater than 5 Å of SiO. It is clear from Fig. 6 that there must be at least one and no more than two disordered monolayers.

We note two further facts concerning the interpretation of these results. The number of disordered layers is not necessarily a quantized number, but has been taken so for simplicity. A layer appears non-registered to the channeled beam depending on its displacement relative to the shadow cone radius. In this case a displacement of ~.15 Å would correspond to non-registration. Secondly a more complex analysis might include an oxide coverage dependent strain of the Si. Such an effect could also give rise to the Si vs. O behavior at low coverage.

ACKNOWLEDGMENTS

Valuable discussions with J. R. MacDonald and J. A. Davies and technical assistance of M. L. Shombert are acknowledged. One of us (T.E.J.) held a postgraduate scholarship from the National Research Council of Canada.

REFERENCES

1. R. L. Kauffman, L. C. Feldman, P. J. Silverman and R. A. Zuhr, Appl. Phys. Lett. 32, 93 (1978).

2. L. C. Feldman, R. L. Kauffman, P. J. Silverman, R. A. Zuhr and J. H. Barrett, Phys. Rev. Lett. 39, 38 (1977), and references therein.

3. T. W. Sigmon, W. K. Chu, E. Lugujjo, and J. W. Mayer, Appl. Phys. Lett. 24, 105 (1974); W. K. Chu, E. Lugujjo, J. W. Mayer, and T. W. Sigmon, Thin Solid Films, 19, 329 (1973).

4. G. Della Mea, A. V. Drigo, S. Lo Russo, P. Mazzoldi, S. Yamaguchi, and G. G. Bentini, Appl. Phys. Lett. 26, 147 (1975).

5. W. F. van der Weg, W. H. Kool, H. E. Roosendaal and F. W. Saris, Rad. Eff. 17, 245 (1973).

6. T. M. Buck and J. M. Poate, J. Vac. Sci. Technol. 11, 289 (1974).

7. Williams, J. S. (1975) Ion Beam Surface Layer Analysis, Plenum Press, New York.

8. J. F. Ziegler and W. K. Chu, At. Data Nucl. Data Tables 13, 463 (1974).

9. J. A. Applebaum and D. R. Hamann, Surface Science, to be published.

TRANSMISSION ELECTRON MICROSCOPY OF MICRO-STRUCTURAL DEFECTS IN Si-SiO$_2$ SYSTEMS
-- Si CLUSTERS IN SiO$_2$ FILM --

Jen-Jon Chen and Takuo Sugano
Department of Electronic Engineering,
The University of Tokyo, Bunkyo-Ku, Tokyo 113, Japan.

ABSTRACT

Transmission electron microscopy at 100KV and selective area electron diffraction at 1000KV were carried out for studying micro-structural defects in Si-SiO$_2$ systems. It has been found that unoxidized Si clusters, whose diameter is 15-50Å, exist within 10-15Å from the interface. The area density is about $6 \times 10^{11} cm^{-2}$.

INTRODUCTION

Micro-structural defects in thermally grown SiO$_2$-Si systems such as unoxidized Si clusters remaining in SiO$_2$ film were pursued by transmission electron microscopy and selective area electron diffraction using JEM100B at 100KV and JEM1250 at 1000KV, respectively.

EXPERIMENTAL PROCEDURES

Phosphorus-doped Si wafers with a resistivity of 5 to 8Ω-cm and surface of <100> orientation were cleaned by rinsing in tri-chlorethylene, aceton and methanol successively with ultrasonic agitation and boiling in a mixed solution of 1:1 H$_2$SO$_4$:H$_2$O$_2$ (in volume ratio). Just before oxidation in dry oxygen, the wafers were etched by 25μm in 1:25 HF:HNO$_3$ (in volume ratio) solution. Oxygen gas was dried by a cold trap containing dry ice and methanol.

Temperature and time of oxidation and annealing, and symbols for samples were listed in Table 1.

The area, to be observed, of a wafer was thinned to 1μm or less by electrochemical etching of Si substrate in 2.5 percent HF solution from the unoxidized side, and the area was localized by illuminating with 4mW He-Ne laser. The thickness of remaining Si in the etched area measured by the transmitted laser beam and controlled electronically. The control circuit was illustrated in Fig. 1.

TABLE 1   Temperature and Time of Oxidation and
Annealing, and Symbols for Samples

oxidation \ annealing	as grown	800°C for 30 min. in N_2	500°C for 30 min. in H_2	oxide thickness
1100°C 40 min.	D10TS			ca. 770A
1100°C 20 min.	D10	D10N2	D10H2	ca. 500A
1150°C 15 min.	D15	D15N2	D15H2	ca. 500A
1200°C 10 min.	D20	D20N2	D20H2	ca. 500A

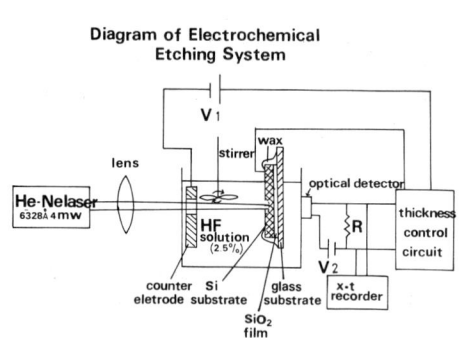

Fig. 1. The diagram of electrochemical
etching system.

Photo. 1. Tramsmission electromicrograph of unoxidized Si
substrate.

EXPERIMENTAL RESULTS

Since no change of images was found during the observation, effects of
heating, damaging and contaiminating the samples by the electron beam were
not significant.

Transmission Electromicrograph of Unoxidized Wafers

In order to separate effects of the preparation of both surfaces of Si
substrate before oxidation and the electrochemical etching on transmission
electromicrograph, the transmission electromicrograph of unoxidized samples
were taken as shown in Photo. 1. and confirmed different from those of
micro-structural defects induced by oxidation.

Small Si Clusters in SiO$_2$ Film

Photo. 2. is a typical transmission electromicrograph at 100KV of SiO$_2$ film. Such transmission electromicrograph cannot be observed for too thick SiO$_2$ film and/or with too high acceleration voltage of the electron microscope, because the transmitted electron beam is scattered randomly in too thick SiO$_2$ film and cannot make clear images, and too high acceleration voltage results bright background and therefore weak contrast between the images and the background.

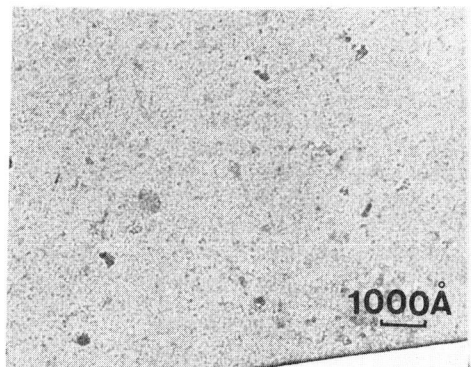

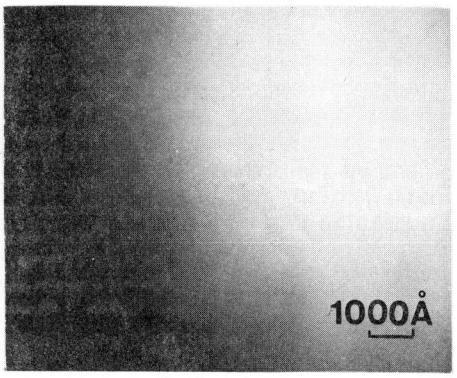

Photo. 2. Si clusters in SiO$_2$ film of D15H2.

Photo. 3. Dark-field image, on (200) reflections of Si clusters in SiO$_2$ film of D10TS.

The big black dots (200Å in diameter) in Photo. 2. were due to big Si clusters remaining on SiO$_2$ film after electrochemical etching process, which were identified by selective area electron beam diffraction.

The small black dots (15-30Å in diameter) were due to Bragg reflection of electrons by small Si clusters in amorphous SiO$_2$ film. This was confirmed by the following. In order to remove the big Si clusters, the samples were boiled in 8cc:17cc:3g water:ethylene-diamine:pyrocatechol solution for 3 minutes. After the big Si clusters were removed completely, the transmission electromicrographs of SiO$_2$ film were taken with JEM100B at 100KV for various tilting angle from <100> orientation. The diameter of the small black dots is 15-30Å and the density is $4-6\times10^{11}$cm^{-2}. The particles, which give the images as small black dots, are surely existing in SiO$_2$ film and crystalline, because they can be seen in bright field. Those particles were identified by the selective area diffraction technique, where the diameter of the area was 600Å, using JEM1250 at 1000KV. Ring diffraction patterns of Si were obtained and compared with ASTM as listed in Table 2. on the next page. The spacings of the measured rings coincide with those of ASTM within error of ±5 percent. Furthermore, the transmission electromicrograph of the SiO$_2$ film was taken in dark field on (200) reflections of Si crystal as shown in Photo. 3. By reoxidizing this SiO$_2$ film at 1150°C for 10 minutes, these small black dots have disappeared as shown in Photo. 4. and the ring diffraction patterns too. These two experimental results indicate that those small black dots are due to small unoxidized Si clusters in SiO$_2$ film.

TABLE 2   Interplanar Spacing of Si in ASTM and Experiment

ASTM			Experiment		
hkl	$d(\overset{\circ}{A})$	I/I_1	$d(\overset{\circ}{A})$	error	intensity
111	3.138	100	3.25	+3.5%	very strong
220	1.920	60	1.81	-5 %	weak
311	1.638	35	1.629	-0.6%	very strong
422	1.108	17	1.1	-0.6%	strong

Average diameters of small unoxidized Si clusters remaining in SiO_2 film in terms of temperature and time of oxidation and annealing for samples were listed in Table. 3.

TABLE 3   Average Diameters of Unoxidized Si Clusters in
terms of Temperature and Time of Oxidation and
Annealing for Samples

annealing / oxidation	as grown	800°C for 30 min. in N_2	500°C for 30 min. in H_2
1100°C 40 min.	15–30$\overset{\circ}{A}$		
1100°C 20 min.	15–30$\overset{\circ}{A}$	20–30$\overset{\circ}{A}$	
1150°C 15 min.	20–40$\overset{\circ}{A}$		20–40$\overset{\circ}{A}$
1200°C 10 min.	40–50$\overset{\circ}{A}$		40–50$\overset{\circ}{A}$

In order to find the location of these small unoxidized Si clusters in SiO_2 film, the SiO_2 was etched by 10-15$\overset{\circ}{A}$ from the previous interface with $NH_4F \cdot HF$ solution.   Then black dots have not been observed in SiO_2 film for any tilting angle from <100> orientation.   Therefore, the small unoxidized Si clusters exist within 10-15$\overset{\circ}{A}$ from the Si-SiO_2 interface.

DISCUSSION

The existence of unoxidized Si clusters in SiO_2 film is suggesting that Si surface is not oxidized uniformly, but non-uniformly along local defects, probably caused by the tension due to the expansion of SiO_2 in volume when it has been formed, and unoxidized parts of Si near the interfaces have been included in SiO_2 as the small Si clusters.   This model of oxidation is illustrated in Fig. 2.

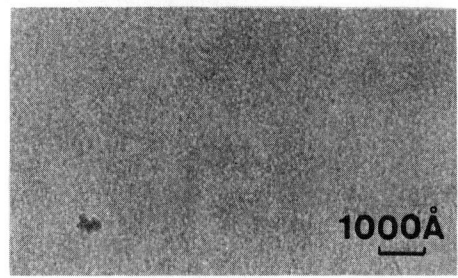

Photo. 4. Transmission electromicrograph of SiO_2 film of D10TS
reoxidized at 1150°C for 10 min.

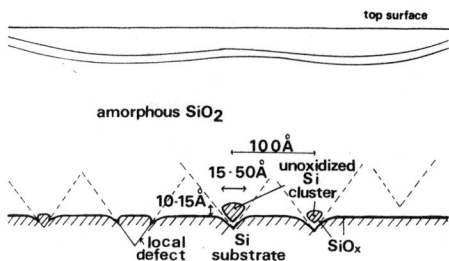

Fig. 2. Model of Si-SiO_2 interface
morphology, showing Si clusters in
SiO_2 film due to non-uniform
oxidation process.

COMPARISON WITH OTHER EXPERIMENTAL RESULTS

Ion scattering (ISS) experiment indicated 20% excess Si in a layer of 15 to
20Å thick in front of the SiO_2 interface. (Ref. 1) The amount of excess Si
atoms has been calculated as 1.4×10^{14} atoms cm^{-2} per monomolecular layer.
Our experiments show that the area density of unoxidized Si clusters is
about $6 \times 10^{11} cm^{-2}$ and the diameter is 15-50Å. This means that the amount of
excess Si atoms is about 6×10^{13} atoms cm^{-2} per monomolecular layer. It
seems somewhat small in comparision with ISS experimental data. This
discrepancy may be due to excess Si atoms, in SiO_x ($1 \leq x \leq 2$) transition layer,
which were not taken into account in our experiment.   Auger electron
spectroscopy revealed that unoxidized Si clusters whose diameter is 5-15Å
exist within 5-10Å from the SiO_2-Si interface. (Ref. 2) The result is
consistent with ours.

CONCLUSION

It has been confirmed for Si-SiO_2 systems made by oxidizing phosphorus doped
(100) Si wafers with a resistivity of 5-8Ω-cm in dry oxygen that unoxidized
Si clusters, whose diameter is dependent on the oxidation temperature and
annealing conditions and ranged from 15 to 50Å, exist within 10-15Å from the
interface and the area density is about $6 \times 10^{11} cm^{-2}$.

ACKNOWLEDGEMENTS

The authors wish to thank Prof. T. Homma, Institute of Industrial Science,
Prof. H. Yanagida, Department of Industrial Chemistry, Messrs. K. Adachi,
H. Nishizawa and H. Zunagawa, Engineering Research Institute, the University
of Tokyo and Prof. Y. Tanabe, Department of Industrial Chemistry, Tokyo
Metropolitan University for many helpful discussion.

REFERENCE

(1) W. L. Harrington, R. E. Honig, A. M. Goodman and R. Williams, Low-
     energy ion-scattering spectrometry (ISS) of the SiO_2/Si interface,
     Appl. Phys. Lett. 27, 644 (1975)

(2) J. S. Johannessen, W. E. Spicer and Y. E. Strausser, An Auger analysis
of the SiO_2-Si interface, J. Appl. Phys. 47, 3028 (1976).

A HIGH RESOLUTION ELECTRON MICROSCOPY STUDY OF THE Si - SiO$_2$ INTERFACE

Ondrej L. Krivanek
Dept. of Mats. Sci. & Min. Eng., U. of Ca., Berkeley, Ca. 94720

and D. C. Tsui, T. T. Sheng and A. Kamgar
Bell Laboratories, Murray Hill, New Jersey   07974

ABSTRACT

The structure of (100), (111) and (911) interfaces between Si and thermally grown, dry SiO$_2$ has been studied in cross-section by high resolution electron microscopy (HREM), at about 3Å resolution. The three types of interfaces have been found to be rather similar. The Si crystal transforms into the amorphous oxide quite abruptly - within about 3Å. One plane high atomic steps exist on the Si surface, separated by typically 20-40Å. They produce a surface roughness of about 4Å over distances of around 50Å. Most interfaces also show a longer range modulation of a height of ~ 4-8Å, with a wavelength of 200 - 500Å. No evidence of a transition region of a nonstoichiometric oxide has been found, and the 10Å-wide dark band reported by us after a preliminary investigation is shown to be an imaging artifact. However, the sensitivity of the HREM technique to changes in the composition of the amorphous oxide is relatively poor.

INTRODUCTION

The structure of the Si-SiO$_2$ interface has been studied by many different techniques. A transition region of nonstoichiometric oxide has always been found, but its width Δ tends to vary, roughly in proportion to the resolution of the technique used. Thus Δ = 600Å from Rutherford back-scattering[1], 20Å from low energy ion back-scattering[2], 35Å and later 10Å by Auger spectroscopy[3,4], and 20Å by electron spectroscopy for chemical analysis[5]. Additional results on the interface structure have been deduced from surface free energy measurements[6], and from the measurement of the electrical properties of the interface[7].

High resolution electron microscopy has, since several years ago, been able to directly resolve the 3.2Å Si crystal lattice. It should give results of similar resolution on the interface structure. Blanc, Buiocchi, Abrahams and Ham[8] have recently used the EM technique, but their resolution was only about 10Å. They found the (100) interface to be smooth and free of Si protusions within this limit. Improving the resolution to 3Å yields several further results, as this study will demonstrate.

SPECIMEN PREPARATION AND THE EXPERIMENTAL TECHNIQUE

The oxides were between 1000 and 1500Å thick, and were grown on 10-30 Ωcm p-type Si wafers in dry air at 1100°C, and subsequently annealed in H$_2$ at 380°C. Typical peak mobilities at 4.2°K on MOSFETs made with identical oxides were 10,000 - 15,000 cm^2/Vsec on the (100) surface, 5000 cm^2/Vsec on (911), and about 1000 cm^2/Vsec on (111).

The thin cross-sections were prepared by cutting a slice normal to a [011] direction lying in the plane of the interface, thinning the slice mechanically to about 0.1 mm, ion beam milling till perforation at 7kV, and finally by milling at 2kV and a low incidence angle to remove any surface damage. For protection of the surface oxide against the milling ions, several MOS devices were stacked on top of each other prior to the slicing, and bonded together with epoxy.

In the immediate vicinity of the hole through the MOS cross-section, the interface is only 100 - 200Å thick, and therefore ideally suited for HREM. A low magnification image of the prepared sample is shown in Fig. 1a.

The interfaces were examined in a Siemens 102 EM at 125kV and electron-optical magnifications of 3000x to 500,000x, using a doubly-tilting specimen holder to allow the proper orientation of the interface to be achieved. Mainly because of the poor coherence of the electron beam from our thermionic electron gun, the Si crystal lattice was only very poorly visible directly on the micro-scope screen and the finer details of the interface structure could only be studied on micrographs recorded with typically 6 sec. exposure. Planned improvements of the microscope performance should however enable us to observe the interface detail directly on the fluorescent screen.

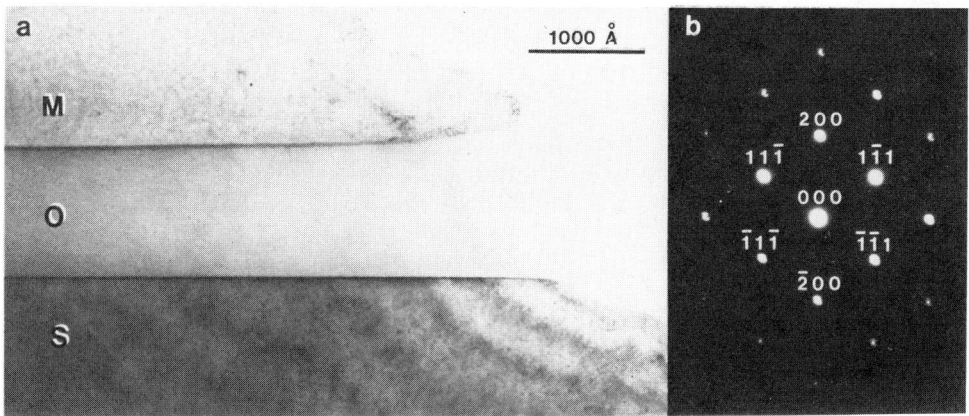

Fig. 1. a) Low magnification image of the finished cross-section through a (100) MOS. b) Electron diffraction pattern from the Si crystal.

IMAGE INTERPRETATION

The high energy electrons are scattered by the Si crystal into Bragg beams and by the oxide into diffuse rings. The electron microscope brings the diverging beams and the diffuse scattering back into coincidence. The resulting interference patterns between overlapping Bragg beams have the periodicity of the original lattice. A perfect imaging system would bring all the Bragg beams into the optimum coincidence for phase contrast imaging, and there would be a one-to-one correspondence between the image and the object. In practice, however, the coincidence can only be controlled by defocussing (changing the current through the electromagnetic objective lens), and by tilting the main beam with respect to the optic axis. The optimum coincidence can therefore be achieved only for a strictly limited number of Bragg beams. Moreover, poor coherence of the incident electron beam frequently makes the lattice fringes invisible by destroying the interference altogether.

In most of the experiments to be described here, the interface was viewed along the [011] direction, and we have used axial illumination and defocus values of about -1400 to -1900Å. This brings the 111 diffracted beams (Fig. 1b) into optimum coincidence with the main beam[9,10,11], and allows us to view the (111) planes as well as (200) and (220) planes with the minimum distortion possible in a microscope of the Siemens 102 type. Higher order planes are not visible due to the effects of limited coherence and microscope aberrations. Tilting the illumination so that the optic axis is half-way between the main beam and one of the Bragg beams reduces the coherence requirements, but the resulting image is likely to contain artifacts due to imperfect coincidence of the other beams.

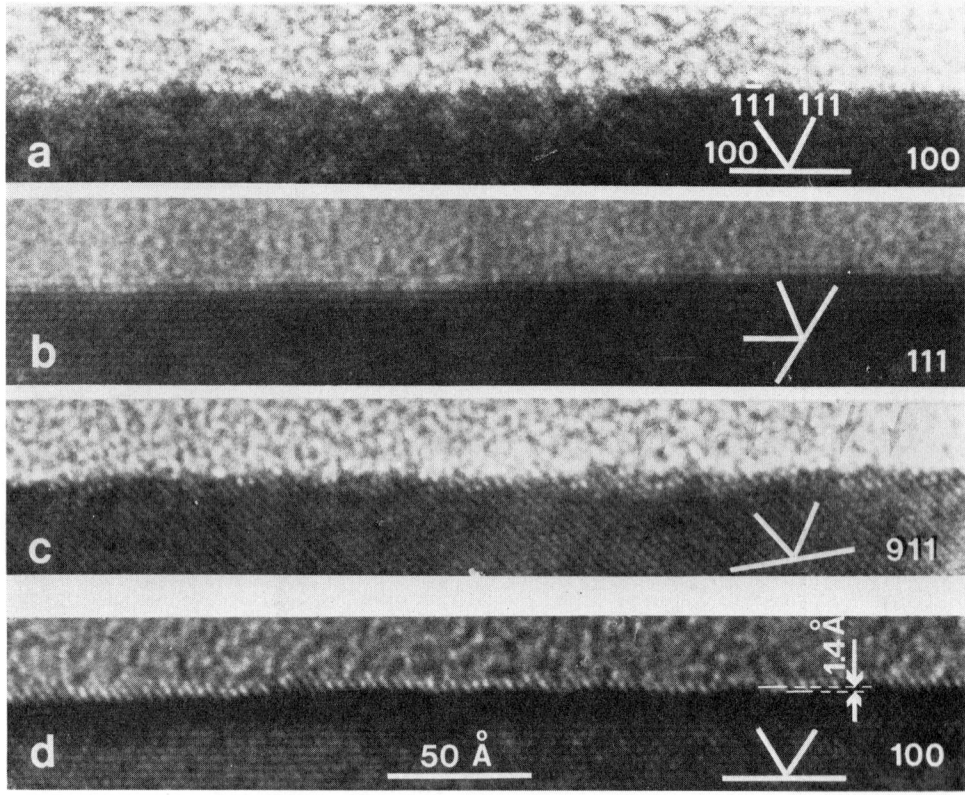

Fig. 2. High magnification images of (100), (111) and (911) interfaces. The major planes in the Si crystal are indicated. The best way to see the steps on the Si crystal surface is to rotate the page by 90° and view it from a low angle.

STRUCTURE OF (100), (111) AND (911) INTERFACES

Fig. 2 shows four interfaces, imaged edge-on parallel to [011]. {111} planes are resolved by interference between the 111 beams and the main beam,

{200} and {220} planes by interference among the 111 beams. One or the other type of lattice fringes may predominate depending on the specimen thickness and the exact imaging conditions.

The transition from the Si crystal into the oxide is quite abrupt, as can be best seen in a) and c), which were taken with axial illumination. The faint fringes above the crystal in b), and the dark band as well as the short fringes above it in d) are artifacts due to the use of tilted illumination. The tilted illumination however makes it easier to see the 3.2Å high (d_{111}) steps on the (111) surface, and the 1.4Å high (d_{400}) steps on (100). The images shown here are typical of 15 different interfaces examined, though the (111) interface shown in b) is perhaps slightly rougher than average.

The surface steps are typically 20-40Å apart, and can therefore be seen directly only in the thinnest parts of the cross-section. However, since regions thinner than 100Å could be susceptible to damage from the ion milling and the electron irradiation, the images selected for Fig. 2 are from regions about 100-200Å thick (as measured from thickness fringes[12] and the variation in lattice fringes' contrast[13]), and therefore show the superposition of several steps. The ~4Å roughness is however clearly visible up to thickness of ~400Å, where projection smooths the roughness out.

Fig. 3. Images of 900Å long segments of (100), (111) and (911) interfaces. The departures from the reference markers show how the interface undulates. The Si crystal peaks are marked with arrows. The black line below the interface in the lower two images arises from the variation of the Si crystal's scattering with thickness (thickness fringe).

LONG-RANGE MODULATION

On most interfaces we have found, in addition to the surface steps spaced by 20-40Å, a longer periodicity (Fig. 3). Its observation however requires that the ion milling produces a thin (less than 400Å) cross-section at least

1000Å long, and as this does not always happen, we are as yet unable to define precisely the conditions under which the modulation arises. So far it appears to be a general phenomenon, the modulation height being about 4-8Å, but the periodicity appears to vary from 200-300Å (911) to 500-600Å (111). Similar modulation of the same height, as well as the one-plane-high surface steps have been also observed in a preliminary investigation of much thicker wet oxides.

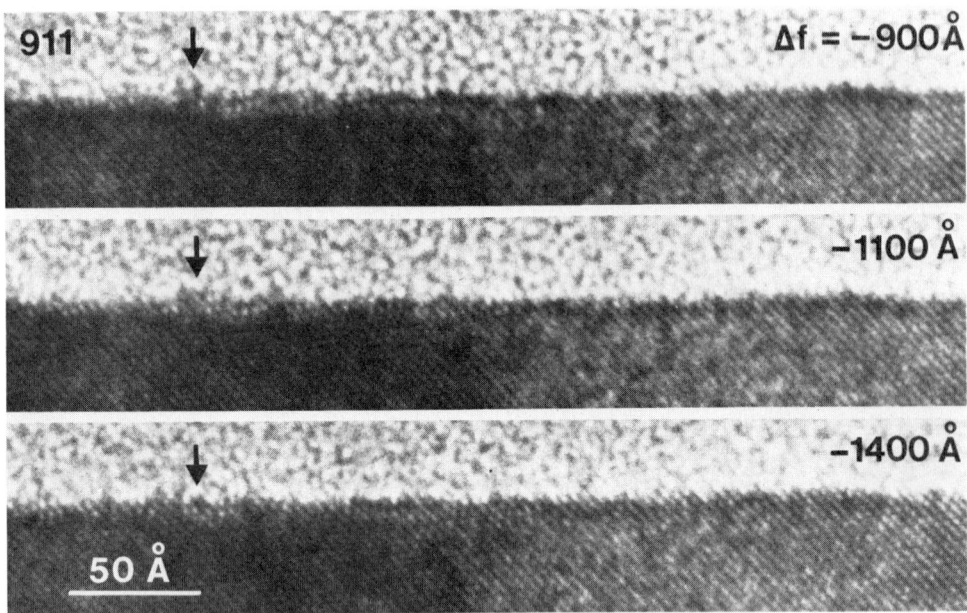

Fig. 4. A through-focal series of high magnification images of (911) interface. The arrowed feature does not disappear with a defocus change, but it is too small to be identified with a cluster of Si atoms.

TRANSITION REGION

The structures seen in the image of the amorphous oxide are produced by the superposition of many atoms throughout the depth of the material, and cannot, except under extremely favorable circumstances[14], be interpreted in terms of the detailed oxide structure. Moreover, the image of the amorphous region is very strongly dependent on defocus[9], as shown in Fig. 4. Though the figure also shows some features at the interface that are relatively defocus-independent, ascribing these directly to small (~5Å diameter) inhomogeneities such as clusters of Si atoms cannot be justified - the image is a projection of a 200Å thick oxide, and random superposition of atoms from various specimen depths is known to produce seemingly non-random image features[14]. The absence of stationary features larger than the arrowed one however shows that inhomogeneities much larger than 5Å are not present in the oxide.

Average changes in the oxide composition should be reflected in its average scattering power, and this might give rise to broad contrast features such as dark or bright bands near the interface. However, our calculations show that the difference in elastic scattering of SiO and SiO_2 never exceeds 10%

at any point in reciprocal space. Any such features would therefore be extremely faint. The dark band previously observed by us[15] in a tilted illumination image, obtained with a microscope of somewhat lower resolution than the one used here, is therefore almost certainly an artifact arising by a failure to achieve exact coincidence of the Bragg beams, and the same probably applies to another independent observation[16] of the "transition region" by EM.

CONCLUSIONS

We have shown that the transition from the Si crystal to the amorphous oxide happens quite abruptly (within about 3Å) . One plane high atomic steps separated by 20-40Å occur on all the interfaces, and most interfaces also appear to show a long range modulation of 4-8Å height, and a wavelength of 200-500Å. Overall, the structure of the three types of interfaces examined here appears to be quite similar.

We are grateful to G. Kaminsky for preparing the MOS wafers. One of the authors (O.L.K.)wishes to acknowledge NSF (Contract #DMR 77-24022) and the Division of Materials Sciences, Office of Basic Energy Sciences, U. S. Department of Energy for financial support and for the provision of laboratory facilities, and Prof. G. Thomas for continued encouragement.

REFERENCES

1. T. W. Sigmon, W. K. Chu, E. Lugujjo, and J. W. Meyer, Appl. Phys. Letts. 24, 105 (1974).

2. W. L. Harrington, R. E. Honig, A. M. Goodman, and R. Williams, Appl. Phys. Letts. 27, 644 (1975).

3. J. S. Johannessen, W. E. Spicer, and Y. E. Strausser, J. Appl. Phys. 47, 3028 (1976).

4. C. R. Helms, C. M. Garner, J. Miller, Z. Lindau, S. Schwarz, and W. E. Spicer, Proc. 7th Int. Vacuum Congress, (Vienna, 1977) p. 2241.

5. R. A. Clarke, R. L. Tapping, M. A. Hopper, and L. Young, J. Electrochem. Soc. 122, 1347 (1975).

6. R. Williams and A. M. Goodman, Appl. Phys. Letts. 25, 531 (1974).

7. B. E. Deal, J. Electrochem.Soc. 121, 198C (1974).

8. J. Blanc, C. J. Buiocchi, M. S. Abrahams, and W. E. Ham, Appl. Phys. Letts. 30, 120 (1977).

9. K.-J. Hanszen, Adv. Opt. Electron Micr. 4, 1 (1971).

10. O. L. Krivanek, Ph.D. Thesis, University of Cambridge (1975).

11. J. M. Cowley and S. Iijima, Physics Today, March 1977, p. 31.

12. P. B. Hirsch, H. Howie, R. B. Nicholson, D. W. Pashley, and M. J. Whelan, Electron Microscopy of Thin Crystals (Butterworths: London) 1965.

13. P. Rez and O. L. Krivanek, Proc. 9th Int. EM Congress (Toronto) to be published.

14. O. L. Krivanek, P. H. Gaskell, and A. Howie, Nature 262, 454-457 (1976).

15. O. L. Krivanek, T. T. Sheng, and D. C. Tsui, Appl. Phys. Letts. 32, 437 (1978).

16. S. Mahajan, A. K. Sinha, T. T. Sheng, unpublished results.

STRUCTURE OF THE Si–SiO$_2$ INTERFACE BY INTERNAL PHOTOEMISSION

T. H. DiStefano

IBM T. J. Watson Research Center, Yorktown Heights, NY 10598

ABSTRACT

Details of the SiO$_2$ conduction band near the Si–SiO$_2$ interface were examined by the technique of field-dependent internal photoemission. From a measurement of the threshold $\Phi(E)$ for internal photoemission for fields up to 8.17×10^6 V/cm, the conduction band bottom $\phi(x)$, as a function of position x in the SiO$_2$ was determined to within approximately two lattice units of the interface. In calculating the effective potential, corrections were made for the effects of the photon induced tunneling. The results of the threshold measurements show only small deviations from the flat band model, about 0.06 eV ± 0.03 eV, up to a distance 4.5Å from the silicon surface. There is no indication of a large ionic charge or of a greatly reduced SiO$_2$ bandgap to within about two lattice units of the interface.

INTRODUCTION

The interface between a given crystallographic surface of Si and an amorphous SiO$_2$ layer may not be precisely smooth on an atomic scale.[1] Evidence [2, 3] based on He backscattering data suggests that SiO$_2$ which has been thermally grown on silicon contains excess silicon near the Si–SiO$_2$ interface. However, ion backscattering measurements are known to be somewhat insensitive and difficult to interpret because of the background from the substrate, even if channeling techniques are employed [4]. Several sputter-etch profiling techniques have been used to examine the Si–SiO$_2$ interface, including Auger [5] and ESCA [6] measurements. Data from the sputter-etch measurements show some irregularities near the Si–SiO$_2$ interface which have been variously interpreted as excess Si in the SiO$_2$, a graded bandgap interface Si–SiO$_x$–SiO$_2$, free silicon inclusions in the SiO$_2$, or an undulating interface. However, the data are difficult to interpret unequivocally due to nonuniformities in the microscopic sputter-etch rate in SiO$_2$, which cause the interface to appear distorted because some portions of the silicon surface are exposed by the sputter-etch process before others. Additionally, these profiling techniques are complicated by details of the electron escape mechanism and by the physical damage induced by the sputtering process itself.

In this paper, we provide information on the local microscopic structure of the Si–SiO$_2$ interface obtained by the field-dependent internal photoemission technique [7]. The electric field dependence of the threshold for the internal photoemission is measured and used to determine the conduction band energy in the insulator as a function of distance away from the interface. In this way, the SiO$_2$ conduction band was probed from 4.5Å to 12Å from the Si–SiO$_2$ interface, and the curvature of the conduction band was determined [8].

FIELD DEPENDENT INTERNAL PHOTOEMISSION

The Si–SiO$_2$ interface has been extensively studied by internal photoemission measurments [9]. The interface barrier is determined by finding the threshold for photoemission over the barrier from spectral photoresponse measurements. The threshold is determined by extrapolating a power law fit to the photoresponse, where the best fit is obtained with power of 2 or 3 [10]. Typically, the threshold Φ determined from a third power fit is 0.15-0.2 eV lower than that obtained from a second power fit.

The field dependence of the internal photoemission from Si into SiO_2 was used to determine the variation of the conduction band potential ϕ with distance from the interface of the SiO_2 with the Si. Any curvature of the SiO_2 conduction band near the interface would result in a deviation from simple Schottky barrier reduction. This deviation $\Delta\phi$ is used to determine the conduction band potential $\phi(x)$ as a function of the distance x. The potential $\phi(x)$, together with the image potential and the applied field, produce a total barrier which is approximately the photoemission threshold $\Phi(E)$. At one particular value of field, the position x of the maximum of the barrier and the potential $\phi(x)$ can be obtained from the dependence of the photoemission threshold $\phi(E)$ upon electric field. An increase in the applied electric field moves the position of the barrier maximum closer to the interface and lowers the total barrier height. By using this field dependence of the photoemission barrier, the SiO_2 band curvature was measured as a function of distance away from the silicon surface.

MEASUREMENTS

The measurements were performed on samples formed by the oxidation of 10 ohm-cm n-type Si (100) in dry O_2 at 1150°C. The thickness of the SiO_2, determined by the ellipsometric measurements, was 1090 Å. A semitransparent electrode of 160Å of aluminum, 0.3 cm in diamter, on top of the SiO_2 was used to apply the electric field and to collect the photocurrent. Photoemission measurements were made up to the highest electric field at which the sample was stable. Three independently fabricated samples were measured, with similar results. The thresholds were determined by extrapolation from a plot of the half-power of the photoyield at a given applied field. At the higher electric fields, the photoyield near threshold was complicated by a component due to the photoemission of electrons from the conduction band of the inverted silicon surface. The second power, rather than the third power, was chosen because it facilitated a separation of the photoyield due to emission from the Si conduction band from that from the valence band.

The photoelectric threshold $\Phi(E)$ for emission from Si into SiO_2 is shown as a function of electric field in Fig. 1.

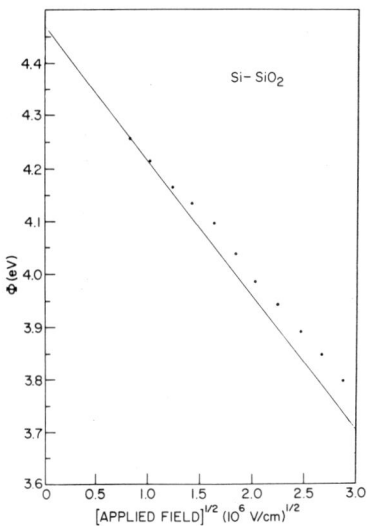

Fig. 1. The threshold for internal photoemission from Si into SiO_2 as a function of electric field.

The experimental points were determined from a fit of the spectral photoemissive yield to a parabolic dependence on photon energy. When extrapolated to zero electric field, the threshold indicates an interface barrier of about 4.47 eV $\pm$ 0.03 eV. This is somewhat higher than the 4.35 eV found from a third power fit, but this shift does not significantly influence the band bending determination. The barrier Φ is fit to a simple Schottky barrier reduction at the low field, using $\varepsilon_\infty = 2.15$ for SiO_2. A very small correction due to photon assisted tunneling is not included; this correction is -0.025 eV at the highest fields. Also, because of an uncertainty in the parameters, no corrections are included at this point for the effects of field penetration into silicon, imperfect screening at the silicon surface, and phonon scattering energy loss.

The potential of the conduction band bottom in the SiO_2, determined from the electric field dependence of $\Phi(E)$, is shown in Fig. 2. The band curvature is small in comparison to the uncertainties in the measurement and to the corrections which have not been included. However, the band bending in the SiO_2 near the silicon surface is small, only about 0.06 eV for distances up to 4.5Å from the silicon surface.

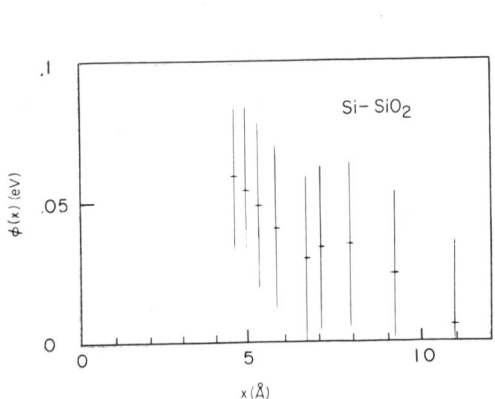

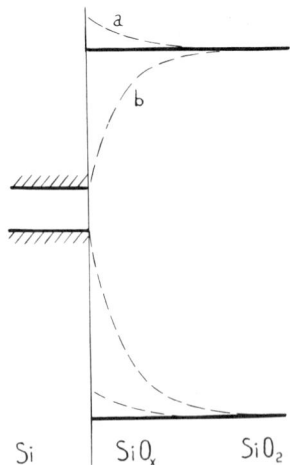

Fig. 2. The potential at the bottom of the SiO_2 conduction band as a function of distance from the interface.

FIg. 3. Schematic represenation of the electronic bands near the Si-SiO_2 interface: (a) conic Si in SiO_2; and (b) graded Si-SiO_x-SiO_2 bandgap.

DISCUSSION

The lack of any significant bending of the conduction band of the SiO_2 near a silicon surface indicates that any substantially graded bandgap does not extend beyond about 4Å from the silicon surface. If there were a non-stoichiometric layer of SiO_2 at the interface, the layer would influence the curvature of the conduction bond, as illustrated in Fig. 3. The dashed curve "a" shows the influence of excess ionic silicon, while curve "b" shows a graded bondgap due to a layer of covalently bonded SiO_x. Since no significant band bending is seen, there is no direct evidence of a non-stoichiometric layer beyond 4Å from the interface.

The corrections due to field penetration, screening, phonon scattering and photon assisted tunneling would change the calculated band curvature somewhat. Inclusion of the corrections would also change the calculated shape of the conduction band, but they would not introduce the sort of gross curvature expected from a non-stoichiometric layer.

REFERENCES

1   S. T. Pantelides and M. Long, Continuous-Random Network Models for the Si-SiO$_2$ Interface, *Proceedings of the International Topical Conference on the Physics of SiO$_2$ and its Interfaces* (1978).

2   (a) J. A. Davies, J. Denhartog, L. Eriksson, and J. W. Mayer *Can. J. Phys.* 45, 407 (1967)

     (b) T. M. Buck and G. H. Wheatley, *Surface Science* 33, 35 (1972)

     (c) W. F. van der Weg, W. H. Kool, H. E. Rosendaal, and F. W. Saris *Radiat. Eff.* 17, 245 (1973).

3   T. W. Sigmon, W. K. Chu, E. Lugujjo, and J. W. Mayer, *Appl. Phys. Lett.* 24, 105 (1974).

4   W. K. Chu, J. W. Mayer, M. A. Nicolet, T. M. Buck, G. Amsel, and F. Eisen, *Thin Solid Films* 17, 1 (1973)

5   J. S. Johannessen, W. E. Spicer, and Y. E. Strausser, *J. Vac. Sci. and Tech.* 13, 849 (1976); C. M. Garner, I. Lindau, J. N. Miller P. Pianetta, and W. E. Spicer, *J. Vac. Sci. Tech.* 14 372 (1977)

6   S. I. Raider, R. Flitch, and M. J. Palmer, *J. Electrochem. Soc.* 122, 413 (1975); S. I. Raider and R. Flitsch, *J. Vac. Sci. Tech.* 13, 856 (1976).

7   C. G. Wang and T. H. DiStefano, *Crit. Rev. Sol. State Sci.* 5, 327 (1975).

8   T. H. DiStefano, *J. Vac. Sci. Technol* 11, 996 (1976)

9   R. Williams, "Injection by Internal Photoemission," Chapter 2 in *Semiconductors and Semimetals*, Vol. 6, (Ed. R. K. Willi ardson, Academic Press, New York, 1970)

10   B. E. Deal, E. H. Snow, and C. A. Mead, *J. Phys. Chem. Solids* 27, 1873 (1966).

AUGER SPUTTER PROFILING STUDIES OF THE Si–SiO$_2$ INTERFACE[*]

C. R. Helms,[†] N. M. Johnson,[††] S. A. Schwarz,[†] and W. E. Spicer[†]

ABSTRACT

In this paper, we present results of new studies of thermally oxidized Si using Auger electron spectroscopy and ion etching. We will describe the results obtained from a "matrix" of samples for oxides grown on (100) silicon that have a range of oxide thickness of from 6.0 to 250 nm and growth temperatures from 800°C to 1150°C. The 10%/90% width of the transition region between SiO$_2$ and Si is independent of oxide thickness. Results on the variation of the transition width on growth temperature will also be discussed.

INTRODUCTION

The controversy that exists concerning the morphology and chemistry of the Si–SiO$_2$ interface can be related primarily to two factors: the limitations of the techniques applied to these studies and variations that exist between the samples studied.

Samples studied to date include silicon on sapphire (SOS), thermal oxides ranging in thickness from a few Å to a few thousand Å grown under a variety of conditions of temperature and oxidizing environment (dry O$_2$, wet O$_2$, etc.). There is no a priori reason to believe that the interface morphology and chemistry will be the same for all of these different types of samples.

Techniques applied to these studies include internal photoemission spectroscopy (Ref. 1), Auger sputter profiling (ASP) (Refs. 2–6), x-ray photoelectron spectroscopy (XPS) (Refs. 7 and 8), and ion thinning transmission electron microscopy (Ref. 9). All of these techniques have limited applicability, and any knowledge to be gained in studies of the SiO$_2$/Si interface must be obtained from interpretation based on each technique's inherent limitations.

In our ASP studies of the Si/SiO$_2$ interface formed by thermal oxidation in dry O$_2$, there are two factors inherent in the technique which limits the depth resolution and therefore our ability to study the interface. These are electron escape depth and ion knock-on broadening. These effects limit the ASP

[*] This research was funded by the Advanced Research Projects Agency, Order No. 2397, through the National Bureau of Standards Semiconductor Program Contract No. 5-35944 and is not subject to copyright.

[†] Stanford Electronics Laboratories, Stanford University, Stanford, CA 94305.

[††] Xerox Palo Alto Research Center, Palo Alto, CA 94304.

technique to a 10%/90% interface width resolution of about 15 Å (escape depth alone gives a 30 Å broadening factor in XPS measurements).

When our results are viewed in this framework, they are consistent with a model of the Si-SiO$_2$ interface shown in Fig. 1. Within the context of this model, it is convenient to define two measures for the interface width. First is the width of the total transition region, on one side of which is pure Si and the other is pure SiO$_2$, and second the width of the connective region between Si and SiO$_2$ at any point. Values or limits on the values of the parameters that come from our present work are stated as follows:

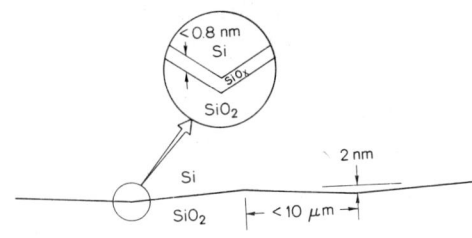

Fig. 1. Model for the Si-SiO$_2$ interface.

  (1) The width of the transition region is ~2.0 nm corrected for electron escape depth and ion knock-on broadening effects.

  (2) This width is probably due to an undulating interface, the period of which is certainly less than 100 μm (the distance over which the electron beam is rastered) which gives an angle of not more than 10^{-3} degrees.

  (3) The SiO$_x$ phase present is in a connective region no greater than 8 Å wide (subsequent work would place a 4 Å limit on this region).

EXPERIMENTAL DETAILS

Our results were obtained using a Varian Model 2730 spectrometer. The experiments were performed at base pressures of about 5×10^{-10} torr. The electron beam used was a few microns in diameter operated at 4.5 KeV with a beam current of 1 μA. Of critical importance in this study was the use of a 200 μm × 200 μm raster on the electron beam to minimize total sample dose (Ref. 3).

This rastered electron beam was directed to the center of the sputtering crater created by a rastered ion beam operated at 1 KeV in argon with an ion flux distribution of about 5 mm FWHM. Rastering the ion beam serves to average out any nonuniformities in the beam itself as well as broaden the distribution so that any aiming errors in the electron or ion beam will cause only minimal effects.

Samples were prepared for two complementary Auger profiling studies. For the first study, single crystal silicon wafers were oxidized in dry O$_2$ at a fixed temperature of 1000 ± 1°C to obtain samples with a range of oxide thicknesses. The wafers had a (100) crystal orientation and were p-type (boron doped) with a resistivity of 11 Ω-cm. The samples received a post-oxidation anneal in flowing N$_2$ for 10 min at the temperature of oxidation. Samples were prepared with four different oxide thicknesses over the range from 0 to 100 nm. Oxide thickness was measured by ellipsometry.

The second study involved the examination of oxides grown at the nominal
extrema of the temperature range typically used in MOS technology. Samples
were prepared from (100)-oriented p-type silicon wafers. One set of wafers
was oxidized in dry O_2 at 800 ± 1°C for 26 hr to obtain an oxide thickness of
44 nm. Another set was oxidized at 1150 ± 1°C for 7 min which produced 47 nm
oxides. The samples in this second study did not receive a high temperature
anneal.

In both studies, specimens for Auger profiling received no further treatment
after oxidation and high-temperature annealing. This eliminated any degradation of spatial depth resolution arising from ion milling through a polycrystalline metallic layer and/or a metal-oxide transition layer.

LIMITATIONS OF AUGER SPUTTER PROFILING

There are a number of factors which tend to distort or broaden the measured
Auger sputter profile. These effects can be divided into two categories.
First, there are effects due to the electron beam used in Auger spectroscopy.
These include electron escape depth broadening and electron stimulated reaction and desorption. Second, there are effects due to the ion beam used in
the sputter profiling. These include ion knock-on mixing, preferential sputtering, and ion beam nonuniformities.

The effect of electron escape depth as well as ion knock-ons is to broaden
the profile obtained. The measured 10%/90% width W_M can be written approximately as

$$W_M = \left\{ W^2 + (2.2\delta L)^2 + (2r)^2 \right\}^{1/2}$$

where W is the actual width, L is the escape depth, δ is a detector
geometry factor (0.75 for our instrument), and r is the ion range. For the
silicon LVV transition with L = 5 nm and Ar^+ ions at 1 KeV, we obtain a
broadening of a step interface of about 1.5 nm.

The effect of electron and ion irradiation on the chemical state of the materials present may mask the presence of SiO_x-type phases present at the Si/SiO_2
interface; this will be discussed in more detail below.

The other effects, electron stimulated desorption and preferential sputtering,
are unimportant in the present study due to the choice of experimental parameters.

RESULTS

Of primary interest in this study was the effect of oxide thickness on the
interface resolution because, in numerous other investigations for other systems, the resolution of the technique was proportional to the thickness of
the removed material. In our study, we prepared thermal oxides as described
above under identical conditions except that the growth time was varied to
give thickness ranging from about 250 to 1000 Å, the range of thicknesses

important for MOS technology.

The width of the interface was taken as that distance over which the Si LVV transition at 92 eV, characteristic of the free Si, in the derivative spectra increased from 10% to 90% of its final value. Subsequent work has shown that very little error is made compared to using the total transition strength in the undifferentiated spectra (Ref. 6). For comparison purposes, the oxygen KLL profile was also measured, with the interface width defined in a similar way. It should be noted that the electron escape depth for this transition is about a factor of two larger than for the Si LVV transition. This, coupled with longer knock-on distances for oxygen versus silicon, cause the resolution of the interface to be poorer viewed with the oxygen transition as versus the silicon LVV transition.

A typical profile is shown in Fig. 2 where the peak-to-peak heights of the Si LVV and oxygen KLL are shown as a function of depth for the 95.5 nm thick oxide sample. A set of Si LVV spectra corresponding to such a profile are shown in Fig. 3.

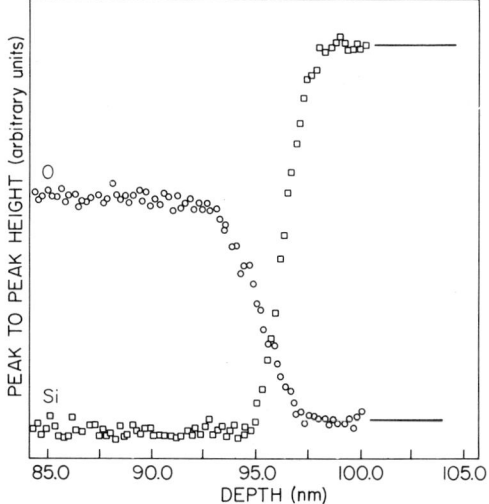

Fig. 2. High resolution Auger sputter profile of a thermally grown Si-SiO2 interface for a 95 nm thick oxide, prepared as described in the text. The oxygen profile is from the KLL transition and the silicon profile is from the LVV transition.

The 10%/90% width of the interface is about 2.5 nm as determined by the Si LVV transition and 3.35 nm as determined by the oxygen KLL transition. There are three factors that may be responsible for the apparent discrepancy between the two values for interface width: escape depth, preferential ion knock-on effects, and the presence of an oxygen containing phase of silicon other than SiO_2.

We have corrected the data for electron escape depth as described above and find that the values for interface width are 2.35 nm for the Si LVV and 2.9 nm for the oxygen KLL. One or both of the other two mechanisms must therefore be important. In subsequent work, Helms, Strausser, and Spicer (Ref. 6) have found an SiO_x phase present in the Auger spectra at the interface; however, it is unclear whether or not this is completely responsible for the 0.55 nm difference in interface widths observed here.

Irrespective of the cause of this discrepancy, we will take the Si LVV 10%/90% width as a measure of the interface width since it appears to be least affected by escape depth, knock-on effects, or ion induced chemical effects. In Table 1, we show measured interface widths as a function of oxide thickness for oxides grown at 1000°C in dry O_2 followed by a 10 min post growth anneal in N_2, as described above. The interface widths are remarkably similar, the average values being

TABLE 1   Dependence of the width
of the Si-SiO$_2$ transition region
on oxide thickness

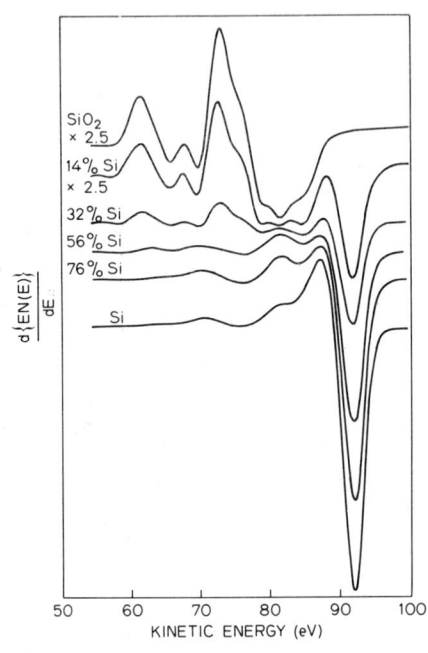

Fig. 3.   Silicon derivative spectra of
the LVV transition in going from
SiO$_2$ to Si. The percentages listed
correspond to the ratio of the 92 eV
peak to that of the pure silicon.

Sample (nm)	Oxygen KLL Width (nm)	Silicon LVV Width (nm)
24	3.16	2.37
24	3.90	2.74
29.5	3.55	2.50
31.0	3.20	2.31
31.0	3.20	2.45
58.5	3.30	2.42
95.5	3.14	2.58
Average	3.35 ± 0.25	2.50 ± 0.15
Average corrected for Escape Depth	2.90 ± 0.5	2.35 ± 0.25

shown in the bottom row.

An estimate of the limit of any broadening contribution that might be proportional to thickness can be obtained by dividing the rms deviation, 0.15 nm by
the average thickness of the samples ~40 nm. This gives an upper limit of
0.4% for broadening proportional to oxide thickness. This value is the level
at which contributions from the nonflat crater bottom must be taken into consideration and probably represents the limit of our present capabilities
(note, this would still give an ~50 Å interface width for a 10,000 Å thick
oxide).

The implications of these findings with respect to device processing and electrical properties will be discussed below. Our findings indicate that, over
this range of thicknesses, the interface morphology is independent of growth
time and, therefore, thickness. In addition, the electrical properties of
these interfaces, as measured by quasistatic CV studies, are also identical,
which is consistent with this finding.

In the next part of the study, we prepared samples with differing electrical properties. This was done by varying the oxide growth temperature. Although we have indicated above that there is little effect of the oxide thickness on interface widths, the growth times were chosen to obtain samples of similar thickness but varying in growth temperature from 800°C to 1150°C. As an example of the variation in electrical properties, Q_{SS} ranged from 2.1×10^{11} cm^{-2} at 800°C to 1.4×10^{10} cm^{-2} at 1150°C. Values for the interface width obtained from four runs each were 2.35 ± 0.04 nm for the 800°C oxide and 2.75 ± 0.08 for the 1150°C oxide. Note, the value for the 1000°C oxide was 2.5 ± 0.15 nm which is in between these two values. The difference between the 800°C and 1150°C oxide is 0.4 nm, a small enough difference to make detailed interpretation inappropriate; the data does, however, indicate an inverse relationship between growth temperature and interface width.

CONCLUSIONS

The value of the transition width of approximately 2.0 nm obtained in this work is consistent with recent ion thinning/TEM experiments where Blanc et al (Ref. 9) find an interface roughness of about 2.0 nm with a periodicity of about 10^3 nm.

Ion knock-ons not only broaden an interface but can also affect the chemical nature of the damage region they create. In general, we would expect the chemistry of a compound material to be "scrambled" by the ion beam. Due to the uncertainty in the effect of the ion beam on SiO_x-type materials, their presence may only be detectable in Auger data taken during sputtering before the interface is reached.

From an analysis of the experimental parameters, we estimate that an SiO_x region thinner than 0.8 nm would be undetectable in our experiments. In fact, neither in the present work or in that of Johannessen et al has Auger chemical shift evidence for the presence of SiO_x in the interface region been found, suggesting that the SiO_x region is thinner than 0.8 nm.

Raider and Flitsch (Ref. 7), on the other hand, using ESCA have evidence for an additional chemical form of Si in the interface region of very thin oxides and suggest an SiO_x thickness somewhat greater than 0.8 nm.

Other studies, however, appear consistent with the interpretation of a connective layer less than 0.8 nm thick. For example, in recent internal photoemission studies, DiStefano (Ref. 1) has found no evidence for an SiO_x region or significant excess charge extending farther than 0.4 nm from the interface.

In interpreting data from the various techniques, however, it must always be kept in mind that all have a limited applicability, each with different experimental limitations. Any model of $Si-SiO_2$ interfaces must be constructed with this in mind.

Our recent results, as well as those of others, all appear consistent with a model for the Si/SiO_2 interface sketched in Fig. 1. The relationship between the parameters of this model and the electrical properties of the interface at this time still remain unclear.

REFERENCES

1. T. H. DiStefano, J. Vac. Sci. Tech. 13, 856 (1976).

2. J. S. Johannessen, W. E. Spicer, and Y. E. Strausser, J. Vac. Sci. Tech. 13, 849 (1976).

3. J. S. Johannessen, W. E. Spicer, and Y. E. Strausser, J. Appl. Phys. 47, 3028 (1976).

4. C. R. Helms, C. M. Garner, J. Miller, I. Lindau, S. Schwarz, and W. E. Spicer, Proc. 7th Intern. Congr. and 3rd Intern. Conf. Solid Surfaces 3, 2241 (Vienna, 1977).

5. C. R. Helms, W. E. Spicer, and N. M. Johnson, "New Studies of the Si-SiO$_2$ Interface Using Auger Sputter Profiling," Solid State Comm., in press.

6. C. R. Helms, Y. E. Strausser, and W. E. Spicer, to be published.

7. S. I. Raider and R. Flitsch, J. Vac. Sci. Tech. 13, 58 (1976).

8. F. J. Grunthaner and J. Maserjian, to be published.

9. J. Blanc, C. J. Buiocchi, M. S. Abrams, and M. E. Ham, Appl. Phys. Lett. 30, 120 (1977).

AUGER ANALYSIS OF THE SiO$_2$/Si INTERFACE OF ULTRATHIN OXIDES*

J. F. Wager and C. W. Wilmsen
Department of Electrical Engineering, Colorado State
University, Fort Collins, Colorado 80523

ABSTRACT

The chemical state and width of the SiO$_2$/Si interface have been investigated by Auger analysis of ultrathin oxides 23 to 45Å thick (as estimated ellipsometrically). The low energy Si (LVV) line was recorded at various depths into the surface. By comparing measured and numerically synthesized Auger lines, it was determined that the interface is composed primarily of SiO$_2$ and Si and that the width of a fully developed interface is approximately 20Å. These results concur with those previously obtained for thick oxides using the Si KLL line. A slight excess oxygen buildup is evident directly at the surface of the ultrathin oxides but the Auger lines at a given distance from the substrate are otherwise the same regardless of the initial oxide thickness.

INTRODUCTION

Over the past few years there has been increased interest in determining the chemistry and width of the SiO$_2$/Si interface. Auger,[1-4] ESCA,[5-8] TEM,[9] ISS,[10] and He[11,12] back scattering have been previously used to investigate this interface. Auger analysis has indicated that in the interface silicon exists in only two distinct bonding states; Si-Si or Si-O$_2$ bonding.[1-3] However, recent work by Grunthaner and Maserjian[5] suggest an additional bonding state. In addition, several impurities (C, H, OH, and N$_2$)[4,13,14] have been found at the interface and appear to be bonded to the interfacial silicon. The exact width of the interface is also in doubt but it appears that the interface is 20Å or less, even for very thick SiO$_2$ layers.[2-5,6,7]

This paper presents the results of an Auger analysis of ultrathin oxide films grown on silicon. A method of quantitatively analyzing the Si LVV lines is used to determine the distribution of SiO$_2$ and Si in the interface. This quantitative information is then used to compare the interfaces of 23, 32, and 45Å thick oxides.

EXPERIMENTAL METHODS

Oxides were grown at 700°C in dry oxygen on 1.5 Ω-cm, <100> oriented p-type silicon wafers. The wafers were first vapor degreased in xylene, etched in 48% HF, rinsed in ultrapure DI water for an extended period of time, and then blown dry with nitrogen. Oxidation times ranged from 15 minutes to 4 hours. The wafer oxidation boat was loaded at different intervals in such a manner

*Supported by DOE, Contract EG-77-S-02-4518.A000

as to allow all wafers to be removed from the furnace simultaneously. Each
wafer was then broken in half with one wafer half immediately placed under
ultrahigh vacuum of the Auger spectrometer. The oxide thicknesses of the
remaining wafer halves were then measured with a Gaertner model L117 ellipso-
meter. Since oxide thickness is an important parameter of this investigation,
the ellipsometric measurements require additional comment.

The nature of the ellipsometric equations are such that for films less than
50Å thick, either the film thickness or index of refraction must be known.
Since a measure of the oxide thickness is the important unknown, the index of
refraction was assumed to be that of SiO_2. This is obviously incorrect since
much of the measured film thickness is the interfacial region containing both
SiO_2 and Si. At 6328Å, the SiO_2 has an index of refraction of 1.462 and Si -
3.855-j.024. Thus the ellipsometric measurements yield a larger thickness than
the actual film. Even though these thickness measurements are incorrect, they
provide a useful indication of the relative thickness of the film. As measured
in the above manner, the oxides were found to be 45, 32, and 23Å thick. The
wafer directly from the water rinse had a measured oxide thickness of 10Å.

The Auger measurements were obtained with a Physical Electronics 540 Thin Film
Analyzer with a 3kV, 5μA electron beam defocused to a diameter of 50μm. The
surface was sputtered with an 1KV ion beam in 5×10^{-5} Torr of Argon. Based on
the ellipsometric measurements, the resulting sputter rate was 3 to 4Å/min.
The sputter rate through the SiO_2 appeared to be the same as through the interfacial region.

AUGER ANALYSIS

There are two major Auger transitions of silicon: the LVV and the KLL. For
thin oxides, the KLL line can not be used because the mean escape depth of the
KLL electrons is greater than the oxide film thickness. On the other hand, the
LVV line shape is disturbed by the sputter beam. Auger lines of Fig. 1
illustrate the changes in the silicon LVV line as a thick SiO_2 film is sputter
etched at the same conditions used to profile the thin SiO_2 samples. It is
seen that the positive peak at 66eV remains unchanged but a negative peak at
92eV appears and tends to wash out the negative 78eV peak.

The peak to peak profiles and silicon LVV lines at various depths of the 23,
32, and 45Å thick oxides are shown in Fig. 2. Note that the profiles have the
same shape through the interfacial region and that the line shapes at a given
depth are the same for all three oxide thicknesses.

Quantitative information was extracted from this Auger data by the following
method. It was assumed that the Auger signal was primarily a mixture of pure
SiO_2 and pure Si signals. These pure signals were digitized and placed in the
memory of an hp9830 desk top calculator. The measured signals from the thin
oxide samples were also digitized and stored in the calculator. The calculator
then added the pure SiO_2 and Si lines in various proportions until the best
RMS fit between measured and synthesized curves was obtained. This resulted
in only a fair fit of the two curves. This was thought to be caused by the
effects of sputtering on the SiO_2 line shape. Since the pure SiO_2 signal
changes in a predictable manner when sputtered, the various sputtered lines
from the thick SiO_2 sample were substituted for the pure undisturbed SiO_2 line
used previously. Thus, when synthesizing the silicon line measured after a
given time of sputtering, the corresponding line from the sputtered thick
SiO_2 was used. Synthesizing the lines in this manner resulted in very close

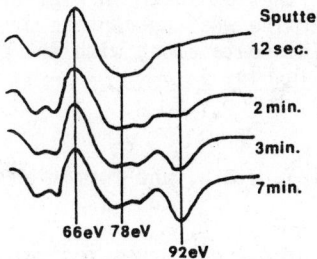

Figure 1. Changes which occur in the Si LVV Auger line due to sputtering a thick SiO_2 film.

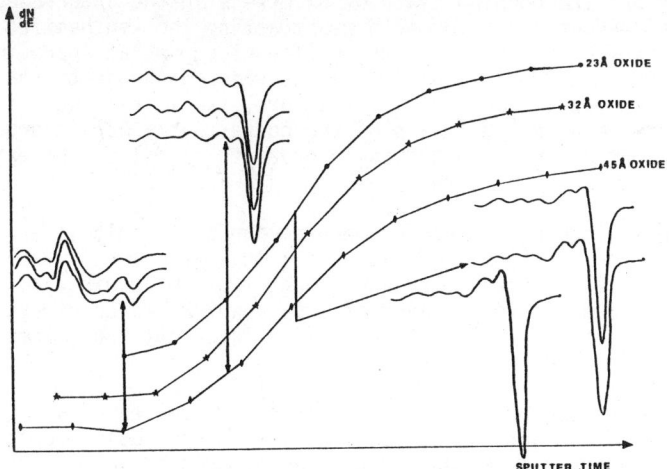

Figure 2. Auger dN/dE peak to peak plots and line shapes for 23, 32, and 45 Å oxides.

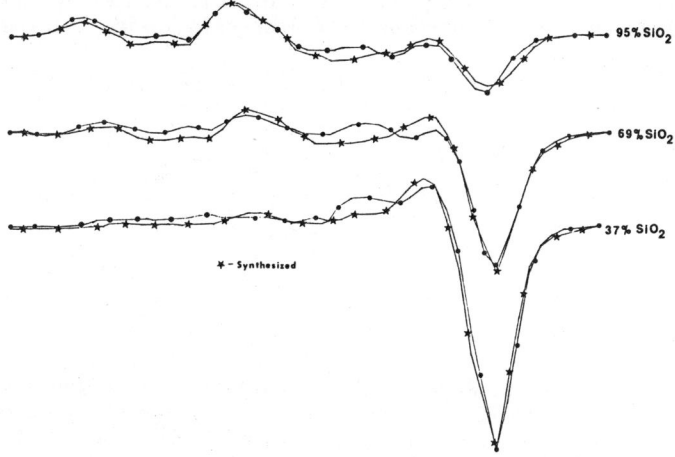

Figure 3. Comparison of the measured and synthesized Auger lines at different depths into the interface of a 45 Å thick oxide.

fits between the measured and calculated lines as shown in Fig. 3. From the synthesized curves it is possible to determine what percentage of SiO_2 and Si exists at each point in the interface. This data along with the peak to peak dN/dE points for the 32Å oxide are given in Fig. 4.

DISCUSSION

The above analysis sheds light into bonding, width, and formation of the SiO_2/Si interface.

The close fit between the measured and synthesized curves indicate that the bonding throughout the interface is made up primarily of Si-Si and Si-O_2 bonding. If a third bonding state of Si exists in the interface, it should be evident in an "Auger line" formed by subtracting the synthesized curve from the measured. The similarity of the different curves at approximately 50% SiO_2/50% Si for the 23, 32, and 45Å oxides (Fig. 5) suggests that such a third state may indeed exist. However, it is more likely that there is a small systematic error in the synthesis of the curves but a more thorough analysis of the different curves is required in order to establish the existance or non-existance of a third state.

The interface width corrected for mean escape depth only is approximately 20Å in agreement with many of the previous investigations. The interface appears to be fully formed after 25 to 30Å of oxide growth. All of the samples grown at 700°C had a surface layer of SiO_2. Wafers taken directly from the water bath did not have a 100% SiO_2 surface layer but had evidence of free silicon on the surface.

The effects of oxygen knock ons was apparent in samples of all thicknesses but was observed to increase monotonically with increasing depth into the oxide. An apparent saturation of knock on effects is seen in the thick oxide where the removal of the surface by sputtering is presumably in equilibrium with knock on effects. The difference between the oxygen peak to peak curve and the calculated SiO_2 at zero concentration (see Fig. 4) is taken as a measure of the magnitude of the knock on effect. These differences are listed in Table I and are observed to increase with increasing oxide thickness.

TABLE I   The Difference Between the O_2 Zero Intercept and that of the Calculated SiO_2.

Oxide Thickness	$X_{O_2} - X_{SiO_2}$
10Å	7.6Å
23	12.0
32	18.6
45	19.4

CONCLUSIONS

The method of analyzing the Auger lines by comparison with synthesized curves has been shown to be useful in assigning a quantitative value to concentration of SiO_2 as a function depth. Using this method it was shown that:
1. the interface is fully developed after 25 to 30Å of oxide has been grown,
2. the interface is primarily free silicon and SiO_2,
3. the interface width is 20Å or less.

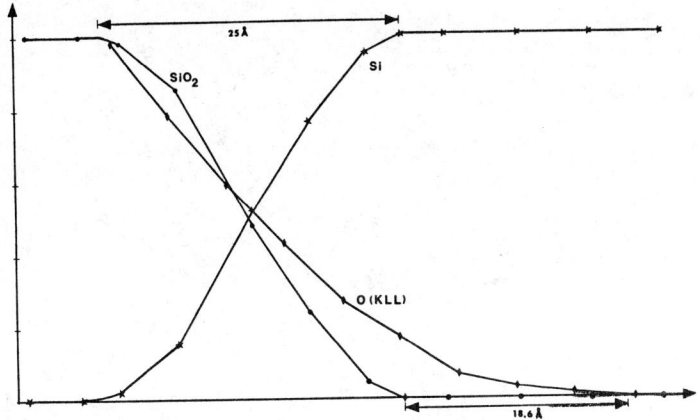

Figure 4. Calculated SiO_2 and Si composition and oxygen KLL line for a 32 Å oxide.

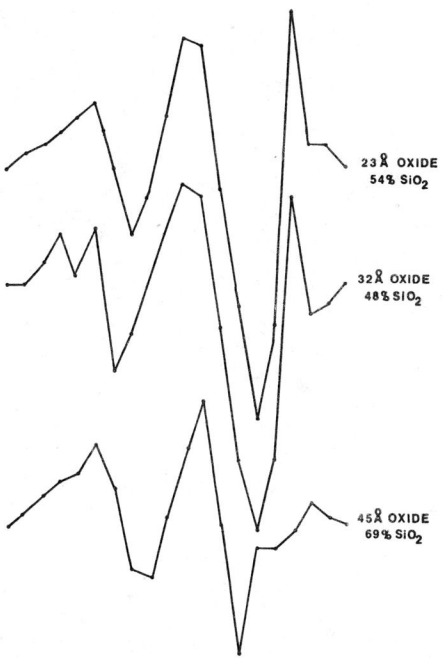

Figure 5. Difference between the measured and synthesized Auger lines at approximately 50% SiO_2 /50% Si for 23, 32, and 45 Å oxides.

ACKNOWLEDGMENT

The authors wish to express their gratitude to Mr. Stephen Goodnick for his help with the ellipsometric measurements and the computer program.

REFERENCES

1. J. S. Johannessen, W. E. Spicer, and Y. E. Strausser, Phase separation in silicon oxides as seen by Auger electron spectroscopy, Appl. Phys. Letters 27, 452 (1975).

2. J. S. Johannessen, W. E. Spicer, and Y. E. Strausser, An Auger analysis of the SiO_2-Si interface, J. Appl. Phys. 47, 3038 (1976).

3. C. R. Helms, W. E. Spicer, and N. M. Johnson, New studies of the Si-SiO_2 interface using Auger sputtering profiles, Solid State Comm. (In press)

4. G. A. Haas and H. F. Gray, Oxidation of Si surfaces, J. Appl. Phys. 46, 3885 (1975).

5. F. J. Grunthaner and J. Maserjian, Chemical structure of the transitional region of the SiO_2/Si interface, J. Vac. Sci. Tech., 15, July/Aug. (1978).

6. R. Flitsch and S. I. Raider, Electron mean escape depths from x-ray photoelectron spectra of thermally oxidized silicon dioxide films on silicon, J. Vac. Sci. Tech. 12, 305 (1975).

7. S. I. Raider, R. Flitsch, and M. J. Palmer, Oxide growth on etched silicon in air room temperature, J. Electrochem. Soc. 122, 413 (1975).

8. R. A. Clarke, R. L. Trapping, M. A. Hopper, and L. Young, An ESCA study of the oxide at the Si-SiO_2 interface, J. Electrochem. Soc. 122, 1347 (1975).

9. J. Blanc, C. J. Buiocchi, M. S. Abrahams, and W. E. Ham, The Si/SiO_2 interface examined by cross-sectional transmission electron microscopy, Appl. Phys. Letters 30, 120 (1977).

10. W. L. Harrington, R. E. Honig, A. M. Goodman, and R. Williams, Low-energy ion-scattering spectrometry (ISS) of the SiO_2/Si interface, Appl. Phys. Letters 27, 644 (1975).

11. T. W. Sigmon, W. K. Chu, E. Lugujjo, and J. W. Mayer, Stoichiometry of thin silicon oxide layer on silicon, Appl. Phys. Letters 24, 105 (1974).

12. P. Offermann, Thickness evaluation of Si/SiO_2 interfaces by He-backscattering experiments, J. Appl. Phys. 48, 1890 (1977).

13. E. Kooi, J. G. van Lierap and J. A. Appels, J. Electrochem. Soc. 123, 1117 (1976).

14. A. G. Revesz, On SiOH and SiH group in SiO_2 films on silicon, J. Electrochem. Soc. 124, 1811 (1977).

STUDIES OF Si/SiO$_2$ INTERFACES AND SiO$_2$ BY XPS

Takeo Hattori and Tatsushi Nishina
Department of Electrical Engineering
Faculty of Engineering
Musashi Institute of Technology
1-28-1, Tamazutsumi, Setagaya-ku, Tokyo, Japan

ABSTRACT

The method to determine thickness distribution in thin films by using X-ray excited photoelectron spectroscopy (1), (2) has been modified for the determination of the thickness distribution of thermally grown silicon oxide films and the thickness of Si-SiO$_2$ interfacial transition layer. Thickness distribution functions and the thickness of the interfacial layer are determined for silicon oxide films made by wet and dry oxidations.

INTRODUCTION

The Si-SiO$_2$ system has been studied extensively because of its importance in the semiconductor device technology (3), (4). However, the details of the chemical structure of the Si-SiO$_2$ interface and its relation to the electronic structure are little understood. Because of this the Si-SiO$_2$ interface has been studied by various surface sensitive techniques such as Auger electron spectroscopy(AES) (5), (6), X-ray excited photoelectron spectroscopy (XPS) (7), high and low ion backscattering (8), (9), scanning electron microscopy (10), and contact angle measurements (11). Flitsch and Raider (7), employing XPS, have identified nonstoichiometric interfacial transition regions in thermally grown ultra thin silicon oxide films. In the present paper Si-SiO$_2$ interfaces and SiO$_2$ were studied by XPS combined with sputter-etch profiling technique and the experimental data were analyzed by considering the non-uniformity of thickness in silicon oxide films.

EXPERIMENTAL DETAILS

The specimens examined in this study were prepared as follows: The substrates used for the thermal growth of silicon oxide have <111> orientation with p-type resistivity of nearly 50 Ωcm. After the substrates were chemically etched by 600 A at the etching speed of nearly 1A/sec, oxide films with their thickness being in the range from 50 to 395 A were grown thermally at 800°C in dry oxygen in one case and in water vapor in another case. The cooling from 800°C to room temperature takes place approximately in 5 minutes. The specimens were not annealed after the oxidation was done. The thickness of the oxide films were measured with an accuracy of 1 A by ellipsometry. The experimental procedure to determine the thickness distribution in silicon oxide film is essentially the same as that described previously (1). The numbers of photoelectrons produced by X-ray excitation from silicon and oxygen in the oxide film were measured as a function of the distance z, which is equal to the sputter etching rate multiplied by the etching time. The measurements were performed using DUPONT ESCA 650B spectrometer. The ion beam for the sputter etching was incident on the film at nearly $20°$ to the surface of the oxide film. In order to obtain uniform sputter etching condition, a small area of nearly 4 mm^2 was used for the measurements. The experimental data described in the present paper were obtained by using 1.5 keV argon ions at a current density of approximately 30 μA/cm^2. The present measurements were performed without breaking vacuum.

THICKNESS DISTRIBUTION IN SILICON OXIDE FILM

In general, an ultra thin silicon oxide film is not uniform in thickness, i.e. the thickness varies from position to position on the substrate surface. This can be described by using thickness distribution function $f(x)$ of thickness x of silicon oxide film. If we assume that a) thickness distribution function is continuous, b) the thickness of the oxide film decreases uniformly by z as a result of uniform sputter etching of SiO$_2$, Si-SiO$_2$ interfacial layer and Si at the same etching speed, c) the thickness of the interfacial layer under the silicon dioxide film is uniform and is independent on the non-uniform thickness distribution in silicon dioxide film, while the thickness of the transition layer for the zero thickness region of silicon dioxide film takes values from zero to a, the detected numbers of photoelectrons NS(z), NO(z) and NT(z) from silicon in silicon substrate, silicon in silicon dioxide and silicon in the interfacial layer, respectively, can be expressed by the following equations.

$$F(z) = NS(z)/[K\sigma_s n_s F\Lambda_s]$$

$$= A + \int_0^z f(x)dx + \int_0^a f(x+z)\exp(-x/\Lambda_t)dx$$

$$+ \int_0^\infty f(x+z+a)\exp(-x/\Lambda_o)\exp(-a/\Lambda_t)dx \qquad (1)$$

$$G(z) = NO(z)/[K\sigma_o n_o F\Lambda_o]$$

$$= \int_0^\infty f(x+z+a)[1 - \exp(-x/\Lambda_o)]dx \qquad (2)$$

$$H(z) = NT(z)/[K\sigma_t n_t F\Lambda_t]$$

$$= \int_0^a f(x+z)[1 - \exp(-x/\Lambda_t)]dx$$

$$+ \int_0^\infty f(x+z+a)\exp(-x/\Lambda_o)[1 - \exp(-a/\Lambda_t)]dx \qquad (3)$$

Here, a is the thickness of the interfacial layer. Λ_o and Λ_t are inelastic mean free path for electrons in SiO$_2$ and that in interfacial layer, respectively. σ_s, σ_o and σ_t are the photoelectric cross section for the excitation of 2p photoelectrons from silicon in silicon substrate, that in silicon dioxide and that in the interfacial region, respectively. n_s, n_o and n_t are the concentration of silicon in silicon substrate, that in silicon dioxide and that in the interfacial region in terms of atom per unit volume, respectively. F is the X-ray flux. K is constant. Assuming $\Lambda_o = \Lambda_t \equiv \Lambda$, the following equations can be obtained (2) from eqn. (1) with Dirac's δ-function.

$$f(z) = + A\delta(z-\varepsilon) - \Lambda\partial^2 F/\partial z^2 + \partial F/\partial z \qquad (4)$$

$$\Lambda = [<x> - \int_0^\infty \{1 - F(z)\}dz]/[1 - F(0)] \qquad (5)$$

Here, ε is chosen to be negligibly small compared with Λ. There seems to be the deviation of the measured value of the thickness of the oxide film determined by ellipsometry from the expected value for oxide mean film thickness smaller than 12 A (12). In the present paper we discuss Si-SiO$_2$ system for oxide mean film thickness larger than 50 A. The following expression is used for the mean film thickness <x> measured by ellipsometry.

$$<x> = \int_0^\infty xf(x)dx \qquad (6)$$

EXPERIMENTAL RESULTS AND DISCUSSION

The photoelectron spectra of the Si 2p core level for silicon oxide film made by wet oxidation is shown in Fig. 1 with z as a parameter. The Si 2p spectra can be separated into three lines as shown in Fig. 2: high binding energy line located at 103.7 eV is associated with silicon dioxide; the middle binding energy line located at 102.3 eV may be associated with the interfacial layer; low binding energy line is associated with silicon in the substrate. The Si 2p dioxide line intensity decreases with an increase in z, while the Si 2p substrate line intensity increases. The transition layer line intensity takes its maximum value if z is approximately equal to the mean film thickness of the oxide. The Si 2p dioxide line shape can be approximated by Gaussian function. An in Si 2p dioxide line width(full-width, half-maximum — FWHM)from 1.7 to 2.0 eV is observed with a initial sputter etching of 8.0 A. The Si 2p dioxide line width does not change with further increase in z. This is also confirmed from the measurements on thick oxide film having mean film thickness of 395 A. The Si 2p transition layer line shape can be approximated also by Gaussian function. The Si 2p substrate line shape is somewhat complicated. Without sputter etching the Si 2p substrate line shape for chemically etched silicon substrate can be approximated by Gaussian function with FWHM of 1.4 eV. With an increase in z photoelectrons having higher binding energy increase to change the line shape asymmetric with respect to 99.3 eV. The Si 2p substrate line shape for any z can be deduced from that for large enough value of z. By the same procedure the Si 2p

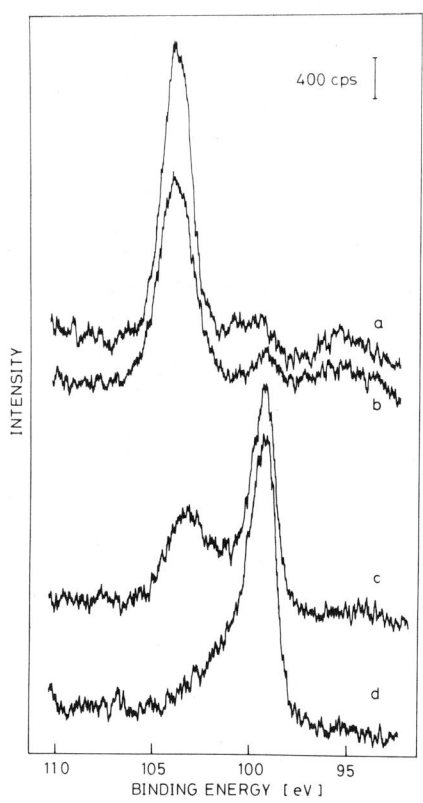

Fig. 1 Si 2p core level spectra for silicon oxide film made by wet oxidation with z as a parameter: <x> = 100 A. a) z = 0, b) z = 8.0 A, c) z = 88 A, d) z = 200 A

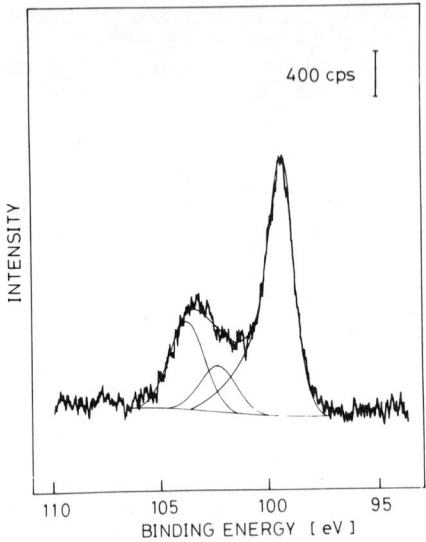

Fig. 2 Separation of photoelectron spectrum c in Fig. 1.

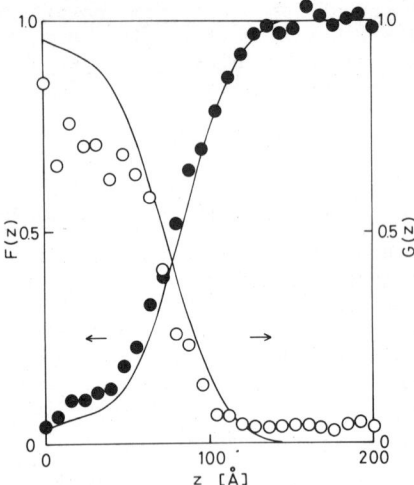

Fig. 4 Dependence of F(z) and G(z) on z for the same specimen as that shown in Fig. 1. ●, $NS(z)/NS(\infty)$: ○, $NO(z)/NS(\infty)$

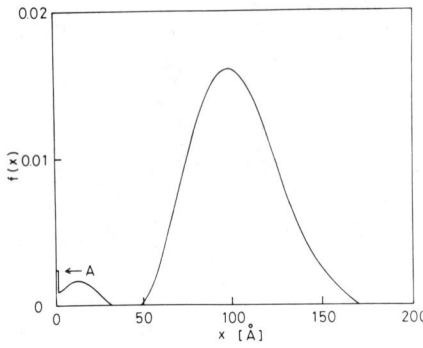

Fig. 3 Thickness distribution function for the same specimen as that shown in Fig. 1. A is explained in the text.

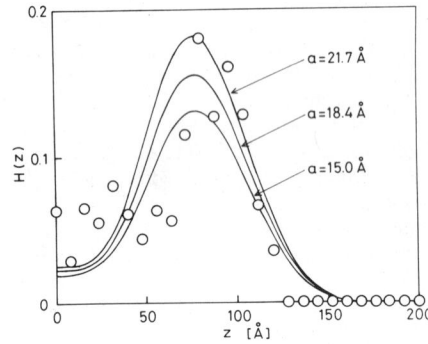

Fig. 5 Dependence of H(z) on z for the same specimen as that shown in Fig. 1. ○, $NT(z)/NS(\infty)$

substrate line shape in Fig. 2 is determined. Fig. 3 shows the thickness distribution function determined for silicon oxide film made by wet oxidation. The inelastic mean free path, Λ, determined in this case is 24.3 A. A in Fig. 3 shows the area of the zero thickness region of silicon oxide film (2). Fig. 4 shows the calculated dependence of F(z) and G(z) on z. The experimental data shown in this figure are $NS(z)/NS(\infty)$ and $NO(z)/NS(\infty)$. Fig. 5 shows the calculated dependence of H(z) on z for silicon oxide film made by wet oxidation using thickness distribution function in Fig. 3. The experimental data shown in Fig. 5 correspond to the ratio of $NT(z)/NS(\infty)$. If we assume that $\sigma_t = \sigma_s$ and $\Lambda_t = \Lambda_o$, the following relation is obtained.

$$H(z)/F(\infty) = \alpha[NT(z)/NS(\infty)], \quad \alpha = [n_o/n_t][G(0)NS(\infty)]/[F(\infty)NO(0)]$$

In the present case α is calculated to be 1.01 with the assumption that density of silicon dioxide and that of the interfacial transition layer is nearly equal and the composition of the interfacial layer is $SiO_{1.36}$ (13). Then, $H(z)/F(\infty) \simeq NT(z)/NS(\infty)$ is satisfied to give 18.4 A for a value of a from Fig. 5. Almost the same value is obtaied for silicon oxide film made by dry oxidation. Rather large value for a obtained may be due to the following reasons: a) there is ambiguity about the determination of the Si 2p substrate line shape and the Si 2p interfacial layer line shape, b) silicon oxide films were not annealed and cooled down to room temperature not slowly after the oxidation was done, c) undulating interface (6) is not considered here, d) the equality $\Lambda_t = \Lambda_o$ is not necessarily satisfied, e) atomic mixing within the range of the collision cascade is neglected.

CONCLUSION

The Si-SiO$_2$ interfaces and SiO$_2$ were studied by XPS combined with sputter etch profiling technique and the experimental data are analyzed by considering non-uniformity of thickness in the silicon oxide film. The thickness of the interfacial layer is estimated to be nearly 20 A for silicon oxide film made by wet and dry oxidations. Further improvements in the separation of three line shapes are necessary to determine the details of the chemical structure of the Si-SiO$_2$ interface.

The authors are grateful to H. Yamauchi of Shimadzu Seisakusho Ltd. for giving us an opportunity to use 1.5 keV argon ion etching system.

REFERENCES

1) T. Hattori, Japan J. appl. Phys. 16, 635 (1977); Thin Solid Films 46, 47 (1977); Proc. 7th Intern. Vac. Congr. & 3rd Intern. Conf. Solid Surfaces, Vienna, 1977, p.2157.
2) T. Hattori, Proc. 7th Intern. Vac. Congr. & 3rd Intern. Conf. Solid Surfaces, Vienna, 1977, p. 2259.
3) B. E. Deal, J. Electrochem. Soc. 121, 198C (1974).
4) R. Williams, J. Vac. Sci. Technol. 14, 1106 (1977).
5) J. S. Johannessen, W. E. Spicer and Y. E. Strausser, J. Appl. Phys. 47, 3028 (1976).
6) C. R. Helmes, C. M. Garner, J. Miller, I. Lindau, S. Schwartx and W. E. Spicer, Proc. 7th Intern. Vac. Congr. & 3rd Intern. Conf. Solid Surfaces, Vienna, 1977, p.2241.
7) S. I. Raider and R. Flitsch, J. Vac. Sci. Technol. 13, 58 (1976).
8) T. W. Sigmon, W. K. Chu, E. Lugujjo and J. W. Mayer, Appl. Phys. Lett. 24, 105 (1974).
9) W. L. Harrington, R. E. Honig, A. M. Goodman and R. Williams, Appl. Phys. Lett. 27, 644 (1975).
10) J. Blanc, C. J. Buiocchi, M. S. Abrahams and W. F. Ham, Appl. Phys. Lett. 30, 120 (1977).
11) R. Williams and A. M. Goodman, Appl. Phys. Lett. 25, 531 (1974).
12) S. I. Raider, R. Flitsch and M. J. Palmer, J. Electrochem. Soc. 122, 413 (1975).
13) T. Hattori and T. Nishina, to be published.

X-RAY PHOTOELECTRON SPECTROSCOPY OF SiO_2-Si INTERFACIAL REGIONS

S. I. Raider[*] and R. Flitsch
IBM System Products Division, Hopewell Junction, N.Y.

ABSTRACT

X-ray photoelectron spectroscopy (XPS) was used to characterize the interface formed between thermally-grown oxide films and single crystal Si substrates. Core level binding energies, photoelectron line intensities, and photoelectron linewidths, each varied over an initial oxide film thickness of about 25Å and then remained constant. Analysis of the XPS data indicate the presence of a nonstoichiometric transition region that is composed of tetrahedral groups, Si-$(O)_x(Si)_{4-x}$, that is nonlinearly graded in composition, and that is ≤ 10Å thick. The transition region width remains invariant with changes in oxidation processing conditions, in oxidant, in oxide film thickness, or in high temperature annealing but is affected by changes in substrate orientation. Si substrates with $<100>$ orientations have narrower transition regions that do substrates with $<111>$ orientations.

INTRODUCTION

The interface between an amorphous, thermally-grown SiO_2 film and a crystalline Si substrate has been intensively studied because of its technological importance. This paper reviews some of our XPS studies of the SiO_2-Si interface (1-3).

EXPERIMENTAL

P- or n-type, 1-2 ohm-cm, $<100>$ and $<111>$ oriented Si substrates were thermally oxidized. Oxidation conditions, oxidant, oxide film thicknesses, post-oxidation annealing, and substrate orientation, were varied (2). Ultrathin oxide films <30Å thick were nondestructively examined using XPS. Thicker (≤ 1000Å) films were first chemically etched to ultrathin film thicknesses.

Ellipsometry was used as the primary measure of oxide film thickness, d. The measured value of Δ and an assumed film index of refraction, n, of 1.46, corresponding to SiO_2, were used to evaluate film thicknesses.

Samples were excited with Mg Kα (1253.6 eV) radiation to obtain XPS spectra. Silicon (Si 2p) and oxygen (O 1s) binding energies, linewidths, and line intensities, were recorded as a function of oxide film thickness. Peak areas were measured to obtain XPS intensity data.

*Present address: IBM T. J. Watson Research Center, Yorktown Heights, NY 10598

Photoelectron Binding Energies and Intensities

The binding energy of an electronic level, referenced to the Fermi level, is $E_B = h\nu - E_K - \phi$, where $h\nu$ is the x-ray excitation radiation energy, E_K is the measured kinetic energy, and ϕ is a spectrometer constant. E_B(Si 2p)$_{oxide}$ linearly increases with oxygen content of the homogeneous reference Si films of Si, SiO, and SiO$_2$ (3).

The ratio of photoelectron intensity, I, from a film of thickness, d, to the intensity of pure bulk material is given by Eq. (1),

$$I(film)_d/I(film)_\infty = 1 - \exp(-d/\lambda) \qquad (1)$$

where λ, the mean escape depth, is the distance for which 1/e of the photoelectrons are not inelastically scattered. The bulk film intensity is

$$I(film)_\infty = \int KD\exp(-d/\lambda) = KD\lambda, \qquad (2)$$

where K is constant for a particular kinetic energy and electron level, and D represents an atom density that varies with film composition. The intensity ratio for photoemitted electrons from a two-layer structure with photoelectron lines arising from the same atomic level and close in kinetic energy is

$$[I_2(\text{Si 2p})_{Si})/(I_1(\text{Si 2p})_{ox}] = [(I_{2,\infty})/(I_{1,\infty})][(\exp(-d/\lambda_1)/(1 - \exp(-d/\lambda_1)] , \qquad (3)$$

$$I_2(\text{substrate})_\infty/I_1(\text{film})_\infty = D_2\lambda_2/D_1\lambda_1 . \qquad (4)$$

Mean escape depths, λ, were evaluated (1) from Eqs. (3) and (4). The obtained values of λ for Si 2p photoelectrons emitted from the oxide and from the substrate are 25Å and 23Å, respectively, when using a Mg Kα x-ray source. The λ values were derived from Si-oxide structures with oxide films >20Å thick. For films <20Å thick, Eq. (3) is used to derive oxide film thicknesses from photoelectron intensity data.

Photoelectron linewidths (FWHM-full width, half maximum) provide a basis for characterizing a film if the linewidths vary with film composition. Reference oxide linewidth data are obtained from SiO (2.3 eV) and from SiO$_2$ (1.8 eV) samples (3).

RESULTS AND DISCUSSION

Binding Energy Data

In XPS spectra of SiO$_2$-Si structures, the Si 2p substrate binding energy, E_B(Si 2p)$_{substrate}$, is constant at 99.3 eV and is used as an internal reference. E_B(Si 2p)$_{oxide}$ from oxide films >20Å thick is constant at 103.7 eV. For thinner oxide films, E_B(Si 2p)$_{oxide}$ decreases as the substrate is approached. The O 1s oxide binding energies, although scattered, remain constant in the oxidized Si films. Oxygen atoms are presumably bonded as Si-O-Si groups in the partially and completely oxidized Si films.

Si 2p oxide binding energy shifts and Si 2p intensity ratios obtained from spectra of oxide films formed using various oxidation processing conditions (2) are plotted in Fig. 1. These data associate changes in oxide film composition with changes in ultrathin oxide film thickness. The open circles and triangles correspond to nondestructively characterized samples. The filled data points correspond to oxide films $\leq$1000Å thick that were chemically etched to ultrathin film thicknesses and then characterized with XPS.

An ideal SiO_2-Si interface possesses one layer of Si tetrahedra with Si atoms in asymmetric chemical environments which will shift $E_B(Si\ 2p)_{oxide}$. The Si 2p data were analyzed to derive a transition region width by a) determining the composition of the first 2.5Å oxide layer from Fig. 1, b) assuming a linear gradation in oxide composition between the first layer and the SiO_2 film, c) synthesizing Si 2p profiles for transition regions of varying widths by incremental addition of oxide layers to the substrate, and d) comparing experimental with calculated data. A single iteration was used to adjust the Si 2p intensities for oxide nonstoichiometry by altering D_1 in Eq. (4).

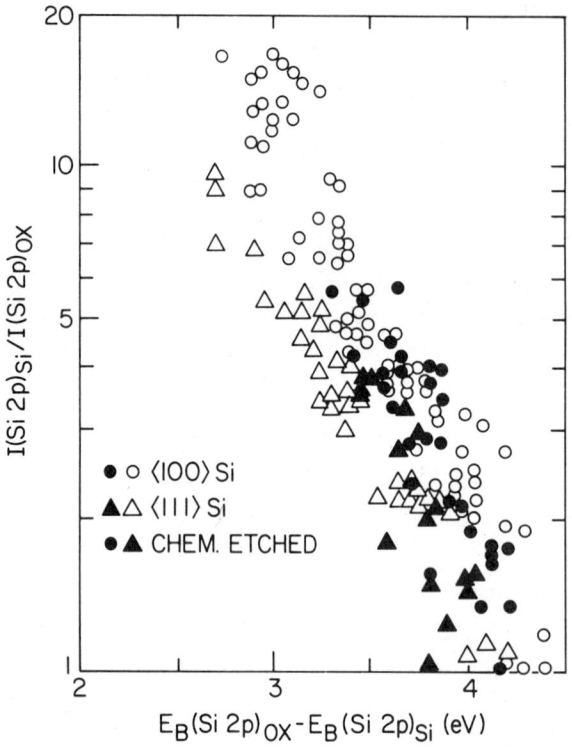

Fig. 1. Experimental Si 2p data for oxide films grown on <111> and <100> oriented substrates.

The data in Fig. 1 are scattered but form two distinct groups that are associated with differences in substrate orientation but are not affected by changes in oxidation temperature, oxidation time, oxide film thickness, post-oxidation annealing, or oxidant. The Si 2p oxide peaks are not resolvable into Si and SiO_2 photoelectron lines but are best represented by a distribution of tetrahedral configurations ranging from Si-(Si)$_4$ in the substrate, to Si-(O)$_4$ in stoichiometric SiO_2. The <100> data extrapolate to a 2.5Å thick layer with an average composition of $SiO_{1.4}$. The transition region width on the <100> Si is 7.5Å thick with a standard deviation of ± 1.4Å. Similarly, the <111> data extrapolate to an initial layer 2.5Å thick with an average composition of $SiO_{1.2}$ and a transition region width about 10Å thick. The transition between Si and stoichiometric SiO_2 in nonlinear in composition.

Linewidth Data

Linewidth was assumed to change linearly with change in oxygen content in the oxide films to estimate linewidth variations with oxide film thickness. Both O 1s and Si 2p oxide linewidths (FWHM) obtained from oxide films grown on <100> oriented substrates increased with decrease in film thickness for films <25Å thick (Fig. 2). For thicker oxide films, a constant minimum photoelectron linewidth is detected for both of these lines. The Si 2p substrate linewidth remains constant and is independent of film thickness. From Fig. 2, it appears that the linewidth and E_B(Si 2p) data are in qualitative agreement with regard to the transition region width.

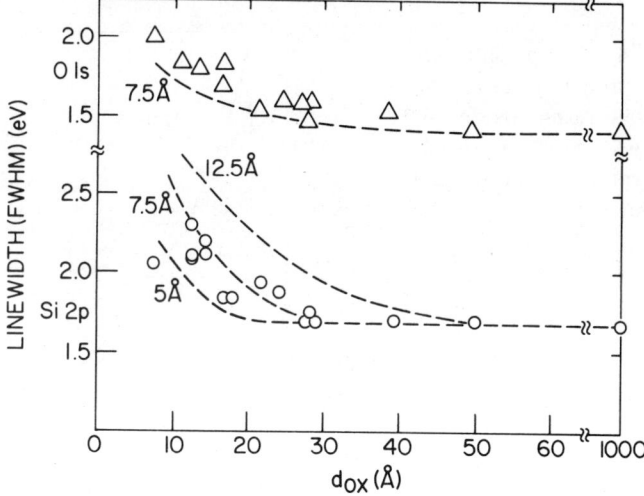

Fig. 2. Oxide Si 2p and O 1s linewidths (FWHM) dependence on oxide film thickness for films on <100> oriented substrates.

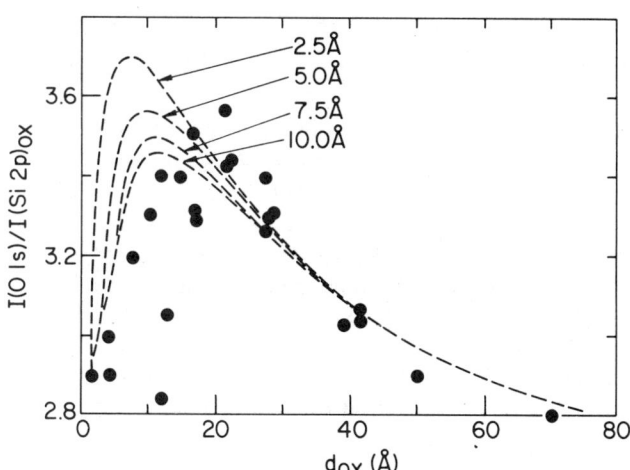

Fig. 3. Oxide O 1s/Si 2p intensity ratio dependence on oxide film thickness for films grown on <100> oriented substrates.

Oxide O 1s and Si 2p Intensity Ratio Data
In Fig. 3, the oxide O 1s/Si 2p intensity ratio is plotted versus oxide film thickness. Oxygen and Si intensities are determined from Eqs. (1) and (2). A mean escape depth curves was calculated from the intensity data for oxide films >20Å thick with I_1(O 1s)$_\infty$/I_2(Si 2p)$_\infty$ = 2.7, λ_1(O 1s) = 16Å, λ_2(Si 2p)$_{oxide}$ = 25Å, and D_1 and D_2 adjusted to account for changes in oxide composition. A one layer interface on a <100> oriented Si substrate is represented as 2.5Å of $SiO_{1.4}$. Curves are also shown for 5.0Å, 7.5Å, and 10.0Å thick transition regions. The oxide intensity data

388

indicates the presence of a transition region whose width appears somewhat greater than that obtained from E_B analysis. This difference can be attributed to nonoxygenated surface impurities.

Fig. 4. Si 2p photoelectron intensity ratio data from ultrathin oxide films versus ellipsometrically determined oxide film thicknesses.

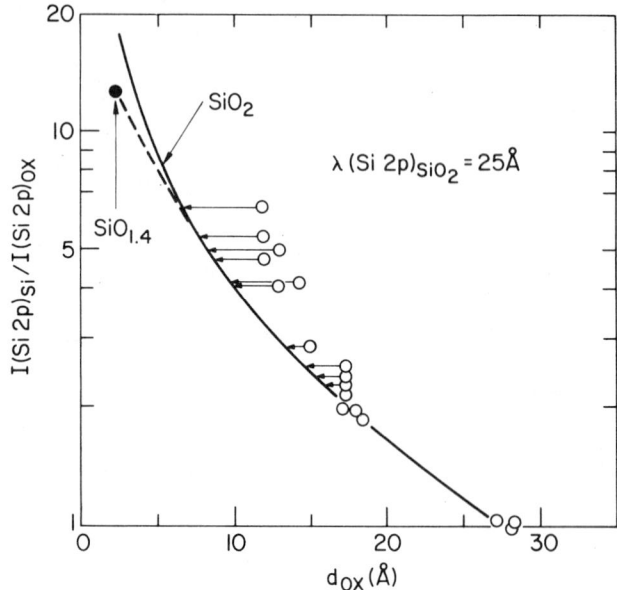

Si 2p Intensity Data:

The mean escape depth curve obtained from Eq. (3) for oxide films <20Å thick is plotted versus film thicknesses in Fig. 4. A point corresponding to a film 2.5Å thick with a composition of $SiO_{1.4}$ was calculated by adjusting D in Eqs. (3) and (4). The dashed curve represents the mean escape depth curve for a graded nonstoichiometric Si oxide film about 7.5Å thick. The thicknesses of oxide films <20Å thick do not fall upon the mean escape depth curve. The required adjustments in film thickness increase as the Si substrate is approached. Decreasing the oxide film ellipsometric thickness corresponds to increasing the refractive index, n, of the oxide film. Increases in n are consistent with decreases in oxygen content of the oxide films, and therefore, with the presence of a nonstoichiometric SiO_2-Si transition region.

SUMMARY

The XPS Si 2p binding energy data indicate that a 7-10Å transition region extends between SiO_2 and a crystalline Si substrate. This conclusion is supported by agreement obtained from analyses of binding energy, linewidth and photoelectron intensity data. No single alternative interpretation is consistent with these XPS data.

REFERENCES

1. R. Flitsch and S. I. Raider, J. Vac. Sci. Technol., **12**, 305 (1975).
2. S. I. Raider and R. Flitsch, IBM J. Res. Dev. (1978).
3. S. I. Raider and R. Flitsch, J. Electrochem. Soc., **123**, 1754 (1976).

CHEMICAL STRUCTURE OF THE TRANSITIONAL REGION OF THE
SiO$_2$/Si INTERFACE*

F. J. Grunthaner and J. Maserjian
Jet Propulsion Laboratory, California Institute of Technology
Pasadena, California 91103

Abstract

XPS and chemical depth profiling are used to determine the chemical structure
of the SiO$_2$/Si interface. The transitional layer is approximately one monolayer
thick and consists of Si(I), Si(II), and Si(III) states. A strained region of
SiO$_2$ exists near the interface (10-40 A thick) and and can be modeled by a
distribution of 3,5, and 7 membered tetrahedral rings while the bulk is best
modeled by 6,4, and 8 membered rings. Theory and experiment are correlated to
develop this model structure based on the local chemical environment.

INTRODUCTION

Highly perfect SiO$_2$/Si interfaces can be routinely prepared by thermal oxidation
of monocrystalline silicon and intensive research has been devoted to the opti-
mization of this structure for microelectronic technology. As noted by Deal (1),
a number of questions concerning the dynamic interaction of the oxide with charge
and the nature of the SiO$_2$/Si interface remain unanswered. In a review of the
chemical nature of this interface, Revesz (2) has suggested that the high degree
of perfection of the interface is largely due to the non-crystallinity of the
thermally grown film. A variety of experiments have indicated that this oxide
is amorphous but substantially similar to α-quartz. These observations have
been interpreted by Revesz in terms of short range order (preservation of the
basic tetrahedral SiO$_{4/2}$ unit) in the structure. A network or polymer structure
is suggested consisting of a variety of rings composed of n tetrahedral units
where n can vary from 3 to 9. In this model, Revesz has proposed that pπ-dπ
bonding increases with ring size and contributes to the observed properties of
the bulk oxide as well as the SiO$_2$/Si interface.

Recently, it has become experimentally possible to study the SiO$_2$/Si interface
in considerable chemical detail through the use of surface analytical techniques
such as x-ray photoelectron spectroscopy (XPS) (Ref. 3,4), Auger electron spectroscopy (Ref. 5), and ion scattering spectroscopy (Ref. 6). In this paper we
report the results of a study using XPS to determine i) the stoichiometry of
thermally grown SiO$_x$ films, ii) the chemical structure of the transitional layer

*This paper presents the results of one phase of research performed at the Jet
Propulsion Laboratory, California Institute of Technology, sponsored by the
National Aeronautics & Space Administration under Contract No. 7-100, and by
the Defense Advanced Research Projects Agency through the National Bureau of
Standards, Order No. 611377.

at the SiO_2/Si interface, and iii) variations in the structure of the oxide as a function of distance from the interface. Finally, we interpret our data in terms of a model of bulk SiO_2 and the SiO_2/Si interface emphasizing the importance of the local atomic environment and the charge distribution in the bridging Si-O-Si bond.

EXPERIMENTAL TECHNIQUES

The XPS spectrometer used in this work has been described elsewhere (Ref. 3). It is comprised of a modified Hewlett-Packard 5950A ESCA Spectrometer connected to an environmental chamber used for sample preparation. This chamber maintains a high purity N_2 atmosphere (<1 ppm H_2O) and is connected via interlock to a furnace for oxide growth under "in situ" conditions. For these experiments, the spectrometer was operated at a high resolution (better than 0.35 eV) and at a pressure of less than 5×10^{-10} torr. The standard deviation of measured peak positions is 0.024 eV. N_{ss} measurements were taken using an automated digital quasi-static method. Depth profiling experiments have been carried out using a newly developed wet-chemical etching technique which minimizes the oxide-solvent reaction. The topographical uniformity of this procedure was verified through SEM micrographs; within the positional resolution of the SEM technique, no thickness variation was observed over the etched area. For oxide thicknesses greater than 50 Å, the film was measured ellipsometrically. Thickness below this value was generally determined by observed oxide/element ratios calculated from the XPS spectra.

RESULTS AND DISCUSSION

In a previous investigation of the electrical effect of carbon contamination at the SiO_2/Si interface (Ref. 3), 60 Å SiO_2 films were grown "in situ" at 850° C on <100> silicon surfaces intentionally contaminated with monolayers of carbonaceous films. The remaining carbon at the interface was determined by

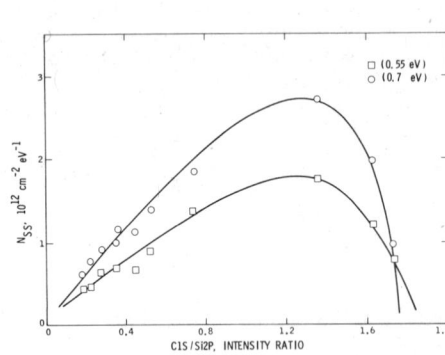

Fig. 1. N_{ss} vs carbon contamination. Carbon signal strength normalized to the intensity of the Si 2p line corresponding to the silicon substrate. Intensity ratio is not atom ratio at interface.

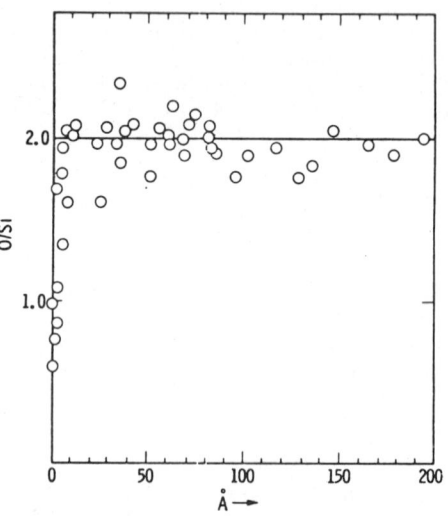

Fig. 2. Normalized oxygen/oxidized Si ratio vs. oxide film thickness.

measurement of the carbon 1s photoelectron line through the grown film and was normalized to the intensity of the silicon 2p line arising from the substrate. Duplicate samples were metallized through a shadow mask to prepare an array of MOS capacitors for N_{ss} measurements. Fig. 1 gives a plot of N_{ss} vs. the interfacial carbon 1s intensity. For these extremely thin, unannealed capacitors, reasonable values of N_{ss} are observed ($\approx 10^{12} cm^{-2} eV^{-1}$). N_{ss} initially increases with carbon contamination, but then decreases, suggesting the formation of a new interfacial layer of different stoichiometry.

Considerable controversy has arisen recently concerning the stoichiometry of the SiO_x film thermally grown on silicon. XPS spectra permit independent observation of oxygen and oxidized silicon signals arising from a sample. A series of samples were prepared with dry thermal oxides of 3 to 200 A thickness. Growth temperatures ranged from 850° C to 1050° C. Several of the 100–200 A samples were profiled by the wet chemical etching procedure previously mentioned. The signal strength of the oxygen and oxidized silicon lines were determined for this range of film thicknesses. The ratios of these intensities is plotted in Fig. 2 as a function of sample film thickness. The mean ratio for film thicknesses greater than 30 A has been normalized to a value of 2. The film composition changes rapidly approximately 5 A from the silicon interface. This transitional layer consists of reduced oxides (SiO_x, $0.5 < x < 2$).

These observations indicate that 1) the interfacial sites can be chemically substituted and thereby cause an electrically observable change in N_{ss}, and 2) the stoichiometry of SiO_2 as determined by XPS is essentially uniform except for a transitional zone within ≈ 5 A of the interface. These observations prompted us to make a controlled modification of the silicon surface, grow a thin oxide structure and determine the subsequent chemical and electrical characteristics of the resulting interface. A series of 28 wafers (N-type <100>) were prepared following a standardized cleaning procedure. The samples were then immersed in a deionized water bath at 90° C for 30 minutes followed by a rinse in high purity water (25° C) to 18 megohm resistivity. Fourteen samples were stripped in HF solutions, and rinsed again to 18 megohm resistivity. The two sample sets were oxidized in HCl-cleaned quartz tubes with dry oxygen at 1000° C with no anneal. Several of these samples were used for chemical

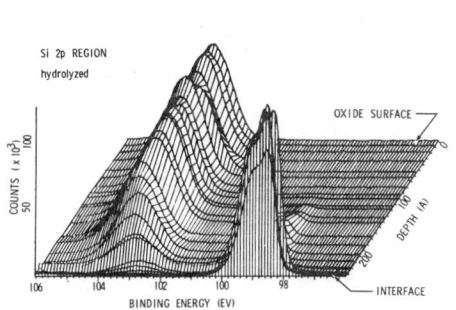

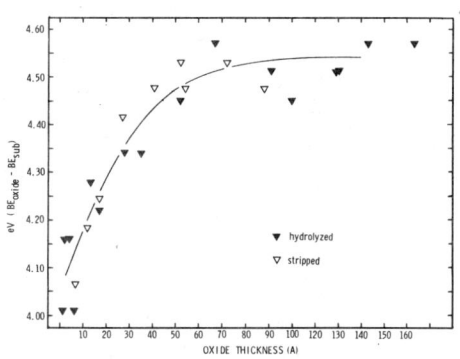

Fig. 3. Composite 3-D plot of Si 2p
XPS spectra vs. film thickness.
Viewing direction is from SiO_2/Si
interface out to original oxide surface.

Fig. 4. Plot of binding energy
difference between centroid of oxide
peak and substrate maximum vs. thickness

profiling experiments and the remainder were metallized and patterned into MOS capacitors for electrical measurements. Initial XPS spectra of the samples just prior to oxidation showed a hydroxylated surface oxide (≈15 A) on the non-stripped sample, while the HF stripped sample had an oxide film (≈6 A) at lower binding energy.

Following oxidation, the final oxide thickness was 700 A for the hydroxylated surface and 800 A for the stripped sample. In order to determine whether this difference was due to a sustained change in interfacial growth kinetics or to some initial transition, a series of oxides were grown at the same temperature for shorter times. The results (Ref. 3) indicate a sustained difference in growth rate, more characteristic of a stabilized structural difference at the SiO_2/Si interface. Electrical measurements of N_{ss} and flatband voltage V_{FB} showed no significant differences before irradiation. However, the hydroxylated wafers show somewhat greater change ΔV_{FB} after irradiation with 2.5 MeV electrons. N_{ss} was in the low $10^{11} cm^{-2} eV^{-1}$ range for all samples.

Three replicate samples of each of the two cleaning processes were depth pro-filed in a series of 40 intervals by the wet chemical proceedure described previously. The observed XPS spectra of the silicon 2p region are plotted as a function of oxide thickness for the hydroxylated sample in Fig. 3. The peak at 103.2 eV corresponds to SiO_2 and that at 98.9 eV is due to the silicon sub-strate. On this scale Si_2O would occur at 100 eV, SiO at 101 eV, and Si_2O_3 at 102 eV. The elemental peak can be seen far into the oxide because of the mag-nitude of the electron mean-free path. Equipotential lines are drawn at 0.1 eV spacings. The z-axis scale is calibrated in Angstroms from the outer surface. From this figure, two major points are evident: i) the chemical composition of the oxide is uniformly SiO_2 down to the interfacial region and ii) the centroid of the oxide peak shifts to lower binding energy as the interface is approached. This latter feature is emphasized in the data of Fig. 4. Here the difference in binding energy between the centroid of the oxide peak and the maximum of the substrate line is plotted vs. film thickness for an hydroxylated and a stripped sample. Note that at approximately 60 A the relative energy position of the oxide peaks begins to change from a mean value of 4.53 eV to a new value of 4.08 eV at about 5 A. Although the magnitude of the shift, 0.45 eV is the same for the two samples within the experimental uncertainty, the observed spectra

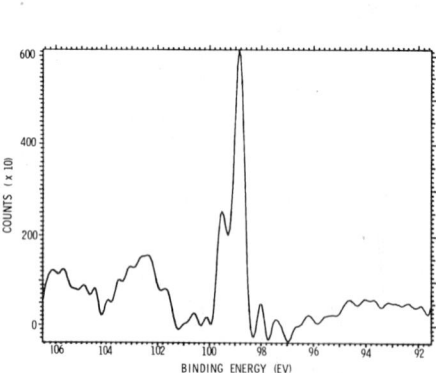

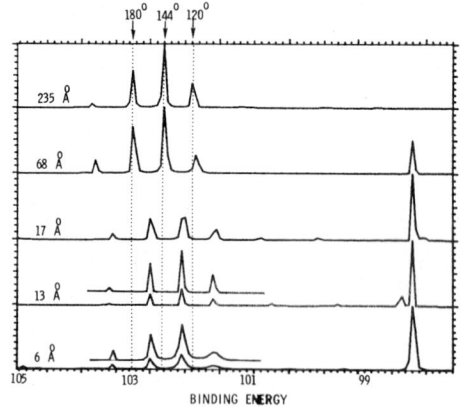

Fig. 5. Difference spectrum of 4 A film less substrate signal with no measureable oxide.

Fig. 6. Resolution enhancement of Si 2p spectra giving posi-tional distributions vs. energy

show different peak shapes for the two samples. The determined mean free-path or electron escape depth was substantially less in the case of the stripped sample as compared to the hydroxylated sample. To demonstrate the observation of the reduced oxides of silicon, Fig. 5 gives the difference spectrum obtained by subtracting the silicon substrate spectrum from that spectrum corresponding to a 4 A film on the substrate. The predominant peak (assigned to Si(I)) is 0.6 eV higher binding energy than the substrate peak. The oxide peak is about 0.9 eV lower in binding energy than SiO_2 and is attributed to Si(III). This total integrated signal strength is a factor of 1000 times weaker than the primary silicon line and is consistent with a monolayer coverage.

RESOLUTION ENHANCEMENT AND CHARGE CORRELATION

Recently, we have developed a number of resolution enhancement proceedures which are capable of detecting the position of component peaks within complex peak envelopes. These methods detect the position of the centroid with considerable reliability and can approximate the intensity distribution. No indication of the original line shape can be preserved. These methods are based on the solution of a linear prediction filter by means of the Yule-Walker equations for a maximum entropy constraint (Ref. 7). In Fig. 6, the spectra resulting from a deconvolution of the original signal with the observed silicon substrate line is given for the hydroxylated sample at a variety of thicknesses. The peak at 98.3 eV is due to the substrate. The oxide envelope is shown to consist of a series of three primary peaks (235 A) at 103.02, 102.55, and 102.05 eV. Closer to the interface the positions of the main oxide peaks shift to lower binding energy (102.75, 102.2, 101.7 eV), and additional peaks are observed at 100.7 and 99.5 eV. Nucho and Madhukar (Ref. 8) have calculated the effective electronic charge removed from the central silicon atom of the $SiO_{4/2}$ tetrahedron in α-quartz using a tight-binding approximation. They have also calculated the charge transfer from silicon to oxygen as a function of the Si-O-Si bridging bond angle. No d-orbital interactions were included in their calculations. The angles chosen for the calculation included the literature values for 6-tetrahedra (144°) and 4-tetrahedra (120°) rings for comparison with the cited non-crystalline network SiO_2 model (Ref. 2). If their value for charge transfer from SiO_2 at 144° (α-quartz) is linearly scaled to the observed chemical shift

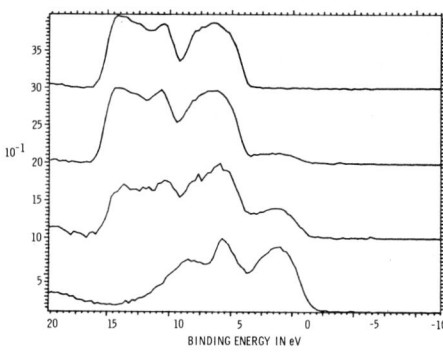

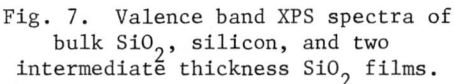

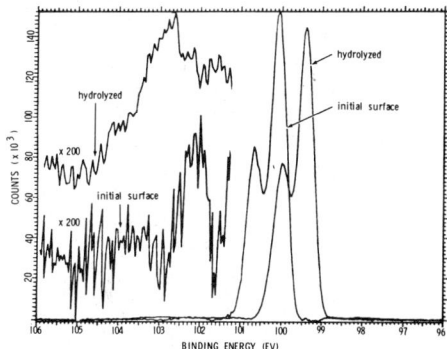

Fig. 7. Valence band XPS spectra of bulk SiO_2, silicon, and two intermediate thickness SiO_2 films.

Fig. 8. High resolutions Si 2p spectra of n-type <100> silicon before and after H_2O exposure at 25° C.

between the elemental substrate and the peak at 102.55 eV, a binding energy
scale can be estimated as a function of Si-O-Si angle. The peak at 102.55
was used for this scaling since this is the most intense peak of the observed
manifold over the range of oxide thicknesses. Figure 6 shows the positions of
the peaks for β-cristobolite (180°), α-quartz (144°), and 4-membered rings
(120°), based on this scaling. This remarkable agreement suggests that bulk
SiO_2 is composed of 4, 6, and 8 membered rings with a substantial population
of 180° Si-O-Si bond angles. Approaching the interface the ring size distribution shifts to smaller numbers, and the bond angle shifts to smaller values
from the 144° optimum, as a result of the strain on the network induced by
the structural mismatch of the oxide and element lattices. It should be
noted that a Si-O-Si bond angle of 97° will permit substitution of a bridging
oxygen for a Si-Si bond in the elemental lattice based on a Si-O bond length
of 1.6 Å. Hence, the structural transition between SiO_2 and elemental silicon would require a change in bond angle consistent with the experimental
observation.

In Fig. 4, a 0.45 eV shift was observed in the position of the oxide centroid
when proceeding from the bulk to the near interface. By this charge-vs-angle
scaling, this shift is equivalent to a bond angle reduction from 144° to 120°.
In a ring formalism, the distribution of ring sizes has changed from 6, 4, 8
to 5, 7, 3 at the near interface. The calculation further indicates that
the charge removed from the silicon atom per Si-O bond is independent of
the total number of Si-O bonds. In terms of 4-coordinate silicon, the charge
removed from a silicon atom with four Si-O bonds is 4 times greater than a
silicon with one Si-O and three Si-Si bonds. Consequently, with a 144° bond
angle, Si^{+1} would be predicted at 99.3 eV, Si^{+2} at 100.4 eV, and Si^{+3} at 101.5
eV. Therefore, near the interface, the peak at 101.7 eV could be assigned to
Si^{+3}, that at 100.7 eV to Si^{+2}, and that at 99.5 eV to Si^{+1}. Additional details concerning spectral interpretation and scaling of charge shifts will be
published elsewhere (9).

This observation of intermediate oxidation states near the interface is further
emphasized by Fig. 7, which gives a series of XPS spectra for the valence
region of the spectrum. The uppermost curve is that of the bulk oxide, and
the bottom curve is due to the clean silicon substrate. The second spectrum
from the top of the figure is due to a ~40 Å thick film on the substrate,
while the remaining curve is due to a 10 Å film. The spectra are presented
as recorded. Note that the edge of the oxide valence band does not change
position with thickness and that the 40 Å and 10 Å spectra show occupied
states at 4 eV not present in either the substrate or bulk oxide spectra. It
would appear that states are filled above the top of the SiO_2 valence band
and that their relative spectral contribution is of the same order of magnitude as the intermediate state observed in the case level spectra.

Recent instrumentation developments at our laboratory have substantially increased the resolving power of the spectrometer and this is illustrated in
Fig. 8, which gives the Si 2p spectra of an n-doped <100> silicon sample.
The curve labeled initial is the spectrum of the sample after cleaning by a
standard procedure which leaves no detectable impurities (C, N, Na, O, etc.)
on the surface. The spectrum labeled hydrolyzed is that of the same sample
after exposure to deionized water (25°C) for one minute. Note that the oxide
signal in both cases is so weak that a 200-fold gain increase is required to
show the structure. Yet the substrate signal has shifted nearly 0.6 eV and
substantially different oxide species are observed. These spectra require
a signal-to-background ratio of better than 1000 to 1. Using the charge and

bond angle criteria developed above, the initial sample has a set of surface species corresponding to Si^{+2} (120° bond angle) and a small amount of Si^{+3} (144°). The hydrolyzed sample shows predominantly Si^{+3} (144°) with lesser amounts of Si^{+2} (144°), Si^{+3} hydrogen bonded and Si^{+4} (144°). The increased surface states after H_2O exposure are essentially Si^{+3} and have caused a band bending of approximately 0.6 eV. Finally, it should be noted that we have successfully synthesized Si^{+1} compounds for energy referencing purposes (10).

CONCLUSIONS

The structure of silicon dioxide can be characterized as comprised of four-coordinate silicon tetrahedra and a variety of Si-O-Si bond angles. Variation in bridging bond angle causes substantial changes in the charge density on silicon. The bulk oxide consists of ring networks with bond angles of 144, 180, and 120° or alternatively, with 6, 8, and 4 tetrahedra. Near the interface, ($\sim 10 - 40$ Å) the bond angle changes substantially due to strain and the sizes of the rings reduce to 3, 5, and 7 members. The final transition layer at the interface is approximately one monolayer thick and consists of Si^{+1}, Si^{+2}, and Si^{+3}. Intermediate oxidation states of silicon can be observed in the valence band spectra.

REFERENCES

1. B. E. Deal, J. Electrochem. Soc. (1977).

2. A. G. Revesz, J. Non-crystalline Solids 11, 309 (1973).

3. F. J. Grunthaner and J. Maserjian, IEEE Trans. Nucl. Sci. NS-24, 2108 (1977).

4. R. Flitsch and S. I Raider, J. Vac. Sci. Technol. 12, 305 (1975).

5. J. S. Johannessen, W. E. Spicer, and Y. E. Strausser, Appl. Phys. Lett. 27, 452 (1975).

6. W. L. Harrington, R. E. Honig, A. M. Goodman, and R. Williams, Appl. Phys. Lett. 27, 644 (1975).

7. J. J. Barton, J. D. Klein, and F. J. Grunthaner, to be published.

8. R. B. Nucho and A. Madhukar, this conference.

9. F. J. Grunthaner and A. Madhukar, to be published.

10. F. J. Grunthaner, J. A. Wurzbach, and J. Maserjian, submitted to J. Vac. Sci. Tech.

ACKNOWLEDGMENTS

The authors wish to thank J. Wurzbach for the carbon/silicon interface experiments, C. Butler, who performed the XPS profile experiments, B. Lewis, who engineered the instrumental system, R. Vasquez, who performed the data reduction, P. Grunthaner, who prepared the manuscript, and A. Madhukar for stimulating discussions.

MOS SOLAR CELL AS A TOOL TO STUDY THE TRANSITION REGION ASSOCIATED WITH ULTRA THIN FILMS OF SiO$_x$*

R. Singh, K. Rajkanan and J. Shewchun
Department of Engineering Physics, McMaster University
Hamilton, Ontario L8S 4M1, Canada

ABSTRACT

The oxide thickness in silicon MOS solar cells is comparable to the non-stoichiometric transition region associated with thermally grown thick oxides. The oxide thickness dependence of open circuit voltage below a particular thickness has been interpreted to give information regarding the physical properties of the transition layer. By using the Maxwell-Garnett relation, the complex dielectric constant of the transition region has been estimated as a function of its thickness. Our study indicates there is a graded interface and the width of transition layers is about 13-14 Å.

INTRODUCTION

Recent experimental work on the interface of thermally grown thick silicon oxide on silicon reveals the existence of a transition region with composition of SiO$_x$, where x varies from one to two across the transition region. As concluded by most observers, the non-stoichiometric region extends to about 20 Å from the interface. In case of MOS devices involving ultra thin oxides (10-20 Å), such as MOS solar cells, the oxide thickness is comparable to the non-stoichiometric transition region and the device properties are expected to be dependent on the thickness and the composition of the oxide. In addition, since the ultra-thin oxide of 10-20 Å consists of only few atomic layers, it can be heavily defected with large pin-hole density, therefore the device performance will also depend on the quality of the oxide.

In this paper, we shall report on the use of Al-SiO$_x$-(p-type)Si solar cells to study the transition layer associated with a thermally grown SiO$_x$ layer. The oxide thickness dependence of open-circuit voltage V$_{oc}$, below a particular oxide thickness can be interpreted to give information regarding the properties of the transition layer. In particular by using the Maxwell-Garnett relation, the complex dielectric constant of the transition layer has been estimated as a function of its thickness.

---

*Work supported by the National Research Council of Canada and the Department of Energy under contract No. E(04-3-1203)

EFFECT OF OXIDE THICKNESS ON OPEN CIRCUIT VOLTAGE

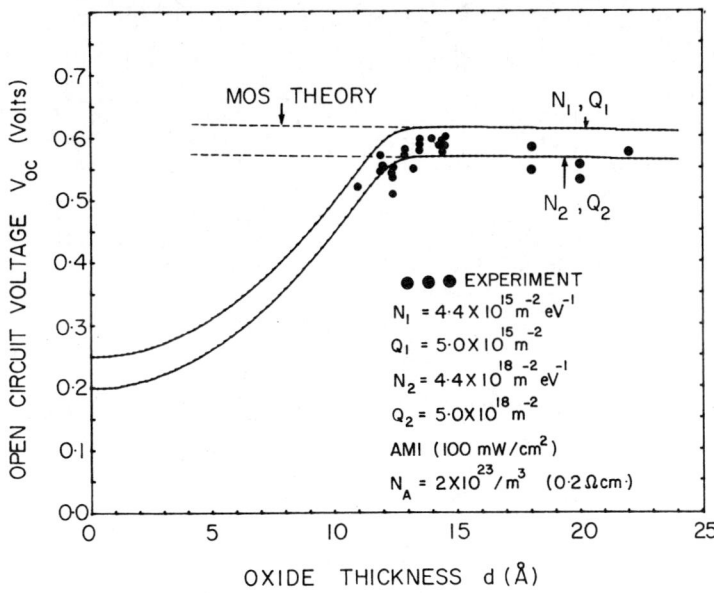

Fig. 1   Experimental and theoretical open circuit voltage
of MOS solar cell as a function of oxide thickness

Figure 1 shows experimentally observed V_{oc} as a function of oxide thickness
in the range of 10-20 Å. The data shows a decrease in V_{oc} with increasing
thickness at approximately 1 mV/Å above about 12 Å and a rapid increase
below 12 Å. In order to explain the observed oxide thickness dependence of
V_{oc}, we examined whether our theory of MOS solar cells (1) could give a similar
functional behaviour. For this purpose, we varied the oxide and interface
parameters within suitable upper and lower limits, to account for the nonstoichiometry of ultra thin oxide layers. The effective mass of carriers, band
gap, electron affinity and dielectric constant of SiO_x do not affect the V_{oc}
in the range of 10-20 Å by any significant amount. The surface state density
N and the oxide charge Q were varied from 4.4×10^{15} m^{-2} eV^{-1} and 5×10^{15}
m^{-2} to 4.4×10^{18} m^{-2} eV^{-1} and 5×10^{18} m^{-2}. On increasing N and Q by three
orders of magnitude, V_{oc} is decreased by about 30 mV. Also the theoretical
curve of V_{oc} versus oxide thickness gives only a linear dependence with a
negative slope of about 1 mV/Å. These results are plotted in Fig. 1. A comparison of experimental results with these theoretical curves reveals a distinct
deviation from theory below about 12-13 Å. The experimental data indicates
that below oxide thickness of about 13 Å, the barrier height of the device
is considerably reduced causing a low V_{oc}.

In the light of above discussion, it seems that a mere variation in the oxide
and interface parameters is not sufficient to explain the positive slope of
V_{oc} versus oxide thickness at low thicknesses. One possible way of accounting
for this positive slope may be the presence of pin holes associated with ultra
thin oxides. It is a well known fact that as the thickness of the oxide is
reduced, it becomes more and more patchy i.e. pin hole density increases and
the variation in oxide thickness across the surface increases. In principle,

thick oxides are also not totally pin hole free, but the effect of pin holes will be seen only on devices with ultra thin oxides because of their large surface density which may even cause patches in the oxide layer.

This situation can be modelled as if there are several solar cells in a parallel combination, some of them having a large barrier height and others with a small barrier height. It should be noted, however, that it is difficult to assign a particular distribution for the variation of oxide thickness and hence of the barrier height across the entire surface due to pin holes. To overcome this problem, we have, somewhat arbitrarily at present, assumed only two barrier heights. One corresponds to the ideal MOS structure for the region with minimum pin holes and the other for a "Schottky" diode in a region where either no oxide exists or the pin hole density is so large that there is virtually no oxide. The exact nature of the metal-semiconductor contact is not well understood, but a survey of the literature reveals that for Si Schottky solar cells, the maximum open-circuit voltage under normal sunlight intensities is about 0.2 - 0.3 volt. Under the same level of illumination, the theoretical V_{oc} for an Al-SiO$_x$-(p-type)Si solar cell of 13 Å oxide thickness is about 0.65 V (1). Therefore, the parallel combination of an ideal MOS and "Schottky" diode will give a V_{oc} between 0.2 - 0.3 V and 0.65 V. To account for the fact that an increase in oxide thickness results in a reduction of pin hole density and an increase in barrier height, we have assumed a Gaussian distribution of pin holes as given below.

$$\rho_{pin} \sim [\exp-(\frac{d}{2do})^2 + \rho_o] \qquad (1)$$

Here d is the oxide thickness and d_o is the Si-O bond length equal to 1.62 Å. ρ_o takes into account the fact that even for very thick oxides, there are some pin holes present. There are reports in the literature (2-4), which indicate that fractional pin hole area is about 10^{-7}. In our model we have taken this fractional area as ρ_o. The fractional pin hole area obtained by equation 1 is plotted against the oxide thickness in Figure 2.

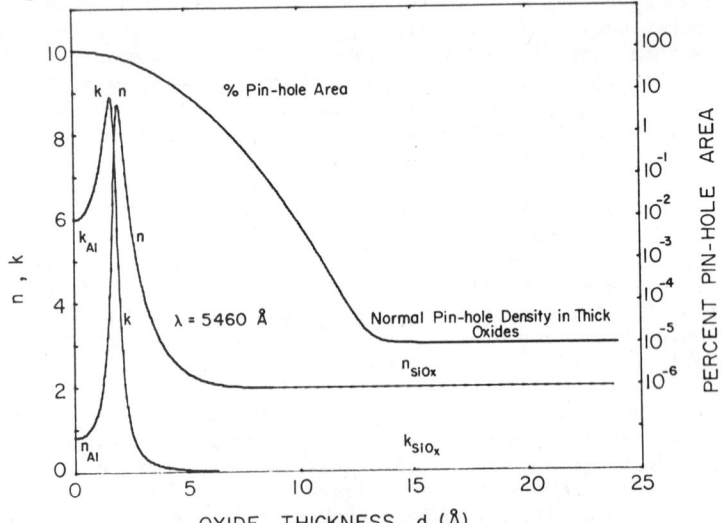

Fig. 2   Calculated n,k of SiO$_x$ as a function of oxide thickness. Percent pin hole area is calculated from Eq. (1)

To perform the actual calculations, we have assumed that for both the "MOS" and "Schottky" parts, the series and shunt resistance effects are negibile and carriers are generated in a similar manner. The diode ideality factors were taken same for both cases. Also, we have assumed that if the fractional area of pin holes is x, then out of a total light generated current I_L, (xI_L) would be due to pin holes and $(1-x)I_L$ due to ideal MOS parts. The total open-circuit voltage is given by (6)

$$V_{oc}^{total} = \frac{nKT}{q} \ln\left[\frac{1}{\dfrac{\{\exp-(\frac{d}{2do})^2 + \rho_o\}}{\{\exp \frac{qV_{oc}^S}{nKT} - 1\}} + \dfrac{\{1 - \exp-(\frac{d}{2do})^2 - \rho_o\}}{\{\exp \frac{qV_{oc}^M}{nKT} - 1\}}} +1\right] \quad (2)$$

Here V_{oc}^M varies with oxide thickness as

$$V_{oc}^M (d) = V_{oc}^M (o) - \alpha d \quad (3)$$

where α is a parameter depending on the interface properties and base material. A more general form of Eq. (2) is

$$V_{oc}^{total} (d) = \frac{nKT}{q} \ln\left[\frac{1}{\displaystyle\int_o^d \dfrac{\rho(t)\ dt}{\exp \frac{qV_{oc}(t)}{nKT} -1}} + 1 \right] \quad (4)$$

where $\rho(t)$ is the distribution function for estimating ρ_{pin}. Equation (2) combined with equation (3) is plotted in Fig. 1. We have varied V_{oc}^M and V_{oc}^S to account for the possible changes due to variation in surface states, oxide charges etc. in the MOS regions and work function variation in the Schottky regions. α in equation (3) has been taken as 1mV/Å, the value obtained by our theoretical calculations on MOS solar cells. Ideality factor, n has been taken to be unity, which is the value obtained by the theory of MOS solar cells. The experimental values are usually slightly larger than unity, but not much different. It should be noted that for each combination of V_{oc}^M and V_{oc}^S, these curves peak or saturate at about 13-14 Å of oxide thickness, much as the experimental data points. Most of the experimental points lies between the curves for N $= 4.4 \times 10^{15}$ m^{-2} eV^{-1} and N $= 4.4 \times 10^{18}$ m^{-2} eV^{-1}. This is in agreement with our hypothesis that the observed variation in the V_{oc} with oxide thickness is due to pin holes in ultra thin oxide. The positive slope below about 12 Å is mainly due to pin holes, although some contributions due to stoichiometric variation will be present, primarily around the peak in Figure 1. It is worth mentioning here that the peak around 13-14 Å is roughly the width of transition region as obtained by Raider and Flitsch (5).

DIELECTRIC PROPERTIES OF THE INTERFACE

The above conclusion also leads to another result concerning the dielectric constant of the thin oxide layer in MOS structures. Because of the pin holes in the oxide layer, the top layer metallic particles will be embedded in the silicon oxide film in a similar density as the pin holes. This will lead to modifications in the dielectric constant of the oxide layer. The effective dielectric constant of a "mixture" is given by the Maxwell-Garnett relation.

$$\frac{\varepsilon_e - \varepsilon_h}{\varepsilon_e + 2\varepsilon_h} = q \, \frac{\varepsilon_f - \varepsilon_h}{\varepsilon_f + 2\varepsilon_h} \tag{5}$$

where ε_e is the effective dielectric constant, ε_h is the dielectric constant of the host material (SiO_x in our case), ε_f is the dielectric constant of foreign particles (Al in our case for the Al-SiO_x-Si structure) and q is the fraction of foreign particles in the mixture (fractional density of pinholes). We have plotted results for the Al-SiO_x-Si structure in Fig. 2. In principle the dielectric constant of an intermediate layer could be deduced from ellipsometric measurements on a complicated MOS structure. The curves of Fig. 2 predict a sharp rise in n and k values around 2 Å of oxide thickness with a return to values associated with SiO_x at about 5 Å oxide thickness. The peak. position is governed by the dipole-dipole interaction between metal and oxide and will therefore change for different metals. At about 5 Å oxide thickness the pin hole area is about 10% and this thickness represents the growth of about two monolayers. At this point the oxide layer begins to be effective in separating the silicon substrate from the metal layer. Therefore, it is expected that after about 5 Å, the n and k values will saturate, as observed in Fig. 2.

CONCLUSION

The experimental results of V_{oc} as a function of oxide thickness for Al-SiO_x-(p-type)Si solar cells has been interpreted to give information regarding the physical properties of the transition layer. By using the Maxwell-Garnett relation, the complex dielectric constant of the SiO_x layer for Al-SiO_x-Si structure has been estimated.

REFERENCES

1. J. Shewchun, R. Singh and M.A. Green, "Theory of Metal-Insulator Semiconductor Solar Cells", J. Appl. Phys., **48**, 765 (1977).

2. S.W. Ing, Jr., R.E. Morrison, and J.E. Sandor, "Gas Permeation Study and Imperfection Detection of Thermally Grown and Deposited Thin Silicon Dioxide Films", J. Electrochem. Soc., **109**, 221 (1962).

3. E.F. Duffek, E.A. Benjamini, and C. Myeroie, "The Anodic Oxidation of Silicon in Ethylene Glycol Solutions", Electrochem. Technol. **3**, 75 (1965).

4. V.I. Prokhorov, and L.M. Sorokin, "Inheritance of Defects in Silicon by a Growing Oxide Film", Inorg. Matter., **9**, 28 (1973).

5. S.I. Raider, and R. Flitsch, "Stoichiometry of SiO_2/Si Interfacial Regions I:Ultra-thin Oxide Films", J. Vac. Sci. Technol. **13**, 58 (1976).

6. K. Rajkanan, R. Singh and J. Shewchun (to be published).

CHAPTER VIII
INTERFACE PROPERTIES

INITIAL STAGES OF SiO$_2$ FORMATION ON Si (111)[*]

R. S. Bauer and J. C. McMenamin

Xerox Palo Alto Research Center, Palo Alto, California 94304

and

H. Petersen[†] and A. Bianconi[‡]

Stanford Synchrotron Radiation Laboratory, Stanford, California 94305

ABSTRACT

Using novel UHV sample preparation techniques and photoelectron spectroscopies, the detailed atomic coordination, spatial characteristics, and growth kinetics of Si-SiO$_2$ interfaces have been elucidated. The kinetics of *in situ* oxide formation were studied by dosing cleaved Si (111) surfaces with O$_2$. The unique tunability of synchrotron radiation allows interpretation of the microscopic bonding from the antibonding states measured by constant-final-state (CFS) photoyield, where oxygen-induced Si (2p) core shifts provide insufficient information. Two intermediate oxidation states of Si are manifested by the 2p core electron shifts previously reported at 1.6 and 2.5 eV. These oxides are shown to be uniquely coordinated by our observation of spin-orbit splitting in the shifted peaks. However, these peaks are only precursors of SiO$_2$ formation since the three inner-well resonances characteristic of SiO$_4$ tetrahedra are not observed in the CFS spectrum. By varying the photoelectron kinetic energy and hence escape depth, these oxide states are found to form below the Si (111) surface double-layer. We understand this and the excess oxygen coordination upon annealing to be caused by O$_2$ molecules covalently-bonded as peroxy radicals to the surface Si. By dosing with atmospheric pressure of O$_2$, by using "excited" oxygen, or by heating the Si substrate, the activation barrier is overcome and the subsurface Si can become tetrahedrally coordinated with oxygen as in SiO$_2$. Since interatomic core-level transitions are limited to near-neighbor atoms, we have used them as a probe to show that an atomically abrupt Si-SiO$_2$ interface can be formed only by high temperature (~1000C) oxide growth. Even in this case, isolated Si-O bonds remain in the connective region and are characterized by the 1.6 eV Si (2p) chemical shift. A model is presented for interface growth which distinguishes between intra- and inter-layer bonds.

I. Introduction

Microscopic properties of the interface formed by SiO$_2$ with Si have recently been investigated extensively by electron emission spectroscopies. Techniques have included UV-stimulated valence-band photoemission,[1,2] x-ray-excited core-level photoemission,[3-6] electron-loss spectroscopy,[7,8] and Auger electron spectroscopy.[9,10] Such measurements have had the common feature of probing within tens of Å from the free surface. In order to relate these studies to practical MOS interfaces, two main experimental approaches have been utilized. In the first, Si is oxidized by standard device techniques and subsequently etched *in situ* by ion sputtering[9] or by chemical stripping.[6] The chemical state of the surface is then monitored with the relevant surface spectroscopy. The second approach has been to produce a clean Si surface and subsequently build up an oxide by *in situ* low-pressure exposure to oxygen at room temperature.[3,4,7,8]

We have studied with several different and novel electron spectroscopies the *in situ* growth of oxides on Si as a function both of substrate temperature and of gas pressure. By tuning synchrotron radiation, the Si (2p) core electron escape depth is varied to allow non-destructive profiling of the Si-O coordination in the first ~10 Å below the surface. The local configurations of SiO$_4$ tetrahedra are most sensitively monitored by the absorption[11] to antibonding states. Here again we use the unique tunability of the synchrotron radiation to measure photoemission partial yield.[12] This technique also provides direct evidence for the abruptness of the Si-SiO$_2$ interface. Since the core exciton transitions[13,14] are local in nature, the observation of inter-atomic absorption from a SiO$_2$ initial state to a Si final state provides an experimental limit of 5-10 Å on the width of the connective region between these two bulk stoichiometrics.

II. Experimental Techniques

A. *In Situ Dosing*

The formation of the Si/SiO$_2$ interface is controlled by the sample surface condition and temperature, the total exposure to oxygen as well as the pressure of the oxygen and its state upon arrival at the Si surface. As shown schematically in Fig. 1, all these parameters could be varied in our study. The Si was cleaved *in situ* at pressures of 2×10^{-10} Torr or less and could be indirectly heated to above 1050°C as measured at the crystal clamping point. This allows both high temperature oxide growth and annealing of intermediate oxidation states into more highly oxygen-coordinated Si. To better control the purity of the gas and to reduce the number of oxygen molecules necessary to obtain a given coverage, the gas was emitted in short bursts (<10 sec. duration) in a collimated stream onto the Si (111) surface ~ 1 cm away. By employing this dosing technique,[15] very high total exposures of 10^6L are possible. The peak pressure at the crystal surface is around 10^{-3} Torr while peak chamber ambient remains below 10^{-5} Torr. Compared to back-filling the chamber, this exposure allows better control, lower oxygen contamination, and much lower exposures of the chamber to the gas. The O$_2$ molecular state was not identified in the studies. However, strong, reproducible effects on the growth kinetics were observed when the oxygen was known to be "excited" by the presence of an ion gauge. Whether this controlled excitation resulted in the long-lived (~ 15 minute) molecular O$_2$ excited state discussed in GaAs oxidation studies[16] or in atomic oxygen, the extra energy imparted to the adsorbate is a key feature in the surface reaction as previously reported.[3,17]

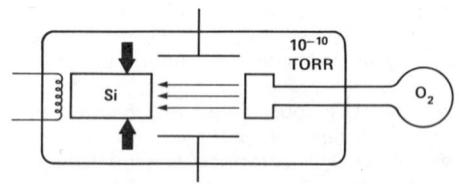

Fig. 1 Schematic diagram of Si oxidation control. For the gas dosing technique, oxygen molecules are passed through a dispersing head and impinge uniformly on the surface of the cleaved Si crystal. The exposure is calculated from the known spray pattern and the volume times pressure of the reservoir. With this method, large exposures up to 10^4 Langmuir can be made without increasing the peak ambient chamber pressure above 10^{-6} Torr.

B. *Core Level Photoemission Using Synchrotron Radiation*

The formation of Si oxides was monitored by measurement of the Si (2p) electron energy state using continuum synchrotron radiation as shown schematically in Fig. 2; the photons were monochromatized in the soft x-ray region by the grasshopper monochromator on the 4° beam line at the Stanford Synchrotron Radiation Laboratory. The Si (2p) electrons photoemitted thereby were analyzed using a PHI 15-255G, double-pass CMA at 75° to the optical axis and 20° from the sample normal. By studying core emission at kinetic energies around $E_K \approx 60$ eV, we

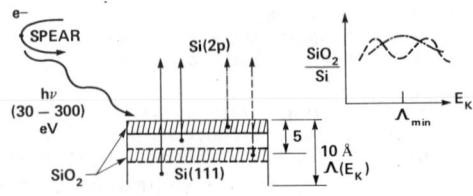

Fig. 2 Schematic diagram illustrating how the tunability of synchrotron radiation provides a spatial probe of interface formation from core-level photoemission. For SiO$_2$ on the surface of the crystal, the variation of the SiO$_2$/Si ratio is shown *vs.* kinetic energy of emitted electrons E_K by the dash-dot curve. If the SiO$_2$ layer is growing beneath a surface Si region, the "M" shaped curve (dashed) is obtained.

are sensitive to the environment of the atoms within ~ 6 layers of the free (111) surface.

The chemical bonding and composition as manifest by core level chemical shifts are probed as a function of depth below the growing interface by tuning the photoelectron escape depth, $\Lambda(E_K)$, by varying the photon energy $h\nu$. While this can be simulated in XPS measurements by tilting the sample relative to the analyzer, angular-dependent contributions of the final states can significantly alter the photoemitted electron energy distribution. Over the kinetic energy range of 10 to 170 eV available with adequate resolution, the probed depth varies by approximately a factor of two with a minimum of ~ 5 - 7 Å. With the aid of Fig. 2 we consider the behavior of the photoemitted electron intensity from an oxide region of one or two layers formed within the first four or five layers of the free Si surface. If such an SiO_2 interface grows inward from the surface, then the partial yield of Si (2p) electrons chemically-shifted by the oxygen relative to elemental Si will reach a maximum at the $h\nu$ for minimum escape depth. As noted in Fig. 2 by the dashed-dotted lines, the SiO_2 emission is constant, while the unshifted Si (2p) intensity is a minimum at Λ_{MIN}. On the other hand, if the SiO_2 forms below a region of elementally bonded Si, the dashed lines depict the relative minimum which will occur in the SiO_2 emission. Eventually, the unshifted Si emission increases at high and low E_K. This general "M-shape" for SiO_2:Si emission is independent of the exact energy where Λ_{MIN} occurs; however, knowledge of the value for Λ_{MIN} (~5-7Å) allows a quantitative determination of the growth depth.

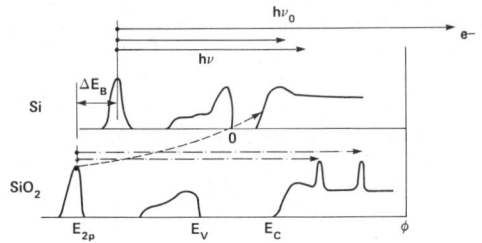

Fig. 3 Schematic diagram depicting how the tunability of synchrotron radiation can be used to probe unoccupied states below φ (by CFS yield, ref. 12) (dash-dot) and interface abruptness by the range (<10Å) of interatomic transitions (dashed).

The unoccupied states can be a more local probe of oxygen coordination than the core level chemical shift measured by photoemission. Such complementary information is obtained by measuring the Si (2p) core electron absorption with high surface sensitivity. As depicted schematically in Fig. 3, normal photoemission is measured at a constant photon energy $h\nu_0$ great enough to excite the electrons over the threshold φ. For unoccupied states below φ, we measure the secondary yield, at a constant final state (CFS) energy E_f, as $h\nu$ is swept through the range for core level transitions. When strong structure is present, characteristic energies such as those shown as dashed-dotted can be used as fingerprints for the local bonding complexes.

The core-electron absorption also can be used as a spatial probe of the interface. Consider transitions to the final state at the bottom of the elemental Si conduction band. The photon energy for exciting a Si(2p) electron in an oxygen environment to this state will be greater than the bulk Si $L_{2,3}$ absorption[14] by the core-level chemical shift ΔE_B. Fortunately this occurs not only below the $L_{2,3}$ threshold for SiO_2, but also in a structureless local minimum of the Si soft x-ray absorption. The lineshape observed in the CFS yield should be a displaced replica of the Si $L_{2,3}$ edge since the initial densities-of-states are identical. The spatial information comes from understanding the excitonic state of the excited solid. Estimates for the binding energy of this localized state range from[22] 140±200 meV to[23] 0.9±0.4 eV. Our measurements[14,24] find a *final-state* shift of 0.6±0.2 eV. In any event the corresponding exciton radius varies from half a Si bond length to 10Å maximum. The observation of strong interatomic transitions from SiO_2 to elemental Si final states then places this experimental upper limit on the separation of Si and SiO_2 coordinations. Clearly then the occurrence of such absorption across the interface is indicative of an atomically abrupt change in stoichiometry since a connective region of width[25] comparable to this range would only be present in this case.

III. Oxide Growth on Si (111) 2 x 1

To understand the inital stages of the SiO_2 formation, we oxidize in steps from coverages below an equivalent monolayer. The oxidation of Si produces three major distinguishable shifts of the Si (2p) core electrons. While each has been identified separately in previous work (ΔE_B = 3.7 eV in ref. 3 and 5, ΔE_B = 2.5 eV in ref. 3 and ΔE_B = 1.6 eV in ref. 6 and 18), we find that these can all occur simultaneously under appropriate growth conditions. Fig. 4 depicts the initial formation of Si-O compounds as those bonds shifted from bulk Si by $\Delta E_B \simeq 1.6$ eV. As the doses increase, there is a dramatic reduction in growth of the intermediate oxide characterized by $\Delta E_B \simeq 2.5$ eV. While the O (1s) photoemission is too weak to obtain adsorbate uptake data, it may be that (a) the impinging oxygen covalently bonds to the surface over this range and thus does not cause a chemical shift of the Si (2p) electrons or (b) a coverage of ~0.6 monolayers has been reached and the sticking coefficient changes dramatically as reported previously.[17,18] The intermediate oxide of ΔE_B = 2.5 eV corresponds to a distinct coordination of Si and O judging by the observed 0.6 eV spin-orbit splitting of the Si $2p^{3/2}$ and $2p^{1/2}$ components in both the elemental and oxidized core-photoemission.

Energy distributions such as those in Fig. 4 can be decomposed into their component structures. One then can measure the partial yield for the four distinctive features (i.e., the area under each of the structures) and obtain a relative amount of each oxidation state by normalizing to the elemental Si (2p) emission. As the photon energy is varied from threshold emisison to high kinetic energies, the changes determine where the growth occurs relative to the free surface by the method described in Fig. 2. We obtain the "M-shape" behavior for each oxygen shifted Si (2p) feature (including that for SiO_2) indicating completely subsurface oxide formation. A precise decomposition of the Si(2p) core emission could determine the depth variation of the oxygen bonding from the ordering of the relative maxima. The minimum escape depth of $6\pm2\text{Å}$ is predicted by this work to occur at $E_K \sim 60$ eV. This is consistent with the universal behavior of electron-electron scattering[26] and recent studies of other semiconductors.[27]

The oxide compounds must be forming below at least the first (111) *double* layer of Si in order for the first single layer of surface atoms to retain their bulk Si bond character. It is then important to ask whether the surface layer is clean Si free of all oxygen. Annealing experiments help resolve this issue. The sample dosed 80 times with 10^4L per exposure in Fig. 4 was heated and allowed to equilibrate at temperatures to 750°C. The SiO_2 precursor state with ΔE_B = 1.6 eV is converted into more highly oxygen-coordinated Si, resulting in an increased 2.5 eV shifted peak. This general trend of higher O coordination for Si and reduced interface bonds upon annealing is also seen for conversion of the 2.5 eV state to SiO_2 for samples cleaved at room temperature in atmospheric pressure of ground-state molecular oxygen. A disproportionate increase of Si atoms surrounded by oxygen is measured. This is due to a source of oxygen *molecules* existing at the surface which are *covalently* bonded. Such a surface oxide which does not cause a substantial 2p core shift has recently been found to tie up the filled Si surface state.[18] Further, the peroxy radical model first predicted theoretically[28] and then reported in XPS studies[20] probably describes this chemisorption bonding.

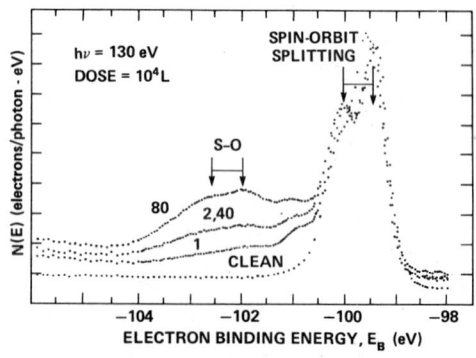

Fig. 4 Energy distributions normalized to constant partial yield for the unshifted Si (2p) core photoemission. The five sets of raw data are for ground-state oxygen dosing by the indicated total number of 10^4 L exposures on room-temperature Si (111). Those Si (2p) electrons chemically shfited by 2.5 eV retain the 0.6 eV S-0 splitting indicating suboxide compound formation.

IV. Si (111)-SiO$_2$ Interface Abruptness

The soft x-ray absorption measured by CFS partial yield at kinetic energies near Λ_{MIN} provides a distinctive fingerprint of local SiO$_2$ bonding. As presented in Fig. 5, the structure appearing in conjunction with a 3.6 eV deep Si (2p) chemically-shifted peak agrees remarkably well with the soft x-ray bulk SiO$_2$ absorption energies.[21] Since the precise energy of the core-level chemical shift is not representative of the local Si-O bonding,[5] these Si 3d-derived inner-well resonances characteristic of SiO$_4$ tetrahedra[11] are a valuable monitor of SiO$_2$ formation. We only observe these transitions (and a Si (2p) peak of $\Delta E_B > 3eV$) when oxidizing at atomospheric pressure, with excited molecular oxygen, or for high Si surface temperatures.

The high temperature oxidized Si (1000 LO$_2$ at 1000°C) exhibits a strong transition having the characteristic L$_{2,3}$ lineshape displaced from the bulk Si L$_{2,3}$ edge by the chemical shift of the Si (2p) core electrons. While this may be caused by distinctive absorption for a suboxide such as SiO, its shape and position suggest at this time an identification as the interatomic transition noted by dashes in Fig. 3. This is only possible if Si atoms four-coordinated with oxygens are within 10Å or less of those Si atoms four-coordinated with other silicons. This would imply an atomically abrupt change in stoichiometry, previously shown to be theoretically possible.[25,29] In such an instance, there would be Si-O bonds forming the connective region which were neither bulk Si nor SiO$_2$ in character. The inability to remove through annealing all of the Si (2p) states which are shifted by 1.6 eV, suggests that these are the Si oxidation states composing the abrupt interface layer. Room-temperature oxidation either with high pressure or excited oxygen does not result in SiO$_2$ to Si absorption structure (see Fig. 5). We conclude from this that such growth conditions do not yield an atomically abrupt interface between the Si and SiO$_2$.

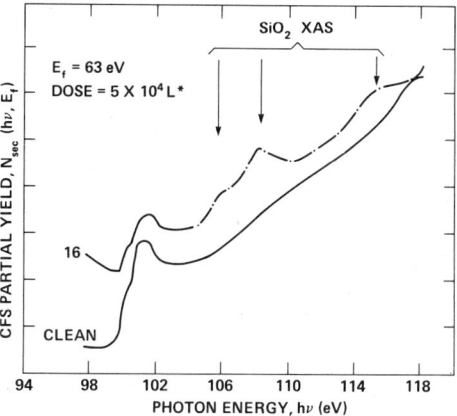

Fig. 5 CFS partial yield for Si (2p) electrons of clean Si and after oxidation at room temperature with 16 exposures of 5 X 10^{4}L excited (*) oxygen. The arrows mark the three peaks characteristic of Si (2p) absorption in SiO$_2$ from ref. 21. The transitions responsible for the oxygen-induced structure are shown by the same symbol as in Fig. 3.

V. Summary and Conclusions

We have studied the initial stages of oxide formation on Si by high pressure oxygen dosing over a wide range of substrate temperatures. The spatial composition of these oxide interfaces was nondestructively probed with core level photoemission by employing the unique tunability of synchrotron radiation. By varying the growth conditions any of the three previously observed oxidation states of Si could be formed as the dominant species and each was found to grow below the first double layer, ~2 Å below the free surface. In all cases annealing experiments indicated the additional presence of a layer of molecular oxygen covalently bonded to the surface. We can sensitively monitor the existence of fully formed SiO$_2$ bonding by observing the characteristic Si resonances for SiO$_4$ tetrahedra in the CFS partial yield spectrum. With this technique we have demonstrated that the activation barrier for SiO$_2$ formation is only overcome by high pressure, high temperature, or "excited" oxidations. An atomically abrupt transition from Si to SiO$_2$ (as measured by inter-atomic soft x-ray absorption across the interface) occurs only for oxide growth temperatures in the 1000°C range. Whenever fully formed SiO$_2$ is present, a constant amount of Si is observed to exist with an intermediate oxidation state (~1.6 eV core level shift). These Si atoms are the interconnecting atoms of the abrupt interface. The 1.6 eV shift is only consistent with an interface bond constructed from the Si (111) inter-layer bond.

ACKNOWLEDGMENTS

We acknowledge stimulating discussions with R. Z. Bachrach on the dosing technique and the experimental collaboration of T. F. Gustafsson and L. Johansson. We are grateful for the support and cooperation of the SSRL staff.

REFERENCES

*Aspects of this work were performed at the Stanford Synchrotron Radiation Laboratory, which is supported in part by NSF Grant No. DMR 73-07692 in cooperation with the Stanford Linear Accelerator Center and the United States Department of Energy.

†Present address: Istituto di Fisica, Universita di Camerino, Camerino, Italy.

‡Present address: 9 Kollaukamp, D-2000 Hamburg 61, West Germany.

1. L. Wagner and W. Spicer, Phys. Rev. Lett. 28, 1381 (1972); Phys. Rev. B 9, 1512 (1974).
2. D. E. Eastman and W. D. Grobman, Phys. Rev. Lett. 28, 1378 (1972).
3. C. Garner, I. Lindau, J. Miller, P. Pianetta and W. Spicer, J. Vac. Sci. Technol. 14, 372 (1977).
4. I. T. McGovern, A. W. Parke, and R. H. Williams, (Daresbury Rpt. DL/SRF/P97).
5. G. Hollinger, Y. Jugnet, and T. M. Duc, Sol. State Comm. 22, 277 (1977).
6. F. J. Grunthaner and J. Maserjian, these proceedings; IEEE Trans. Nucl. Sci. NS-24 (1977).
7. H. Ibach and J. E. Rowe, Phys. Rev. B 9, 1951 (1974); B 10, 710 (1974). See also M. Nishijima, M. Miyamura, Y. Sakisaka, and M. Onchi, Solid State Comm. 25, 457 (1978).
8. A. Koma and R. Ludeke, Phys. Rev. Lett. 35, 107 (1975).
9. C. R. Helms, W. E. Spicer, and N. M. Johnson, Sol. State Comm. 25, 673 (1978).
10. J. F. Wager and C. W. Wilmsen, these proceedings.
11. A. Bianconi, R. S. Bauer, and J. C. McMenamin, to be published.
12. R. S. Bauer, R. Z. Bachrach, S. A. Flodström, and J. C. McMenamin, J. Vac. Sci. Technol. 14, 378 (1977) and cited references.
13. R. S. Bauer, D. J. Chadi, J. C. McMenamin, and R. Z. Bachrach, Proc. 7th Int. Vac. Congr. & 3rd Int. Conf. Sol. Surf. (Vienna, 1977), p. A-2699.
14. J. C. McMenamin and R. S. Bauer, J. Vac. Sci. Technol. 15, #4 (1978); R. S. Bauer, R. Z. Bachrach, J. C. McMenamin, D. J. Chadi, and H. Petersen, Proc. 5th Int. Conf. Vac. UV Rad. Phys., Montpellier, Sept. 5-9, 1977.
15. J. T. Yates, Jr., T. E. Madey, G. B. Fisher, NBS Report NR 056-573, Jan. 31, 1976
16. P. Pianetta, I. Lindau, C. M. Garner, and W. E. Spicer, Phys. Rev. B (to be published).
17. R. J. Archer and G. W. Gobel, J. Phys. Chem. Sol. 26, 343 (1965).
18. C. M. Garner, I. Lindau, C. Y. Su, P. Pianetta, J. N. Miller, and W. E. Spicer, Phys. Rev. Lett. 40, 403 (1978).
19. G. Hollinger, Y. Jugnet, P. Pertosa, L. Porte, and T. M. Duc, Proc. 7th Int. Vac. Congr. & 3rd Conf. Sol. Surf. (Vienna, 1977), p. 2229.
20. J. E. Rowe, G. Margaritondo, H. Ibach, and H. Froitzheim, Sol. State Comm. 20, 277 (1976).
21. O. A. Ershov, D. A. Goganov and A. P. Lukirskii, Soviet Physics - Solid State 1, 1903 (1966); O. A. Ershov and A. Lukirskii, Soviet Physics - Solid State 8, 1699 (1967).
22. W. Eberhardt, G. Kalkoffen, C. Kunz, D. Aspnes and M. Cardona, Proc. 5th Intl. Conf. Vac. UV Rad. Phys. 11, 70, Montpellier, Sept. 5-9, 1977
23. G. Margaritondo and J. E. Rowe, Phys. Lett. 59A, 464 (1977).
24. R. Bauer, R. Bachrach, J. McMenamin, and D. Aspnes, Il Nuovo Cim. 39B, 409 (1977).
25. S. Pantelides, J. Vac. Sci. Tech. 14, 965 (1977); M. Long and S. Pantelides, these proc.
26. I. Lindau and W. E. Spicer, J. Electron. Spectrosc. 3, 409 (1974).
27. R. S. Bauer and J. C. McMenamin, J. Vac. Sci. Technol. 15, #4 (1978) and cited references.
28. W. A. Goddard, A. Redondo, and T. C. McGill, Sol. State Comm. 18, 981 (1976).
29. F. Herman and R. V. Kasowski, these proceedings.

THE Si-SiO$_2$ INTERFACE AND LOCALIZATION IN THE INVERSION LAYER

M. Pepper*
Cavendish Laboratory, Cambridge, U. K.
*On leave of absence from The Plessey Co., Allen Clark
 Research Centre, Caswell, Northants, U. K.

INTRODUCTION

At low temperatures the inversion layer of a Si MOSFET exhibits a metal-insulator transition as the carrier concentration in increased. For carrier concentrations below the critical value the inversion layer conductance decreases with decreasing temperature; increasing the carrier concentration decreases the activation energy for conduction and when the carrier concentration is at the critical value metallic conduction is exhibited. It was first suggested by Mott (1), and Stern (2), that this behaviour, initially observed by Fong and Fowler (3), indicated the existence of Anderson localization (4,5) arising from disorder at the Si-SiO$_2$ interface. Consequently the device can be used as a convenient vehicle for the investigation of the Anderson transition. As the Fermi level can be adjusted at will, by changing the gate voltage of the device, the system possesses advantages compared to other systems, for example, the impurity band of a doped semiconductor. However, the disadvantage of the inversion layer is that there is little definite knowledge of the disorder at the Si-SiO$_2$ interface, whereas in the semiconductor impurity band the disorder is known to arise from the randomness in the impurity distribution and the presence of compensating ions.

In this paper evidence is summarised which suggests that the low temperature device conductance is a very sensitive probe of the random fluctuations in potential at the Si-SiO$_2$ interface, and may be imformative on the nature of the charged defects at the interface.

ELECTRICAL CHARACTERISTICS OF THE ANDERSON TRANSITION

When the Fermi energy, E_F, is below the mobility edge, E_c, conduction is thermally activated. There are two transport mechanisms in parallel.
a) Current is due to transport of carriers at the mobility edge and the conductance σ varies as $\sigma_{min} \exp -(W/kT)$ where σ_{min} is the minimum metallic conductance.
b) The variable range hopping of carriers from one localized state to another, the conductance varies, in two dimensions, as $\exp -(B/T)^{1/3}$.

In two dimensions the numerical calculations of Licciardello and Thouless (6) suggest that σ_{min} is a universal constant $\sim 0.1 \ e^2/\hbar$, $(3.10^{-5}\Omega^{-1})$. Experiments have confirmed these predictions but two types of deviation have been found (7), these are

a) The device conductance obeys the law $\sigma = \sigma_{min} \exp -(W/kT)$ but σ_{min} is greater than $0.1 \ e^2/\hbar$.

b) σ varies as $\sigma_0 \exp -(W/kT)$ where σ_0 increases as W decreases and the final value of σ_{min} is much greater than $0.1 \ e^2/\hbar$

INTERFACIAL CHARGE AND LOCALIZATION

Typically the charge at the $Si-SiO_2$ interface determined by C-V or device threshold, is between 10^{11} and $\sim 2.10^{10}$ positive charges cm^{-2}. However up to $2.10^{12} \ cm^{-2}$ electrons or holes can be localized; the minimum concentration of localized electrons achievable appears to be in the range $1-2 \times 10^{11} \ cm^{-2}$, a greater value is found for holes. In general, when the carrier concentration is small the value of $|E_c - E_F|$ obtained (~ 7 meV) is considerably less than that for electrons when Na^+ ions are present at the $Si-SiO_2$ interface (~ 18 meV) (8). However by pushing carriers close to the interface with a substrate bias the activation energy at low values of carrier concentration can be increased close to the value for binding to discrete coulombic centres.

These results suggest that the measured interfacial charge is the result of the compensation of much larger amounts of positive and negative charge. Furthermore, the rapid increase in activation energy as carriers approach the interface, (when the carrier concentration is low), indicates the positive and negative centres are in the form of pairs so giving a dipolar potential. This weak potential is why the negative centres are not observed at higher temperatures by more conventional techniques. When carriers are far from the interface binding is weaker than that produced by a coulombic potential, the increase in activation energy with decreasing distance from the interface is interpreted as indicating that the charges are separated by less than ~ 40 Å (7). Mott (9) has discussed the similarity between these charge pairs and those in the chalcogenide glasses. There are two additional effects associated with pulling carriers away from the interface,

a) An increase in the number of localized carriers,

b) Quite often an increase in the value of σ_{min} and the appearance of a carrier concentration dependent pre-exponential factor when conduction is by excitation to E_c.

These effects can be observed for both n and p type inversion layers and are consistent with the potential fluctuations becoming longer range as the carriers are pulled away from the interface. At low values of carrier concentration, the activation energy decreases as carriers move away from the interface. Thus, when carriers are close to the interface, the wavelength of the fluctuations is too small to allow localization and the diffusive, Ioffe-Regal, regime persists to very low carrier concentrations. Evidence that pushing carriers closer to the interface results in a reduction in the mean free path is obtained from device conductance measurements at 77 K and also by the Shubnikov-de Haas effect (10). In figures 1 and 2 the magneto-resistance oscillations for an n channel device on the (100) Si surface are shown for two different values of substrate bias. The principal point here is that when carriers are close to the interface, ($V_{SUB} = -25$ volts), the resolution of the oscillations is considerably poorer than when they are further from the interface. Thus, clearly the rate of scattering has increased, but measurements in the absence of the magnetic field show the localization to have decreased when electrons are closer to the interface. Another aspect of this effect is that as the carriers move away from the interface localized carriers can describe cyclotron orbits, as if the fluctuations are increasing in range, i.e. longer mean free path. As the net inter-

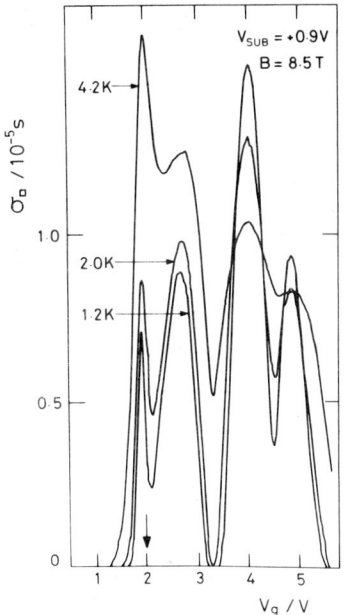

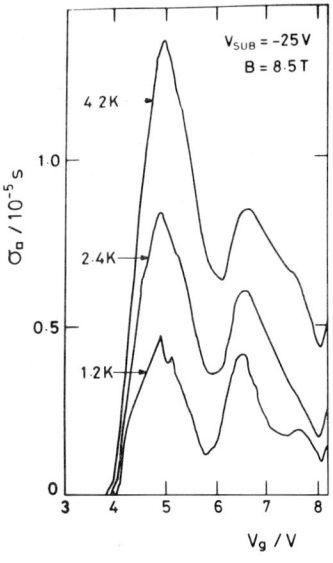

Fig. 1 Conductance versus gate voltage when the Fermi energy is in the zeroth Landau level. The arrow shows the gate voltage required to induce metallic behaviour in the absence of the magnetic field of 8.5 Tesla. The substrate bias is +0.9 volts so pulling electrons away from the interface.

Fig. 2 The same device as for Fig. 1 except that the substrate bias is −25 volts and forces electrons towards the Si-SiO$_2$ interface. Both Fig. 1 and Fig. 2 are from reference 10.

facial charge is progressively made more positive the substrate bias has a smaller effect on the conductance - carrier concentration relation. Initially results suggested that, when the net interfacial charge is low, pushing carriers towards the interface always decreased the localization (7,11). This, however, is not so and will now be discussed.

RADIATION HARDNESS, SLOW STATES AND INTERFACE ENGINEERING

We now discuss radiation sensitivity due to the trapping of holes, close to the Si-SiO$_2$ interface, after photo-generation of electrons and holes in the SiO$_2$. The dependence of charge build up on the oxidation and annealing conditions has been discussed by Derbenwick and Gregory (12). The essential point is that reducing the temperature of the post-oxidation inert gas annealing decreases hole trapping close to the Si-SiO$_2$ interface. Using p channel devices, it was found that, at 4.2 K, forcing holes closer to the interface in a "hard" device results in a reduced conductance, whereas, as previously discussed, the opposite effect is found in a "soft" device (13,14). This is illustrated in Fig. 3.

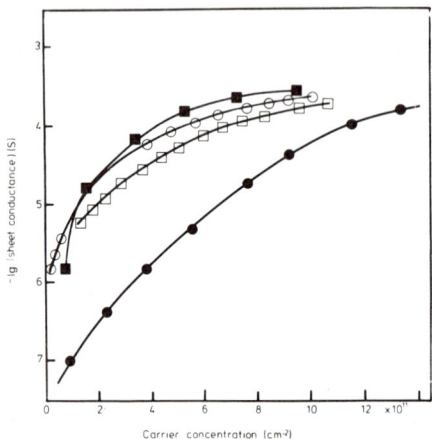

Fig. 3 Here we compare the conductance of a "hard" p channel device with a "soft" device, the temperature of measurement is 4.2 K. Points ● and o are for the "soft" and "hard" devices respectively without the application of substrate bias, ■ and □ are for the "soft" and "hard" devices respectively when the holes are pushed towards the interface, from reference 13.

This difference in behaviour suggests that the total charge at a "hard" interface is less than at a soft interface, and that, as carriers approach the interface, the potential fluctuations do not change in form, but merely become slightly deeper so enhancing localization. It appears as if the closely separated positive and negative pairs have been removed by the lower temperature of the anneal, i.e. an exercise in interfacial engineering is being performed. The reduction in hole trapping on irradiation is then merely due to the reduced concentration of negative centres, the net charge is similar to soft interfaces, and at temperatures greater than ∿30 K there is no difference between "hard" and "soft" p channel devices prior to irradiation. At present similar experiments have not been performed on "hard" n channel devices. Experiments, which are reported in greater detail elsewhere (13,14), show that if a small concentration of holes is trapped (less than ∿3.10^{11} cm^{-2}) near the $Si-SiO_2$ interface then a reduction in localization is observed in p channel devices. This result strongly suggests that negative centres are trapping holes in SiO_2, as a result of this trapping inversion layer holes are less strongly bound and localization decreases. It also appears that the hole traps are distributed into the SiO_2, about half being within ∿40 Å of the SiO_2, the other half being sufficiently far from the Si to have a small effect on the localization. Slow state discharging can alter the extent of the localization, and, in a similar way to radiation effects, information on the nature of the slow states can be obtained. The effect first found by Breed (15) has been investigated in detail (7); here, by annealing in oxygen at ∿800°C, the interfacial (net) charge is made increasingly positive but the concentration of slow states is similarly increased. Experiments on p channel devices showed that discharging the slow states increased the positive oxide charge but initially decreased the localization. In a similar manner to hole trapping after irradiation this result suggests that the slow states are negative centres, and that the increase in positive oxide charge is due to their discharging to form neutral centres.

A model for D^+ and D^- centres as non bridging oxygen ions in SiO_2 has been proposed by Mott (9), on the basis of the localization dependence on the slow states it is most likely that the D⁻ centres close to the Si are the slow states. The energy of these centres could differ between the Si rich interfacial regime of the SiO_2 and the bulk of the SiO_2. In this respect it is to be emphasised that the localization is most sensitive to charges close to the Si.

Experiments by the author and P. Vohralik (16) have shown in detail that the value of σ_{min}, characteristics of the localization, and deviations from $T^{-1/3}$ behaviour are dependent on the post-oxidation annealing treatments received by the interface. Thus the range of the fluctuations, and the characteristics of the localization, can be adjusted by changing the distance between carriers and the interface (substrate bias) and also by changing the annealing treatments received by the interface. In this respect the inversion layer is one of the most sensitive probes of the interface and it is anticipated that further, detailed, information on interfacial defects and disorder will be obtained from such experiments.

ACKNOWLEDGEMENTS

I should like to thank Professor Sir Nevill Mott F.R.S. for very many discussions on this work. Support was provided by the Science Research Council and the AFWL, Air Force Systems Command, United States Air Force under grant AFOSR-77-3398.

REFERENCES

1.   N. F. Mott, Electronics and Power 19, 321 (1973).

2.   F. Stern, Phys. Rev. 9, 2762 (1974).

3.   F. F. Fang and A. B. Fowler, Phys. Rev. 169, 619 (1968).

4.   P. W. Anderson, Phys. Rev. 109, 1492 (1958).

5.   N. F. Mott, Metal-Insulator Transitions, Taylor and Francis, London (1974).

6.   D. C. Licciardello and D. J. Thouless, J. Phys. C 8, 4157 (1975).

7.   M. Pepper, Proc. Roy. Soc. A353, 225 (1977).

8.   A. Hartstein and A. B. Fowler, Physics of Semiconductors, F. G. Fumi, 741 (1976).

9.   N. F. Mott, Adv. Phys. 26, 363 (1977).

10. M. Pepper, Phil. Mag. 37, 83 (1978).

11. A. B. Fowler, Phys. Rev. Lett. 34, 15 (1975).

12. G. F. Derbenwick and B. L. Gregory, IEEE Trans. Nucl. Sci. 22, 2151 (1975).

13. M. Pepper, J. Phys. C 10, L445 (1977).

14. M. Pepper, Amorphous and Liquid Semiconductors, Ed W.E. Spear, 477 (1977).

15. D. J. Breed, Solid-St. Electron 17, 1229 (1974).

16. P. Vohralik, Ph.D. Thesis, Cambridge (1978).

METASTABILITIES AT THE Si-SiO$_x$ INTERFACE

C. T. White[+] and K. L. Ngai
Naval Research Laboratory, Washington, D. C. 20375

ABSTRACT

We summarize some of the results of our investigation of the effects of dang-ling/weaker and stronger bonds on the electronic structure of the Si-SiO$_2$ interface. These dangling bonds give rise to a new type of interfacial com-pensating charged state and have an interesting dynamic character with impli-cations for device characteristics. It is pointed out how these states can be severely modified by external perturbations when they are in contact with e.g. the p-channel inversion layer (IL) which could represent a sensitive probe of their nature and distribution.

INTRODUCTION

At the Si-SiO$_x$ interface the covalently bonded semiconductor Si is joined to the dielectric SiO$_2$; hence, it is reasonable to expect that this inter-face's electronic structure is such that not all the possible bonded states are occupied and some that are can be easily altered. Inherent in the nature of such states is their ability to mediate an effective local pairing inter-action between electrons, a feature employed by Anderson (Ref. 1) in his discussion of certain amorphous glasses. Here we will primarily be concerned with perhaps the most straightforward example of a dynamic electron pairing center that might well be present at the Si-SiO$_x$ interface: an Si dangling bond. We will discuss how the p-channel Si IL can be used as a probe of the nature and distribution of these states and how their metastable charac-ter could give rise to measurable alterations in the device characteristics after the external perturbations causing these changes have been removed.

THE MODEL AND RESULTS

To begin we envision an Si atom at the interface bonded to three neighboring silicons and/or oxygens, leaving a dangling hybrid (DH). One would then expect that the average number of electrons in this hybrid would deviate significantly from one. The origin of this effect can be understood by observing that a lowering of the system energy from what is obtained when the nonbonded hybrid is singly occupied can usually be achieved by essen-tially transferring (donating) an electron to (from) this hybrid from (to) one of the other available occupied (unoccupied) states of the system and simultaneously distorting the atom possessing the dangling bond so to weaken (strengthen) its associated backbonds, while concomitantly lowering (raising) the DH level. Distortions would not be advantageous if the DH were con-strained to be singly occupied since, although they should produce changes in the DH and associated back levels linear in the displacement, these changes would tend to cancel leaving an elastic-like energy to dominate.

This situation is entirely different if the DH occupancy is allowed to vary since e.g. if the DH was essentially unoccupied, then raising its level would have little energetic consequence. To specify and quantify this picture further one can employ (Ref. 2) the following Hamiltonian

$$H = \sum_{m\sigma} \varepsilon_o \, n_{m\sigma} + \sum_{\sigma jm} V_{jm} a^\dagger_{j\sigma} a_{m\sigma} + \sum_{\sigma} E_b n_{b\sigma},$$ (1)

$$- \lambda u(n_{b\uparrow} + n_{b\downarrow} - 1) + cu^2/2 + V_o \sum_\sigma (c^\dagger_{b\sigma} a_{b\sigma} + a^\dagger_{b\sigma} c_{b\sigma})$$

to describe a dangling bond in contact with the IL. The first two terms on the right hand side of (1) comprise a tight binding Hamiltonian, H_{TB}, taken to represent the IL where $a^\dagger_{m\sigma}$, $a_{m\sigma}$ create and annihilate electrons of spin σ in the Wannier state $|m\sigma\rangle$ centered at the site m and $n_{m\sigma} = a^\dagger_{m\sigma} a_{m\sigma}$ is the usual number operator. The remaining terms entering (1) refer entirely to the Si atom with the DH with the exception of the last which couples the dangling hybrid to the IL. The operators $c^\dagger_{b\sigma}$, $c_{b\sigma}$ create and destroy electrons in the dangling hybrid orbital $|b\sigma\rangle$ and u represents a local displacement of the atom possessing the dangling bond from its position if this bond were singly occupied with energy E_b. The fifth term appearing on the right hand side of (1) is an elastic energy while the fourth is a dehybridization energy which can be obtained (Ref. 3) by applying the valence bond theory (Ref. 4).

The free energy corresponding to (1) can be expressed for low temperatures as

$$F(u) = K + \lambda u + cu^2/2 + \frac{1}{\pi} \sum_\sigma \text{Im} \int_{-\infty}^{E_f} dE \int_{-\infty}^{E} \frac{(1 - V_o^2 G'_\sigma(\tilde{E}^+)) d\tilde{E}}{\tilde{E}^+ - E_b + \lambda u - V_o^2 G_\sigma(\tilde{E}^+)} \quad , $$ (2)

where E_f is the Fermi level, $G_\sigma(E) = \langle \sigma m | [Z - H_{TB}]^{-1} | m\sigma \rangle$, $\tilde{E}^+ = \lim_{s \to 0^+} (\tilde{E} + is)$, $\tilde{G}'_\sigma = \partial G_\sigma / \partial \tilde{E}$ and K is independent of u.

Both experimental (Ref. 5) and theoretical results (Ref. 6) concerning freshly vacuum-cleaved Si (111) surfaces imply the existence of a split surface dangling bond band in the vicinity of the valence band edge. It is generally agreed (Refs. 3,5) that the observed splitting of this band can in effect be ascribed to a dehybridization energy mediated attractive interaction amongst the surface dangling bond band electrons similar in form to that employed here. This data leads one to expect that the centers of reconstruction, E_b, of these ideal dangling bonds lie ∿.15 eV above the valence band edge with typical values of λ^2/c of ∿.15 eV. Since the coupling of the DH to the IL should not be large, one can show from (2) that those DHs with centers of reconstruction that lie not too far from E_f have a very interesting property, that is for those the free energy (2) exhibits two local minima associated with distinctly different occupancy of the DH. In one instance, effectively two electrons are strongly self-trapped by one another in the DH which lies λ^2/c below E_b, while in the other case, the two electrons lie in the IL and the DH state appears at λ^2/c above E_b. Grossly, if $E_f > E_b$, the system prefers to put two electrons in the DH. However, if $E_f < E_b$, it prefers two holes in the DH. Thus, although a pair of electrons (holes) may be strongly trapped (e.g. .1 eV) below (above) E_f, this pair can be eventually broken by a small change in the chemical potential

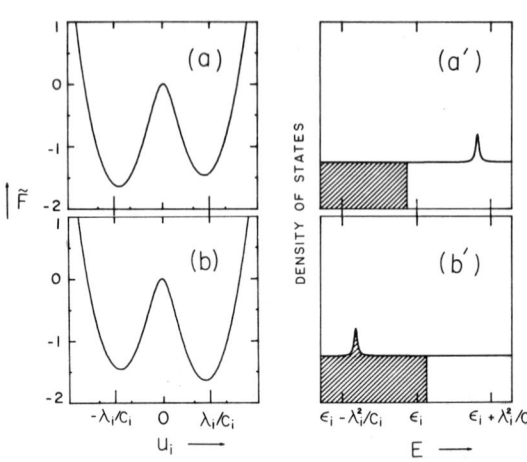

Fig. 1. The behavior of $F(u_i)$ [(a), (b)] for zero temperature and $2\lambda_1^2/\pi c_i > 1$ if the inversion layer DOS, ρ_0, is taken as constant. Also shown a schematic of the effective one-electron picture of the DOS [(a'),(b')] corresponding to an absolute minimum in F and the Fermi level below [(a),(a')] and above [(b), (b')] the center of reconstruction ε_i. Shaded areas represent occupied states. In (a), (b), $\varepsilon_i=0$ and $\lambda_i/c_i= .75$ with $\lambda_i= 9$ in units taken so that $V_0^2\pi\rho_0= 1$.

(e.g. 1 meV). In Fig. 1 we summarize this discussion where the results obtained there have been found in assuming that the IL density of states (DOS) is independent of energy. We have obtained essentially similar results by taking this DOS as elliptic, rectangular, etc.

The above properties of these dangling hybrids when in contact with the IL leads us to suspect that application of external perturbations could result in a rearrangement of the available holes over the dangling hybrid and IL states and these rearrangements might maintain themselves after the perturbations are removed. Assuming such is the case, measurable changes in the IL behavior could result. In particular, because of the disordered nature of the interface, the IL exhibits a mobility edge (Ref. 7) which sharply separates localized from extended states. This latter feature gives rise to an activated behavior for p-channel conduction when E_f lies above E_c and since meV changes in the activation energy, E_A, are detectable, one could then possibly measure metastable changes in the number of holes in the IL at constant gate voltage by observing changes in E_A. Of course for any such changes to be easily observed in this manner it is necessary that the DOS of the dangling hybrids be comparable with and overlap to some extent the IL. For vacuum cleaved Si(111) surfaces the number of dangling bond states is $\sim 10^{14}$ per cm^2 with centers of reconstruction presumably peaked at about .15 eV above the valence band edge. How many of these states are eliminated with the formation of the interface is difficult to estimate, as is the degree of the expected spreading of the centers of reconstruction. Note though that because of the metastable character of these states one should be able to sample not only those with centers of reconstruction right at E_f, but any others that can be put into improperly reconstructed states.

We now detail several examples of how such changes could be produced. (a) Mechanical stress: Application of uniaxial stress will shift the IL band as well as the centers of reconstruction of the pair states and such shifts should not be too strongly correlated. Thus a redistribution of electrons and holes is expected. Upon removal of the stress this state could persist as a strongly self-trapped metastable one whose existence could be reflected in changes in the activation energy from the prestressed situation. (b) External magnetic field: At low carrier densities magnetic fields of

sufficient strength can be generated so that there is essentially only one
spin species of IL hole states. It then becomes energetically favorable
to transfer some electrons from the IL to the dangling hybrids resulting
in a strong pairwise trapping of these electrons below E_f. Removal of the
field could thus leave the system in what then is a very low lying self-
trapped excited state which will exhibit an activation energy reduced from
what was found prior to application of the field. Although this example
is amusing, such effects would probably not be easily observable. (c)
Annealing: Heating the device and then rapidly cooling it could easily
freeze in different rearrangements of the available holes. Annealing could
also produce a significant reduction in the pair state DOS. (d) Negative
bias stress: Under negative voltage bias at room or elevated temperatures,
negatively charged electron pair states can be activated to an excited state
of one electron on the Si dangling bond which is now neutral, with the other
electron transferred to the Si substrate. Subsequently the remaining electron is also transferred across to the Si substrate and a positively charged
hole pair state results, changing the nature of the surface charge Q_{ss} and
hence perhaps the energetic position of the mobility edge. (e) Radiation:
Photo excitation of electrons from the IL to the unoccupied dangling bond
hole band could result in a condensation of these electrons at least strongly
reconstructed dangling bonds effectively producing a nonradiative recombination process eventually taking the initially photoexcited electrons below
the Fermi level. Such a process could actually change the device behavior
from activated to conducting.

Implicit in some parts of our above discussion is the assumption that the
mobility edge does not change drastically with alterations in the occupancy
of the dangling hybrids. Such an assumption is quite reasonable since these
hybrid states upon reconstruction should lie either far above or below the
mobility edge. Indeed we have carried out calculations assuming firstly,
a tight binding model of the IL, and, secondly, that the dangling hybrids
randomly occupy 10% of the lattice sites and thirdly, that the one electron
potentials and centers of reconstruction of the pair states both obey Anderson probability distributions of width W_1, W_2 respectively. It is found
for a physically reasonable choice of the parameters of this model, that
the assumption of the essential constancy of the mobility edge is indeed
verified. Furthermore, within this model we have obtained an example of
a rather severe rearrangement of the holes over the system which actually
changes the behavior of the device from activated to conducting but represents a low lying strongly self-trapped excitation of the system.

CONCLUDING REMARKS

The p-channel MOSFET was chosen here to illustrate some of the important
effects that could be produced by Si dangling bonds at the Si-SiO$_x$ interface.
These dynamic pairing states not only contribute to Q_{ss} if E_f is above or
below E_b but also can provide a source of interface "fast" states N_{st}, each
of which can be thought of as lying essentially at a center of reconstruction. Other pairing states that arise from weaker and stronger (in the
spirit of Anderson) Si-Si interfacial bonds have also been studied and characterized by the present authors and this work will be reported elsewhere.
We propose that these pairing states when taken altogether can account partially for the interface "fast" states which are commonly observed but seldom
understood. It is important to remark that irrespective of whether our
identification of the pair states with N_{st} is correct or not, experiments
on inversion layers have invariably been carried out with the assumption

that the interface states can be neglected. No one seems to have measured N_{st} inside the conduction or valence bands although as the associated band edges are approached from midgap the density of N_{st} states is rapidly rising and it is conceivable that it is sufficiently high near the subband energies that it should be accounted for especially in the localized low carrier density regime of both n and p-channel MOSFETs.

In closing, we make an important prediction of the frequency dependence of the dielectric response of the interface. The density of pair states could well be smooth, continuous and occupied to E_f (Ref. 1). Fast hopping/transition of interface carriers induced by an external field switches on a sudden potential followed by the emission of electron pairs across E_f in transient response (Ref. 8). We have discovered the existence of an infrared divergence in the pair state excitation response and predict an interface dielectric loss $\chi''(\omega) \sim \omega^{n-1}$ which is exactly the "universal" Curie-von Schweidler law of dielectric response of almost all solid dielectrics correlated by Jonscher (9). We have also shown that the infrared divergence of the pair states could naturally be responsible for this "universal" dielectric behavior of solids. Details will be reported elsewhere.

REFERENCES

+ NRC-NRL Resident Research Associate

(1) P. W. Anderson, Possible consequences of negative U centers in amorphous materials, J. Physique C4, 341 (1976).

(2) C. T. White and K. L. Ngai, Random attractive electron-electron interaction in two dimensional systems, Surface Sci. 73 (to be published).

(3) W. A. Harrison, Surface reconstruction on semiconductors, Surface Sci. 55, 1 (1976).

(4) C. A. Coulson, (1961) Valence, Oxford University Press, New York.

(5) Winfried Mönch, On the correlation of geometrical structure and electronic properties at clean semiconductor surfaces, Surface Sci. 63, 79 (1977).

(6) M. Schlüter, J. R. Chelikowsky, S. G. Louie and M. L. Cohen, Self-consistent pseudo potential calculations Si(111) unreconstructed and (2x1) reconstructed surfaces, Phys. Rev. Letts. 34, 1385 (1975).

(7) See for compilation of references, J. J. Quinn and P. J. Stiles (1976), Electronic of Quasi-Two-Dimensional Systems, North Holland, New York.

(8) J. J. Hopfield, Infrared divergence, x-ray edges and all that, Comm. Solid State Phys. 11, 40 (1969).

(9) A. K. Jonscher, The "universal" dielectric response, Nature 267, 673 (1977).

PHOTOCAPACITANCE PROBING OF Si-SiO$_2$ INTERFACE STATES

Emil Kamieniecki and Ryszard Nitecki
Institute of Physics, Polish Academy of Sciences
Al. Lotników 32/46, 02-668 Warsaw, Poland

ABSTRACT

Results of studies of the interface states in n-type MOS structures by the low-temperature photocapacitance technique are reported. From the comparison of the photocapacitance and the quasi-static $C-V$ measurements the existance of two types of interface states, optically active (OAS) and optically inactive (OIS), is inferred. It has been established that the density of OAS decreases towards the band edges, while the density of OIS increases towards the band edges. The photocapture cross-section of OIS is a few orders of magnitude smaller than that of OAS.

INTRODUCTION

The most common probing techniques of Si-SiO$_2$ interface states are based on electrical measurements (Ref. 1). The electrical techniques detect effects associated with the *thermal excitation* of carriers from the interface states into the bulk of a semiconductor. The results of such measurements are generally interpreted in terms of a continuous distribution of the interface states over the band-gap of a semiconductor with a broad minimum near midgap and an increasing concentration towards the band edges.

In this paper we report the results of studies of the interface parameters by a low-temperature photocapacitance technique based on the *photoemission* of carriers from the interface states into the bulk of a semiconductor. We present the results of studies of the energy distribution of the interface states and their photocapture cross-section over nearly entire band-gap of Si in the state-of-the-art clean Al - polycrystalline Si - SiO$_2$ - Si capacitors. The details of the experimental technique have been given elsewhere (Refs. 2, 3).

METHODS

The investigated n-type MOS capacitor is first cooled down to the liquid nitrogen temperature and biased from the accumulation to the deep depletion conditions. This biasing depopulates, by the thermal emission, the interface states lying near the conduction band edge (Ref. 4). At the liquid nitrogen temperature the thermal emission depopulates the interface states at energies $E_c - E_t < E_Q \simeq 0.2$ eV (Ref. 4). The deeper interface states remain occupied. After the saturation of the thermal depopulation the sample was illuminated with I-R radiation ($E_Q < h\nu < E_G$) depopulating the filled interface states.

At the fixed gate voltage the change of the interface states charge induces the change of the semiconductor space charge and subsequently the change of the total MOS capacitance. Thus the change of the interface states charge due to the illumination can be determined from the change of the total MOS capacitance (Ref. 2).

Photocapacitance Saturation

For the incident photon energy $h\nu < E_G/2$, the photo-induced hole emission is impossible, and the change of the interface charge is due only to the optical depopulation of the interface states (Ref. 4). Long enough illumination of the sample leads, at this photon energy, to the complete depopulation of the interface states with energies $E_c - E_t$ ranging from E_Q to $h\nu$. The total capacitance of the sample per unit area saturates then at the value C_{sat} . A subsequent illumination with light of the photon energy $h\nu + \Delta h\nu < E_G/2$ leads to the depopulation of deeper states having energies ranging from $h\nu$ to $h\nu + \Delta h\nu$ and the total capacitance of the sample achieves a new saturation value $C_{sat} + \Delta C_{sat}$. If $\Delta C_{sat}/C_{sat} \ll 1$, then the average density of the interface states having energies ranging from $h\nu$ to $h\nu + \Delta h\nu$ is,

$$ N_{ss} = (N\epsilon_s C_{ox}/C_{sat}^3)(\Delta C_{sat}/\Delta h\nu) , \tag{1} $$

where N is the doping concentration of a semiconductor, ϵ_s is the dielectric permittivity of a semiconductor, C_{ox} is the oxide capacitance per unit area (Ref. 3).

Initial Photocapacitance Time Response

As it has been discussed the interface states at the energies $E_c - E_t > E_Q \simeq$ 0.2 eV are fully occupied before the illumination. Therefore, just after the illumination is on the photo-induced hole emission is impossible (when $h\nu < E_G/2$), or negligibly small (when $E_G/2 < h\nu < E_G$). The change of the interface states charge is then dominated by the optical depopulation of the filled interface states (Ref. 2). The initial rate of change of the filled states density at the energy E_t , $dn_t/dt|_o$, is proportional to the density of the interface states at this energy, $N_{ss}(E_t)$, times the product of the photocapture cross-section of these states, $K(E_t,h\nu)$, by the incident photon flux, $I(h\nu)$.

Since the photons of energy $h\nu$ depopulate states in the energy range from $E_c - E_Q$ to $E_c - h\nu$, the initial change of the interface states charge can be obtained by integrating $dn_t/dt|_o$ over all energies of the states involved in the process (Ref. 3), yielding,

$$ \left.\frac{dQ_{ss}}{dt}\right|_o = -qI(h\nu) \int_{E_c - E_Q}^{E_c - h\nu} N_{ss}(E_t)K(E_t,h\nu)dE_t \quad . \tag{2} $$

This method of determination of the energy distribution of the interface states density requires the knowledge of the dependence of the photocapture cross-section both on E_t and on $h\nu$.

RESULTS

A typical energy distribution of the interface states density determined from the saturation of the photocapacitance is presented in Fig. 1 (crosses).

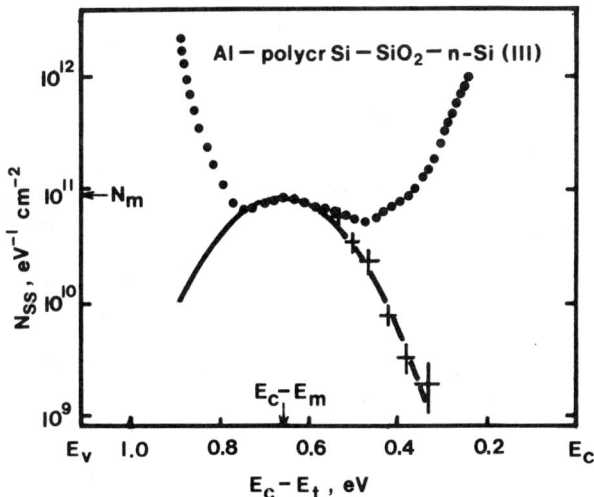

Fig. 1. Energy distribution of the interface states density; from the photocapacitance saturation (crosses); from the initial rate of the photocapacitance change (solid line); from the quasi-static $C-V$ measurement (dots).

Note that the density of the interface states decreases towards the conduction band edge. We observed that at $h\nu \lesssim 0.3$ eV the illumination had no effect on the photocapacitance. From this we could infer that the density of the interface states for energies $E_c - E_t \lesssim 0.3$ eV is lower than about $3 \cdot 10^8$ eV^{-1} cm^{-2}.

The initial rate of change of the interface states charge as described by Eq. (2) accounts best for the measured values, for $K(E_t, h\nu)$ given by the Lucovsky relationship (Ref. 5), when $N_{ss}(E_t)$ is Gaussian with the maximum at $E_c - E_t = 0.66$ eV and the standard deviation $\sigma = 0.11$ eV (solid line in Fig. 1). The effective field ratio in the Lucovsky relationship was assumed to be the same for all the interface states which is equivalent to the assumption that the maximum of $K(E_t, h\nu)$ vs. $h\nu$ for a given E_t, $K_{max}(E_t)$, is proportional to $1/(E_c - E_t)$. The fitting of Eq.(2) to the experimental values of $dQ_{ss}/dt|_o$ does not determine the absolute magnitude of the interface states density, $N_{ss}(E_t)$, but it determines the factor $N_m E_t K_{max}(E_t)$, where $N_m = N_{ss}(E_m)$. The knowledge of $N_{ss}(E_t)$ from measurements of the photocapacitance saturation enables one to determine N_m and hence the factor $E_t K_{max}(E_t)$. For the sample of Fig. 1 , $K_{max}(E_t) = 8 \cdot 10^{-20}/(E_c - E_t)$ cm^2. The value of $K_{max}(E_t)$ of the same order of magnitude was also obtained for other samples.

SUMMARY

The interface states have been shown (Ref. 3) to be responsible for the extrinsic photocapacitance effect reported in this paper. The interface states density determined by the photocapacitance technique near $E_c - E_t = 0.7$ eV is in a good agreement with the quasi-static $C-V$ measurements (dots in Fig. 1). However, the interface states density determined from the photocapacitance measurements *decreases* towards the band edges whereas the density determined from the quasi-static measurements *increases* towards the band edges. This suggests the existence of two types of interface states: "optically active" and "optically inactive". The comparison of the photocapacitance and the quasi-static results shows that the photocapture cross-section of the optically inactive states at $E_c - E_t \simeq 0.25$ eV is at least four orders of magnitude smaller than the photocapture cross-section of the optically active states.
   The optically active states in clean MOS structures are most probably due to some typical impurities or imperfections at the interface. The optically inactive states probably originate from the charge transfer stimulated electron transitions (ion-electron interface states) associated with the fixed charges in the oxide interior (Ref. 6).

REFERENCES

(1) A. Goetzberger, E. Klausmann and M. J. Schulz, Interface states on semiconductor/insulator surfaces, CRC Critical Reviews in Solid State Sciences Januar 1976 p. 1 .
(2) E. Kamieniecki, Low-temperature photocapacity measurement in MOS structures, Solid-State Electronics 16, 1487 (1973).
(3) E. Kamieniecki, R. Nitecki and A. Swiątek, Optically active interface states in MOS structures, to be published.
(4) R. F. Pierret, Photo-thermal probing of Si-SiO$_2$ surface centers, Solid-State Electronics 19, 577 (1976).
(5) G. Lucovsky, On the photoionization of deep impurity centers in semiconductors, Solid State Commun. 3, 299 (1965).
(6) E. Kamieniecki, Would-be interface states in MOS structures, to be published.

TRANSIENT CAPACITANCE MEASUREMENTS OF ELECTRONIC STATES AT THE SiO$_2$-Si INTERFACE

N. M. Johnson and D. J. Bartelink
Xerox Palo Alto Research Center, Palo Alto, California 94304

M. Schulz,* Institute for Applied Solid State Physics, Freiburg, Germany

ABSTRACT

Deep-level transient spectroscopy (DLTS) has been applied to measure the energy distribution and capture cross-section of electronic trapping centers at the Si-SiO$_2$ interface. The experimental and analytical techniques for DLTS analysis of interface states in MOS structures are summarized, and results from combined capacitance-voltage and DLTS measurements of a characteristic discrete level at the oxidized silicon surface are presented.

INTRODUCTION

The interface between thermally-grown SiO$_2$ and Si is generally found to possess a continuous distribution of electronic trapping centers which extends throughout the Si forbidden energy band, with the density monotonically increasing with energy toward the band edges from a minimum near midgap (1). The metal-oxide-semiconductor (MOS) structure has been extensively used to study these states. From conductance measurements on MOS capacitors (2,3) and charge transfer loss measurements on charge-coupled devices (4) it is found that the cross section for electron capture by fast surface states is constant over the midgap region, but decreases rapidly with energy near the Si conduction band. Deep-level transient spectroscopy (DLTS) (5) is an alternative technique for studying interface states which features high sensitivity and complements the above techniques for measuring dynamic interface properties in that it is not affected by surface potential fluctuations, which arise from the random spatial distribution of fixed positive charge in the oxide. In addition, the technique has been used to measure bulk defects introduced by ion implantation in the near-surface region of MOS structures (6). Here we summarize the experimental and analytical techniques for the DLTS analysis of interface states and present combined capacitance-voltage (C-V) and DLTS measurements of a characteristic discrete level at the Si-SiO$_2$ interface.

DLTS MEASUREMENTS ON MOS STRUCTURES

DLTS is a transient capacitance technique which was developed by Lang (5) to measure deep levels in bulk semiconductors. A block diagram of the apparatus is shown in Fig. 1. The

---

* Present address: Institute for Applied Physics, University of Erlangen-Nurnberg, Germany; partially supported by ARO.

measurement system consists of a capacitance bridge with fast transient response, a pulse generator for rapidly changing sample bias, a dual-gated signal integrator and X-Y recorder, and a variable temperature cryostat. For studying interface states in MOS structures, the system in Fig. 1 includes a feedback loop which maintains the capacitance of an MOS device at a constant value by varying the gate voltage. Thus, the monitored signal is a voltage transient. In the next section we show that the constant-capacitance DLTS, or CC-DLTS, technique offers significant advantages for the analysis of interface states in MOS structures.

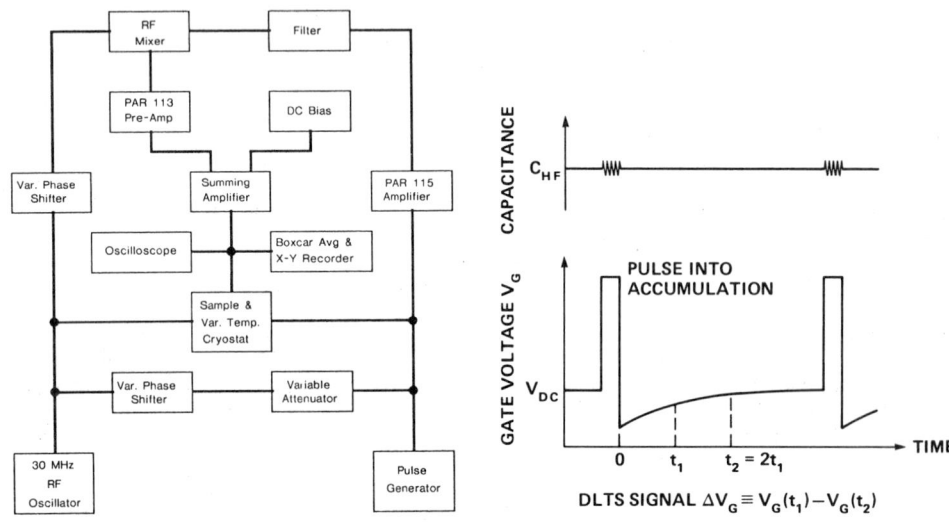

Fig. 1. Block diagram of apparatus for constant-capacitance DLTS (CC-DLTS) measurements of interface states in MOS structures.

Fig. 2. Schematic diagrams of the capacitance variation and voltage transient in a CC-DLTS measurements.

The experimental procedure involves first biasing an MOS device into depletion. Superimposed on the dc bias is a charging pulse which drives the semiconductor surface into strong accumulation in order to populate the interface states with majority carriers. After a charging pulse, the gate voltage varies with time as the occupation of the interface states returns to its equilibrium distribution.

The voltage transient is schematically illustrated in Fig. 2. At low temperatures (<300 K) and short times (~1 msec) the transient arises from the emission of majority carriers and consists of a superposition of many exponentially decaying signals, since the interface traps are distributed in energy and hence emit their charge at different rates. With the DLTS technique these signals can be deconvolved. The CC-DLTS signal, ΔV_G, is obtained by forming the difference of the gate voltages measured at two delay times t_1 and t_2 after a charging pulse, as shown in Fig. 2. The signal is measured in a temperature scan, and good noise discrimination is obtained by time averaging over many cycles with a boxcar averager or on-line computer.

THEORY OF CONSTANT-CAPACITANCE DLTS MEASUREMENTS

For majority carrier emission from interface states, the CC-DLTS signal is readily shown to be proportional to the change with time of the net charge in interface states after a charging pulse.

Since the depletion capacitance of the semiconductor is held constant during the transient response, the change in gate voltage required to maintain a constant capacitance appears only across the oxide layer, and the emission signal ΔV_G is related to the net charge (per unit area) in interface states, Q_{is}, as follows:

$$\Delta V_G = (1/C_{ox})[Q_{is}(t_1) - Q_{is}(t_2)], \tag{1}$$

where C_{ox} is the oxide capacitance per unit area. The gate-voltage transient is thus linearly proportional to the difference in the net charge in interface states at the two delay times t_1 and t_2 after a charging pulse. This linear dependence obtains for all interface trap densities, and the proportionality factor is independent of substrate doping and temperature. By contrast, in DLTS transient-capacitance measurements with fixed gate voltages, linearity is lost at high interface-state densities, and the proportionality factor depends on substrate doping and varies with temperature.

For electron emission from a continuous distribution of interface traps, the CC-DLTS signal is

$$\Delta V_G = (1/C_{ox}) \int qN_{is}(E) \left[\exp(-t_1/\tau_n) - \exp(-t_2/\tau_n)\right] dE, \tag{2}$$

where τ_n is the electron emission time constant which from considerations of detailed balance may be expressed as

$$1/\tau_n = \sigma_n v_n N_c \exp(-E/kT). \tag{3}$$

The quantities appearing in the above equations are defined as follows: q is the electronic charge, $N_{is}(E)$ is the interface-state density at an energy E below the conduction-band minimum, σ_n is the capture cross-section for electrons, v_n is the mean thermal velocity for electrons, N_c is the effective density of states in the Si conduction band, k is the Boltzmann constant, and T is the absolute temperature. In Eq. (2) it is assumed that the traps are completely filled during a charging pulse. The exponential terms in the integrand of Eq. (2) form a function which for $t_2 = 2t_1$ is peaked at an energy $E_0 = kT \, ln(\sigma_n v_n N_c t_1/ln2)$. For a constant capture cross-section this function is sharply peaked, and the integral can be solved by assuming that N_{is} varies slowly over an energy interval of order kT. We obtain

$$\Delta V_G = q \, kT \, ln2 \, N_{is}(E_0)/C_{ox}. \tag{4}$$

Thus, the emission signal, divided by temperature, is directly proportional to the interface-state density at the energy E_0. The energy interval over which interface traps contribute to the DLTS signal is $\Delta E = kT \, ln2$. Both the width of the interval, ΔE, and its location in energy, E_0, increase linearly with temperature. The energy resolution is therefore greatest at low temperatures where the sampled interval is closest to the conduction band and the interface-state density is expected to vary most rapidly. The larger energy intervals obtained at high temperatures provide enhanced sensitivity for detecting low densities of interface states which are typically found near midgap.

RESULTS AND DISCUSSION

CC-DLTS Spectra in an N-Type MOS Structure

Several qualitative features of CC-DLTS measurements on n-type MOS capacitors can be illustrated by examining Fig. 3. This figure shows DLTS spectra obtained with delay times t_1, t_2

= 1,2 msec for different fixed depletion capacitances and also one spectrum for t_1, t_2 = 10,20 msec. In n-type MOS capacitors, the DLTS signal arises from the emission of electrons from interface states in the upper half of the Si bandgap. Below room temperature the DLTS signal is independent of the device capacitance, provided that the capacitor is biased far enough into depletion so that under equilibrium conditions the Fermi level intersects the interface below the Si midgap. In this temperature regime the signal is dominated by majority carrier emission. The rapid drop in the emission signal at temperatures below ~130 K may arise from a dependence of the capture cross-section on energy near the Si conduction band. Above 300 K the DLTS signal is strongly dependent on the depletion capacitance and increases rapidly as the device is biased further into depletion. The peak at ~335 K and the sharp increase in the signal at higher temperatures occur when the capacitor is biased near inversion and is ascribed to the onset of minority carrier emission and capture processes at the interface.

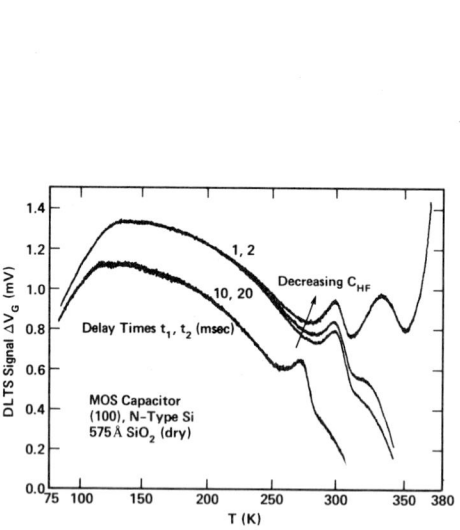

Fig. 3. CC-DLTS spectra for an n-type MOS capacitor for different bias conditions and delay times

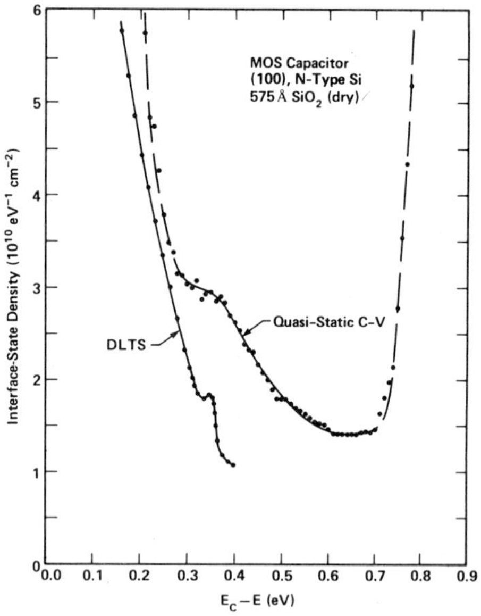

Fig. 4. Interface-state distributions for the capacitor of Fig. 3 obtained by the quasi-static C-V technique and from a DLTS analysis, with a constant capture cross-section of 10^{-18} cm^2.

The analysis presented in the last section is applicable over the region of the spectrum where the signal is dominated by electron emission and the capture cross-section can be assumed to be a constant. The interface-state distribution at midgap cannot be determined from DLTS data with the analysis presented here due to the onset of minority carrier effects at high temperatures. Also, contrary to a previous report (7), the analysis is not sufficient to determine the possible existence of a thermally activated capture cross-section. The emission spectra in Fig. 3 can be used to compute both σ_n and the energy scale for the interface-state distribution. The shift of the spectrum with delay time over the temperature range from ~130 K to room temperature is consistent with a capture cross-section that is constant or varies slowly with energy. The data can be adequately fitted with a σ_n of 10^{-18} cm^2. In Fig. 4 is shown the interface-state distribution obtained from the DLTS analysis. For comparison the distribution as measured by the quasi-static C-V technique (8) is also shown. In both distributions the density increases with energy from midgap toward the Si conduction band. The peak at ~300 K in the DLTS spectrum in Fig. 3 corresponds to the shoulder in the C-V distribution.

A Characteristic Discrete Interface Level in a P-Type MOS Structure

Certain processing and testing procedures introduce discrete levels at the Si-SiO$_2$ interface. In clean, unimplanted MOS structures, discrete interface levels appear when the device is electrically biased at elevated temperatures (9). In particular, bias-temperature treatments produce a discrete level of high density at an energy between 0.35 and 0.4 eV above the valence-band maximum (9,10). We have observed a similar level in *unannealed* specimens of oxidized Si, *without* the use of bias-temperature treatments. This peak is located at E$_V$+0.35 (±0.01) eV and is readily removed by a low-temperature anneal (450 C).

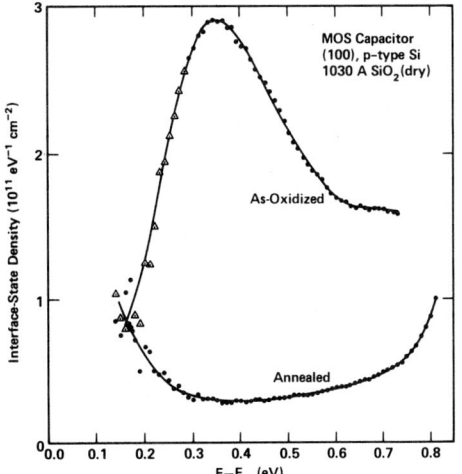

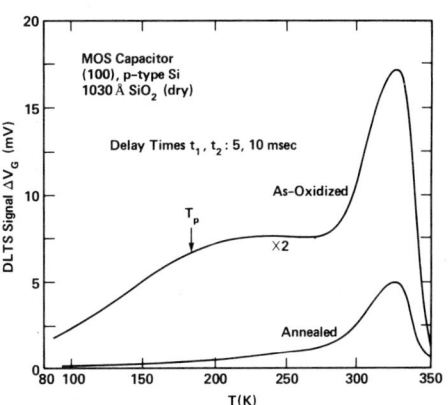

Fig. 5. Interface-state distributions in as-oxidized and annealed p-type MOS capacitors, as measured by the quasi-static C-V technique.

Fig. 6. CC-DLTS spectra for as-oxidized and annealed p-type MOS capacitors.

The interface-state distributions for both as-oxidized and annealed specimens, measured by the quasi-static C-V technique, are shown in Fig. 5. As-oxidized samples display a broad, prominent peak centered at E$_V$+0.35 eV. The full width of the peak at half maximum is approximately 0.3 eV, and the peak density is 2.9×10^{11} eV^{-1} cm^{-2}. In the annealed samples this peak is completely absent; remaining is the generally-observed continuum of interface states, with a minimum density near midgap of 3×10^{10} eV^{-1} cm^{-2}.

DLTS spectra for both as-oxidized and annealed specimens are shown in Fig. 6. Both spectra were recorded with the same delay times and fixed depletion capacitance. The only difference in preparation between the two specimens is a low-temperature, post-metallization anneal (450 C, 60 min, H$_2$/N$_2$). Both spectra display a large peak at ~330 K which we ascribe to combined majority and minority carrier processes. These peaks are not the DLTS signatures of the discrete level shown in Fig. 5, although the relative magnitudes of the peaks do reflect the difference in the interface-state densities of the two specimens. The broad interface-state peak is responsible for the large emission signal in the as-oxidized specimen at temperatures below ~300 K. Although the peak is associated with a discrete level, it must be treated as a continuous trap distribution in the DLTS analysis since its density varies slowly with energy on a scale of kT. In annealed specimens the hole emission signal is strongly attenuated due to the reduced density of the discrete interface level. The interface-state distribution (vs temperature) obtained from the emission spectrum for the as-oxidized specimen displays a peak at a temperature T_p; this temperature is noted on the emission spectrum in Fig. 6. The interface-state density at the peak is 2.5×10^{11} eV^{-1} cm^{-2}, in close agreement with the peak density as measured by the quasi-static C-V technique (Fig. 5). From the peak temperature the capture cross-section for holes is found to be approximately 7×10^{-15} cm^2.

DLTS emission spectra for an annealed specimen are shown in Fig. 7. The spectra were recorded with enhanced sensitivity and with two different pairs of delay times. Two features are clearly evident. First, the emission signal monotonically increases with temperature over the entire range, in marked contrast with results for electron emission from interface states in n-type MOS structures (Fig. 3). In MOS capacitors on p-type Si, the signal is due to hole emission from interface states near the Si valence band. However, the interface-state distribution which is obtained from the DLTS spectrum indicates that the density of states in the continuum increases with energy from the valence band toward midgap, in disagreement with results from quasi-static C-V measurements on annealed specimens (Fig. 5). This feature has been previously noted (11); it was suggested that the DLTS signal arises from hole emission from interface states in a conduction-band tail which extends into the lower half of the Si bandgap. The second feature pertains specifically to the discrete interface level. Even in the annealed specimens, a residual peak, superimposed on the continuum, is detectable in the DLTS emission spectra (Fig. 7). From the shift of the peak with delay time, the peak energy is found to be $E_v+.34(+.15,-.08)$ eV. Further, we estimate that the density of discrete levels at the peak is ~6×10^8 eV^{-1} cm^{-2}. As is evident in Fig. 5, a residual peak with this density is not detectable by the quasi-static C-V technique. The minimum detectable interface-state density in the DLTS measurements presented in Fig. 7 is ~2×10^8 eV^{-1} cm^{-2}.

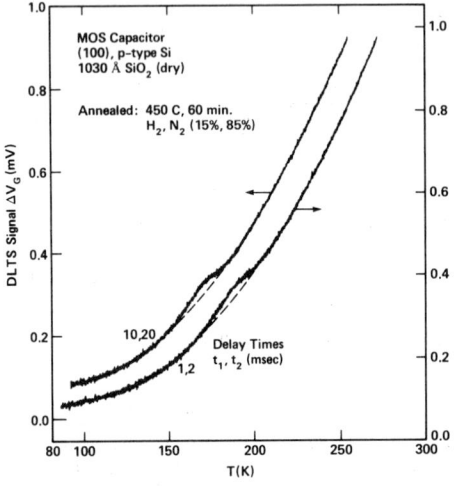

Fig. 7. CC-DLTS emission spectra, recorded on an expanded scale, for an annealed p-type MOS capacitor

We propose that the discrete interface level at $E_v+0.35$ eV is due to a characteristic defect at the Si-SiO$_2$ interface. Its presence in clean, as-prepared MOS structures and the effect of a low-temperature anneal, as demonstrated here, as well as its response to bias-temperature treatment (9,10) support this identification. Additional information further suggests that this defect is a Si dangling bond. Recent theoretical studies by Laughlin, Joannopoulos, and Chadi (12) reveal that Si dangling bonds at the Si-SiO$_2$ interface introduce a discrete interface level in the lower half of the Si bandgap near midgap, while neither oxygen dangling bonds nor strained Si-O bonds (i.e., bond angle distortions) contribute discrete levels in the bandgap. Since the thermal oxidation process favors an excess of Si in the Si-SiO$_2$ transition layer (13), which has been experimentally observed (14), Si dangling bonds may be considered a characteristic defect of the thermally oxidized surface. Even an atomically abrupt interface may possess a surface density of dangling bonds due to the Si-SiO$_2$ lattice mismatch (15). Further, with electron spin resonance Si dangling bonds have been identified at the Si-SiO$_2$ interface in freshly oxidized wafers and in processed wafers after bias-temperature treatment (16). In the present study the surface density of discrete levels in as-oxidized specimens (Fig. 5) would correspond to ~10^{11} cm^{-2} Si dangling bonds, or of the order of one dangling bond for every 10^4 surface atoms. These dangling bonds would be readily hydrogenated by a low-temperature anneal in a hydrogen-rich ambient; on a free Si surface this removes the discrete level from the Si bandgap (17). The surface density of discrete interface levels in the annealed specimens (Fig. 7) is estimated to be $<5\times10^7$ cm^{-2}.

ACKNOWLEDGEMENT

The authors express their appreciation to D. J. Chadi, J. P. McVittie, and R. Tremain for helpful discussions and to C. Healy, T. Kloffenstein, R. Lujan, and B. Pugh for technical assistance during the course of this work.

REFERENCES

1. A. Goetzberger, E. Klausmann, and M. Schulz, CRC Critical Review, 1 (Jan. 1976), and references listed therein.
2. E. D. Nicollian and A. Goetzberger, Bell Syst. Tech. J. *46*, 1055 (1967).
3. H. Deuling, E. Klausmann, and A. Goetzberger, Solid State Electron. *15*, 559 (1972).
4. R. J. Kriegler, J. Shappir, and T. F. Devenyi, IEEE Trans. Elect. Devices *ED-24*, 1206 (abs.) (Sept. 1977).
5. D. V. Lang, J. Appl. Phys. *45*, 3014 & 3023 (1974).
6. K. L. Wang and A. O. Evwaraye, J. Appl. Phys. *47*, 4574 (1976).
7. M. Schulz and N. M. Johnson, Solid State Commun. *25*, 481 (1978).
8. C. N. Berglund, IEEE Trans. Electron Devices *ED-13*, 701 (1966).
9. A. Goetzberger, A. D. Lopez, and R. J. Strain, J Electrochem. Soc. *120*, 90 (1973).
10. K. Saminadayar and J. C. Pfister, Solid-State Electron. *20*, 891 (1977).
11. M. Schulz and N. M. Johnson, Appl. Phys. Lett. *31*, 622 (1977).
12. R. B. Laughlin, J. D. Joannopoulos, and D. J. Chadi, (this conference).
13. B. E. Deal, M. Sklar, A. S. Grove, and E. H. Snow, J. Electrochem. Soc. *114*, 266 (1967).
14. W. L. Harrington, R. E. Honig, A. M. Goodman, and R. Williams, Appl. Phys. Lett. *27*, 644 (1975).
15. S. T. Pantelides, J. Vac. Sci. Technol. *14*, 965 (July/Aug. 1977).
16. E. H. Poindexter, E. R. Ahlstrom, and P. J. Caplan, (this conference).
17. J. A. Applebaum and D. R. Haman, Rev. Mod. Phys. *48*, 479 (1976).

INTERFACE STATES RESULTING FROM A HOLE FLUX INCIDENT ON
ON THE SiO_2/Si INTERFACE

J. M. McGarrity, P. S. Winokur, H. E. Boesch, Jr.,
and F. B. McLean
US Army Electronics Research and Development Command
Harry Diamond Laboratories, Adelphi, Maryland 20783

ABSTRACT

Interface states are produced in MOS capacitors following exposure to pulsed
e-beam irradiation. Data is presented here which show that, in addition to hole
transport to the SiO_2/Si interface, a positive field must be maintained across
that interface to observe the long-term time-dependent buildup of interface
states in wet-oxide capacitors. Results of detailed experiments on the time
dependence of the interface-state buildup in wet-oxide MOS capacitors suggest
that some chemical species in a metastable state in the SiO_2/Si interface
region must be removed in order to produce an interface state. Processes
occurring at the interface that may be responsible for the buildup of interface states in wet- and dry-oxide capacitors are discussed.

INTRODUCTION

Ionizing radiation effects in CMOS and MOS devices in satellite space environments degrade circuit performance. The physics of the radiation response of
SiO_2 is being investigated in order to provide a basis for improving device
performance. In this paper new results are presented on the effects of a rad-
idation-generated hole flux on the Si/SiO_2 interface, and the time dependence
of the interface-state buildup in thermally grown SiO_2 is explored in more
detail. It was demonstrated previously [1,2] that it was not direct interaction of radiation at the interface which produced interface states, that it was
necessary to transport holes to the interface to observe a buildup, and that
the buildup in some oxides took a long time after the holes reached the interface. It is the long-term buildup which is examined here.

SAMPLES AND EXPERIMENTAL TECHNIQUES

MOS capacitor samples were supplied by Hughes Aircraft Corp. The wet-process
samples consisted of SiO_2 thermally grown in a pyrohydrolytic H_2O ambient at
950 C to 965 Å thickness. The dry-process samples consisted of SiO_2 thermally
grown at 1000 C to 907 Å thickness. Both processes produced samples which
exhibited minimal hole trapping following exposure to ionizing radiation.

Both charge displacement and capacitance-voltage (C-V) measurements were performed on the samples following irradiation with 4-μs 13-MeV electron pulses
from an electron linear accelerator (LINAC). The radiation-induced current
through the samples was integrated during the LINAC pulses using a fast elec-
trometer-input operational amplifier. The C-V characteristics of the sample
were measured, at logarithmically increasing intervals after irradiation, by
momentarily interrupting the applied bias and applying a 50-ms voltage ramp.

During the ramps the sample capacitance was monitored with a phase-sensitive detector operating at 1 MHz, and the resulting C-V curves were recorded on film as oscilloscope traces. The fast C-V system was programmed to change the applied bias at selected times after the radiation pulses. The C-V curves were digitized and analyzed by the Terman technique. The resulting distributions of interface states were integrated from midgap to flat-band to provide single values, N_{ss} (cm^{-2}), of interface states in the samples as functions of the experimental parameters (3).

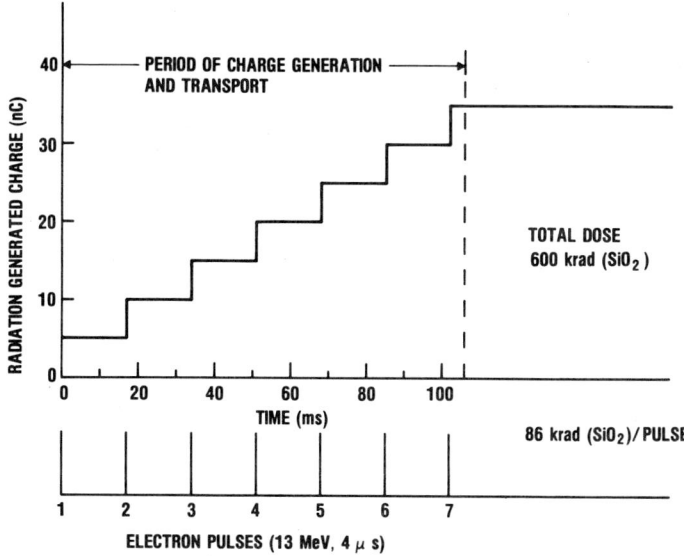

Fig. 1 Radiation generated charge (integrated current) in a wet-oxide MOS capacitor resulting from a series of LINAC pulses. The total dose delivered by the seven pulses is 600 krad(SiO$_2$).

EXPERIMENTAL RESULTS

Figure 1 shows the integrated current in a wet-oxide MOS capacitor resulting from exposure to a series of LINAC pulses at an applied field of 4 MV/cm. The charge is seen to increase in steps corresponding to each of the seven LINAC pulses (delivered at a rate of 60 pulses per second) used to obtain a total dose of 600 krad(SiO$_2$) while avoiding space charge problems. After the last pulse the integrated current is constant, indicating that at the higher field (4 MV/cm) the hole transport is over in milliseconds. The total measured charge of 35 nC was in agreement with a predicted value of 39 $\pm$ 9 nC based on the absorbed dose in SiO$_2$. The uncertainty in the predicted value was due to uncertainties in dosimetry for the shot and in the G-value for electron-hole pair creation in SiO$_2$ (4).

The time dependence of the interface-state buildup was reported previously (2,3). Here this buildup is examined by varying the field across the oxide at critical times during the period of interface-state growth. The results for the pyrogenic wet oxide are presented in Fig. 2.

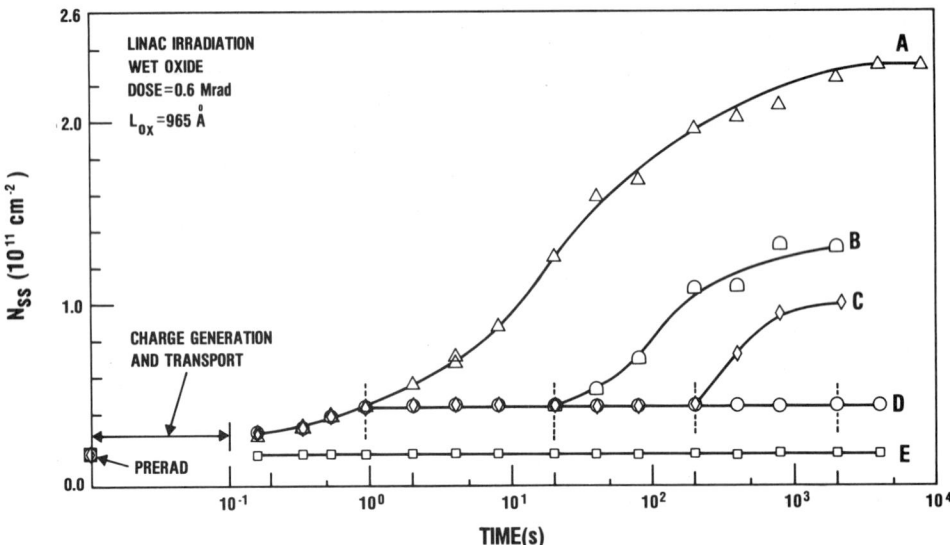

Fig. 2 Integrated interface-state density for wet-oxide capacitors as a function of time following a 600 krad(SiO$_2$) LINAC irradiation. Curves A-E present results for varying conditions of applied field across the oxide during both the initial hole transport and the subsequent long-term interface-state buildup.

Curve A shows the buildup of interface states under a constant positive applied field of 4 MV/cm as a function of time following LINAC irradiation. The major fraction of the buildup is seen to take place over several thousand seconds, a time scale considerably longer than the time it takes holes to transport to the SiO$_2$/Si interface. The saturation value of the interface-state density is 2.3 x 10^{11} cm^{-2}. Curve E shows the result when the bias is kept negative during and following the irradiation. No measurable buildup of interface states is observed.

To examine the slow process by which interface states are produced under a positive field following irradiation (curve A), the holes were moved to the interface under a positive bias for 0.8 s and then the bias was reversed for varying time intervals. For curves B, C, and D a positive field of 4 MV/cm was applied for 0.8 s, at which time the bias polarity was reversed for 20, 200, and 2000 s, respectively, and then returned to positive polarity. For all three curves a small increase in the interface-state density is observed during the initial 0.8 s when the bias is positive; when the bias is reversed no additional increase in the interface-state density is observed. Only when the bias is returned to positive polarity is any further increase in N$_{ss}$ observed, and the magnitude of that subsequent buildup is smaller the longer the bias is held negative. For the case where the bias is held negative for 2000 s (curve D), there is no increase in N$_{ss}$ when the bias is returned positive.

DISCUSSION

From the results presented here, the following features of the interface-state buildup in the wet-oxide MOS capacitors may be identified. (a) Under 4 MV/cm positive field, hole transport through the oxide to the interface is completed in milliseconds, well before interface-state buildup is observed. (b) Interface-state buildup takes place over a long time period (to 2000 s) and only progresses while a positive field is present. (c) After the holes are transported to the interface, and even after the buildup commences, a change to negative bias immediately halts any further increase of interface states. (d) If the bias is made negative and then reversed to positive again only a portion of the saturated interface-state density is actually observed.

In contrast with the results reported in this paper on the wet oxide, when hardened dry oxides are irradiated there is an increase in the interface-state density but no time-dependent buildup (3). In experiments on the thermally grown dry oxides, it is found that under positive bias the postirradiation increase in the interface-state density at 4 MV/cm was only about 10 percent of the final saturation value observed in wet oxides. Also, in the dry oxide the interface-state increase was observed at the earliest time of measurement following the radiation pulse. When a negative bias was applied during and after the radiation pulse, no interface states were produced.

Our proposed explanation for the experimental results is based on models by Revesz (5), Sah (6), Jeppson and Svensson (7), and Pepper (8). The buildup of interface states following irradiation is not caused by direct interaction of radiation at the SiO_2/Si interface but by holes that either transport through or are trapped at deep sites in the interface region (1,2). These latter processes can result in the release of an appreciable amount of energy ($\sim$5 eV). This energy release breaks relatively weak H (5,7,9) or OH bonds (6) with trivalent silicon, or strained Si-O bonds formed during the processing.

In the wet oxides it seems plausible that the interface state buildup begins by the breaking of a Si-H or Si-OH bond. To explain the time dependent buildup which follows the breaking of the bond we suggest that a released ion from a broken Si-H or Si-OH bond is transferred to a nearby site (in the direction of the ion drift in a positive field) where it interacts weakly with the "broken bond" site (trivalent silicon). This weak interaction or metastable configuration, through some compensating effects, prevents the appearance of an interface state. When a positive field is maintained at the interface, the weakly bound ion eventually diffuses away from the "broken bond" site and is removed from the interface region; at this time an interface state associated with the "broken bond" site appears. When a negative field is maintained at the interface, the potential barrier for the reverse reaction, i.e., recombination of the ion and broken bond, is significantly reduced and the recombination process becomes very efficient relative to the drift diffusion process. Thus, negligible interface-state buildup occurs and the number of weakly interacting trivalent silicon-ion centers (or potential interface states) is reduced.

In the dry oxide the weak bonds broken by the transporting holes are strained Si-O bonds that are formed during the oxide growth process. When these bonds are broken by the transporting holes the interface states are formed immediately. The broken bond results in a trivalent Si and a nonbridging oxygen, and there are no diffusing ions.

In summary, the transport of holes to the Si/SiO$_2$ interface results in the breaking of bonds and a subsequent increase in interface-state density. In wet oxides this increase is a slow process which takes place under positive bias and is probably caused by a field-induced diffusion of ions out of the interface region. In the dry oxides the broken bonds are probably strained bonds, and the new interface states appear immediately upon the arrival of the holes. The observation of a time-dependent buildup in any particular case would then depend on how many states were tied up by weak S-H or Si-OH bonds during the processing.

REFERENCES

1. P. S. Winokur and M. M. Sokoloski, Comparison of Interface-State Buildup in MOS Capacitors Subjected to Penetrating and Non-Penetrating Radiation, Appl. Phys. Lett. 28, 627 (1976).

2. P. S. Winokur, J. M. McGarrity, and H. E. Boesch, Jr., Dependence of Interface-State Buildup on Hole Generation and Transport in Irradiated MOS Capacitors, IEEE Trans. Nucl. Sci. NS-23, 1580 (1976).

3. P. S. Winokur, H. E. Boesch, Jr., J. M. McGarrity, and F. B. McLean, Field- and Time-Dependent Radiation Effects at the SiO$_2$/Si Interface of Hardened MOS Capacitors, IEEE Trans. Nucl. Sci. NS-24, 2113 (1977).

4. G. A. Ausman, Jr. and Flynn B. McLean, Electron-Hole Pair Creation Energy in SiO$_2$, Appl. Phys. Lett. 26, 173 (1975).

5. A. G. Revesz, The Irradiation Behavior of Noncrystalline SiO$_2$ Films on Silicon: A Review, IEEE Trans. Nucl. Sci. NS-24, 2102 (1977).

6. C. T. Sah, Origin of Interface States and Oxide Charge Generated by Ionizing Radiation, IEEE Trans. Nucl. Sci. NS-23, 1563 (1976).

7. K. O. Jeppson and C. M. Svensson, Negative Bias Stress of MOS Devices at High Electric Fields and Degradation of MNOS Devices, J. Appl. Phys. 48, 2004 (1977).

8. M. Pepper, Electron and Hole Localization at the Si-SiO$_2$ Interface, Second International Conference on Electronic Properties of Two-Dimensional Systems Part 1, 58 (1977).

9. E. H. Nicollian, C. N. Berglund, P. F. Schmidt, and J. M. Andrews, Electrochemical Charging of Thermal SiO$_2$ Films by Injected Electron Currents, J. Appl. Phys. 42, 5654 (1971).

THE INFLUENCE OF THE pH ON THE SURFACE STATE DENSITY
AT THE SiO_2-Si INTERFACE

N.F. de Rooij and P. Bergveld
Twente University of Technology, P.O. Box 217,
Enschede, The Netherlands

ABSTRACT

Quasistatic C-V measurements have been carried out on $Si-SiO_2$-electrolyte
structures. The C-V curves show a shift in voltage when the pH of the electrolyte is varied. This is due to a variation in the potential difference
across the electrolyte-SiO_2 interface. The shape of the C-V curves also
changes; from the variation in the minimum of these curves the change in the
surface state density near midgap is derived.
It is found that a decrease in pH leads to a decrease in the surface state
density at the $Si-SiO_2$ interface. This decrease is believed to be caused by a
hydrogen bearing species released at the SiO_2-electrolyte interface and transported to the $Si-SiO_2$ interface.
The possibility of application in ISFETs is discussed.

INTRODUCTION

Recently the system Si-thermally grown SiO_2-electrolyte has gained interest
as a consequence of its application in unreferenced Ion Sensitive Field
Effect Transistors (ISFETs) (1,2,3).
Especially, the influence of variations in the electrolyte composition on the
surface state density at the $Si-SiO_2$ interface has been discussed (2,3). In
order to investigate whether and how the surface state density depends on
changes in the electrolyte composition, e.g. a pH change, quasistatic capacitance-voltage (C-V) measurements (4) were carried out on $Si-SiO_2$ electrolyte
structures.
In this paper the results of the C-V measurements will be presented and the
consequences of the application of surface state modulation in ISFETs will be
discussed.

EXPERIMENTAL

P-type (111) silicon wafers of 10 Ωcm resistivity were oxidized at a temperature of 1200°C. The oxide thickness was about 500 Å.
Afterwards aluminum was evaporated on the oxide and on the back of the wafer,
followed by a heat treatment at 450°C with wet N_2 during 10 minutes. The
aluminum on the oxide was etched away just before the experiment.
The $Si-SiO_2$ structures were mounted in a sample holder, in such a way that
an electrolyte could be brought in contact with the oxide and also with the
reference electrode. The exposed area of the oxide was in the order of 0.2 cm^2.
The quasistatic C-V measurements were carried out in accordance with Ref. 4.
After recording a C-V curve with an electrolyte of a given pH, the electrolyte was replaced by one with a different pH value and a C-V recording repeated. The pH was varied in the range from 1 to 13 and also in the reverse
direction.

434

RESULTS

The measured C-V curves at three different pH values are shown in Fig. 1.

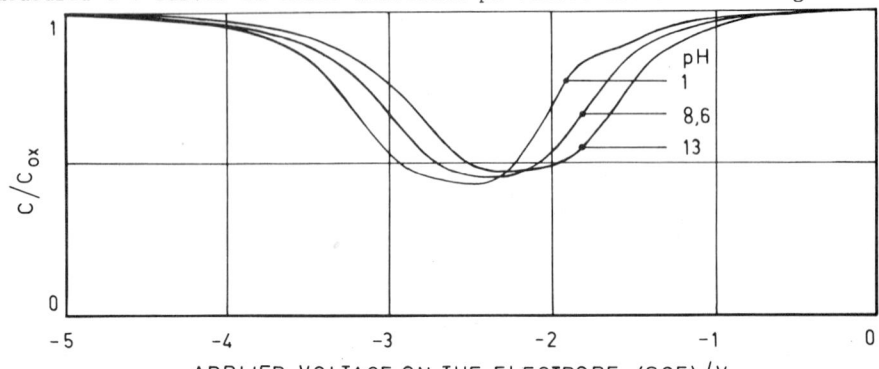

Fig. 1. Measured C-V curves at different pH values (exposed oxide area 1.96×10^{-1} cm^2; oxide thickness 500 Å; $N_A = 1.5 \times 10^{15}$ cm^{-3}).

The shift of these curves along the voltage axis is very pronounced.
This observed shift can be explained by a change in the electric potential difference across the SiO$_2$-electrolyte interface, due to a change in the net charge, which exists in a layer of hydrolysed material near the SiO$_2$-electrolyte interface. The silanol groups of this layer respond to the proton activity in the electrolyte according to

$$SiOH \rightleftarrows SiO^- + H^+ \tag{1}$$

An increase in the proton activity leads to a higher hydrogen content in the SiO$_2$.
The total voltage shift between the curves at pH 1 and pH 13 is about 400 mV, which is considerably less than predicted by the Nernst equation (59 mV per pH unit at room temperature). The deviation from the Nernst equation has been discussed by several authors (5,6).
It can also be seen from the curves of Fig. 2 that the minimum in the capacitance ratio $(C/C_{ox})_{min}$ depends on the pH.
This dependency manifested itself after exposing the SiO$_2$ to the electrolyte for at least a few hours. It appears that the observed shift in $(C/C_{ox})_{min}$ is reversible.
The small change in $(C/C_{ox})_{min}$ can be explained by a change in the surface state density $N_{ss}(E)$.
According to the method outlined by Van Overstraeten et al. (7), $N_{ss}(E)$ near midgap is obtained from $(C/C_{ox})_{min}$ using the following formula

$$\frac{qN_{ss}(E)}{C_{ox}} = \frac{(C/C_{ox})_{min}}{1-(C/C_{ox})_{min}} - \left(\frac{C_{Si}}{C_{ox}}\right)_{min} \tag{2}$$

Here q is the elementary charge, C_{ox} the oxide capacitance, C the measured capacitance and C_{Si} the theoretical semiconductor capacitance. All capacitances are per unit area. A value for $(C_{Si}/C_{ox})_{min}$ is obtained from Fig. 3 of Ref. 8 for a given oxide thickness and doping concentration. Values of $N_{ss}(E)$ at three pH values are given in Table 1.

TABLE 1   Surface state density $N_{ss}(E)$ near midgap at different pH values using $(C_{Si}/C_{ox})_{min}$ = 0.25

pH	$(C/C_{ox})_{min}$	$N_{ss}(E)/10^{11}cm^{-2}eV^{-1}$
1	.43	2.1
8.6	.45	2.4
13	.47	2.7

It is found, that $N_{ss}(E)$ near midgap increases about 30 percent in the pH range from 1 to 13.

The C-V measurements show that the surface state density is influenced by changes in the electrolyte composition. Which processes actually take place at the two interfaces is uncertain at the present time.

As pointed out by Revesz (4) the most realistic process will be the transport of (1) hydrogen in the form of hydroxyl, (2) a fast moving H-related species, (3) H_2 molecules. Also complexes involving Na^+ ions are amongst the possibilities.

These species are thought to be either released or captured at the SiO_2-electrolyte interface. They are transported through the SiO_2 and will then interact with the surface states at the $Si-SiO_2$ interface.

EFFECT OF VARIATIONS IN THE SURFACE STATE DENSITY ON THE CHANNEL CONDUCTIVITY OF ISFETs

In order to find a value for the change in the channel conductivity of an ISFET as a function of a variation in the surface state density the following calculations have been carried out.

Without an external gate the semiconductor space charge (Q_{sc}) is exactly compensated by the fixed oxide charge (Q_{ss}) and the mobile oxide charge (Q_o) as well as the charge in the surface states (Q_{st}) or

$$Q_o + Q_{ss} + Q_{sc} + Q_{st} = 0. \tag{3}$$

All charges are per unit area. The charge Q_{st} in the surface states is equal to

$$Q_{st} = -q \int_E \frac{N_{ss}(E)}{1 + \exp\{\frac{E-E_F}{kT}\}} dE \tag{4}$$

while Q_{sc} can be written as

$$Q_{sc} = 2qn_i L_D F(u_s, u_b). \tag{5}$$

Here n_i is the intrinsic carrier concentration, L_D the intrinsic Debye length and $F(u_s, u_b)$ the Kingston-Neustadter function (8).

Only acceptor type states are considered because these states can interact with the conduction band in the inverted silicon, as a consequence of their position near the conduction band edge.

It is assumed that the distribution of $N_{ss}(E)$ around the maximum state density $N_{ss}(E_{ss})$ is Gaussian as expressed in the following equation:

$$N_{ss}(E) = N_{ss}^o \frac{1}{\sigma\sqrt{2\pi}} \exp\{-\frac{(E-E_{ss})^2}{2\sigma^2}\} \tag{6}$$

where N_{ss}^o denotes the total number of acceptor states per unit area and σ denotes the standard deviation of the surface state density.
Combining equations (4) and (6) gives:

$$Q_{st} = -q \frac{N_{ss}^o}{\sigma\sqrt{2\pi}} \int_E \frac{\exp - \{\frac{(E-E_{ss})^2}{2\sigma^2}\}}{1+\exp \{\frac{E-E_{ss}}{kT} - u_{ss}\}} \, dE \qquad (7)$$

where $u_{ss} = \frac{E_F - E_{ss}}{kT}$.

Now Q_{st} can be solved numerically as a function of u_{ss} for given values of N_{ss}^o and σ, resulting in $Q_{st}(u_{ss}, N_{ss}^o, \sigma)$.
Moreover, because E_{ss} is fixed with respect to the intrinsic level E_i, it follows that

$$u_{ss} - u_s = n \qquad (8)$$

where n is constant.
We will consider the case in which only the number of acceptor states changes. The gaussian distribution (σ) and also the position of the maximum in $N_{ss}(E)$ are not influenced by any chemical process.
By combining equations (3), (5), (7) and (8) we obtain

$$Q_o + Q_{ss} + 2qn_i L_D F(u_s, u_b) = -Q_{st}(n+u_s, N_{ss}^o, \sigma) \qquad (9)$$

The left and right-hand part of equation (9) can be represented graphically as a function of u_s. The values of n (-16) and σ (100 mV) are based on the data as given in Fig. 9 of Ref. 9 and the value of u_b (-11.51) follows from the doping concentration ($N_A = 1.5 \times 10^{15}$ cm^{-3}). The intersections of the curves representing the right-hand part of equation (9) for different values of N_{ss}^o with the curves representing the left-hand part with $Q_o + Q_{ss}$ as a parameter, yield values of u_s from which Q_{sc} can be calculated according to equation (5). N_{ss}^o is varied between 2×10^{11} and 2×10^{12} states cm^{-2} and ($Q_o + Q_{ss}$) between 1×10^{12} and 2×10^{12} elementary charges cm^{-2}.
For a certain value of u_s the inversion charge Q_n can be calculated. The conductivity G of the inversion channel is given by

$$G = -\mu_n Q_n \qquad (10)$$

where μ_n is the electron mobility ($\mu_n = 600$ cm^{2}V^{-1}s^{-1}) (1). It is found that the change in the conductivity as a function of a change in the total number of states $(\frac{dG}{dN_{ss}^o})$ is approximately -5×10^{-17} S cm^2. For an ISFET, the change in the drain current (ΔI_d) due to a change in $N_{ss}^o (\Delta N_{ss}^o)$ is equal to

$$\Delta I_d = \frac{W}{L} (\frac{dG}{dN_{ss}^o}) \Delta N_{ss}^o V_{ds}. \qquad (11)$$

If $N_{ss}(E)$ increases with the same percentage over the whole energy range as near midgap and we take 1×10^{12} states cm^{-2} for N_{ss}^o at pH 1, ΔN_{ss}^o is about 3×10^{11} states cm^{-2}.
According to equation (11) the decrease in drain current (ΔI_d) for an ISFET with a $\frac{W}{L}$ value of 50 and a V_{ds} of 0.1 Volt, is about 75μA in the pH range 1 to 13, which is a measurable quantity.

CONCLUSION

It can be concluded from the C-V measurements that the surface state density at the Si-SiO$_2$ interface is influenced by a change in the electrolyte composition, although the mechanism involved is uncertain.
Modulation of the surface state density at the Si-SiO$_2$ interface due to variations in the composition at the ambient (liquid or gas) will be an interesting subject for further investigation in order to find applications in Chemically Sensitive Semiconductor Devices (3).

ACKNOWLEDGEMENT

This work has been supported by the Netherlands Organization for the Advancement at Pure Research (Z.W.O.) and the Hogeschoolfonds of the Twente University of Technology.

REFERENCES

(1) P. Bergveld, Development, operation and application of the Ion Sensitive Field Effect Transistor as a tool for electrophysiology, IEEE Trans. on Biomed. Eng. BME-19, 342 (1972).

(2) A.G. Revesz, On the mechanism of the ion sensitive field effect transistor, Thin Solid Films 41, L43 (1977).

(3) Jay N. Zemel, Chemically sensitive semiconductor devices, Research/Development 28, 38 (1977).

(4) M. Kuhn, A quasi-static technique for MOS C-V and surface state measurements, Solid-St. Electron. 13, 873 (1970).

(5) S. Levine and A.L. Smith, Theory of the Differential Capacity of the Oxide/Aqueous Electrolyte Interface, Discuss. Faraday Soc. 52, 290 (1971).

(6) J.W. Perram, Structure of the Double Layer at the Oxide/Water Interface, J. Chem. Soc. Faraday II, 993 (1973).

(7) R. van Overstraeten, G. Declerck and G. Broux, Graphical Technique to Determine the Density of Surface States at the Si-SiO$_2$ Interface of MOS Devices Using the quasistatic C-V Method, J. Electrochem. Soc. 120, 1785 (1973).

(8) R.H. Kingston, S.F. Neustadter, Calculation of the Space Charge, Electric. Field and Free Carrier Concentration at the Surface of a Semiconductor, Journ. of Appl. Physics 26, 718 (1955).

(9) B.E. Deal, The Current Understanding of Charges in the Thermally Oxidized Silicon Structure, J. Electrochem. Soc. 121, 198c (1974).

THE Si-SiO$_2$ INTERFACE: OXIDE CHARGE, ELECTRON AFFINITY AND FAST SURFACE STATES

L. A. Kasprzak and A. K. Gaind
IBM Corporation, East Fishkill, New York

Thermal oxidation of silicon results in a Si-SiO$_2$ interface which exhibits interface charge and fast electronic states in the bandgap. We believe that these properties are related to the oxidation process. The purpose of this paper is to describe results of an abrupt Si-SiO$_2$ interface created by chemical vapor deposition (CVD) of SiO$_2$ on Si at 1000°C in H$_2$ carrier gas using CO$_2$ as the oxidant for SiH$_4$. Recently we have described results using this technique which clearly identify Si as the source of fixed oxide charge Q_{ox} in CVD SiO$_2$ [1]. The CVD SiO$_2$ system allowed control of the magnitude and distribution of Q_{ox}. Q_{ox} is made up of Si-SiO$_2$ interface charge (Q_{ic}) and bulk charge distributed in the oxide itself (Q_{dc}). Figure 1 is a plot

$$V_{FB} + \phi_n = \frac{\phi_{MS}^*}{q} - \frac{Q_{ic}}{\varepsilon_{ox}}d_{ox} - \frac{\rho_{ox}}{2\varepsilon_{ox}}d_{ox}^2 \text{ for n-silicon and}$$

$$V_{FB} - \phi_p = \frac{\phi_{MS}^*}{q} - \frac{Q_{ic}}{\varepsilon_{ox}}d_{ox} - \frac{\rho_{ox}}{2\varepsilon_{ox}}d_{ox}^2 \text{ for p-silicon; where } V_{FB} \text{ is the MOS flat-}$$

band voltage (see Fig. 2 for band diagrams), $\phi_n \equiv E_c - E_F$, $\phi_p \equiv E_V - E_F$, ϕ_{MS}^* = $\phi_m - \chi_{Si}$, ϕ_m = metal work function, χ_{Si} = silicon electron affinity, q = electronic charge, d_{ox} = oxide thickness, and $\rho_{ox} = \frac{Q_{dc}}{d_{ox}}$ = bulk oxide charge.

The dashed curves in Fig. 1 indicate distributed bulk charge. The straight lines are the same points corrected for a uniform fixed distributed charge as indicated in Table 1, which also shows similar data for (111) silicon. Figure 3 is uncorrected data for the same deposition conditions except for the addition of HCl to the system. Note the absence of the curved lines indicating no distributed bulk oxide charge (Table 2). Also note the absence of interface charge for the (100) surface (Fig. 3) indicated by the zero slope. Finally it should be pointed out that Q_{ic} is also reduced on (111) Si because of the HCl. Along with the existing data on thermal SiO$_2$ (see Ref. in Table 3) and the known chemisorption phenomena on Si, [2,3] we have formulated a model of the Si-SiO$_2$ interface which explains the discrepancies in Q_{ox} and metal silicon work function (ϕ_{MS}) commonly encountered in the literature. The importance of the interdependence of these two parameters is both fundamental and practical. The model is that the degree of hydrogen chemisorption controls the electron affinity (Table 3) and through ϕ_{MS} causes anomalous values of Q. The fundamental relationship between fixed interface charge Q_{ic} and fast surface states N_{fs} is explained via the role of hydrogen chemisorption and annealing behavior [2,4]. Parametric dependencies of N_{fs} density near mid-gap as observed in the CVD SiO$_2$ system (Fig. 4) indicate that the N_{fs} density depends primarily on the abruptness of the interface and the degree of hydrogen chemisorption [4]. The specific

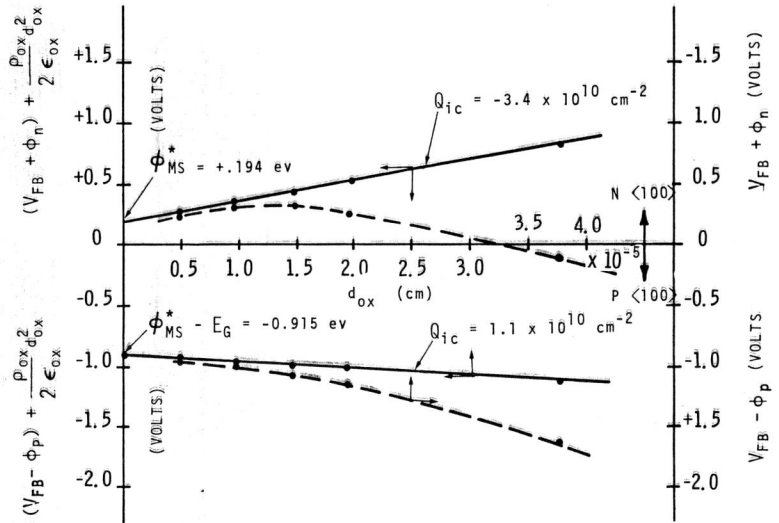

Fig. 1. $V_{FB} + \phi_n$ (or $V_{FB} - \phi_p$) vs. dox dashed curves, $V_{FB} + \phi_n + \frac{1}{2}$ ρox $\frac{dox^2}{tox}$

or $V_{FB} - \phi_p + \frac{1}{2}$ ρox $\frac{dox^2}{tox}$ vs. dox solid lines CVD-ox deposited

without HCl, <100> Si

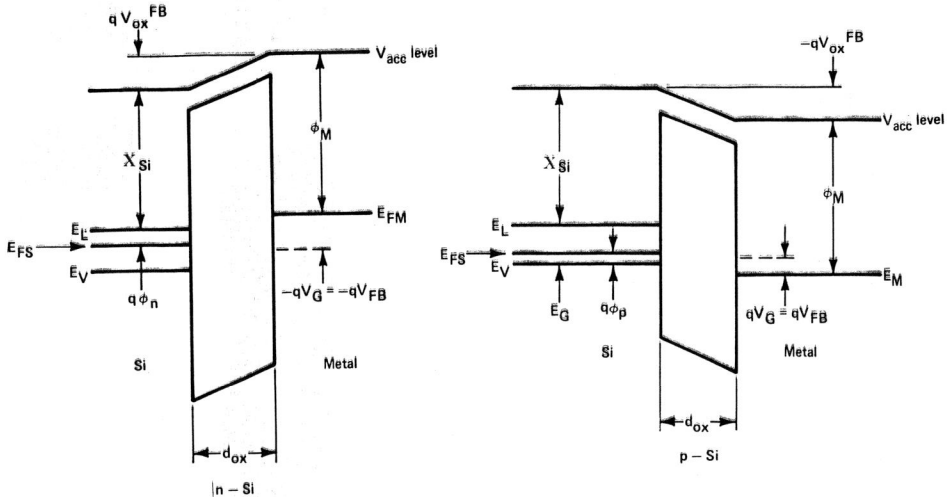

Fig. 2.   Band diagrams

TABLE 1 The interface charge (Q_{ic}), the distributed charge density, and ϕ_{MS} from A. O. S. capacitors, fabricated with CVD-SiO$_2$ deposited without HCl.

Si-type & Orientation	$\phi_{MS}{}^*$ or $\phi_{MS}{}^* - E_G$ (ev)	$\dfrac{\rho_{ox}\dagger}{2}$ x 10^{15} (No./cm^3)	$Q_{ic}\dagger$ x 10^{10} (No./cm^2)
N <100>	+.194 ± .02	1.37	−3.4
N <111>	+.103 ± .01	0.94	+5.8
P <100>	−.915 ± .03	0.75	+1.1
P <111>	−.955 ± .02	0.62	+9.4
MEAN	—	0.92	—

† Error ≤ ±10%

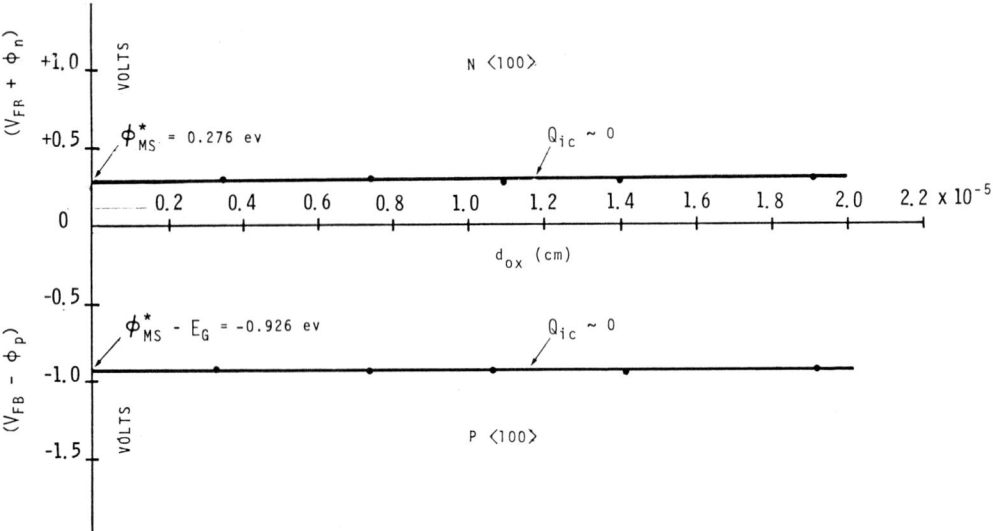

Fig. 3. $V_{FB} + \phi_n$ (or $V_{FB} - \phi_p$) vs. dox for CVD-ox deposited with HCl <100> Si

evidence for these phenomena is the location, height and FWHM (full width at half max) of the H chemisorption peak in N_{fs} (Table 4). The total mid-gap fast state density (peak height x FWHM) for the two surface orientations were found approximately proportional to the ratio of bonds on the respective surfaces. The presence of HCl during the deposition of these films had the effect of doubling the total mid-gap fast state density while decreasing Q_{ic} and eliminating Q_{dc}. This behavior is opposite to that observed in thermal-SiO$_2$ where proportionality between N_{fs} and Q_{ic} has been observed [5]. In conclusion, we contend that hydrogen chemisorption at the Si-SiO$_2$ interface affects χ_{Si} and the fast surface state density. Finally, the CVD technique reported herein allows one to approach an abrupt, near ideal, Si-SiO$_2$ interface.

TABLE 2   The interface charge, (Q_{ic}), the distributed charge density and ϕ_{MS}^*
(Dep'n with Hcl)

Si-type & Orientation	ϕ_{MS}^* or $\phi_{MS}^* - E_G$ (ev)	$Q_{ic} \times 10^{10}$ (No./cm^2)
N <100>	+.276 ± .02	∽ 0
N <111>	+.101 ± .01	4.6
P <100>	−.926 ± .01	∽ 0
P <111>	−.965 ± .01	7.5

TABLE 3   Change in X_{Si} as a Function of Atomic Hydrogen

Reference	Condition & Comments	X_{Si} (ev)	X_{Si} (ev) Relative
W.F. Werner S.S. Elec., **17**, 769 (1974)	Dry O_2 (No Atomic Hydrogen)	3.77	0.0
Present Work	CVD−SiO_2 (Some Atomic Hydrogen)	3.87	+0.10
S. Kar S.S. Elec., **18**, 169 (1975)	H_2O (Large Atomic Hydrogen)	4.21	+0.44

Table 4   Fast Surface State Data

	Peak height (cm^{-2}eV^{-1})	Peak width (V)	Total N_{fs} (cm^{-2})
100 no HCl	1.2×10^{10}	0.20	0.24×10^{10}
111 no HCl	1.8×10^{10}	0.32	0.57×10^{10}
100 HCl	2×10^{10}	0.35	0.64×10^{10}
111 HCl	3×10^{10}	0.45	1.2×10^{10}

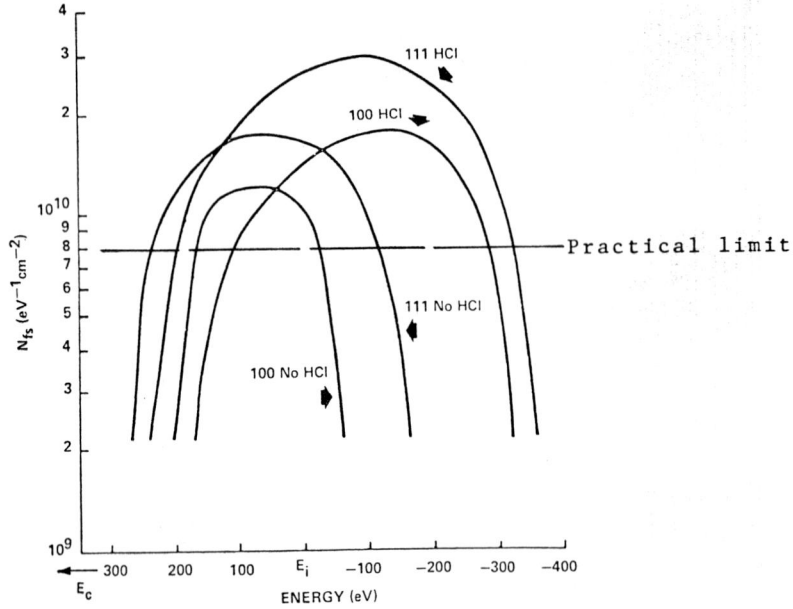

Fig. 4. Distribution of N_{fs} near mid-gap

SUMMARY

1. The fixed flat band charge is distributed in CVD-oxides.

2. The distributed charge is virtually eliminated by the use of HCl during deposition.

3. On <111> surface, CVD depositions result in 2x decrease in Q_{ic} compared to thermal SiO_2.

4. HCl decreases Q_{ic}.

5. ϕ_{MS}* differs from thermal oxide due to hydrogen chemisorption at the Si–SiO_2 interface.

REFERENCES

1. A. K. Gaind and L. A. Kasprzak, "Determination of fixed-charge distribution in CVD SiO_2 and its virtual elimination by use of HCl," Electrochemical Society Fall Meeting, Atlanta, Ga. (1977). Abs. #334.

2. T. W. Hickmott, "Parallel between surface states at the Si–SiO_2 interface and the B_2 center in irradiated SiO_2," J. Vac. Sci. Tech. 9, 311 (1972).

3. T. Sakurai and H. D. Hagstrom, "Chemisorption of atomic hydrogen on the silicon (111) 7 x 7 surface," Phys. Rev. B14, 1593 (1976).

4. L. A. Kasprzak and A. K. Gaind, "The influence of hydrogen and hydrogen chloride on the fast state density distribution at the Si–SiO_2 interface," Electrochemical Society Spring Meeting, Seattle, Wash. (1978). Abs. #278.

5. B. E. Deal, "The current understanding of charges in the thermally oxidized silicon structure," J. Electrochem. Soc., 121, 198C (1974).

LATERAL NONUNIFORMITIES (LNU) OF OXIDE AND INTERFACE STATE
CHARGE*

N. Zamani and J. Maserjian
Jet Propulsion Laboratory, California Institute of Technology
Pasadena, California 91103

ABSTRACT

A method is described for determining the distribution density of charge LNU
in MOS capacitors. The method involves freezing the initial occupation of
interface states with rapid C-V measurements at low temperatures. The charge-
distribution density is then readily deconvolved from the C(V) data using an
FFT computer analysis. Results obtained for a radiation "soft" MOS capacitor
with nearly ideal initial characteristics before irradiation show—after 10^5
rads—large charge LNU with a half-width spread of about $3 \cdot 10^{12}$ cm^{-2}. The
apparent increase of interface state density near mid-gap is in excess of
10^{12} cm^{-2} eV^{-1} when LNU is not considered. It is shown that when the LNU is
taken into account, an alternate explanation can be simply the contribution
of interface states near the band edges. This ambiguity raises serious questions on the proper interpretation of the apparent phenomena of interface
state generation during radiation and electrical stress.

INTRODUCTION

The possibility of lateral nonuniformities (LNU) of charge seen at the SiO_2/Si
interface is clearly of concern because it would effect the interpretations of
various interface measurements. However, in spite of some interest in this
problem, it has been largely ignored because of the lack of a convenient method
of measuring charge LNU.

Recently, there has been considerable attention given to the problem of interface state buildup during irradiation and electrical stressing of MOS devices.
However, if charge LNU also builds up significantly during these experiments,
it may not be possible to properly measure the true interface state density
$N_{ss}(E)$. Castagné and Vapaille (1) analyzed the effect of an assumed LNU on
different kinds of measurements of $N_{ss}(E)$ and showed how erroneous values would
be generated. Lopez and Strain (2) have also considered this problem and have
emphasized the need for a reliable method of estimating charge LNU.

The measurement of charge LNU was first treated by Nicolian and Goetzberger (3)
with their ac conductance-measurement technique. However, this technique is

*This paper presents the results of one phase of research performed at the
Jet Propulsion Laboratory, California Institute of Technology, sponsored by
the National Aeronautics & Space Administration under Contract No. 7-100.

only useful for estimating small fluctuations in surface potential. Recently, Chang (4) studied the effect of LNU in detail and introduced a simple approximate procedure for calculating the flatband distribution, but which is useful only for low values of $N_{ss}(E)$. His general treatment inspired the method developed in this work, which is useful for any values of $N_{ss}(E)$.

The method described here is based on the ability to freeze the interface state charge and the following consequences: (1) the deviation of the measured C–V curve from the ideal—other than parallel shifts—is due to charge LNU only; and (2) this C–V curve results from the convolution of the ideal capacitance with a function which depends on the charge LNU.

THEORY AND TECHNIQUE

The analysis is based upon a model of the nonuniform capacitor as a parallel array of smaller uniform capacitors, differing from each other only in the magnitude of the effective interface charge. This model is valid if the dimensions of the uniform regions are larger than the width of the depletion region (Ref. 5). The measured capacitance C of the device, in the absence of interface states, is given by (Ref. 3)

$$C(V) = \int_{-\infty}^{\infty} C_o(V - V_{FB}) \ f(V_{FB}) \ dV_{FB},$$

(1)

where $C_o(V - V_{FB})$ is the ideal capacitance of a uniform region, but displaced along the voltage axis by its flatband voltage V_{FB}, and $f(V_{FB})$ is the density function for the flatband distribution.

The above relation is generally valid for high-frequency, quasistatic, and non-equilibrium capacitances. It is also applicable when interface states are present if the occupancy of these states does not change during the measurement. In this case the interface state charge would simply add to the fixed oxide charge.

Equation (1) is also a convolution equation, in which $C(V)$ and $C_o(V)$ are known and $f(V)$ is unknown. We wish to use Fourier transforms to solve for $f(V_{FB})$, but, since the Fourier integrals of C and C_o do not converge, we need first to take the derivative of Eq. (1). We then have

$$C'(V) = \int_{-\infty}^{\infty} C_o'(V - V_{FB}) \ f(V_{FB}) \ dV_{FB},$$

(2)

where the prime signifies the derivative with respect to V. Since these derivatives can be Fourier-transformed, the solution can immediately be written as

$$f(V_{FB}) = \mathscr{F}^{-1}\{\mathscr{F}[C'(V)]/\mathscr{F}[C_o'(V)]\},$$

(3)

where $\mathscr{F}$ and $\mathscr{F}^{-1}$ are, respectively, the Fourier transform operator and its inverse. In practice, one needs to use fast-Fourier-transform (FFT) techniques for numerical processing of the data with the aid of a computer.

Therefore, to find $f(V_{FB})$ we need only to determine C and C_o, and to apply the procedure indicated by Eq. (3). The ideal capacitance C_o can be calculated if the doping profile for the MOS depletion region is known. This doping profile

N(x) can be determined easily and accurately with well-known C-V techniques before subjecting the MOS capacitor to processes which induce charge LNU. The
existence of charge LNU would distort N(x), and thus, the results as well.

The method requires the measurement of the nonequilibrium C-V curve because of
the requirement that the interface state charge remain fixed. This measurement is accomplished at liquid-nitrogen temperatures and by first biasing the
capacitor in strong accumulation to fill all interface states. A data point
C(V) is then obtained by momentarily switching to V and measuring the high-
frequency capacitance. The complete C-V curve is obtained by repeating this
process. This procedure requires an automatic digital data system. The
measurement time for each data point is kept to a few milliseconds to minimize
the number of interface states near the majority band which can change their
occupation. However, we have found that increasing this time by four orders
of magnitude causes virtually no change in the resulting C-V curve. In view
of the fact that we later show that large values of $N_{ss}(E)$ occur near the
majority conduction band, it would appear that the time constants for these
interface states are much larger at liquid-nitrogen temperatures than one
would predict from simple extrapolation of published room-temperature data.
In effect, the interface state charge does indeed appear frozen for the purposes of this measurement.

The digital C(V) data and calculated $C_o(V)$ points are then numerically processed using the FFT techniques outlined above to obtain $f(V_{FB})$. This function
may then be transformed to room temperature or any other temperature by the
dependence

$$f_{T2}(V_{FB}) = f_{T1}(V_{FB} - \Delta E_F),$$

where ΔE_F is the Fermi-level shift between temperatures T1 and T2. One can
also use $f(V_{FB})$ to describe the distribution of effective charge density Q_{ss}
by using the relation

$$V_{FB} = \phi_{MS} + Q_{ss}/C_{ox},$$

where ϕ_{MS} is the metal-semiconductor barrier difference, and C_{ox} is the oxide
capacitance per unit area.

RESULTS AND DISCUSSION

The following example uses an MOS capacitor chosen for its radiation softness
to emphasize the effect of charge LNU. The oxide was grown in wet HCℓ at
$1000^{\circ}C$ to a thickness of 1090 Å and annealed in N_2 at $1100^{\circ}C$ for 30 minutes.
Post-aluminum-metallization anneals were carried out in N_2 at $450^{\circ}C$ for 15
minutes. After initial measurements, the capacitors were irradiated with Co^{60}
to 10^5 rads.

Figures 1 and 2 show the measured nonequilibrium C(V) curves before and after
irradiation, respectively. The ideal $C_o(V)$ curves are also shown for comparison. The initial C(V) curve before irradiation in Fig. 1 shows nearly ideal
behavior and corresponds to a small value of V_{FB} of about 0.2 V and values of
$N_{ss}(E)$ in the low 10^{10} cm^{-2} eV^{-1} range near mid-band. It therefore represents
a gate oxide of excellent device quality. The C(V) curve, after irradiation
in Fig. 2, shows considerable distortion and stretch-out which one normally
attributes to a large increase of $N_{ss}(E)$.

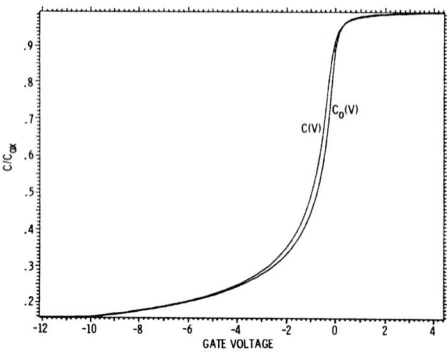

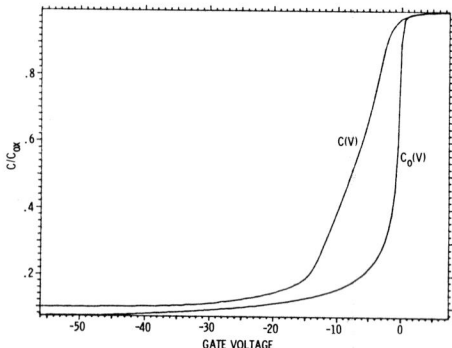

Fig. 1. Nonequilibrium C-V curve
(before irradiation)
C(V) measured, C_o(V) calculated ideal

Fig. 2. Nonequilibrium C-V curve
(after irradiation)
C(V) measured, C_o(V) calculated ideal

Figure 3 shows the derivatives of the curves in Fig. 2. Applying the deconvolution methods described above, $f(V_{FB})$ is obtained and plotted in Fig. 4.
This result shows an enormous spread in V_{FB} which would completely obscure
N_{ss}(E) measurements. It also indicates at least two peaks in the V_{FB} distribution, one at 3.3 V, another near 11.0 V. These double peaks have been
observed consistently in our measurements to date, but we have not yet had an
opportunity to measure a wide range of processes. We might speculate that the
peak at large V_{FB} corresponds to the presence of distinct defect species at the
interface responsible for nucleating large charge clusters, as indicated in
other work related to oxide breakdown (Refs. 6-8).

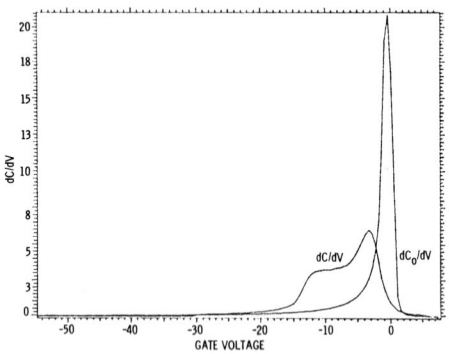

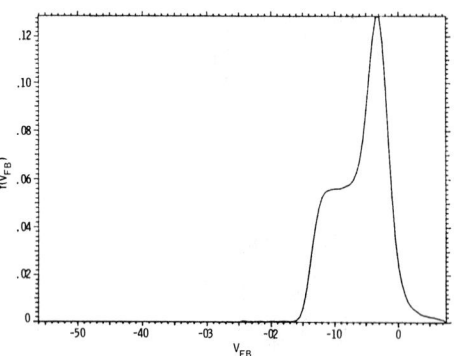

Fig. 3. Derivatives of C-V curve
(after irradiation) dC/dV measured,
dC_o/dV calculated ideal

Fig. 4. Distribution density
for V_{FB}
(after irradiation)

Figure 5 shows the measured post-irradiation high-frequency C(V) curve at room
temperature (equilibrium), along with the ideal C_o(V) curve at room temperature

and a calculated curve labeled C_o*f. The latter is obtained by convolving $C_o(V)$ with $f(V_{FB})$, where V_{FB} is adjusted to room-temperature values. This convolved curve would be the high-frequency C(V) curve if there were no interface states. By noting that there is basically a parallel shift except for the rather prominant tails near the maximum and minimum, one can deduce the presence of high $N_{ss}(E)$ near the band edges. The flatband distribution dominates the behavior of the C(V) curve in the transition between the maximum and minimun and thus obscures the effect of interface states in this region.

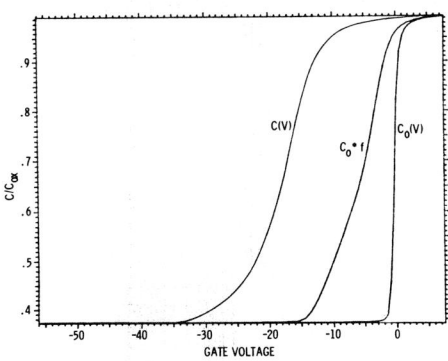

Fig. 5. Equilibrium high frequency
C-V curve (after irradiation)
C(V) measured,
$C_o(V)$ calculated ideal,
C_o * f calculated with LNU

Figures 6 and 7 demonstrate the insensitivity to interface states near midband. For the purpose of this demonstration, we have assumed a Gaussian form for $N_{ss}(E)$ at the band edges for both examples and have added a constant N_{ss} term for the example in Fig. 7. Furthermore, we have ignored the possibility that the interface states themselves could be laterally nonuniform. The dashed curves are obtained by convolving $f(V_{FB})$ with the capacitance calculated by including the assumed $N_{ss}(E)$. The parameters, describing the height and width of Gaussian distributions of $N_{ss}(E)$, were adjusted until a reasonably good fit was obtained, including a constant N_{ss} term (mid-band) of 1×10^{12} cm^{-2} eV^{-1} for the example in Fig. 7. One can clearly see that it is entirely ambiguous whether or not we include an apparent mid-band value for N_{ss}.

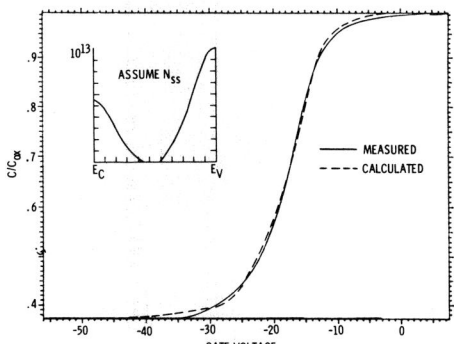

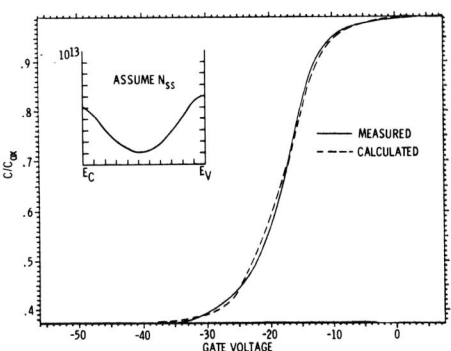

Fig. 6. Comparison of Measured C(V)
with calculated curve
mid-band $N_{ss}(E) = 0$

Fig. 7. Comparison of measured C(V)
with calculated curve
mid-band $N_{ss}(E) = 1 \times 10^{12}$ cm^{-2} eV^{-1}

It is possible in principle to determine $N_{ss}(E)$ uniquely in the presence of charge LNU, if $N_{ss}(E)$ itself is laterally uniform. This could be done by another FFT deconvolution of the high-frequency C(V) curve, using the measured function $f(V_{FB})$. We have attempted to do this but without success. It is not clear at this time whether this was due to the effect of interface state LNU or to the limits of the method.

REFERENCES

1. R. Costagné and A. Vapaille, "Apparent Interface-States Density Introduced by the Spatial Fluctuation of Surface Potential in MOS Structures," Electron. Lett. 6, 671 (1976).

2. A. D. Lopez and R. J. Strain, "MOS Surface Potential and the Gross Non-uniformity," Solid-St. Elect. 16, 507 (1973).

3. E. H. Nicolian and A. Goetzberger, "The Si-SiO$_2$ Interface Electrical Properties as Determined by the Metal-Insulator-Silicon Conductance Technique," Bell Syst. Tech. J. 46, 1055 (1967).

4. C. Chang, "Study of Lateral Nonuniformities and Interface States in MIS Structures," Ph.D. Dissertation, Princeton University, 1976.

5. J. R. Brews, "Admittance of an MOS Device with Interface Charge Nonhomogeneities," J. Appl. Phys. 43, 3451 (1972).

6. R. Williams and M. W. Wood, "Image Forces and the Behavior of Mobile Positive Ions in Silicon Dioxide," Appl. Phys. Lett. 22, 458 (1973).

7. T. H. DiStefano, "Dielectric Breakdown Induced by Sodium in MOS Structures," J. Appl. Phys. 44, 527 (1973).

8. S. P. Li, E. T. Bates, and J. Maserjian, "Time-Dependent MOS Breakdown," Solid-St. Electron. 19, 235 (1976).

TEMPERATURE DEPENDENCE OF RELAXATION OF INJECTED CHARGE AT THE POLYCRYSTALLINE-SILICON–SiO$_2$ INTERFACE

T. W. Hickmott
IBM Thomas J. Watson Research Center
Yorktown Heights, New York 10598

ABSTRACT

Electron injection can occur at the polycrystalline-silicon–SiO$_2$ interface of a polycrystalline-silicon gate MOS capacitor when relatively low negative voltages are applied to the gate. The temperature and voltage dependence of relaxation of injected negative charge has been measured. The amount of charge injected is higher and its relaxation is faster at higher temperatures. The moment of charge can shift towards the substrate silicon-SiO$_2$ interface at negative gate voltages whose absolute magnitudes are less than the flat-band voltage of the capacitor. For positive gate voltages, the activation energy for motion of the negative charge moment is about 1 eV.

INTRODUCTION

Polycrystalline silicon films deposited by thermal decomposition of SiH$_4$, are widely used as gate electrodes in semiconductor technology.(1,2)   Yun and Hickmott (3) have observed hysteresis in high-frequency C–V curves of polycrystalline-silicon–SiO$_2$–Si capacitors at low negative ($\lesssim -1$ V) gate voltages, which they attribute as being due primarily to electron injection at low fields from the polycrystalline-silicon gate into thermal SiO$_2$. We report here measurements of the dependence of the relaxation of injected charge on temperature and on the voltage at which the sample is held during charge relaxation measurements.   The measurements show that rearrangement of injected charge as well as removal of charge to the polycrystalline silicon gate can occur.

EXPERIMENTAL   AND   RESULTS

Details of sample preparation and of measurement procedures are given elsewhere.(4,5) The MOS capacitors used have oxide thicknesses of 380 Å. The substrate is 2 Ω–cm p–silicon; the polycrystalline silicon is n–type, doped to degeneracy. The samples studied here were annealed in N$_2$ at 550 °C after deposition of polycrystalline-silicon on SiO$_2$, an anneal which does not decrease the number of surface states greatly.   Electrons are injected at the polycrystalline-silicon–SiO$_2$ interface by biasing the gate at negative voltages greater than the flat-band voltage, V$_{fb}$. Figure 1 shows a typical quasi-static capacitance-voltage (C–V) curve measured from inversion to accumulation, and high-frequency C–V curves measured for both directions of voltage sweep, for a sample that exhibits significant electron injection. C/C$_{ox}$, the ratio of the capacitance at voltage V to the capacitance in strong accumulation, C$_{ox}$, is plotted as a function of voltage in Fig. 1.   The sample was chosen because it had a large hysteresis in the C–V curves, as shown by the arrows in Fig. 1. C/C$_{ox}$ for the quasi-static curve is greater than 1.0 in accumulation which is often observed when charge injection occurs.(4,5)

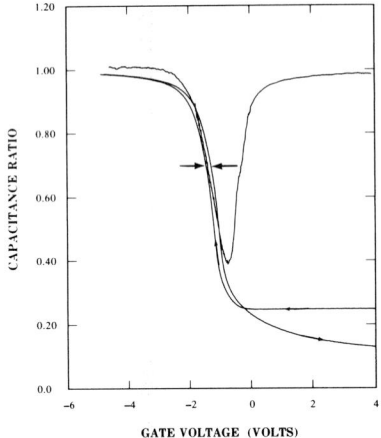

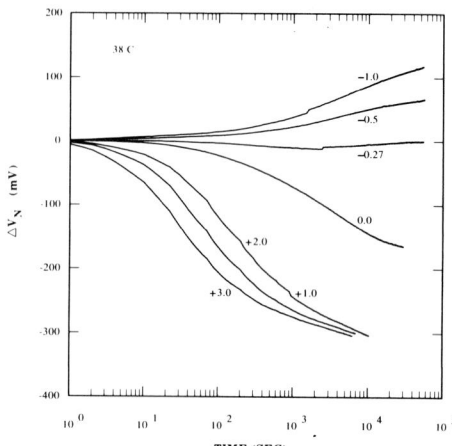

Fig. 1. Quasi–static C–V curve
from inversion to accumulation, and
high-frequency C–V curves for
both directions of voltage sweep,
of a polycrystalline-gate MOS
capacitor.

Fig. 2. Time dependence of the
change of V_N for different
holding voltages at constant
temperature. T=38 °C.

Measurements of relaxation of injected charge have been made by measuring the shift of the high-frequency C–V curve as a function of time at a constant holding voltage after injection of negative charge. A typical charge relaxation measurement sequence is as follows;

1) Apply +8.0 V to the polycrystalline-silicon gate for 30 seconds to reduce the amount of charge injected previously.
2) Apply −10.0 V to the polycrystalline-silicon gate for 30 seconds to inject electrons.
3) Switch the applied voltage to the holding voltage, V_H, the voltage at which a sample is held between measurements of V_{fb}. Periodically, measure the shift of the high-frequency C–V curve by measuring the null voltage, V_N, which is required to make the difference between the capacitance of the sample and the value of a reference capacitance equal to zero. The procedure is equivalent to measuring the change in flat-band voltage, V_{fb}, but the notation V_N is used since the reference capacitance is chosen to be somewhat smaller than V_{fb}. A logarithmic time scale is used.

Figure 2 shows the change in V_N, ΔV_N, as a function of time and holding voltage at 38 °C. For convenience, $\Delta V_N(0)$ is taken as the value of V_N measured 0.5 seconds after the start of the measurement sequence, $\Delta V_N(t) = V_N(t) - V_N(0.5 \text{ sec})$. Positive values of ΔV_N correspond to motion of the negative charge moment away from the polycrystalline-silicon gate, either due to an increase in the number of electrons injected or due to motion of injected negative charge closer to the substrate Si–SiO$_2$ interface. Negative values of ΔV_N correspond to a shift of the negative charge moment at the polycrystalline-silicon–SiO$_2$ interface towards the polycrystalline-silicon gate. The results of Fig. 2 agree with Yun and Hickmott (3) for zero volts applied to the gate. There is a period of little change in ΔV_N followed by a logarithmic decrease. For larger positive holding voltages, the rate of decrease of ΔV_N is larger. For −1.0 V, ΔV_N increases and for about −0.27 V, $\Delta V_N = 0$. $\Delta V_N = 0$ implies that the moment of injected charge is stable. The value of V_H for which $\Delta V_N = 0$, V_C, depends on temperature between 10°C and 90°C; at 10°C it is about −0.20

V, at 60°C V_C reaches a maximum value of −0.48 V, and it is approximately constant and equal to −0.45 V between 60°C and 90°C.

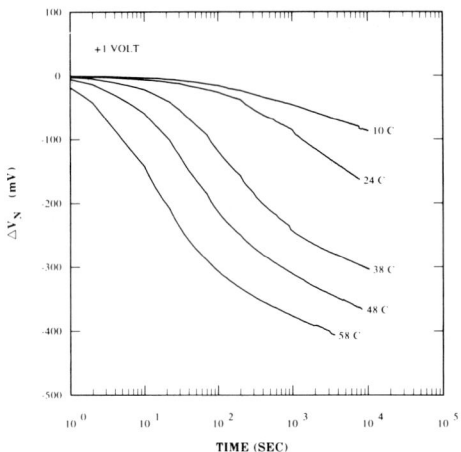

Fig. 3.   Time dependence of the change of V_N for different temperatures at constant voltages. $V_H = 1$ volt.

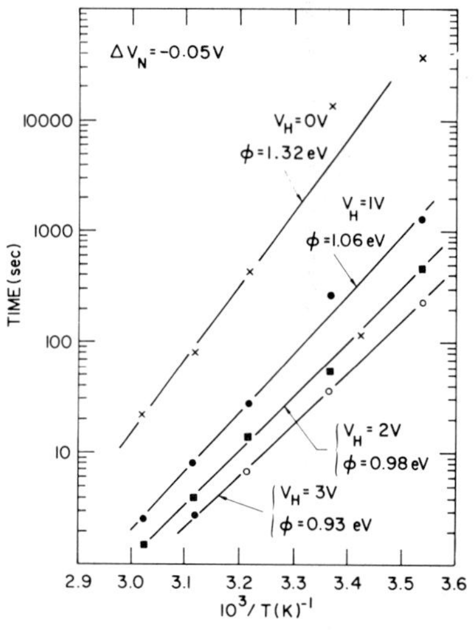

Fig. 4.   Temperature dependence of the relaxation time of the moment of injected charge at different holding voltages.

An alternate way to plot the data of Fig. 2 is at constant holding voltage, V_H, and different temperatures. Figure 3 shows ΔV_N as a function of time for V_H equal to 1 V. The total amount of charge injected is higher at higher temperature, judging both by the change in V_N immediately after application of the injecting pulse and by ΔV_N over longer time intervals. Figure 3 can be used to determine an activation energy for the change of the moment of charge. Assume that the time for V_N to change by a constant amount is of the form, $\tau = \tau_0 \exp(q\phi/kT)$, where ϕ is the activation energy in eV, τ_0 is a preexponential factor and τ is experimentally determined. Figure 4 shows log τ versus $1/T$ for $\Delta V_N = -50$ mV and for four different values of V_H. The activation energy decreases significantly between 0 and 1 V, but is only moderately field dependent at higher fields, with a value about 1 eV.

DISCUSSION

The change in V_N due to injected charge is the same as the change in flat-band voltage, ΔV_{fb};

$$\Delta V_N = -\frac{1}{\varepsilon} \int_0^L x \; \Delta\rho(x) \; dx \quad , \tag{1}$$

where L is the oxide thickness, $\Delta\rho(x)$ is the change in charge density at x after charge injection or charge relaxation, x is the distance from the polycrystalline-silicon gate, and ε is the permittivity of SiO_2. Yun and Hickmott proposed that traps responsible for electron injection were less than 30 Å from the gate-SiO_2 interface. In Fig. 5a, a schematic model of a distribution of traps at the polycrystalline-silicon−SiO_2 interface is given. The precise distribution is, of course, not known. Electron traps are shown in a narrow range of energies with respect to E_F, the fermi level of degenerate polycrystalline silicon, since one of the most striking features of charge injection and relaxation at the polycrystalline-silicon–SiO_2 interface is the low voltage and field at which it occurs. At V_{fb}, the field at both substrate and gate is zero if there is no charge in the insulator. For the sample studied, V_{fb} is between -0.95 V and -1.40 V, depending on the amount of charge injected. Initially charge injection occurs for negative voltages greater than V_{fb}. Once charge injection occurs in the sample it is not possible to tell from the present measurements if changes in ΔV_N are due to charge exchange at the polycrystalline-silicon–SiO_2 interface or are due to motion of charges in SiO_2. When $\Delta V_N = 0$ as a function of time, the net change in the moment of charge is zero.

It is frequently assumed that charge injected at the gate–SiO_2 interface has a small effect on V_{fb} because of the lever arm effect of the charge moment, as given in Eq. (1). In Fig. 5b, the number of traps/cm^2 needed to change V_{fb} by 100 mV is given under the assumption that the trap distribution is a delta function at the indicated distance. The numbers of electrons involved are comparable to values of surface states and surface charge at poorly annealed substrate silicon interfaces. Just as at the substrate–SiO_2 interface, it is necessary to anneal such states at the gate–SiO_2 interface to have stable MOS structures.(4)

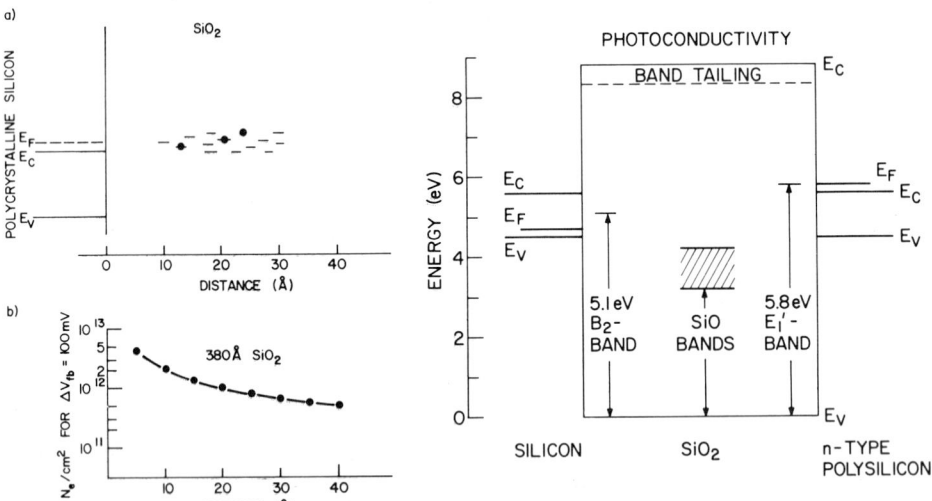

Fig. 5. a) Schematic representation of electron charge trapping centers at the polycrystalline-silicon–SiO_2 interface.
b) Amount of injected charge required for a shift of 100 mV in substrate flat-band voltage if a delta-function off charge is at a given distance from the polycrystalline-silicon–SiO_2 interface

Fig. 6. Schematic energy-band diagram of the silicon-SiO_2– n-type polysilicon–MOS system, showing different optical absorption bands in oxygen-deficient SiO_2. Substrate is 2 Ω-cm silicon, polysilicon is degenerate.

An important problem is to connect surface state phenomena with atomic defects in SiO_2 which are due to oxygen deficiency, impurities, or other causes. Fig. 6 suggests some possible correlations of defects in SiO_2 with surface states, surface charge and charge injection at either SiO_2 interface.(6) Figure 6 assumes that the energy required for photoconductivity in SiO_2, 8.8 eV, is equal to $E_c - E_v$.(7) The barrier height from the valence band of both silicon and polysilicon to the conduction band of SiO_2 is 4.3±0.1 eV.(8) The fermi level of p-silicon is given for a doping level of

7×10^{15} /cm^3; that for polysilicon assumes the doping is 3×10^{20} /cm^3. Two of the defects observed in high purity SiO$_2$ after irradiation by neutrons, electrons, or x-rays are the B$_2$-band and the E$_1$'-band with optical absorption bands at 5.1 eV and 5.8 eV, respectively. The B$_2$-band falls just opposite the forbidden gap of silicon and coincides with the position of a surface state level observed at the substrate silicon−SiO$_2$ interface of some unannealed samples.(4,6) The E$_1$'-band falls directly opposite the fermi level of degenerate n-type polysilicon. The E$_1$'-band is character- ized by an electron spin resonance signal due to a trapped electron; it should be able to exchange charge with degenerate polysilicon even after annealing. Annealing of the E$_1$'-band closely parallels that of surface states.(6,9) For the substrate, it would appear as excess surface charge since the fermi level of p-silicon is pinned below E$_c$ in strong inversion. The trapping level at the polysilicon−SiO$_2$ interface which is responsible for charge injection has been estimated to be less than 0.3 eV above the fermi level of polysilicon from charge relaxation measurements.(3) This is also in good agree- ment with the position of the E$_1$'-band. The model of Fig. 6 differs from that proposed by Mott (10) primarily because it assumes that the band gap of SiO$_2$ is 8.8 eV instead of 10.6 eV. In such a case, the E$_1$'-band lies at the fermi level of Si without assuming the existence of hole states, due to sodium, and located ~1 eV from the valence band edge of SiO$_2$.

Several authors have discussed instabilities and charge relaxation in MOS structures.(11-13) The distinguishing feature of charge relaxation at the polycrystalline-silicon−SiO$_2$ interface is the low fields required to inject charge and remove it. The charge relaxation curves of Figs. 2 and 3 are most similar to those observed by Lundkvist, et al on MNOS structures.(11,12) In both cases a region of no change in ΔV_{fb} with time is followed by a logarithmic dependence of V_{fb} on time. In both cases, charge relaxation is thermally activated. The activation energy for motion of the negative charge moment is ~1 eV, from Fig. 4. The activated process for motion of the moment can be either removal of electrons at the polysilicon-Si−SiO$_2$ interface under positive gate bias or it can be activated electron transport in the oxide. The voltage to inject electrons from polysilicon into SiO$_2$ is less than -1 V from V_{fb}, judging from Fig. 1. Therefore activated electron transport in the oxide probably limits charge decay. The latter process can be either by activation into the conduction band or by activation from traps with a depth of ~1 eV. Trap levels at the fermi level of degenerate n−silicon should be ~3.2 eV below the conduction band. Thus we estimate a trap depth of 1 eV for states in the oxide at the polycrystalline-silicon−SiO$_2$ interface. The distribution of traps at the polycrystalline silicon interface is poorly defined in energy, depth and number. Further work is needed to determine the relative importance of tunneling or thermally activated conduction in charge injection and in changing the location of the charge centroid under bias.

REFERENCES

1) F. Faggin and T. Klein, Solid-State Electronics 13 , 1125 (1970).
2) D. Frohman-Bentchkowsky, Solid-State Electronics 17, 517 (1974).
3) B. H. Yun and T. W. Hickmott, J. Appl. Phys. 48, 718 (1977).
4) T. W. Hickmott, J. Appl. Phys. 48, 723 (1977).
5) T. W. Hickmott, J. Appl. Phys. In Press, May (1978).
6) T. W. Hickmott, J. Vac. Sci. and Tech. 9, 311 (1972).
7) T. H. DeStefano and D. E. Eastman, Phys. Rev. Lett. 27, 1560 (1971).
8) D. J. DiMaria and D. R. Kerr, Appl. Phys. Lett. 9, 505 (1975).
9) T. W. Hickmott, J. Appl. Phys. 42, 2543 (1971).
10) N. F. Mott, Adv. in Physics 26, 363 (1977).
11) L. Lundkvist, I. Lundström and C. Svensson, Solid-State Electronics 16, 811 (1973).
12) L. Lundkvist, C. Svensson and B. Hansson, Solid-State Electronics 19, 221 (1976).
13) B. H. Yun, Appl. Phys. Lett. 23, 152 (1973).

EFFECTS OF ULTRA-THIN SiO$_x$ IN CONDUCTING M-I-S Structures*

T.E. Sullivan, R.B. Childs, J.M. Ruths, S.J. Fonash
Department of Engineering Science and Mechanics, The Pennsylvania
State University, University Park, PA 16802

ABSTRACT

Barrier formation in Schottky barrier-type devices, metal-semiconductor and
metal-thin film insulator-semiconductor, is examined as a function of controlled
surface oxidation. Metalizations used are thin film ($\leq$200A) Ag and Pd.

The evolution of the semiconductor barrier with purposeful oxidation is measured
by reverse C-V barrier height determinations, dark IV barrier height, photoresponse barrier height, and photovoltaic open circuit voltage.

The photovoltaic open circuit voltage for Ag and Pd devices exhibited a correlation to the change in barrier height when operated as solar cells, indicative of majority carrier devices.

INTRODUCTION

There is a resurgent interest in obtaining a better understanding of Schottky
barrier-type (M-S and conducting M-I-S) diodes. In addition there is a strong
interest in Schottky barrier-type solar cells.

In this reported work silicon baseline devices (semiconductor surface etched
and then immediately metalized) and silicon conducting metal-oxide-semiconductor
devices have been fabricated and characterized. The baseline surfaces (no
purposeful oxide) were prepared using a hydrofluoric acid etch. The surfaces
for the M-I-S devices were prepared by oxidizing, after the etch step. The
baseline surface will have some native oxide present. The purposeful, additional oxidizing of this surface as done for the M-I-S devices allows for the
systematic examination of chemical effects and trapping in barrier formation.
Thus the evolution of barrier formation, for different metals, can be followed
as the interfacial layer is varied.

All devices were fabricated on .5Ω-cm single crystal silicon of (100) orientation. All the material was given a standard "new material" clean. Immediately
prior to the device fabrication the chips were soaked in isopropyl alcohol for
three minutes, rinsed in a stream of de-ionized water, then blown dry in N$_2$.

The final preparation of the semiconductor surface consisted of two three
minute hydrofluoric acid (HF) etches. The freshly etched chips were then

*This work was supported in part by NSF Grant No. ENG 76-12874 and in part by
DOE Contract E (04-3)1282.

rinsed in running de-ionized water for 15 to 20 minutes. In the cases of the purposefully oxidized chips the oxidation was performed by insertion into an environment of water bubbled O_2 flowing at .05 SLPM. The oxidation temperature was 540°C. Oxidation time was 15 minutes. Ellipsometry measurements indicate these oxides are ~25-30 Å. The metallizations were performed in a Varian vacuum system at 10^{-8} torr. Both Ag or Pd (front) and Al (back) contact metallizations were performed without breaking vacuum.

These Schottky barrier-type devices, fabricated as outlined above, were then electrically characterized by dark current-voltage (I-V) data, forward bias capacitance-voltage (C-V) measurements, reverse capacitance-voltage data ($1/C^2$ plots), zero bias photoresponse measurements, and photovoltaic behavior. (No attempt was made to optimize this photovoltaic performance.) The I-V data, $1/C^2$ data, and photoresponse each yield (1) a barrier height for characterizing the device. The photovoltaic open circuit voltage also characterizes the barrier formed in the sense that it gauges the strength of the current bucking the photogenerated current (2,3).

DEVICE CHARACTERISTICS

Figure 1 shows typical $1/C^2$ reverse bias data for the Ag Schottky barrier-type devices of Table I. The increase of the barrier height (with purposeful oxidation) as measured by the reverse capacitance technique, $\phi_B(CV)$, is evident. The indicated doping variation is small and is consistent with that expected from wafer to wafer for this material. The $1/C^2$ plots are found to be ideal straight lines for both the baseline (M-S) and oxide (M-I-S) devices.

Dark IV data typically seen for Ag devices are shown in Figure 2. The n or ideality factors for baseline devices are close to unity. The n factors for M-I-S devices are typically ~1.2. The barrier height $\phi_B(IV)$, obtained from the dark current voltage data evolves in the same way as $\phi_B(CV)$ with purposeful oxidation. In general, there is reasonable agreement between the barrier heights measured by $1/C^2$, and dark IV.

Photovoltaic behavior of these Ag devices is summarized in Table I. Again, the evolution of the device behavior with oxidation is apparent i.e., increased $\phi_B(IV)$, $\phi_B(CV)$ and V_{oc} with oxidation. In Ag devices it is seen from Table I that the increased photovoltaic response shown by the increased open circuit voltage is principally attributable to the increase in barrier height; however, the increased n factor provides some additional contribution.

Ag M-I-S devices in general exhibited stable device characterization parameters. The n factors obtained from the dark IV characteristics showed a well defined constant n region. At lower voltages the dark IV's shown some recombination effects i.e., increased n factor in the current range 10^{-8} -10^{-6} amps for an area of .2 cm^2.

Figure 3 shows typical $1/C^2$ reverse bias data for Pd Schottky barrier-type devices of Table I. Palladium devices present a marked departure from the trend of increased barrier heights $\phi_B(CV)$ with purposeful oxidation. Baseline

(M-S) devices have barrier heights higher than the corresponding M-I-S devices. Again the variation (from device to device) in the slope of the $1/C^2$ plots is consistent with the manufacturers indicated doping.

The dark IV characteristics (Figure 4) also show higher barrier heights for the M-S over the M-I-S configuration. Both Pd-SiO$_x$-Si and Pd-Si structures exhibited n or ideality factors of unity. Neither device configuration exhibited appreciable variation in the n factor which might be attributable to either field shaping (redistribution of the voltage developed in the device), appreciable attenuation of transport across the I layer, or recombination effects (2,3) .

DISCUSSION

The reverse bias capacitance ($1/C^2$) technique provides one method to determine the Schottky barrier height, ϕ_B(CV), which does not involve the transport of majority carriers through the interfacial oxide. In this sense the attenuation properties of the oxide film, if any, (2,3) , do not affect the determination of the diffusion potential. However, the effects of localized charge redistributions at the semiconductor surface due both to the surface oxidation and metalization are reflected in the space charge region of the semiconductor and in turn in the measured diffusion potential. Further, if the reverse $1/C^2$ plot is linear, it represents a meaningful technique for obtaining device barrier heights (4,5).

The reverse bias capacitance ($1/C^2$) for Ag shows a clear trend of increased barrier height ϕ_B(CV) with purposeful oxidation. The increased barrier height could have been caused, effectively, by the introduction of negative charge into the interfacial region, thereby, causing the increased shielding seen from the intercept of the $1/C^2$ plot i.e., increased diffusion potential. Since ϕ_B(CV) values correlate with those calculated from the dark IV and correlate with photovoltaic open circuit voltage V_{oc}, the conclusion is that barrier height modification takes place in the M-I-S structure. The correlation of device performance with barrier height indicates that these Ag devices are majority carrier devices.

It is interesting to speculate on the origin of the effective charge necessary in the interfacial region to cause the increase in barrier height. Clearly the effects of chemical preparations, cleaning and etching procedures, may leave residual ions representative of the chemical preparation. These effects though, are more likely dominated by the effects of the surface oxidation and metalization. The bonding rearrangements in oxide formation and the metal film nucleation (and or reaction with the oxide in M-I-S structures) probably are significant.

Palladium Schottky barrier-type devices showed a decreased open circuit voltage between the baseline and M-I-S configurations. This is in marked contrast to the Ag devices which showed improved photovoltaic response with a purposeful interfacial oxide present.

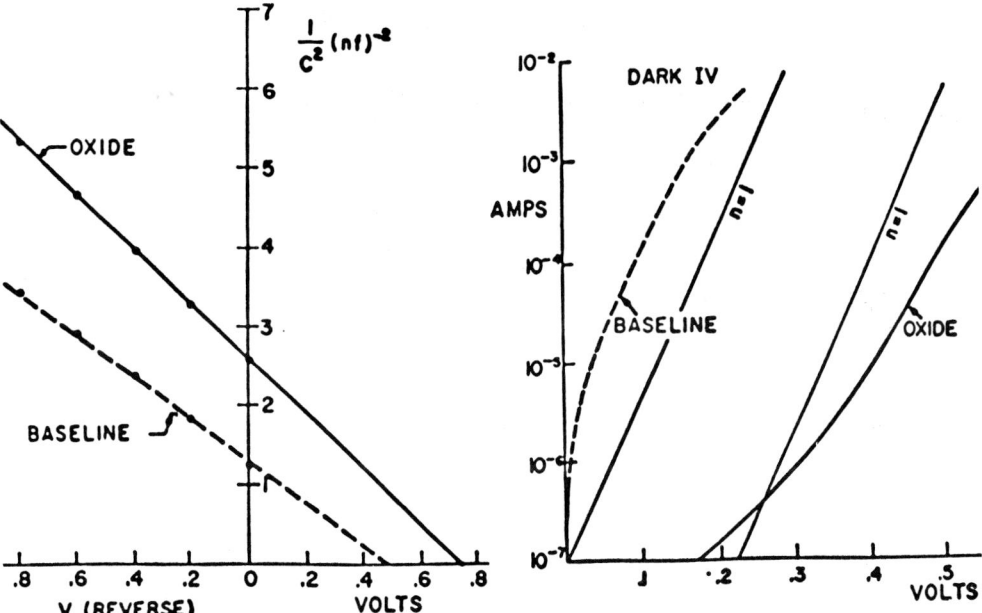

Fig. 1. $1/C^2$ plot for Ag Schottky barrier-type devices. The device area
is 0.20 cm^2.

Fig. 2. Dark IV for Ag Schottky barrier-type devices. Series resistance
can be seen at higher voltages.

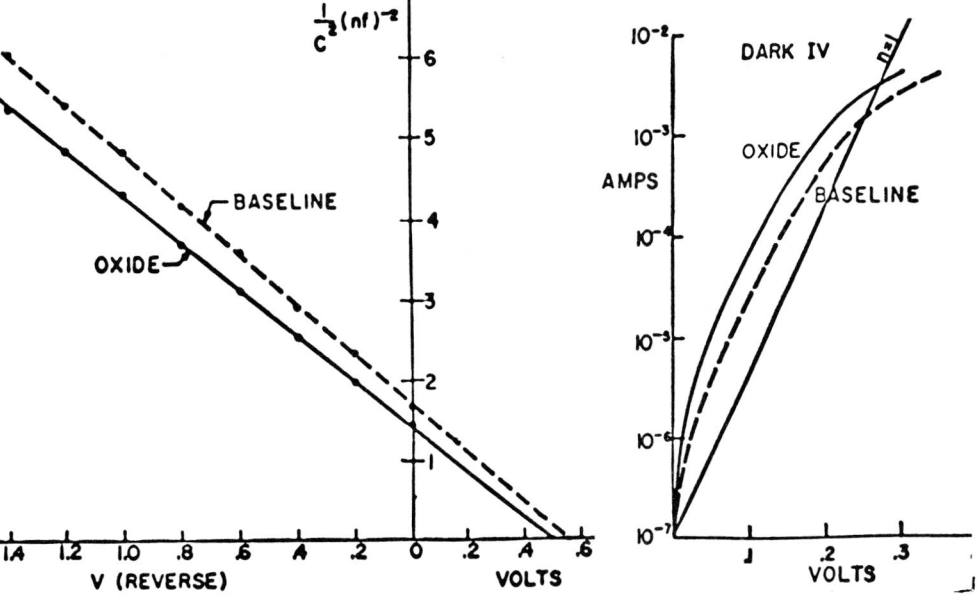

Fig. 3. $1/C^2$ plot for Pd Schottky barrier-type
devices. The device area is 0.20 cm^2.

Fig. 4. Dark IV for Pd Schottky barrier-type devices.

TABLE 1 Electrical Parameters for Various Conducting Metal-Oxide-n Si Devices

Device	Voc	n	ϕ(P.E.)	ϕ(CV)	ϕ(IV)
Ag baseline	.20	~1.0	.69	.73	.65
Ag oxide	.53	1.24		.96	.94
*Ag baseline	.15	~1.0	.62	.69	.71
*Ag oxide	.52	~1.3		.96	.99
Pd baseline	.20	~1.0	.73	.73	.72
Pd oxide	.15	~1.0	.65	.65	.66
*Pd baseline	.23	~1.0	.71	.74	.76
*Pd oxide	.18	~1.0	.62	.69	.73

Asterisk indicates devices shown in Figures 1-4.

The relative change in open circuit voltage between the M-S and M-I-S Pd structures can be seen from Table I to closely follow the change in barrier heights ϕ_B(CV) and ϕ_B(IV). Both baseline and oxide devices had n factors of unity.

In the data presented in this study we have examined the effects of controlled surface oxidation and subsequent metalization on the characteristics of Schottky barrier-type devices. We have seen from the electrical characterization techniques that specific and often detailed macroscopic information regarding the device operation can be obtained. This information reflects the rather complex and little understood interface chemistry involved in barrier formation at semiconductor surfaces.

Considering the devices examined as photovoltaic structures, it has been demonstrated that for these systems (metal and oxide combinations), the change in V_{oc} in going from M-S to the corresponding M-I-S structure is caused principally by barrier height changes. Clearly these structures are majority carrier devices.

(1) J. D. van Otterloo, L. J. Gerritsen, J. Appl. Phys. 49 (1978) 723.
(2) S. J. Fonash, J. Appl. Phys. 46 (1975) 1286.
(3) S. J. Fonash, J. Appl. Phys. 47 (1976) 3597.
(4) S. J. Fonash, J. Appl. Phys. 48 (1977) 3953.
(5) A. M. Goodman, J. Appl. Phys. 34 (1963) 329.

CONFIRMATION OF HYDROGEN SURFACE STATES AT THE
Si-SiO$_2$ INTERFACE

B. Keramati[†] and J.N. Zemel
Moore School of Electrical Engineering, Dept. of
Electrical Engineering and Science, University of
Pennsylvania, Philadelphia, PA 19104

ABSTRACT

Experiments are reported on the hydrogen effect in Pd n-Si Schottky diodes.
Gas ambients were 140 ppm H$_2$ in N$_2$ and pure O$_2$. The thin SiO$_2$ was estimated
at 3-5 nm. In O$_2$, the I-V and C-V characteristics in reverse bias followed
the expected behavior for a moderately good Schottky barrier. Forward bias
data was unreliable due to an epoxy back contact. In hydrogen, the reverse
I-V characteristic degraded substantially and a high density of interface
states associated with the Si-SiO$_2$ interface was observed. This behavior was
quite reversible and suggest that H is responsible for a loss of surface
states below midgap.

INTRODUCTION

In the last several years, there has been considerable work on the properties
of MIS structures based on a palladium gate or contact and a silicon semiconductor (1,2,3). Most of the MIS work has dealt with IGFET structures and
only a limited effort has dealt with the Schottky diode structure. Zemel (4)
initially proposed the Schottky diode as a possible hydrogen sensor and examined the comparative properties of the diode and IGFET structures.(5) The
basic assumption in this and other studies was that the changes in surface
condition was due to a buildup of hydrogen in the Pd at the Pd-SiO$_2$ interface.
While not explicitly stated, it was implicitly assumed that the H atom would
not permeate the Pd-SiO$_2$ interface. Some support for this was obtained from
a study of the Pd-Si Schottky diode when Pd$_2$Si was formed at the interface.(6)
No hydrogen response was observed with these devices. Lundstrom et al (3)
were able to obtain a H response by growing a thin native oxide on the Si
prior to depositing the Pd. This thin oxide acted as a diffusion barrier and
inhibited the formation of the silicide.

We have been systematically studying the properties of the Pd-Thin SiO$_2$-Si
(Pd-i-S) Schottky diode in order to elucidate the device physics that is
applicable. In this preliminary report, we show that the mechanism of sensitivity does not primarily involve changes in the Pd work function as has been
previously assumed and reported, but rather is due to surface state formation

---

[†] This work is in partial fulfillment of
the requirements for a Ph.D. degree in
Electrical Engineering.

460

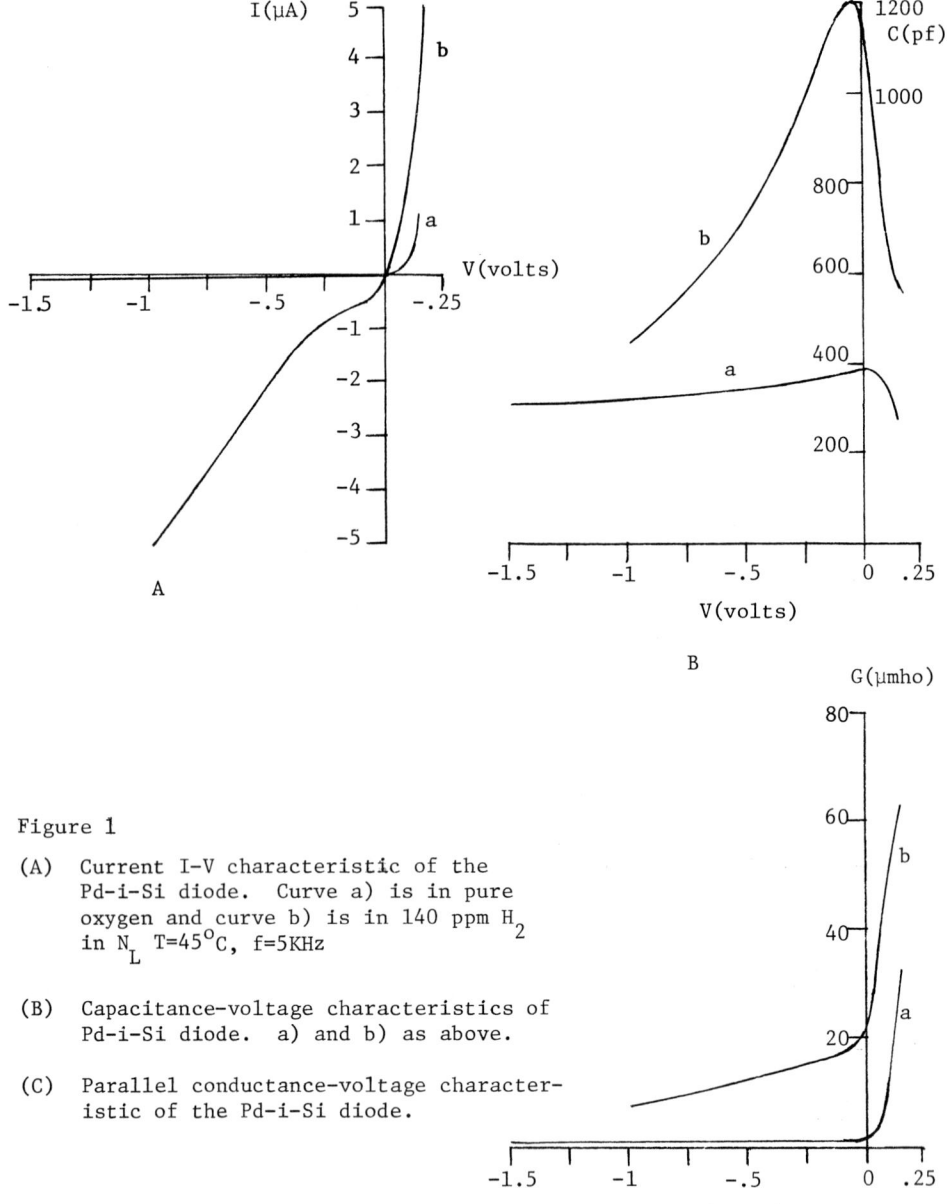

Figure 1

(A) Current I-V characteristic of the
     Pd-i-Si diode. Curve a) is in pure
     oxygen and curve b) is in 140 ppm H_2
     in N_L T=45°C, f=5KHz

(B) Capacitance-voltage characteristics of
     Pd-i-Si diode. a) and b) as above.

(C) Parallel conductance-voltage character-
     istic of the Pd-i-Si diode.

in the thin oxide itself.

The diodes were prepared in our laboratory from our own masks. The area of the diodes were 1.5mm x .6mm approximately. The diode window was lightly oxidized by baking in air for 1/2 hr at 200°C. The 30nm Pd contact was evaporated from an E gun. The silicon was n type and back contact has been made n+ by a phosphorous diffusion from a spin-on dopant. After dicing the wafer, the chips were bonded with a conducting epoxy to standard TO5 headers.

The measurement were carried out on programmable calculator controlled measurement facility. By using an appropriate circuit, it was possible to measure simultaneously C-V, G-V and I-V characteristics of the diode. The data indicated that the back contact was inadequately ohmic under forward bias due to the conducting epoxy. However, the reverse bias data were quite satisfactory. A set of representative I-V, C-V and G-V curves under oxygen and hydrogen ambient (Fig. 1) are presented. The changes in the curves are quite profound. The diode I-V characteristics indicate good rectification with low leakage currents in oxygen. The only question about the diode arises from the decreasing forward bias capacitance and increasing conductance values. The peak value for the capacitance is ~375pf. By contrast, in H ambient, the reverse I-V characteristic goes soft and the peak reverse bias is ~1,200pf, a three fold increase from that in the oxygen ambient. A study of the zero bias admittance as a function of frequency yields the capacitance-frequency curves of figure 2 for oxygen and hydrogen ambients. Not only does the low frequency capacitance increase by a factor of three, the roll off point shifts to lower frequency in the hydrogen ambient. The saturation of the capacitance at ~100pf may be a further manifestation of the non-ohmic back contact. Experiments on eutectically bonded diodes are in progress and will be reported shortly.

There are two possible interpretations for the phenomena observed in our Pd-i-Si diodes. First, the hydrogen could introduce changes in ϕ_{MS}, the metal semiconductor work function. This model provides some explanation for behavior of the I-V characteristics, particularly the high temperature (>70°C) behavior. However, if the H simply shifted ϕ_{MS} to induce a more strongly depleted or inverted surface, the principal contribution to the capacitance would come from the depletion layer in the silicon. The voltage drop across the i region might allow the surface potential to vary slightly but the high conductance of the diode, as evidenced by the reverse bias I-V curve, assure us that this is a small effect. Thus, in this case, a varying ϕ_{MS} can only influence the depletion layer capacitance. Since this is expected to be reasonably constant (at most varying by 40%), it cannot explain the hydrogen effect. On the other hand, introducing surface states provides an excellent accounting for the admittance data and the I-V behavior. These states would be associated with the Si-SiO$_2$ interface rather than the Pd layer. These surface states would also change the surface recombination velocity, thereby introducing a shunt path for the admittance. By increasing the recombination velocity, the reverse bias current is also increased.

We can conclude that H is responsible for a trapping level at the Si-SiO$_2$ interface. We have shown that H enters and leaves sites in the interface in a strikingly reversible fashion. Cycling between H and O$_2$ ambients repeatedly yields substantially identical results. There is some drift which could be accounted for by irreversible site destruction. These conclusions are based not only on our own observation but also on those of other researchers. (2,3) Finally, Pd appears to be an excellent injector of H into the surface of other

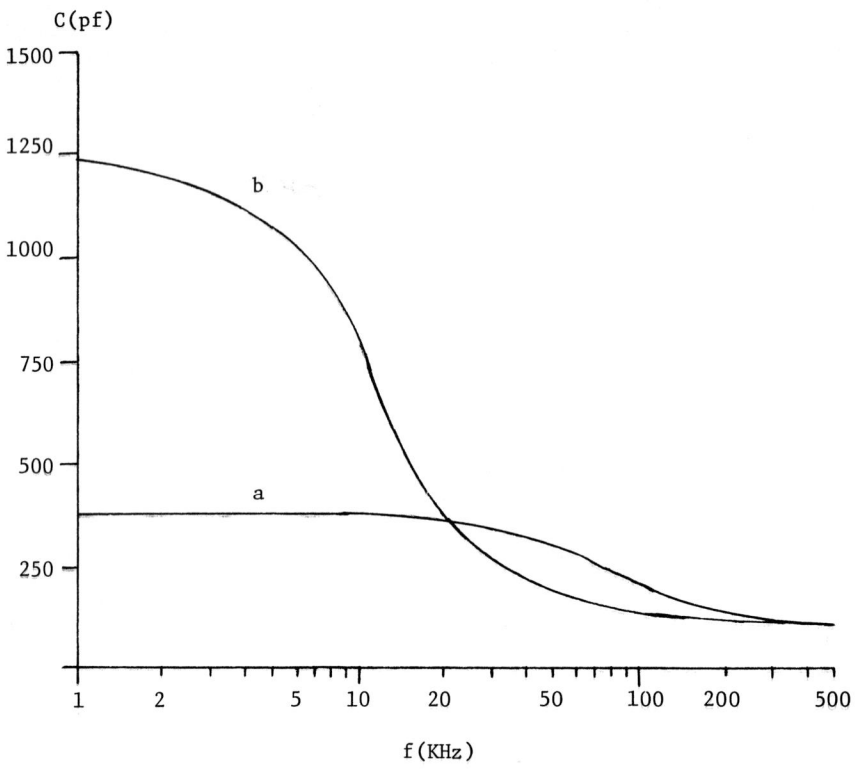

Figure 2

Zero bias capacitance as a function of
frequency of 45°C for oxygen (a) and 140ppm
hydrogen in N_2 (b) saturated Pd-i-Si diode.

materials. This conclusion is based on the work of Steele, et al (7) who find behavior akin to that on the Pd-i-Si structure.

We conclude by observing that Pd or possibly other transition metal contacts provide an intriguing way of studying surface state generation on real semi-conductor surfaces.

ACKNOWLEDGMENTS

The authors are pleased to acknowledge the financial support from The Pennsylvania Science and Engineering Foundation (PSEF-326) and The National Science Foundation (ENG-76-84146).

Facilities of the Laboratory for Research in the Structure of Matter have been utilized.

REFERENCES

(1)   I. Lundström, M.S. Shivaraman, C. Svensson, and L. Lundkvist, "A Hydrogen Sensitive MOS Field-Effect Transistor", _Appl. Phys. Lett._, 26, pp 55-7, (1975).

(2)   I. Lundström, M.S. Shivaraman and C. Svensson, "A Hydrogen Sensitive Pd-Gated MOS Transistor", _J. Appl. Phys._ 46, pp 3876-81, (1975).

(3)   M.S. Shivaraman, I Lundström, C. Svensson and H. Hammarsten, "Hydrogen Sensitivity of Palladium-Thin Oxide-Silicon Schottky Barriers", _Electronics Letters_, 12, pp 483-4, (1976).

(4)   J.N. Zemel, "Ion-Sensitive Field Effect Transistors and Related Devices", _Analytical Chemistry_, 47, pp 255A-268A, (1975).

(5)   J.N. Zemel, "Chemically Sensitive Semiconductor Devices", _IEDM Technical Digest_, 27.7, pp 635-8, (1975).

(6)   B. Keramati, unpublished results.

(7)   M.C. Steele, J.W. Hile, and B.A. McIver, "Hydrogen-sensitive Palladium Gate MOS Capacitors", _J. Appl. Phys._ 47, pp 2537-38, (1976).

ELECTRICAL PROPERTIES OF SiO_2-Si INTERFACE FOR DEFORMED
Si SURFACES

K. Murty, B. Lalevic, B. W. Lee, H. Suga, S. Weissmann
College of Engineering, Rutgers University, New Brunswick, New Jersey

ABSTRACT

Deformation by bending of Si wafers with [100][111] and [112] surface orientation was produced by sputtering tantalum films on Si surface with a thermally grown oxide. Stress on Si surface was determined by measuring radius
of curvature by the ABAC method, as a function of sputtering voltage and Ta
film thickness. Surface defect structure and defect distribution were determined by x-ray topographic method. The following parameters are determined
as a function of the stress on Si surface: surface potential, flat band voltage, surface charge density, surface state distribution, capture cross section and recombination time. I-V characteristics for SiO_2 were measured in
the accumulation region of an MOS capacitor before and after sputtering. The
dispersion at low frequencies changed. The observed changes in mechanical
and electrical properties of SiO_2 and SiO_2-Si interface were correlated to
the distribution of Ta through SiO_2 and Si surface as determined by the Auger
electron spectroscopy.

INTRODUCTION

The effect of stress field on the properties of MOS structure has been studied previously, providing somewhat contradictory results; some suggesting
gross stress effects (1,2,3,4,5) and others indicating negligible or no effect at all (6,7,8). Discrepancies in the reported results can be attributed
to differences in heat treatment and deposition techniques.

In this paper we report on the effects of tantalum deposition by sputtering
on the defect structure and electrical properties of SiO_2 and Si-SiO_2 interface. Ta film deposition on Si wafer, or Si-SiO_2 structure, deformed silicon wafer by bending and induced surface curvature. Using Auger Electron
Spectroscopy (AES) analysis, we have found that sputtered Ta atoms have pene-
trated through SiO_2 layer by an apparent diffusion aided implantation process
and reached Si surface. This process changed the I-V characteristics and
breakdown voltage of the SiO_2 layer. The induced curvature and stresses in
the Si wafer were studied as function of the Ta sputtering voltage for a
given Ta film thickness and as function of Ta film thickness at the given
sputtering voltage. The radii of curvature were determined by the automatic
Bragg angle control scan (ABAC). The microplastic defect structure and spatial distribution of the induced stresses at the silicon surface and in the
bulk were studied using x-ray transmission topographic method. Electrical

characteristics of the SiO_2 and $Si-SiO_2$ interface were studied in the MOS configuration and relevant parameters were determined from the C-V-f and G-V-f measurements before and after deformation of the silicon wafer.

EXPERIMENTAL TECHNIQUES

P-type silicon wafers with [100] and [111] orientation and resistivity of 1.5 ohm-cm and wafers with [112] orientation, 1.4 mm thick with resistivity of 0.0032 ohm-cm were used as substrates in the MOS capacitor structure. These wafers were free of defects as attested by sharpness of the x-ray rocking curves. Wafers were subjected to the standard treatment of polishing and etching and were oxidized at 950°C in the controlled steam atmosphere. Geometrical configuration of the Al ohmic contacts and the wafer areas covered by sputtered Ta were given previously (9) where an arrangement was made to avoid radiation damage under Al electrodes. Ta sputtering on Si and $Si+SiO_2$ structure was performed in an NRC triode sputtering module at 2, 4, 6 and 7 kV peak to peak voltage in 1 μm Ar atmosphere. The thickness of Ta films was determined by the interference method and ranged between 3500-10,000 Å.

The sensitivity of ABAC unit used in the determination of the radii of curvature was such that a change in lattice curvature of ± 1.5 sec. of arc could be detected, thus the stresses in Ta film and Si wafer could be accurately determined. X-ray section topographs were taken, with Ta film deposit at the entrance and exit to locate the sites and distribution of induced defects.

Electrical measurements were made on the MOS capacitor structure using a lock in amplifier (PAR-HR-8) with photometric preamplifier and signal conditioning amplifier and electrometer at room temperature.

RESULTS AND DISCUSSION

It was shown using ABAC method that oxidation of Si and subsequent annealing did not introduce lattice curvature in Si wafer in the limits of ABAC resolution which was about 2000 m in radius of curvature. X-ray topographic scan and rocking curves did not show any defect structure in Si and $Si-SiO_2$ wafers prior to sputtering of Ta films.

Deposition of Ta film by sputtering on Si and $Si-SiO_2$ wafers induced radii of curvature which were functions of the sputtering voltage and Ta film thickness. The stresses in the silicon wafer and Ta film were calculated from the magnitude of radius of curvature as determined by ABAC, Poisson ratio, elastic modulii and substrate and Ta film thickness (10).

Values of the radii of curvature and stresses in Ta films and Si wafers are given in Table 1 as a function of sputtering voltage for Si and $Si-SiO_2$ structure with oxide thickness of 800 Å and Ta film thickness of 3500 Å. Stress dependence of Ta film thickness and sputtering voltage is presented elsewhere (9).

It is of interest to observe in Table 1 that the stresses in Si wafer without oxide are smaller than in $Si-SiO_2$ structure at the same sputtering voltage. Similar effect has been seen in the x-ray topographs.

TABLE 1 Radius of Curvature R, stress σ_s in Si wafer, and stress σ_f in Ta film as a function of Ta sputtering voltage for p-type [100] Si wafer

Sputtering Voltage (kV)	R (m)	σ_s (kg/mm^2)	σ_f kg/mm^2
2 - no oxide layer	9.2	0.63	168
2 - with oxide	6.5	0.89	237
4 - "	5.9	1.00	280
6 - "	4.24	1.38	383

The x-ray topographic scans of the Silicon wafer surface with the Ta deposit at the x-ray exit are shown in Fig. 1 for Si-Ta and Si-SiO$_2$-Ta structures. Thicknesses of Ta film (3500 Å) and SiO$_2$ layer (800 Å) were kept constant, while the sputtering voltage was varied. As mentioned previously, Si-surface prior to Ta deposition was strain-free.

The lattice defects in the surface region of Si are responsible for the periodic elastic strain field at the interface, since they scattered the x-ray beam throughout the Borrmann triangle, resulting in the streaking black and white contrast effect shown in Fig. 1. Density of the induced strain increases with increasing sputtering voltage and at higher voltages large localized strain fields can be observed (Fig. 1-c). It is believed that these large fields are responsible for the observed film flaking at the high sputtering voltages. The concentration of localized high strain fields is also large in Si wafer without SiO$_2$ layer (Fig. 1-a) although the total strain density is lower than in the corresponding case with SiO$_2$ layer (Fig. 1-b). Sectional x-ray topographs have shown that the strain is localized at the Si-Ta or Si-SiO$_2$ interface.

Auger electron spectroscopy was used to analyze chemical composition of SiO$_2$ before and after sputtering of Ta film. The results are shown in Fig. 2. It is apparent that the sputtered Ta atoms penetrate SiO$_2$ layer by a diffusion aided implantation process and reach Si surface. Concentration of Ta atoms decreases approximately exponentially through the SiO$_2$ layer, although the experimental data are not yet sufficient to establish an exact relationship.

The presence of Ta impurities in the SiO$_2$ layer has changed the electrical properties of SiO$_2$. The I-V characteristics of SiO$_2$ are shown in Fig. 3 for evaporated Al electrodes and Ta electrodes sputtered at 4 kV. Initially, as shown in Fig. 3, current increased in both cases as $I \propto \exp \kappa \sqrt{V}$ with κ being larger for Ta electrodes. On further increase in applied voltage a current instability region was reached at the fields of ~2.5x10^6 V/cm for Ta and 5x10^6 V/cm for Al electrode. After this region, the I-V characteristics became quite stable and reproducible with the behavior shown in Fig. 3, with the current for Ta electrode being three orders larger than that for Al electrodes. The dielectric breakdown for Ta electrodes occurred at the field of 5x10^6 V/cm while for Al electrode the field was 1.2x10^7 V/cm.

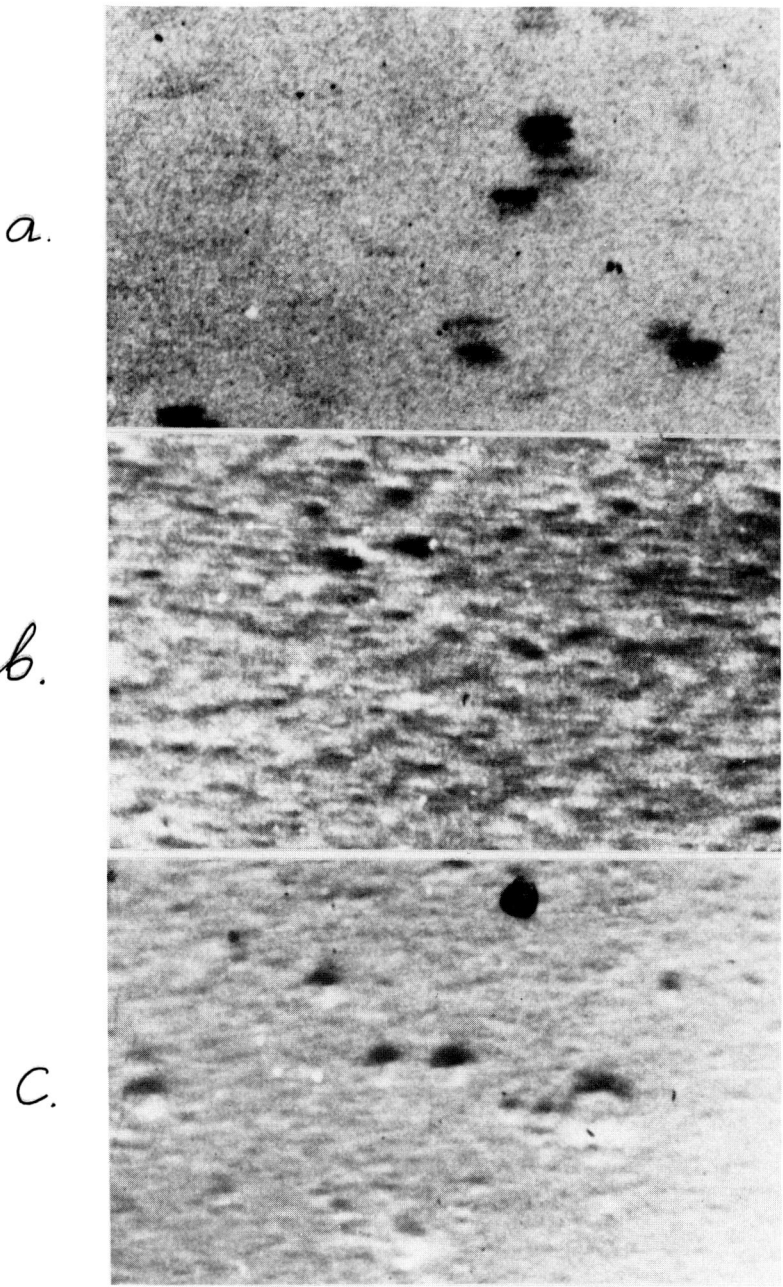

Fig. 1. X-ray topographic scans of a) Si wafer with Ta sputtered at
2 kV; b) Si wafer with SiO_2 and Ta sputtered at 2 kV and c) Si wafer
with SiO_2 and Ta sputtered at 6 kV. Ta and SiO_2 thickness were kept
constant.

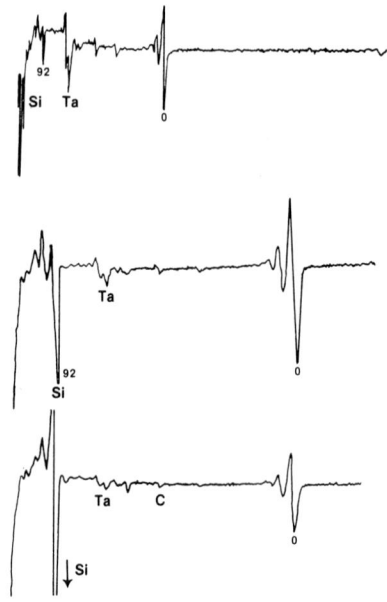

Fig. 2. Auger electron spectroscopic analysis of the sputtered
Ta atom through the SiO_2 layer. A finite concentration of Ta
atoms at the $Si-SiO_2$ interface is apparent from the above analysis.

The electrical properties of the $Si-SiO_2$ were determined from the capacitance
and conductance measurements on the MOS capacitor before and after Ta sputtering. Values of the following parameters of the MOS capacitor were obtained
from the experimental data: surface potential ψ_s(V) and ψ_s°(V=0), flat band
voltage V_{FB}, distribution of the surface state N_{ss} across the semiconductor
energy gap, surface state charge density, capture cross section and recombination time. The values of these parameters are given in Table 2 as function of the Ta sputtering voltage, i.e. induced strain in Si wafer.

TABLE 2 Values of the flat band voltage V_{FB}, surface potential
ψ_s°(V=0), surface charge density Q_{ss}, capture cross section σ_p
and recombination time τ_p as a function of Ta sputtering voltage
for p-type Si wafer with [100] orientation.

Sputtering Voltage (kV)	V_{FB}(V)	Q_{ss} ($\times 10^{11}$ cm^{-2})	ψ_s° (V)	σ_p ($\times 10^{-16}$ cm^2)	τ_p ($\times 10^{-2}$ sec)
0	1.8	7	0.15	1.6	3-4
2	2.45	7.1	0.18	8.7	0.75
4	2.76	7.8	0.25	16	0.4
6	3.4	9.0	0.25	36	0.18

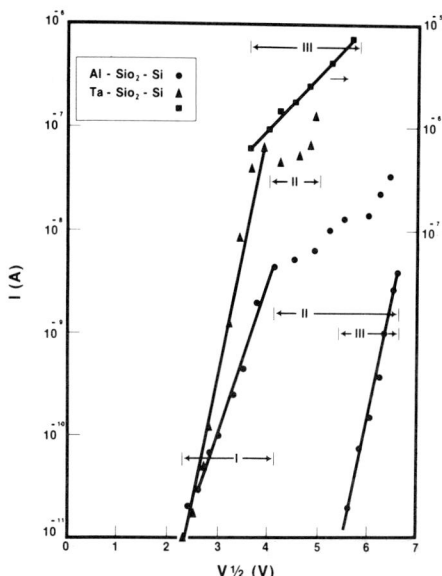

Fig. 3. I-V characteristics of SiO$_2$ with evaporated Al$_2$O$_3$ and sputtered Ta electrodes. Region I represents initial I-V characteristics, region II instability state and III the final and stable characteristics.

As seen in Table 2 large changes occur in the values of the capture cross section and recombination time due to the stresses induced by Ta sputtering. This is in agreement with the x-ray topograph (Fig. 2) where large density of strain fields is shown at the Si surface. The defects causing the observed strain field, do not affect, apparently, the values of other parameters such as Q_{ss} or $\psi_s{}^\circ$ to any significant extent, which is in agreement with some previous measurements (6,7,8) on the effect of stress on these two parameters. The large effect on σ_p and τ_p and relatively insignificant effect on Q_{ss} and $\psi_s{}^\circ$ would tend to indicate that the induced defect could be misfit dislocations. Further investigation is directed toward the identification of the defect structure.

REFERENCES

(1) C. H. Lane, IEEE Trans. Electron Devices, ED-15, 998 (1968).
(2) J. H. Serebrinsky, Solid-State Electron. 13, 1435 (1970).
(3) N. D. Godzhiev, F. D. Kasinow and V. Malinkova, Sov. Phys. Semicond. 5, 835 (1971).
(4) H. Friedrich, Solid-State Electron. 14, 639 (1971).
(5) K. Bulthuis, Philips Res. Rep. 20, 514 (1965).
(6) M. V. Whelan, A. H. Goemans, and L. M. C. Goosens, Appl. Phys. Lett. 10, 262 (1967).
(7) R. S. Wagner, A. K. Sinha, T. T. Sherg, H. J. Levinstein, and F. B. Alexander, J. Vac. Sci. Technol. 11, 582 (1974).
(8) B. C. Wensiewicz and D. V. McCauglan, J. Appl. Physics 44, 5476 (1973).
(9) K. Murty, B. Lalevic, H. Suga and S. Weissmann, J. Vac. Sci. Tech. March-April (1978).
(10) C. Rozgonyi, T. J. Ciesielk, Rev. Sci. Inst. 44, 1053 (1973).

SHEAR STRENGTH OF METAL - SiO$_2$ CONTACTS

Stephen V. Pepper, NASA - Lewis Research Center
21000 Brookpark Rd., Cleveland, Ohio 44135

ABSTRACT

The strength of the bond between metals and SiO$_2$ is studied by measuring the
static coefficient of friction of metals contacting α-quartz in ultrahigh
vacuum. It was found that copper with either chemisorbed oxygen, nitrogen
or sulphur exhibited higher contact strength on stoichiometric SiO$_2$ than
did clean copper. Since the surface density of states induced by these
species on copper is similar, it appears that the strength of the interfacial
bond can be related to the density of states on the metal surface.

INTRODUCTION

The useful electrical properties of the interface between different solids
has led to the fabrication of solid state electronic devices and a parallel
investigation of the physical basis of these properties. The investigation
into the physical nature of the adhesive forces between solids required in
fabrication of electronic devices has not progressed as rapidly. This lack
of progress is partly due to the experimental difficulty of obtaining a
measure of interfacial strength (1, 2) that can be interpreted as due to only
one of the many different forces that can act between solids (3). Adhesion
problems that have arisen in device fabrication have been dealt with rather
empirically, without fundamental understanding in which to act.

A system that should benefit from a fundamental investigation of interfacial
strength is the metal-SiO$_2$ system. This system is considered to be a
"problem" in that noble metals do not adhere well and must be joined to SiO$_2$
with intermediate layers of Ti or Cr.

In this paper changes in the shear strength of Cu-SiO$_2$ contacts induced by
chemisorbed O, S and N on Cu are studied. The shear strength has been shown
to be correlated to thin film adhesion via the popular scratch test (2, 4).
This correlation is the principle justification for associating the results
obtained here with what is generally considered to be adhesion. The use of
O, S and N is motivated by recent work (5) indicating that they induce similar
changes in the Cu surface valence electron density of states. Since it is
the valence electrons of a solid which determine the chemical interaction of
the solid with other species, the contact strength should change in similar
ways due to the presence of those elements on the Cu if in fact the adhesive
bonding between Cu and SiO$_2$ is due to a chemical interaction.

APPARATUS

Contact shear strength is measured in ultrahigh vacuum by pressing a copper sphere onto an SiO_2 flat and then subjecting the contact to a tangential force. The maximum tangential force that the contact can support divided by the normal force yields a static coefficient of friction, μ_s. Changes in μ_s due to chemisorbed 0, S and N on Cu, from the value obtained for the clean surfaces, are the principle experimental results. This method has been used to study metal – Al_2O_3 contact strength (6). Fig. 1 depicts the shear apparatus.

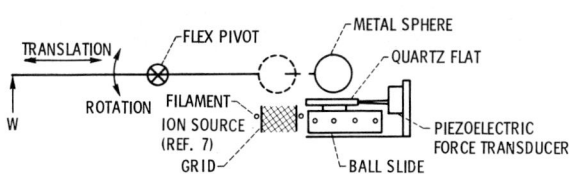

Figure 1. Shear apparatus in ultrahigh vacuum.

The load of the metal on the SiO_2 flat is provided by vertical force W. A differential screw provides tangential motion and tangential force is measured by the piezoelectric force transducer in vacuum. The sphere is retracted (dotted line) to allow specimen cleaning by Ar^+ bombardment of the SiO_2 (ion gun) or of the sphere by the electron bombardment ion source (7). This source also "activates" N_2 which does not chemisorb on Cu in molecular form (5). For "activated" chemisorption the metal sphere is grounded and the source operated at 5×10^{-5} torr N_2 with the ion pump on.

Surface analysis with a single pass cylindrical mirror analyzer with coaxial electron gun was performed in another part of the vacuum chamber, but with the specimens and ion sources removed from the shear apparatus. Auger electron spectroscopy (AES) and energy loss spectroscopy (ELS) for SiO_2 were used to characterize the surfaces after ion bombardment and after exposure to gases. Since the same surface preparation procedures were used both prior to contact and prior to surface analysis, it is assumed that the analysis is valid for the contacting surfaces. Polycrystalline metal and single crystal α-quartz were used.

SURFACE CHARACTERIZATION

After Ar^+ bombardment, the Cu Auger spectrum in Fig. 2 resulted.

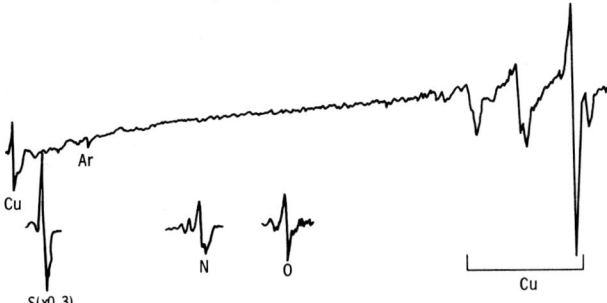

Figure 2. Auger spectrum of clean Cu and adsorbate Auger peaks.

The usual contaminants are absent and there is some Ar. Subjecting this clean
Cu surface to exposure of either 10^3 L O_2 or 10^2 L H_2S yields either the O or
S Auger peaks indicated. Subjecting the clean Cu surface to activated N_2 for
3 min. yields the N Auger peak indicated. Analysis of the fine structure of
the adsorbate peaks showed good agreement with those obtained by Tibbetts et al
(5). Assuming the peak-peak heights are proportional to elemental concentration,
taking into account the linear energy dependence of the sensitivity of the CMA
and the Auger yields of the elements (8), it is found that the elemental concentrations for O, N and S are in the ratio 1:.73:.77. Thus the relative
concentrations are quite similar. The absolute concentrations are probably
on the order of ½ monolayer (5).

The Auger spectrum of Ar^+ bombarded SiO_2 indicated that the surface was free
of impurities and the Si LVV structure is shown in Fig. 3a.

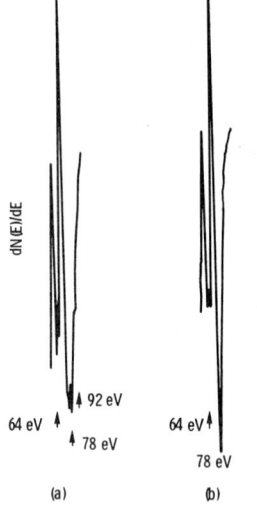

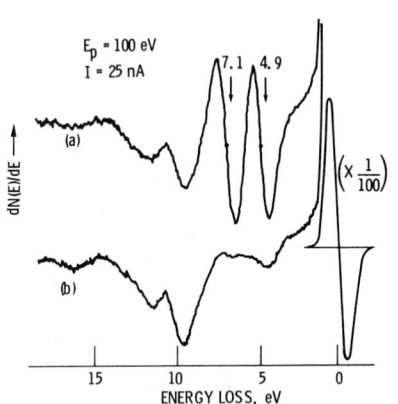

Figure 4. Energy loss spectrum of SiO_2 (a) after Ar^+
bombardment and (b) after O_2^+ bombardment.

Figure 3. Si LVV Auger peak (a) after
Ar$^+$ bombardment and (b) after O_2^+
bombardment.

The virtual absence of the 92eV transition indicates that the surface was not
reduced to elemental Si by the Ar^+. However, the structure at 78eV is absent
after 1 keV O_2^+ bombardment, Fig. 3b. Since the latter surface is probably
stoichiometric SiO_2 (9, Fig. 3), it appears that Ar^+ reduces the surface, but
not all the way to elemental Si. This is corroborated by ELS, Fig. 4. A
comparison with the second derivative ELS of SiO_2 (9, Fig. 5) indicates that
the spectrum of Fig. 4b is close to that of stoichiometric SiO_2, but that
Fig. 4a contains large peaks at 4.9eV and 7.1eV that have been attributed to
the presence of a "surface phase of approximately monoxide composition" (10).
The stoichiometric surface is the better defined of the two and may be easier
to handle analytically in a future theory of the SiO_2 - metal interface. Both
surfaces were investigated because the reduced surface is generated by the
sputter-etch step in thin film deposition and thus the results may be applicable to thin film adhesion.

RESULTS AND DISCUSSION

The results for stoichiometric SiO_2 are presented first. Both surfaces were cleaned by Ar^+ bombardment, the SiO_2 was oxidized by O_2^+ bombardment and finally the Cu was recleaned to prepare a clean Cu - SiO_2 system. Although the static friction coefficient varied from one experimental run to another (due perhaps to differing contact geometrics on the metal sphere), the changes in static friction following the exposures were quite reproducible. In Fig. 5 are depicted force-displacement curves for the clean contact and

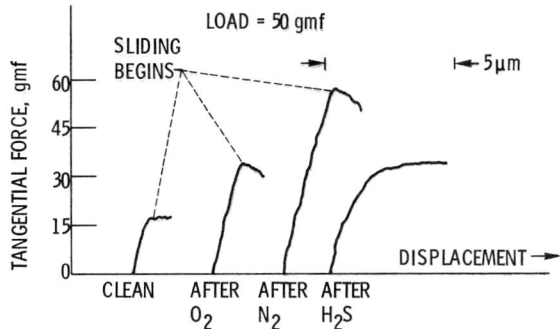

Figure 5. Force-displacement curves for contact on stoichiometric SiO_2.

for contact after exposure to O_2, N_2 and H_2S. The clean contact has μ_s ~ .3 and significant increases over this value are exhibited after all the exposures. The largest increase was due to N_2. Smaller increases were found after O_2 and H_2S, even though the force-displacement curves that illustrate the approach to sliding are (reproducibly) different. The contact strength on the reduced SiO_2 was somewhat different: the clean contact strength and the effect of N_2 was the same as for the stoichiometric SiO_2, but little, if any, increase was observed after the O_2 or H_2S exposures. Motivated by recent work which showed that the density of states induced by O, N and S on Ag were also quite similar to those observed on Cu (11), experiments were performed with a Ag sphere. For this relatively inert metal, all adsorption on the clean surface was by electron beam activation. The results were similar to Cu, although here increased contact strength was observed only on the stoichio‐metric SiO_2 and not on the reduced surface, even for N_2 exposures.

In considering these results, first note that the increased strength due to the presence of these adsorbates is contrary to the view that maximum inter‐facial strength is obtained by removing all foreign species from both surfaces prior to joining. Although this may be the case in joining similar materials as in metal-metal contacts, it is certainly not the case in the situation considered here. Secondly, a previous study of metal - Al_2O_3 contact strength (6) has assumed the need for two separate theories of interfacial strength to understand the effect of adsorbed oxygen on the metal - one for clean metal - Al_2O_3 and one for metal oxide - Al_2O_3. Here, however, we deal not with the copper oxide but with a partial monolayer chemisorbed surface species. Thus one theory of contact strength should be sufficient - for copper and its sur‐face modifications - rather than for the very different entities Cu and Cu_2O in contact with SiO_2.

474

The adsorbates that change the valence electron density of states in similar ways have led to similar changes in contact strength only on stoichiometric SiO_2. A different situation prevailed on reduced SiO_2, where only N_2 increased the contact strength. The contact strength cannot be understood solely in terms of the surface properties of one of the partners and requires a more complicated, as yet undeveloped, understanding of the interaction. An important finding, however, is the effect of adsorbed N. This large effect was not anticipated on grounds other than the induced valence electron density of states and thus lends some confidence to this approach to understanding contact strength.

Finally, the enhanced bonding observed here can have the practical effect of improving thin film adhesion. This has been noted for some time for rather thick intermediate oxide films (2) but results here show that the enhancement may also be possible by monolayer concentrations. This has in fact been observed by the author in preliminary experiments with the $Cu-N-SiO_2$ system by R. F. sputter deposition in a partial pressure of N_2.

REFERENCES
1. Tabor, D.: (1975), in Surface Physics of Materials, vol. II, (J. M. Blakely, ed.), Academic Press, New York.
2. Campbell, D. S.: (1970), in Handbook of Thin Film Technology, (L. I. Maissel and R. Glang, eds.), McGraw - Hill, New York.
3. Krupp, H.: Particle Adhesion Theory and Experiment, Advan. Colloid Interface Sci., 1, III, (1967).
4. Benjamin, P. and Weaver, C.: Measurement of Adhesion of Thin Films, Proc. Roy. Soc., London, A254, 163, (1960).
5. Tibbetts, G. G., et al: Electronic Properties of Adsorbed Layers of Nitrogen, Oxygen and Sulphur on Copper (100), Phys. Rev. B, 15, 3652, (1977).
6. Pepper, S. V.: Shear Strength of Metal-Sapphire Contacts, J. Appl. Phys., 47, 801, (1976).
7. Cuthrell, R. E. and Tipping D. W.: Sandia Laboratory Report SC-RR-72-0783, (1972).
8. Meyer, F. and Vrakking, J. J.: Quantitive Aspects of Auger Electron Spectroscopy, Surf. Sci., 33, 271, (1972).
9. Ibach, H. and Rower, J. E.: Electron Orbital Energies of Oxygen Adsorbed on Silicon Surfaces and of Silicon Dioxide, Phys. Rev. B, 10, 710, (1974).
10. Rowe, J. E.: Photoemission and Electron Energy Loss Spectroscopy of GeO_2 and SiO_2, Appl. Phys. Lett., 25, 576, (1974).
11. Tibbetts, G. G. and Burkstrand, J. M.: Electronic Properties of Adsorbed Layers of Nitrogen, Oxygen and Sulphur on Silver (111), Phys. Rev. B, 16, 1536, (1977).

List of Participants

J. M. AITKEN
IBM RESEARCH CENTER, 13-259
P. O. BOX 218
YORKTOWN HEIGHTS, NY   10598

D. D. ALLRED
OPTICAL SCIENCE CENTER
UNIVERSITY OF ARIZONA
TUCSON, ARIZ   85721

D. A. ANDERSON
GORDON MCKAY LAB
9 OXFORD ST
HARVARD UNIVERSITY
CAMBRIDGE, MA   02138

GORDON WOOD ANDERSON
CODE 5262
NAVAL RESEARCH LABORATORY
WASHINGTON, DC   20375

P. W. ANDERSON
BELL TELEPHONE LABORATORIES
600 MOUNTAIN AVE.
ROOM 1D-268
MURRAY HILL, NJ   07974

T. ANDO
IBM RESEARCH CENTER 26-138
P. O. BOX 218
YORKTOWN HEIGHTS, NY   10598

M. ANTONINI
ISTITUTO DI FISICA
VIA UNIVERSITA', 4
41100 MODENA, ITALY

ALAN APPLETON
DEPT. OF PHYSICS
BIRKBECK COLLEGE
MALET STREET
LONDON W.C. 1, U.K.

B. H. ARMSTRONG
IBM SCIENTIFIC CENTER
1501 CALIFORNIA AVE
PALO ALTO, CALIF   94304

EMIL ARNOLD
PHILIPS LABORATORIES
BRIARCLIFF MANOR, NY   10510

GEORGE W. ARNOLD
SANDIA LABORATORIES - 5112
ALBUQUERQUE, NM   87115

D. E. ASPNES
1C-319
BELL LABORATORIES
MURRAY HILL, NJ   07974

M. AV-RON
IBM SPD EAST FISHKILL
B/300-100
HOPEWELL JUNCTION, NY   12533

A. BALDERESCHI
LABORATOIRE DE PHYSIQUE APPLIQUEE
ECOLE POLYTECHNIQUE FEDERALE
2 RUE RUCHONNET
CH-1003 LAUSANNE, SWITZERLAND

STEPHEN W. BARBER
3806 GLENDALE AVE.
TOLEDO, OH   43614

DIRK BARTELINK
XEROX PARC
3333 COYOTE HILL RD
PALO ALTO, CA   94304

N. M. BASHARA
LABORATORY FOR SURFACE STUDIES
COLLEGE OF ENGINEERING AND TECHNOLOGY
UNIVERSITY OF NEBRASKA
LINCOLN, NEBRASKA   68588

INDER P. BATRA
IBM RESEARCH LABS., K33-281
5600 COTTLE ROAD
SAN JOSE, CA   95193

ROBERT S. BAUER
XEROX PARC
3333 COYOTE HILL RD.
PALO ALTO, CA   94304

GARY D. BENT
DEPT. OF THE ARMY
U. S. MILITARY ACADEMY
WEST POINT, NY   10966

LEN BERENBAUM
IBM EAST FISHKILL FACILITY
DEPT 46K, BLDG 300-095
HOPEWELL JUNCTION, NY   12533

ARNOLD BERMAN
IBM EAST FISHKILL FACILITY
DEPT 265, BLDG 300, ZIP 100
HOPEWELL JUNCTION, NY 12533

JERZY BERNHOLC
IBM RESEARCH CENTER
P. O. BOX 218
YORKTOWN HEIGHTS, NY   10598

JOSEPH BLANC
RCA LABORATORIES
PRINCETON, NJ   08540

W. BLOCK
INST. FUER FESTKOERPERPHYSIK (I)
    DER TECHNISCHEN UNIVERSITAET BERLIN
STRASSE DES 17. JUNI 135
1000 BERLIN 12, GERMANY

NATHAN BLUZER
WESTINGHOUSE ATL MS 3525
P. O. BOX 1521
BALTIMORE, MD   21203

H. BOESCH
BR 280
HARRY DIAMOND LABS.
2800 POWDER MILL RD.
ADELPHI, MD   20783

R. G. BOHN
DEPT. OF PHYSICS AND ASTRONOMY
THE UNIVERSITY OF TOLEDO
2801 W. BANCROFT ST.
TOLEDO, OH   43606

M. A. BOSCH
ROOM 4C-328
BELL LABORATORIES
HOLMDEL, NJ   07733

M. R. BOUDRY
PHILIPS RESEARCH LABORATORIES
REDHILL, SURREY
ENGLAND

KARL BRACK
IBM LABORATORY
SCHOENAICHER STR. 220
7030 BOEBLINGEN
WEST GERMANY

D. CANON BRADLEY
WATT AND CHESTNUT STS.
MT. HOLLY SPRINGS, PA   17065

W. F. BRINKMAN
BELL LABORATORIES
MURRAY HILL, NJ   07974

MARC BRODSKY
IBM RESEARCH CENTER 17-119
YORKTOWN HEIGHTS, NY   10598

WILLEM BROUWER
COULTER INFORMATION SYSTEMS
35 WIGGINS AVE
BEDFORD, MA   01730

KEITH BROWER
ORG. 5111
SANDIA LABORATORIES
ALBUQUERQUE, NM   87115

ELIAS BURSTEIN
DEPT. OF PHYSICS
UNIVERSITY OF PENNSYLVANIA
PHILADELPHIA, PA   19104

F. J. CADIEU
DEPT. OF PHYSICS
QUEENS COLLEGE OF CUNY
FLUSHING, NY   11367

HOWARD C. CARD
DEPT. OF ELEC. ENGINEERING
    AND COMPUTER SCIENCE
COLUMBIA UNIVERSITY
NEW YORK, NY   10027

D. J. CHADI
XEROX RESEARCH CENTER
3333 COYOTE HILL ROAD
PALO ALTO, CA   94304

HU H. CHAO
M/S 134
TEXAS INSTRUMENTS, INC
P. O. BOX 5936
DALLAS, TX   75222

J. R. CHELIKOWSKY
ID-262
BELL LABORATORIES
MURRAY HILL, NJ   07974

JEN-JON CHEN
DEPARTMENT OF ELECTRONIC ENGINEERING
FACULTY OF ENGINEERING
UNIVERSITY OF TOKYO
BUNKYO-KU, TOKYO 113, JAPAN

YUNG CHEN
MCDONNELL DOUGLAS CORP.
5301 BOLSA AVE
MAIL STATION 22-2
HUNTINGTON BEACH, CA   92647

DONALD J. COLEMAN
TEXAS INSTRUMENTS
MAIL STOP 118
BOX 5936
DALLAS, TX 75222

LARRY R. COOPER
CODE 427
OFFICE OF NAVAL RESEARCH
ARLINGTON, VA   22217

ROBERT CRAVEN
MONSANTO CO.
MAIL STOP T1H
800 N. LINDBERGH
ST. LOUIS, MO   63166

WALTER E. DAHLKE
DEPT. OF ELEC. ENGINEERING
PACKARD LABORATORY, BLDG. 19
LEHIGH UNIVERSITY
BETHLEHEM, PA   18015

PHILLIP H. DAVIS
DEPARTMENT OF PHYSICS
SUNY AT ALBANY
1400 WASHINGTON
ALBANY, NY   12222

BRUCE E. DEAL
FAIRCHILD RES. AND DEV.
4001 MIRANDA AVE.
MS 30-402
PALO ALTO, CA   94304

R. F. DE KEERSMAECKER
IBM RESEARCH CENTER, 29-117
P. O. BOX 218
YORKTOWN HEIGHTS, NY   10598

N. F. DE ROOIJ
DEPT. OF APPL. PHYSICS
TWENTE UNIV. OF TECHNOLOGY
P.O. BOX 217
ENSCHEDE, THE NETHERLANDS

JAMES F. DETRY
202 ELECTRICAL ENGIN. RES. LAB.
UNIVERSITY OF ILLINOIS
URBANA, IL   61801

D. J. DI MARIA
IBM RESEARCH CENTER, 30-148
P. O. BOX 218
YORKTOWN HEIGHTS, NY   10598

THOMAS H. DI STEFANO
IBM RESEARCH CENTER, 14-1
P. O. BOX 218
YORKTOWN HEIGHTS, NY   10598

ROBERT H. DOREMUS
DEPT. OF MATERIAL SCIENCE
RENSSELAER POLYTECHNIC INSTITUTE
TROY, NY   12181

J. DOROSTI
SCHOOL OF ENGIN. AND APPL. SCIENCE
UNIVERSITY OF CALIFORNIA
LOS ANGELES, CA

ARTHUR EDWARDS
SHERMAN FAIRCHILD LAB 161
LEHIGH UNIVERSITY
BETHLEHEM, PA   18015

DAVID EMIN
SOLID STATE THEORY DIV. - 5151
SANDIA LABORATORIES
ALBUQUERQUE, NM   87115

JON M ENGELAGE
DEPT OF PHYSICS
LOUISIANA STATE UNIVERSITY
BATON ROUGE, LA   70803

ROGER EVRARD
INSTITUT DE PHYSIQUE
UNIVERSITE DE LIEGE
SART-TILMAN/4000 LIEGE
BELGIUM

FRANK FANG
IBM T. J. WATSON RESEARCH CENTER
YORKTOWN HEIGHTS, NY   10598

ANTHONY J. FARINA
IBM RESEARCH CENTER
YORKTOWN HEIGHTS, NY   10598

FRANCIS P. FEHLNER
SULLIVAN PARK RB3
CORNING GLASS WORKS
CORNING, NY   14830

FRANK J. FEIGL
PHYSICS DEPARTMENT
LEHIGH UNIVERSITY
BETHLEHEM, PA   18015

L. C. FELDMAN
RM. 1E-434
BELL LABORATORIES
MURRAY HILL, NJ   07974

D. K. FERRY
E.E. DEPARTMENT
COLORADO STATE UNIVERSITY
FORT COLLINS, CO   80523

478

RICHARD A. FLASCK
ENERGY CONVERSION DEVICES
1675 WEST MAPLE ROAD
TROY, MI   48084

RICHARD FLITSCH
IBM EAST FISHKILL FACILITY
DEPT 877, BLDG 300-79
HOPEWELL JUNCTION, NY   12533

STEPHEN J FONASH
DEPT OF ENG. SCIENCE AND MECHANICS
PENN STATE UNIVERSITY
UNIVERSITY PARK, PA   16802

ALAN B. FOWLER
IBM RESEARCH CENTER, 29-145
P. O. BOX 218
YORKTOWN HEIGHTS, NY   10598

W. BEALL FOWLER
SHERMAN FAIRCHILD LAB. NO. 161
LEHIGH UNIVERSITY
BETHLEHEM, PA   18015

E. J. FRIEBELE
CODE 5580
NAVAL RESEARCH LABORATORY
WASHINGTON, DC   20375

CLIFF D. J. FUNG
ENGINEERING DESIGN CENTER
BINGHAM BLDG
CASE WESTERN RESERVE UNIV
CLEVELAND, OH   44106

A. K. GAIND
IBM EAST FISHKILL
HOPEWELL JUNCTION, NY   12533

F. L. GALEENER
XEROX PALO ALTO RES. CENTER
3333 COYOTE HILL RD.
PALO ALTO, CA   94304

JOHN C. GARTH
RADS (ETSR - STOP 30)
DEPUTY FOR ELECTRONIC TECHNOLOGY
HANSCOM AFB, MA   01731

CAROLINE GEE
MIT DEPARTMENT OF PHYSICS
ROOM 13-2134
77 MASSACHUSETTS AVE
CAMBRIDGE, MASS   02139

ALVIN M. GOODMAN
RCA LABORATORIES
DAVID SARNOFF RESEARCH CENTER
PRINCETON, NJ   08540

D. GRISCOM
NAVAL RESEARCH LABORATORY
OPTICAL SCIENCES DIVISION
CODE 5580
WASHINGTON, DC   20375

F. J. GRUNTHANER
JET PROPULSION LABORATORY
4800 OAK GROVE DR
PASADENA, CA   91125

DANIEL GUTERMAN
TEXAS INSTRUMENTS
P.O. BOX 1443
MAIL STATION 631
HOUSTON, TX   77001

L. E. HALLIBURTON
OKLAHOMA STATE UNIVERSITY
PHYSICS DEPARTMENT
STILLWATER, OK   74074

W. L. HARRINGTON
RCA LABORATORIES
PRINCETON, NJ   11540

WALTER A. HARRISON
HANSEN LABS.
STANFORD UNIVERSITY
STANFORD, CA   94305

A. HARTSTEIN
IBM RESEARCH CENTER, 17-135
P. O. BOX 218
YORKTOWN HEIGHTS, NY   10598

CHARLES M. HARTWIG
SANDIA LABORATORY
DIV. 8342
P. O. BOX 969
LIVERMORE, CA   94550

TAKEO HATTORI
DEPT. OF ELEC. ENGINEERING
FACULTY OF ENGINEERING
MUSASHI INST. OF TECHNOLOGY
1-28-1, TAMAZUTSUMI
SETAGAYA-KU, TOKYO, JAPAN

C. R. HELMS
STANFORD ELECTRONICS LABORATORY
STANFORD UNIVERSITY
STANFORD, CA   94305

ERNEST HENNINGER
PHYSICS DEPARTMENT
DEPAUW UNIVERSITY
GREENCASTLE, IN   46135

FRANK HERMAN
IBM RESEARCH LABS., K33-281
5600 COTTLE ROAD
SAN JOSE, CA   95193

KARL HESS
ELECTRICAL ENGINEERING RES. LAB
UNIVERSITY OF ILLINOIS
URBANA, IL   61801

T. HICKMOTT
IBM RESEARCH CENTER, 11-263
P. O. BOX 218
YORKTOWN HEIGHTS, NY   10598

M. W. HILLEN
RIJKSUNIVERSITEIT
LAB. VOOR TECH. NATUURKUNDE
ZERNIKELAAN
UNIVERSITEITSCOMPLEX PADDEPOEL
GRONINGEN 8002, THE NETHERLANDS

P. S. HO
IBM RESEARCH CENTER, 20-242
P. O. BOX 218
YORKTOWN HEIGHTS, NY   10598

KOITIRO HOH
COOPERATIVE LABORATORIES
VLSI TECHNOLOGY RESEARCH ASSOC.
4-1-1 MIYAZAKI, TAKATSU-KU
KAWASAKI, JAPAN

RON T. HOLM
NAVAL RESEARCH LABORATORY
WASHINGTON, DC 20375

K. HUEBNER
SEKTION PHYSIK
WILHELM-PIECK-UNIVERSITAT ROSTOCK
25 ROSTOCK
EAST GERMANY

GARY HUGHES
RCA LABORATORIES
PRINCETON, NJ 08540

HAROLD HUGHES
NAVAL RESEARCH LABORATORY
CODE 5216
WASHINGTON, DC   20375

R. C. HUGHES
SANDIA LABORATORIES
ALBUQUERQUE, NM   87115

E. A. IRENE
IBM RESEARCH CENTER, 7-131
P. O. BOX 218
YORKTOWN HEIGHTS, NY   10598

RANDALL ISAAC
IBM T. J. WATSON RESEARCH CENTER
P. O. BOX 218
YORKTOWN HEIGHTS, NY   10598

J. D. JOANNOPOULOS
DEPT. OF PHYSICS (13-2037)
MASS. INST. OF TECH.
CAMBRIDGE, MA   02139

NOBLE M. JOHNSON
XEROX PALO ALTO RES. CENTER
3333 COYOTE HILL RD.
PALO ALTO, CA   94304

COLIN E. JONES
ELECTRONICS & SEMICONDUCTORS
HONEYWELL RESEARCH
10701 LYNDALE AVE. S.
BLOOMINGTON, MN   55420

AVID KAMGAR
BELL LABORATORIES
MURRY HILL, NJ   07974

EMIL KAMIENIECKI
INSTITUTE OF PHYSICS
POLISH ACADEMY OF SCIENCES
AL. LOTNIKOW 32/46
01-668 WARSAW, POLAND

MOTOTAKA KAMOSHIDA
NIPPON ELECTRIC CO.
TAMAGAWA PLANT
1753 SHIMONUMABE
NAKAHARA-KU KAWASAKI
JAPAN

L. A. KASPRZAK
IBM EAST FISHKILL FACILITY
DEPT. 929, BLDG 330-275
HOPEWELL JUNCTION, NY   12533

MARC KASTNER
MIT, RM. 13-2142
CAMBRIDGE, MA   02139

BAHMAN KERAMATI
DEPT. OF ELECTR. ENGIN.
UNIV. OF PENNSYLVANIA
PHILADELPHIA, PA   19104

DONALD KERR
IBM EAST FISHKILL
DEPT 265, BLDG 300, ZIP 100
HOPEWELL JUNCTION, NY   12533

S. M. KIM
AECL
CHALK RIVER
ONTARIO, CANADA K0J 1J0

SANG U. KIM
IBM BURLINGTON
ESSEX JUNCTION, VT   05452

BARRY M. KLEIN
CODE 6484
NAVAL RESEARCH LABORATORY
WASHINGTON, DC   20375

NICHOLAS KLEIN
DEPT. OF ELEC. ENG.
TECHNION
HAIFA 32000, ISRAEL

F. KOCH
PHYSIK-DEPARTMENT
    DER TECHN. UNIVERSITAET MUENCHEN
8046 GARCHING, W. GERMANY

TUNG-SHENG KUAN
IBM RESEARCH CENTER, 16-240
P. O. BOX 218
YORKTOWN HEIGHTS, NY   10598

S. LAHIRI
IBM RESEARCH CENTER
P. O. BOX 218
YORKTOWN HEIGHTS, NY   10598

ANDRAS LAKATOS
XEROX, WEBSTER RESEARCH CENTER
W128
WEBSTER, NY   14580

F. C. LAMAN
DEPT. OF CHEMISTRY
UNIVERSITY OF SASKATCHEWAN
SASKATOON, SASKATCHEWAN S7N 0W0
CANADA

ROBERT B LAUGHLIN
PHYSICS DEPT
M. I. T.
77 MASSACHUSETTS AVE
CAMBRIDGE, MASS   02139

JOHN S. LECHATON
IBM EAST FISHKILL FACILITY
DEPT. 28G, BLDG 320-030
HOPEWELL JUNCTION, NY   12533

HAN-SHENG LEE
DEPT. OF ELECTRONICS
GM RESEARCH LABS.
WARREN, MI   48090

SOOK LEE
DEPT. OF PHYSICS
SAINT LOUIS UNIVERSITY
ST. LOUIS, MO   63103

SHENG S. LI
ELECTRICAL ENGINEERING DEPT.
227 BENTON HALL
UNIVERSITY OF FLORIDA
GAINESVILLE, FL   32611

H. C. LIN
E. E. DEPT.
UNIVERSITY OF MARYLAND
COLLEGE PARK, MD   20903

INGOLF LINDAU
SEL, STANFORD UNIVERSITY
STANFORD, CA   94305

NUNZIO O. LIPARI
IBM WATSON RESEARCH CENTER
YORKTOWN HEIGHTS, NY   10592

M. A. LITTLEJOHN
ARMY RESEARCH OFFICE
P.O. BOX 12211
RESEARCH TRIANGLE PARK, NC

STEVEN LOUIE
IBM RESEARCH CENTER
P.O. BOX 218
YORKTOWN HEIGHTS, NY   10598

W. HOWARD LOWDERMILK
LAWRENCE LIVERMORE LAB
MAIL STOP L-479
P.O. BOX 808
LIVERMORE, CA   94550

T. P. MA
DEPT. OF ENGRG. & APPL. SCIENCE
YALE UNIVERSITY
NEW HAVEN, CT   06520

A. MADHUKAR
DEPT. OF PHYSICS
UNIV. OF SOUTHERN CALIFORNIA
LOS ANGELES, CA   90007

HERMAN MAES
KATHOLIEKE UNIVERSITEIT LEUVEN
KARDINAAL MERCIERLAAN 94
B-3030 HEVERLEE
BELGIUM

A. MANARA
EURATOM CCR
21027 ISPRA (VARESE)
ITALY

J. MASERJIAN
JET PROPULSION LABORATORY
4800 OAK GROVE DR.
PASADENA, CA   91125

SANTOS MAYO
NATIONAL BUREAU OF STANDARDS
BLDG 225 ROOM A327
WASHINGTON, DC 20234

JAMES M. MC GARRITY
HARRY DIAMOND LABORATORIES
2800 POWDER MILL RD
ADELPHI, MD   20783

FLYNN B. MC LEAN
BR 280
HARRY DIAMOND LABS.
2800 POWDER MILL RD.
ADELPHI, MD   20783

MICHAEL MEISSNER
INST. FUER FESTKOERPERPHYSIK (I)
    DER TECHNISCHEN UNIVERSITAET BERLIN
STRASSE DES 17. JUNI 135
1000 BERLIN 12, GERMANY

GENE MELE
DEPARTMENT OF PHYSICS
MIT
77 MASSACHUSSETTS AVE
CAMBRIDGE, MA   02139

DEAN MITCHELL
NATIONAL SCIENCE FOUNDATION
WASHINGTON, DC   20550

A. MORITANI
DEPT. OF ELECTRONICS
FACULTY OF ENGINEERING
OSAKA UNIVERSITY, SUITA CITY
OSAKA 565, JAPAN

S. ROY MORRISON
STANFORD RESEARCH INSTITUTE
MENLO PARK, CA   94025

SIR NEVILL F. MOTT
CAVENDISH LABORATORY
MADINGLEY ROAD
CAMBRIDGE CB3 0HE, ENGLAND

JAMES MURDAY
CODE 6170
NAVAL RESEARCH LABS.
WASHINGTON, DC   20375

SUNIL MURGAI
IBM EAST FISHKILL FACILITY
DEPT. 87C, BLDG 330, ZIP 275
HOPEWELL JUNCTION, NY   12533

CHERRY MURRAY
MIT DEPARTMENT OF PHYSICS
ROOM 13-2073
77 MASSACHUSETTS AVE
CAMBRIDGE, MASS   02139

KRISHNA MURTY
C217, DEPT. OF ELEC. ENGINEERING
RUTGERS UNIVERSITY
NEW BRUNSWICK, NJ   08903

M. I. NATHAN
IBM RESEARCH CENTER, 29-129
P. O. BOX 218
YORKTOWN HEIGHTS, NY   10598

ARTHUR H. NETHERCOT
IBM RESEARCH CENTER, 20-136
P. O. BOX 218
YORKTOWN HEIGHTS, NY   10598

K. L. NGAI
NAVAL RESEARCH LABORATORY
CODE 5277
WASHINGTON, DC   20375

EDWARD H. NICOLLIAN
BELL LABS
MURRAY HILL, NJ 07974

T. H. NING
IBM RESEARCH CENTER, 11-243
P. O. BOX 218
YORKTOWN HEIGHTS, NY   10598

R. NITECKI
INSTITUTE OF PHYSICS
POLISH ACADEMY OF SCIENCES
AL. LOTNIKOW 32/46
02-668 WARSAW, POLAND

PAUL NORTON
HONEYWELL RESEARCH
10701 LYNDALE AVE. S.
BLOOMINGTON, MN   55420

ROGER NUCHO
DEPT OF PHYSICS
UNIV. OF SOUTHERN CALIFORNIA
LOS ANGELES, CA   90007

IWAO OHDOMARI
IBM RESEARCH CENTER
P. O. BOX 218
YORKTOWN HEIGHTS, NY   10598

JEAN OLIVIER
THOMSON-CSF (L.C.R.)
DOMAINE DE CORBEVILLE
91401 ORSAY, FRANCE

HILDING M. OLSON
BELL LABS
2525 N. 11TH ST.
READING, PA   19604

STANFORD R. OVSHINSKY
ENERGY CONVERSION DEVICES, INC.
1675 WEST MAPLE ROAD
TROY, MI   48084

EDWARD PALIK
NAVAL RESEARCH LABORATORY
WASHINGTON, DC 20375

SOKRATES T. PANTELIDES
IBM RESEARCH CENTER 28-146
YORKTOWN HEIGHTS, NY   10598

MARINA PASCUCCI
CASE WESTERN RESERVE UNIV.
WHITE BLDG
CLEVELAND, OH   44106

C. K. N. PATEL
ROOM 1D269
BELL LABORATORIES
MURRAY HILL, NJ   07974

R. LYLE PATTON
UNION CARBIDE
TARRYTOWN TECH. CENTER
TARRYTOWN, NY   10591

MARTIN C. PECKERAR
WESTINGHOUSE ATL MS 3525
P. O. BOX 1521
BALTIMORE, MD   21203

M. PEPPER
CAVENDISH LABORATORY
CAMBRIDGE CB3 0HE
ENGLAND

STEPHEN V. PEPPER
MS 23-2
NASA - LEWIS RESEARCH CENTER
21000 BROOKPARK ROAD
CLEVELAND, OH   44135

ROBERT PFEFFER
8 STANDISH DRIVE
OCEAN, NJ   07712

H. R. PHILIPP
GE CENTER FOR RESEARCH
    AND DEVELOPMENT
SCHENECTADY, NY

EDWARD H. POINDEXTER
US ARMY ELECTRONICS TECHNOLOGY
    AND DEVICES LABORATORY
3 ROSEDALE TERRACE
HOLMDEL, NJ   07733

RAYMOND POIRIER
THOMSON-CSF
LABORATOIRE CENTRAL DE RECHERCHES
DOMAINE DE CORBEVILLE
91401 ORSAY, FRANCE

ROGER POLLAK
IBM T. J. WATSON RESEARCH CENTER
P. O. BOX 218
YORKTOWN HEIGHTS, NY   10598

WILLIAM B. POLLARD
MIT
13-2041
CAMBRIDGE, MA   02139

JOHANNES POLLMANN
IBM RESEARCH CENTER
P. O. BOX 218
YORKTOWN HEIGHTS, NY   10598

STEVE QUIGLEY
IBM CORP.
DEPT. F05 BLDG 052-1
POUGHKEEPSIE, NY   12601

S. I. RAIDER
IBM RESEARCH CENTER 24-136
YORKTOWN HEIGHTS, NY   10598

DENNIS RATHMAN
SHERMAN FAIRCHILD LAB
LEHIGH UNIVERSITY
BETHLEHEM, PA   18015

ASIT RAY
IBM T. J. WATSON RESEARCH CENTER
YORKTOWN HEIGHTS, NY   10598

K. H. RAU
HERAEUS QUARZSCHNELZE
P. O. BOX 463
D-6450 HANAU, GERMANY

A. G. REVESZ
COMSAT LABS.
CLARKSBURG, MD   20734

DAVID RHIGER
SANTA BARBARA RESEARCH CENTER
75 COROMAR DR
GOLETA, CA   93017

ARTHUR RICH
DEPARTMENT OF PHYSICS
UNIVERSITY OF MICHIGAN
ANN ARBOR, MI 48109

V. L. RIDEOUT
IBM RESEARCH CENTER, 11-223
P. O. BOX 218
YORKTOWN HEIGHTS, NY   10598

JACK RIFE
A251 PHYSICS BLDG
NATIONAL BUREAU OF STANDARDS
WASHINGTON, DC   20234

HOWARD K. ROCKSTAD
MICRO-BIT CORP.
40 HARTWELL AVE
LEXINGTON, MA   02173

PETER ROITMAN
NATIONAL BUREAU OF STANDARDS
BLDG. 225, RM. A327
WASHINGTON, DC   20234

JOHN E. ROWE
BELL TELEPHONE LABS.
ROOM 1-C 306
MURRAY HILL, NJ   07974

GARY RUBLOFF
IBM T. J. WATSON RESEARCH CENTER
YORKTOWN HEIGHTS, NY   10598

C. T. SAH
DEPT. OF ELEC. ENGINEERING
UNIVERSITY OF ILLINOIS
URBANA, IL   61801

KRISHNA SAPRU
ENERGY CONVERSION DEVICES, INC
1675 WEST MAPLE ROAD
TROY, MI   48084

SEPPO SARI
OPTICAL SCIENCES CENTER
UNIVERSITY OF ARIZONA
TUCSON, AZ   85721

WILLIAM C. SAUDER
DEPT. OF PHYSICS
VMI
LEXINGTON, VA   24450

ARJUN N. SAXENA
DATA GENERAL CORP.
433 N. MATHILDA AVE.
SUNNYVALE, CA   94086

A. SCHIFRIN
DATA GENERAL CORP.
433 N. MATHILDA AVE.
SUNNYVALE, CA   94086

M. SCHLUTER
BELL LABORATORIES
MURRAY HILL, NJ   07974

MANFRED SCHMIDT
AKADEMIE DER WISSENSCHAFTEN DER DDR
ZENTRALINSTITUT FUER ELECTRONENPHYSIK
INSTITUTSTEIL IV
RUDOWER CHAUSSEE 5
1199 BERLIN-ADLERSHOT
EAST GERMANY

KLAUS SCHWIDTAL
US ARMY ERADCOM
DELET-EDS
FORT MONMOUTH, NJ   07703

C. SENEMAUD
LABORATOIRE DE CHIMIE-PHYSIQUE
11 RUE PIERRE ET MARIE CURIE
75005 PARIS, FRANCE

MORRIS SHATZKES
IBM EAST FISHKILL
DEPT. 265, BLDG 300, ZIP 100
HOPEWELL JUNCTION, NY   12533

EDWARD SIEGEL
MOLECULAR ENERGY RESEARCH INSTITUTE
150 WEST END AVE.
BROOKLYN, NY   11235

GEORGE H. SIGEL, JR.
NAVAL RESEARCH LABORATORY
CODE 6440
WASHINGTON, DC   20390

JAMES H SIMPSON
THE SINGER CO., KEARFOTT DIV.
1150 MCBRIDE AVENUE
LITTLE FALLS, NJ   07424

R. N. SINCLAIR
PHYSICS DIVISION
AERE HARWELL
DIDCOT, OXFORDSHIRE OX11 0RA
ENGLAND

RAJENDRA SINGH
DEPT. OF PHYSICS
MC MASTER UNIVERSITY
HAMILTON, ONT. L8S 4M1
CANADA

G. E. SMITH
MOS DEVICE DEPARTMENT
BELL LABS
MURRAY HILL, NJ 07974

W. LEE SMITH
LAWRENCE LIVERMORE LABORATORY
BOX 808, MAIL CODE L8479
LIVERMORE, CA   94550

LAWRENCE C. SNYDER
ROOM 1A-359
BELL LABS
MURRAY HILL, NJ 07974

PAUL SOLOMON
IBM RESEARCH CENTER, 10-244
P. O. BOX 218
YORKTOWN HEIGHTS, NY   10598

ASHOK SOOD
HONEYWELL ELECTRO-OPTICS CENTER
2 FORBES RD
LEXINGTON, MA   02173

E. J. SOXMAN
SLOAN TECH. CORP.
414 EAST COTA STREET
SANTA BARBARA, CA   93101

J. R. SROUR
NORTHROP RES. AND TECH. CENTER
ONE RESEARCH PARK
PALOS VERDES PENINS., CA 90274

M. STAPELBROEK
MATERIAL SCIENCES DIVISION
NAVAL RESEARCH LABORATORY
WASHINGTON, DC   20375

SUSAN TURNBACH STEIGERWALT
NAVAL OCEAN SYSTEMS CENTER
CODE 9251
SAN DIEGO, CA   92152

RICHARD STEIN
MAIL STOP 119
TEXAS INSTRUMENTS
P.O. BOX 5936
DALLAS, TX 75222

FRANK STERN
IBM RESEARCH CENTER, 29-253
P. O. BOX 218
YORKTOWN HEIGHTS, NY   10598

PHILLIP J. STILES
PHYSICS DEPARTMENT
BROWN UNIVERSITY
PROVIDENCE, RI   02912

ALAN STIVERS
ELECTRICAL ENGINEERING RES. LAB
UNIVERSITY OF ILLINOIS
URBANA, IL   61801

Y. E. STRAUSSER
VARIAN ASSOCIATES
611 HANSEN WAY, MS G-226
PALO ALTO, CA   94303

R. L. STREEVER
U. S. ARMY ELECTRONICS COMMAND
(DRSEL-TL-ESA)
FORT MONMOUTH, NJ   07703

TAKUO SUGANO
DEPT. OF ELEC. ENGINEERING
UNIVERSITY OF TOKYO
7-3-1 HONGO, BUNKYO-KU
TOKYO 113, JAPAN

CHRISTER SVENSSON
RES. LAB. OF ELECTRONICS III
CHALMERS UNIVERSITY OF TECHNOLOGY
FACK, 40220 GOTEBORG, SWEDEN

SIMON SZE
BELL LABS
MURRAY HILL, NJ 07974

MING L. TARNG
RCA LABORATORIES
DAVID SMIRNOFF RESEARCH CENTER
PRINCETON, NJ   08540

JOHN H. THOMAS
RCA LABORATORIES
PRINCETON, NJ   11540

M. F. THORPE
PHYSICS DEPARTMENT
MICHIGAN STATE UNIVERSITY
E. LANSING, MI   48824

JAMES TOPICH
ENGINEERING DESIGN CENTER
CASE WESTERN RESERVE UNIVERSITY
CLEVELAND, OH   44106

D. C. TSUI
BELL LABORATORIES
MURRAY HILL, NJ   07974

DAVID VANDERBILT
MIT DEPT. OF PHYSICS
77 MASS AV
CAMBRIDGE, MA   02139

BOYD W. VEAL
ARGONNE NATIONAL LABORATORY
9700 S. CASS AVE.
ARGONNE, ILLINOIS   60439

J. F. VERWEY
PHILIPS RESEARCH LABORATORIES
EINDHOVEN
THE NETHERLANDS

C. R. VISWANATHAN
SCHOOL OF ENGIN. AND APPL. SCIENCE
UNIVERSITY OF CALIFORNIA
LOS ANGELES, CA

JOHN VITKO
SANDIA LABORATORY
P. O. BOX 969
LIVERMORE, CA   94550

PETER E. WAGNER
DIRECTOR, CENTER FOR ENVIRONMENTAL
   AND ESTUARINE STUDIES
UNIVERSITY OF MARYLAND
P. O. BOX 775
CAMBRIDGE, MD   21613

ROBERT F. WAITES
HEWLETT PACKARD
1501 PAGE MILL RD
PALO ALTO, CA   94304

J. WAKEFIELD
DEPARTMENT OF ELECTRICAL AND
   ELECTRONIC ENGINEERING
THE QUEEN'S UNIVERSITY OF BELFAST
ASHBY BLDG, STRANMILLIS RD
BELFAST BT9 5AH
IRELAND

RODGER M. WALSER
DEPT. OF ELEC. ENGINEERING
439 ENS
UNIVERSITY OF TEXAS
AUSTIN, TX   78712

KANG WANG
GENERAL ELECTRIC R & D CENTER
SCHENECTADY, NY   12301

SAMUEL WANG
HUGHES AIRCRAFT
500 SUPERIOR AVE
NEWPORT BEACH, CA   92663

JOHN A. WEIL
DEPT. OF CHEM. & CHEM. ENGINEERING
UNIVERSITY OF SASKATCHEWAN
SASKATOON, SASKATCHEWAN S7N 0W0
CANADA

ZEEV WEINBERG
IBM RESEARCH CENTER, 29-105
P. O. BOX 218
YORKTOWN HEIGHTS, NY   10598

STUART H. WEMPLE
BELL LABORATORIES
MURRAY HILL, NJ   07974

C. T. WHITE
NAVAL RESEARCH LABORATORY
CODE 5277
WASHINGTON, DC   20375

MARVIN H. WHITE
WESTINGHOUSE ELEC. CORP.
ADVANCED TECHNOLOGY LABS.
BOX 1521
BALTIMORE, MD   21203

CARL W. WILMSEN
DEPT. OF ELECTRICAL ENGINEERING
COLORADO STATE UNIVERSITY
FT. COLLINS, CO   80523

P. WINOKUR
BR 280
HARRY DIAMOND LABS.
2800 POWDER MILL RD.
ADELPHI, MD   20783

M.C. WINTERSGILL
PHYSICS DEPT.
OKLAHOMA STATE UNIVERSITY
STILLWATER, OK   74074

JOE WONG
GENERAL ELECTRIC R & D CENTER
1 RIVER ROAD, K-1 5C8
SCHENECTADY, NY   12301

A. C. WRIGHT
J. J. THOMSON PHYSICAL LABORATORY
WHITEKNIGHTS, READING
RG6 2AF, UNITED KINGDOM

G. B. WRIGHT
STEVENS INST. OF TECHNOLOGY
HOBOKEN, NJ   02030

C. T. WU
IBM RESEARCH CENTER
P.O. BOX 218
YORKTOWN HEIGHTS, NY   10598

EDWARD S. YANG
DEPT. OF ELEC. ENGINEERING
COLUMBIA UNIVERSITY
NEW YORK, NY   10027

JAMES C. YEN
TEKTRONIX INC.
M.S.48-222
BEAVERTON, OR   97077

D. R. YOUNG
IBM RESEARCH CENTER, 29-141
P. O. BOX 218
YORKTOWN HEIGHTS, NY   10598

BOB H. YUN
DEPT. G39
IBM CORP.
ESSEX JUNCTION, VT   05452

STOYAN ZALAR
IBM EAST FISHKILL FACILITY
DEPT 87C, BLDG 330, ZIP 275
HOPEWELL JUNCTION, NY   12533

N. ZAMANI
JET PROPULSION LABORATORY
4800 OAK GROVE DR
PASADENA, CA   91125

PAUL ZAVRACKY
COULTER INFORMATION SYSTEMS, INC.
35 WIGGINS AVE
BEDFORD, MA   01730

JAY N. ZEMEL
MOORE SCHOOL
ELEC. ENGINEERING, RM. 310
UNIVERSITY OF PENNSYLVANIA
PHILADELPHIA, PA   19104

ROBERT ZETO
US ARMY ELECTRONICS RESEARCH
    AND DEVELOPMENT COMMAND
DELET-ESP
FORT MONMOUTH, NJ 07703

AUTHOR INDEX

Ahlstrom, E. R., 227
Allred, D. D., 210
Anderson, G. W., 200
Antonini, M., 316
Appleton, A., 94
Appleton, B. R., 210
Armstrong, B. H., 128
Arnold, G. W., 278
Av-Ron, M., 46
Barber, S. W., 139
Bartelink, D. J., 421
Batra, I. P., 65, 333
Bauer, R. S., 401
Bergveld, P., 433
Bianconi, A., 401
Block, W., 122
Boesch, Jr., H. E., 19, 428
Bosch, M. A., 294
Brower, K. L., 258
Buyers, W. J. L., 314
Cadoff, I. B., 46
Calabrese, E., 70
Caplan, P. J., 227
Chadi, D. J., 55, 321
Chen, J. J., 351
Childs, R., 454
Chiranjivi, T., 94
Clark, G. J., 210
Comas, J., 200
Costa Lima, M. T., 75
Dec, K., 304
DeKeersmaecker, R. F., 189
de Rooij, N. F., 433
DiMaria, D. J., 51, 160, 189
DiStefano, T. H., 46, 362
Dorosti, J., 184
Embree, D., 289
Emin, D., 14
Faulconnier, P., 89
Feldman, L. C., 344
Ferry, D. K., 29
Flitsch, R., 384
Fonash, S. J., 454
Fowler, W. B., 70
Gaind, A. K., 438
Galeener, F. L., 284

Greaves, G. N., 268
Greytak, T. J., 309
Griscom, D. L., 232, 263
Gritsenko, V. A., 40
Grunthaner, F. J., 389
Harrison, W. A., 105
Hartstein, A., 51
Hartwig, C. M., 215
Hattori, T., 379
Helms, C. R., 366
Herman, F., 333
Hickmott, T. W., 449
Hillen, M. W., 179
Houston, J., 99
Hubner, K., 111
Hughes, R. C., 14
Irene, E. A., 205
Jackman, T. E., 344
Jafaripour-Ghazvini, M., 94
Joannopoulos, J. D., 55, 321
Johnson, N. M., 284, 366, 421
Jones, C. E., 289
Kamgar, A., 356
Kamieniecki, E., 417
Kasowski, R. V., 333
Kasprzak, L. A., 438
Keramati, B., 459
Kim, C. J., 155
Kim, S. M., 314
Krivanek, O. L., 356
Lagally, M. G., 99
Lalevic, B., 464
Lam, D. J., 299
Laman, F. C., 253
Laughlin, R. B., 55, 321
Lee, B. W., 464
Lensi, P., 316
Long, M., 339
Madhukar, A., 60
Manara, A., 316
Maserjian, J., 389, 443
Mattern, P. L., 215
McGarrity, J. M., 19, 428
McLean, F. B., 19, 428
McMenamin, J. C., 401
Meissner, M., 122

Meixner, A. E., 85
Mikkelsen, Jr., J. C., 284
Misawa, K., 144
Mogilnikov, K. P., 40
Moore, G., 99
Moritani, A., 144
Mott, N. F., 1
Murday, J. S., 99
Murray, C. A., 309
Murty, K., 464
Nakai, J., 144
Ngai, K. L., 412
Nishina, T., 379
Nitecki, R., 417
Nucho, R. N., 60
Olivier, J., 89
Oona, H., 155
Ovshinsky, S. R., 304
Pantelides, S. T., 80, 189, 339
Pepper, M., 407
Pepper, S. V., 470
Petersen, H., 401
Platzman, P. M., 85
Poindexter, E. H., 227
Poirier, R., 89
Powell, R. A., 195
Raider, S. I., 384
Rajkanan, K., 396
Ramaker, D. E., 99
Revesz, A. G., 222
Rubloff, G. W., 24
Ruths, J., 454
Rzhanov, A. V., 40
Sapru, K., 304
Sari, S. O., 155
Saxena, A. N., 195
Schluter, M., 85
Schmidt, W. A., 200
Schneider, P. M., 70
Schulz, M., 421

Schwarz, S. A., 366
Schwidtal, K., 273
Senemaud, C., 75
Shatzkes, M., 46
Sheng, T. T., 356
Shewchun, J., 396
Shu, K., 155
Siegel, E., 149
Silverman, P. J., 344
Sinclair, R. N., 133
Singh, R., 396
Solomon, P. M., 35
Spicer, W. E., 366
Spitzmann, K., 122
Stapelbroek, M., 263
Stensgaard, I., 344
Suga, H., 464
Sugano, T., 351
Sullivan, T. E., 454
Svensson, C. M., 328
Thorpe, M. F., 116
Tsong, I. S. T., 210
Tsui, D. C., 356
Turner, N. H., 99
Viswanathan, C. R., 184
Vitko, J., 215
Veal, B. W., 299
Wager, J. F., 373
Weil, J. A., 253
Weinberg, Z. A., 24, 51
Weissman, S., 464
White, C. T., 412
White, C. W., 210
Wilmsen, C. W., 373
Winokur, P. S., 428
Wright, A. C., 133
Zamani, N., 443
Zemel, J. N., 459